WILLIAM F. MAAG LIBRARY
YOUNGSTOWN STATE UNIVERSITY

Advances in
INORGANIC CHEMISTRY

Volume 32

ADVISORY BOARD

A. H. Cowley
University of Texas
Austin, Texas

H. B. Gray
California Institute of Technology
Pasadena, California

O. Kahn
Université de Paris Sud
Orsay, France

A. Ludi
Universität Bern
Bern, Switzerland

J. Reedijk
Leiden University
Leiden, The Netherlands

A. M. Sargeson
Australian National University
Canberra, Australia

D. F. Shriver
Northwestern University
Evanston, Illinois

K. Wieghardt
Ruhr Universität Bochum
Bochum, Federal Republic of Germany

Advances in
INORGANIC CHEMISTRY

EDITOR

A. G. SYKES

Department of Chemistry
The University
Newcastle upon Tyne, England

VOLUME 32

ACADEMIC PRESS, INC.
Harcourt Brace Jovanovich, Publishers
San Diego   New York   Berkeley   Boston
London   Sydney   Tokyo   Toronto

COPYRIGHT © 1988 BY ACADEMIC PRESS, INC.
ALL RIGHTS RESERVED.
NO PART OF THIS PUBLICATION MAY BE REPRODUCED OR
TRANSMITTED IN ANY FORM OR BY ANY MEANS, ELECTRONIC
OR MECHANICAL, INCLUDING PHOTOCOPY, RECORDING, OR
ANY INFORMATION STORAGE AND RETRIEVAL SYSTEM, WITHOUT
PERMISSION IN WRITING FROM THE PUBLISHER.

ACADEMIC PRESS, INC.
1250 Sixth Avenue
San Diego, California 92101

United Kingdom Edition published by
ACADEMIC PRESS INC. (LONDON) LTD.
24-28 Oval Road, London NW1 7DX

LIBRARY OF CONGRESS CATALOG CARD NUMBER: 59-7692

ISBN 0-12-023632-X (alk. paper)

PRINTED IN THE UNITED STATES OF AMERICA
88 89 90 91   9 8 7 6 5 4 3 2 1

CONTENTS

PREFACE ix

Dynamics of Spin Equilibria in Metal Complexes

JAMES K. BEATTIE

I.	Introduction	2
II.	Static Properties	4
III.	Techniques	14
IV.	Solution Dynamics	22
V.	Solid-State Dynamics	36
VI.	Summary and Interpretation	39
VII.	Implications	43
	References	49

Hydroxo-Bridged Complexes of Chromium(III), Cobalt(III), Rhodium(III), and Iridium(III)

JOHAN SPRINGBORG

I.	Introduction	56
II.	Structural Considerations	57
III.	Spectroscopic and Magnetic Properties	70
IV.	Formation of Polynuclear Complexes	75
V.	Stability Constants	98
VI.	Acid–Base Equilibria	106
VII.	Kinetics for the Condensation Reaction of Mononuclear Species to Give Dinuclear Species	119
VIII.	Cleavage of Polynuclear into Mononuclear Species	121
IX.	Equilibria between Mono- and Dihydroxo-Bridged Dinuclear Complexes of Chromium(III), Rhodium(III), and Iridium(III) . . .	131
X.	Kinetics of the Hydrolysis of Dihydroxo-Bridged Cobalt(III) Complexes .	141
XI.	Equilibria between Tri- and Dihydroxo-Bridged Complexes . . .	145
XII.	Bridge Cleavage of Mixed Bridge Complexes	149
XIII.	Concluding Remarks	156
XIV.	Abbreviations for Ligands and Solvents	158
	References	160

Catenated Nitrogen Ligands Part II. Transition Metal Derivatives of Triazoles, Tetrazoles, Pentazoles, and Hexazine

DAVID S. MOORE AND STEPHEN D. ROBINSON

I.	Introduction	171
II.	Triazole and Triazolate Complexes	173
III.	Tetrazole and Tetrazolate Complexes	205
IV.	Pentazolate and Hexazine Complexes	230
	References	232

The Redox Chemistry of Nickel

A. GRAHAM LAPPIN AND ALEXANDER MCAULEY

I.	Introduction	241
II.	Steric and Electronic Requirements	242
III.	Probes of Structure	243
IV.	Oxidation of Nickel(II)	245
V.	Reduction of Nickel(II)	281
VI.	List of Abbreviations	288
	References	290

Nickel in Metalloproteins

R. CAMMACK

I.	Introduction	297
II.	Urease	300
III.	Hydrogenase	304
IV.	Methyl-Coenzyme M Reductase	323
V.	Carbon Monoxide Oxidoreductase	326
VI.	Concluding Remarks	329
VII.	Abbreviations	329
	References	330

Nitrosyl Complexes of Iron–Sulfur Clusters

ANTHONY R. BUTLER, CHRISTOPHER GLIDEWELL, AND MIN-HSIN LI

I.	Introduction	336
II.	Synthesis	337
III.	Molecular Structure: X-Ray Crystallography	354

CONTENTS

IV.	Molecular Structure: NMR Spectroscopy	364
V.	Electronic Structure	366
VI.	Chemical Reactivity	373
VII.	Biological Chemistry	384
	References	389

INDEX 395

CONTENTS OF RECENT VOLUMES 407

PREFACE

It was in 1959 that Eméleus and Sharpe initiated this series of reviews. Since that time, under their able guidance, it has become established as one of the leading review series in inorganic chemistry. It is with considerable pleasure, therefore, that I now take over the editorship.

At its inception, Eméleus and Sharpe adopted what they described as a broad definition of inorganic chemistry. As they indicated, the subject depends very much for its existence on the application of physical and physicochemical principles to chemical phenomena. One of their aims was the integration of structural, kinetic, and thermodynamic data with descriptive chemistry. All this and more has, I am quite sure, been achieved. Inorganic chemistry has certainly not become any less broad over the intervening years.

It is intended that the *Advances in Inorganic Chemistry* series will continue to provide a forum for scholarly and critical reviews by recognized experts. It does not seek to catalog each and every event so much as to guide the reader into thinking creatively about a subject, with recent significant developments to the fore.

The response to this series will hopefully remain international. Contributions will continue to be solicited by the editor, who will be assisted by a newly appointed advisory board. However, suggestions for future articles and for more general comments regarding the series will be welcome at any time.

Reviews in a number of rapidly developing areas, including the bioinorganic and solid-state areas, will no doubt feature prominently in future volumes. Topics formerly included in *Advances in Inorganic and Bioinorganic Mechanisms* (of which there are four volumes) will henceforth form a part of this series. Over the years, coverage of radiochemistry has not in fact merited inclusion in the title, and without wishing to exclude any such reviews from future issues, it is timely that this appendage be dropped.

In the present volume, articles are of a varied nature and include recent advances in the redox chemistry of nickel alongside the newly discovered roles for nickel in a number of enzymes. Overall editorial policy will remain unchanged with the aim to provide diverse topics within each volume. This will also, it is hoped, cater more for a cross-fertilization of ideas, which is so very important in the further development of the subject.

I am most grateful to those who have contributed to this volume and to those who have promised contributions to future volumes.

A. G. SYKES

DYNAMICS OF SPIN EQUILIBRIA IN METAL COMPLEXES

JAMES K. BEATTIE

School of Chemistry, The University of Sydney, New South Wales 2006, Australia

I. Introduction
II. Static Properties
    A. Magnetic Susceptibility
    B. Geometric Structure
    C. Electronic Spectra
    D. Vibrational Spectra
III. Techniques
    A. Magnetic Resonance Methods
    B. Temperature-Jump Relaxation
    C. Ultrasonic Relaxation
    D. Photoperturbation
    E. Mössbauer Spectroscopy
IV. Solution Dynamics
    A. Iron(II)
    B. Iron(III)
    C. Cobalt(II)
    D. Nickel(II)
V. Solid-State Dynamics
    A. Lifetime Limits
    B. Measured Rates
VI. Summary and Interpretation
    A. Octahedral $\Delta S = 2$ Equilibria
    B. Octahedral $\Delta S = 1$ Equilibria
    C. Planar–Tetrahedral $\Delta S = 1$ Equilibria
    D. Planar–Octahedral $\Delta S = 1$ Equilibria
VII. Implications
    A. Reaction Mechanisms
    B. Excited States
    C. Porphyrins and Heme Proteins
    References

I. Introduction

Octahedral transition metal complexes with four to seven d electrons and with intermediate ligand-field strengths can exist in high-spin or low-spin electron configurations. In some cases the energies of these two configurations are sufficiently similar such that both are thermally populated at some accessible temperature. Such complexes are described as being in spin equilibrium. The interconversion between the spin states has been termed "spin-crossover." It is closely related to the intersystem crossing process in excited states. It can also be described as an intramolecular electron transfer reaction (45). Examples of octahedral spin-equilibrium complexes generally have been confined to the $3d$ transition metals, with only a few examples reported among the $4d$ elements (31, 34, 83, 142).

Four-coordinate d^8 complexes can display a closely related electronic and geometric equilibrium between paramagnetic tetrahedral and diamagnetic planar isomers. Numerous examples are known in nickel(II) chemistry (80). In this case, as well as with the octahedral complexes described above, there is no change in the coordination number of the metal ion.

A change in the spin state of a metal ion also can accompany a change in coordination number. Again, in some cases conditions may be established in which an equilibrium exists between two complexes with different coordination numbers and different numbers of unpaired electrons. Some of the concepts which are used to describe intramolecular spin equilibria can be extended to the description of these coordination-spin equilibria. Examples include equilibria among four-, five-, and six-coordinate nickel(II) complexes and equilibria involving coordination number changes in iron porphyrin complexes and in heme proteins.

The phenomenon of spin equilibrium in octahedral complexes was first reported by Cambi and co-workers in a series of papers between 1931 and 1933 describing magnetic properties of tris(N,N-dialkyldithiocarbamato)iron(III) complexes. By 1968 the concept of a thermal equilibrium between different spin states was sufficiently well established that the definitive review by Martin and White described the phenomenon in terms which have not been substantially altered subsequently (112). During the 1960s the planar-tetrahedral equilibria of nickel(II) complexes were thoroughly explored and the results were summarized in comprehensive reviews published by Holm and co-workers in 1966 and 1973 (79, 80). Also, in 1968, Busch and co-workers published a review on iron, cobalt, and nickel complexes having anomalous magnetic moments which described, among others, complexes which undergo coordination-spin equilibria (8).

Thus 20 years ago there was an adequate understanding of many of the electronic and structural properties of complexes in spin equilibrium. There was little knowledge, however, of the dynamics of the spin equilibrium. The description of the complexes as spin isomers, with discrete nuclear configurations, and the concomitant observation of separate electronic spectra for the two spin states, set a lower limit for the spin state lifetimes longer than nuclear vibrational periods, i.e., $>10^{-13}$ second. There was some evidence that equilibration between spin states is rapid, for in certain cases averaged NMR spectra were observed in solutions, implying spin state lifetimes $<10^{-4}$ second. Furthermore, it was known that in the solid state some iron complexes display separate Mössbauer signals for the two spin isomers whereas other complexes display an averaged signal, which implies spin state lifetimes both greater than and less than the ^{57}Fe excited nuclear state lifetime of 10^{-7} second. Nevertheless, the only rate constants known had been reported from low-temperature NMR studies on the planar–tetrahedral equilibrium of dihalobis(phosphine)nickel(II) complexes (99, 129). Extrapolation to room temperature gave rate constants of the order of 10^5 sec^{-1} for the spin state interconversion.

Since 1968 there have been numerous studies on the physical and chemical properties of spin-equilibrium complexes. Many additional examples have been discovered or deliberately prepared. Extensive investigations of spin crossover in the solid state have focused on the differences between abrupt and gradual transitions which occur with a change in temperature. Most of these developments have been adequately reviewed (62, 65, 95).

This article examines the dynamics of spin-equilibrium processes, principally from studies in solutions. The properties of the complexes which are relevant to the dynamics studies are first reviewed. Then the techniques used to observe these rapid processes are described. Some aspects of solid-state dynamics are mentioned. Finally, some implications for the description of intersystem crossing processes in excited states and for spin equilibria in heme proteins are described.

Many of the complexes which occur in spin equilibrium possess ligands with complicated structures. Trivial abbreviations are used, with structural formulas given in the table in which the complex first appears. Generally the complexes are of low symmetry, but in the description of their electronic structure idealized symmetries are assumed and the appropriate term symbols are used accordingly.

II. Static Properties

A. MAGNETIC SUSCEPTIBILITY

Anomalous magnetic susceptibility is the characteristic feature of spin-equilibrium complexes. Measurement of magnetic susceptibility in the solid state, including its temperature dependence, is carried out on metal complexes using the classic Gouy and Faraday methods, as well as using the more recent techniques employing vibrating sample magnetometers and the superconducting quantum interference detector (SQUID) (121).

Measurement of magnetic susceptibility in solution is of more direct relevance to the investigation of solution dynamics. The change of state from solid to solution generally results in a change in the position of the spin equilibrium, usually but not always with an increase in the high-spin fraction. Consequently, determination of the magnetic moment in solution is desirable in order to evaluate the equilibrium constant under conditions as close as possible to those in which the dynamics are measured. Although the Gouy method can be used to measure the magnetic susceptibility of solutions, the method of choice is the nuclear magnetic resonance (NMR) technique of Evans (54).

The Evans NMR method has been widely described and applied (20, 36, 55, 87). The method involves the measurement of the chemical shift difference of the resonances of a reference molecule between a solvent and a solution containing the paramagnetic substance. An important criterion is that the reference substance does not interact directly with the paramagnetic solute. This is not a problem for most octahedral spin equilibrium complexes but must be considered in studies of coordination-spin equilibria. High-field NMR spectrometers confer advantages both of a larger intrinsic chemical shift difference due to the higher magnetic field and a twofold increase in the shift difference due to the geometry of the field parallel with instead of perpendicular to the axis of the NMR tube (15).

The Evans method gives excellent results provided adequate care is taken. A most important requirement is that the solution temperature is measured reliably. One effective means of accomplishing this for ^{1}H NMR is to insert into the NMR tube a capillary or additional coaxial sample of an NMR temperature calibrant solvent, usually methanol (158) or ethylene glycol (88). In this way the temperature measurement is made simultaneously with the susceptibility measurement. A second important factor is the variation of the solvent density with temperature (126). Because the density difference between the solvent and solution depends linearly on the concentration of the solute, it is only necessary to measure the concentration dependence of the density at one temperature and to use the temperature variation of the density of the pure solvent to calculate the temperature variation of the concentration (20). A final consideration in the data analysis is the choice of the limiting moments for the high-spin and low-spin isomers. Generally the accessible temperature range is insufficient for both of these to be determined empirically. One or both moments must be treated as adjustable parameters. With judicious choices of these magnetic moments, the data give excellent linear plots of the spin-equilibrium constant with temperature (Fig. 1), from which values of ΔH^0 and ΔS^0 for the spin equilibrium can be evaluated (20). Alternatively, nonlinear behavior is an indication of more complex equilibria (61).

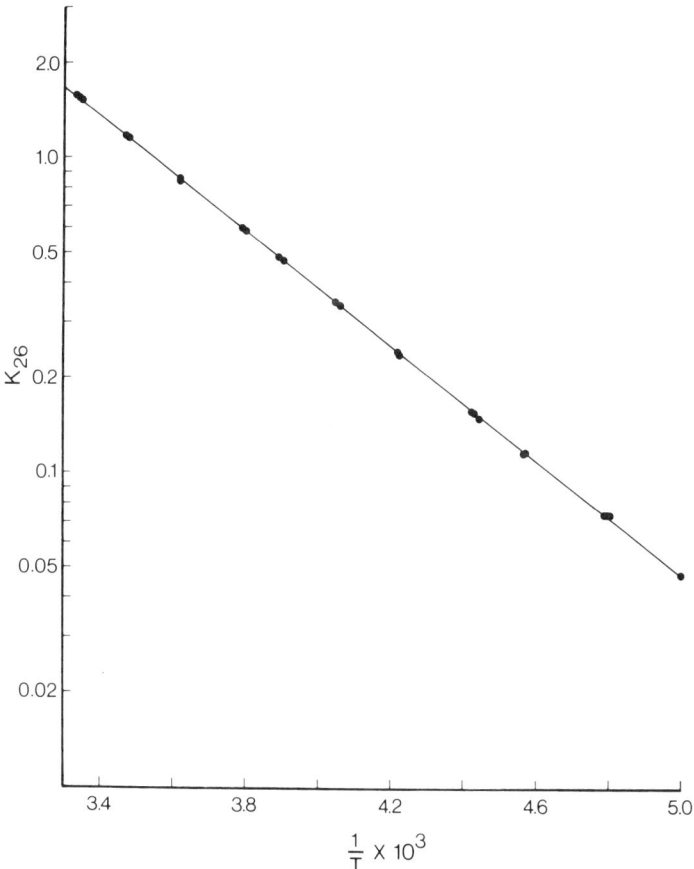

FIG. 1. Linear plot of solution magnetic susceptibility obtained by the Evans method.

Observation of magnetic susceptibility relaxation after perturbation of a spin equilibrium would be the most direct way to measure the dynamics of the equilibrium. This does not appear to have been reported as measured in solution. In principle susceptibility relaxation as a function of frequency could be measured much as dielectric relaxation is examined. The requirement is for a sufficiently strong magnetic field with very sensitive detection. A nonequilibrium magnetic susceptibility has been generated by light at low temperatures in the solid state (39).

B. GEOMETRIC STRUCTURE

A change in spin state among transition metal complexes in spin equilibrium invariably involves a change in the electron population of the σ antibonding $e_g{}^*$ orbitals. This produces a substantial change in the properties of the metal–ligand bonds. This variability in the population of σ^* antibonding orbitals is a conspicuous feature of the complexes of the $3d$ transition metals and accounts for many of their unique properties.

Changes in metal–ligand bond lengths accompanying a spin state transition can be observed with X-ray crystallography in a number of ways. One is the unusual occurrence of obtaining both spin isomers in the same crystal or in closely related polymorphs. The former has been found, for example, for a dibromobis(phosphine)nickel(II) complex, with both the planar and tetrahedral isomers existing together in the same crystal (89, 92). Different spin states may also occur in crystal lattices which differ only in the degree or nature of solvation or in the identity of counterions. This has been observed, for example, in the structures of the tris(α-picolylamine)iron(II) chloride, in which the dihydrate lattice contains a low-spin complex with a facial ligand configuration, while the methanol solvate contains a high-spin complex with the meridional configuration. The difference in spin states is ascribed to differences in hydrogen bonding (64). More commonly, the two spin states are obtained by variation in temperature, although the accessible temperature range is often not wide enough to encompass the two limiting structures. Finally, comparisons can be made among closely related compounds which occur in different spin states.

Some selected data for bond length differences accompanying spin state changes are presented in Table I. For iron(II) complexes there is a substantial change of 20 pm in the iron–ligand bond length between the low-spin and high-spin states. In three cases the ligand donor atoms are all nitrogens. In another case the donor atom set is P_4Cl_2. For the four Fe^{II}–P distances the difference between the high-spin and low-spin

TABLE I
Metal–Ligand Bond Length Differences between Spin States

Complex	M–L bond	Low spin (pm)	High spin (pm)	Δr (pm)	Reference
	Octahedral ⇌ Octahedral				
[Fe(2-pic)$_3$]Cl$_2$·xSol	Fe^{II}–N	201	220	19	*64, 63*
[Fe(2-pic)$_3$]X$_2$	Fe^{II}–N	200	220	20	*90*
Fe[HB(R$_2$pz)$_3$]$_2$	Fe^{II}–N	197	217	20	*125*
FeCl$_2$(dppen)$_2$·2(CH$_3$)$_2$CO	Fe^{II}–P	230	258	28	*29*
	Fe^{II}–Cl	233	236	3	*29*
[Fe$_3$(Ettrz)$_6$(H$_2$O)$_6$](CF$_3$SO$_3$)$_6$	Fe^{II}–N	203	217	14	*159*
Fe(dtc)$_3$	Fe^{III}–S	230	245	15	*30*
[Fe(R$_2$trien)]X	Fe^{III}–N	197	214	17	*143*
	Fe^{III}–O	188	192	4	*143*

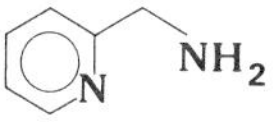

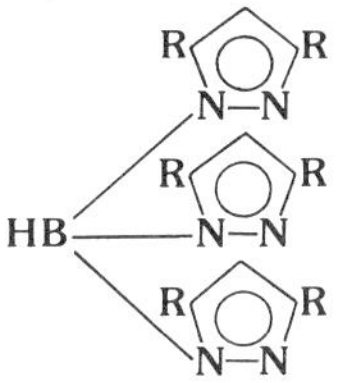

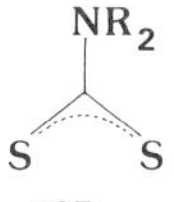

2-pic: pyridine-CH$_2$-NH$_2$

HB(R$_2$pz)$_3$: R = H, Me

dppen: Ph$_2$PCH=CHPPh$_2$

Ettrz: N-ethyl triazole

dtc: S$_2$CNR$_2$

R$_2$trien: RN-H-N-N-H-NR; R = H, Sal, acac, acacCl, 5-OCH$_3$Sal

(*continued*)

TABLE I (*Continued*)

Complex	M–L bond	Low spin (pm)	High spin (pm)	Δr (pm)	Reference
[Fe(Sal$_2$tet)](NO$_3$)	Fe^{III}–N	198	215	17	84
	Fe^{III}–O	188	194	6	84
	R$_2$3,2,3-tet (low spin)	RN–NH–NH–NR, R=Sal			
	R$_2$3,3,3-tet (high spin)	RN–NH–NH–NR, R=Sal			
[Co(terpy)$_2$]X$_2$	Co^{II}–N^a	207	214	7	58
	Co^{II}–N^b	187	208	21	58
	terpy	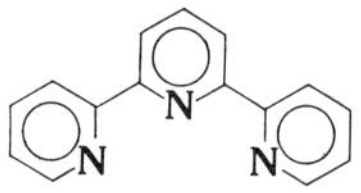			
	Planar ⇌ *Tetrahedral*				
Ni(P(CH$_2$Ph)Ph$_2$)$_2$Br$_2$	Ni–P	226	231	5	92
	Ni–Br	230	235	5	92
	Planar ⇌ *Octahedral*				
[Ni(cyclam)](ZnCl$_4$) ⇌ *cis*-[Ni(cyclam)(H$_2$O)$_2$]X$_2$	Ni–N	192	210	18	7
[Ni(Me$_4$cyclam)]X$_2$ ⇌ [Ni(Me$_4$cyclam)(H$_2$O)$_2$]X$_2$	Ni–N	199	214	15	9
	cyclam (R = H), Me$_4$cyclam (R = Me)	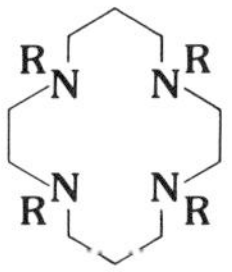			
[NiL$_\alpha$](ClO$_4$)$_2$ ⇌ NiL$_\alpha$(ClO$_4$)$_2$	Ni–N	197	207	10	75
	L$_\alpha$	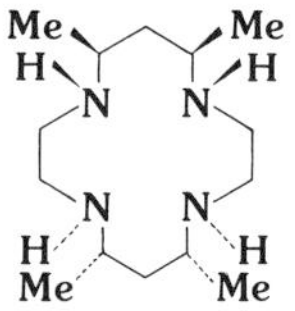			

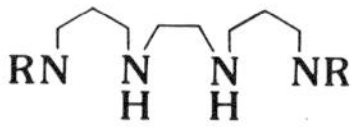

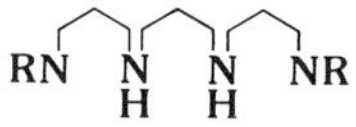

[a] Distal N.
[b] Central N.

states is 28 pm, the largest observed in any octahedral spin-equilibrium complex. The difference in the Fe–Cl distances is only 3 pm, however, giving an average distance for all six ligands of 20 pm, the same as is observed for the nitrogen donor ligands. Only in the case of a complex trinuclear structure, in which just the central iron changes spin state, is the change in Fe^{II}–N distance significantly less than 20 pm.

For iron(III) complexes the differences are somewhat smaller. From a large number of structures of iron(III) dithiocarbamate complexes a difference of 15 pm can be extracted between the limiting low-spin and high-spin structures (30). From the structures of a series of hexadentate ligands derived from triethylenetetramine with N_4O_2 donor atom set, differences are found of 17 pm for the Fe–N distances, but only 4 pm for the Fe–O distances. This gives an average change in the metal–ligand distances of 13 pm.

In both the Fe(II) and Fe(III) cases the spin state change involves a change in the population of the σ antibonding e_g^* orbitals of two d electrons. For the spin equilibrium of the d^7 cobalt(II) complexes the population of the e_g^* orbitals changes by only one electron. From examination of the structures of a series of $[Co(terpy)_2]^{2+}$ salts, a bond length difference between the two Co–N (central) distances of 21 pm was found between the spin states, with a difference of only 7 pm found between the four Co–N (distal) distances. This gives an average difference of 12 pm.

Thus it does appear from the available evidence that the effect of the transfer of an electron from the t_{2g} to the e_g^* orbitals is more substantial in the II oxidation state than in the III oxidation state. The $\Delta S = 1$ transition in Co(II) produces as large a bond length change as the $\Delta S = 2$ transition in Fe(III) complexes.

For nickel(II) complexes involved in planar–tetrahedral equilibria, the difference in nickel(II)–ligand distances is only 5 pm. This relatively small difference is understandable when it is recognized that the t_2 orbitals in tetrahedral complexes are only weakly σ antibonding, in contrast with the strong σ^* character of the e_g^* orbitals in octahedral complexes. There is, of course, substantial rearrangement of bond angles.

In the planar–octahedral equilibria of nickel(II) the d orbital population changes by transfer of one electron from the d_z^2 orbital to the $d_{x^2-y^2}$ σ antibonding orbital. This results in a substantial increase in the nickel–nitrogen distances in the plane. Accompanying this is the formation of new metal–ligand bonds in the axial positions.

A change in metal–ligand distances inevitably means a change in the volume occupied by the complex, although for coordination-spin equilibria the outcome is complicated by the different volumes occupied by the coordinated and uncoordinated ligands. For a nonzero volume change between the isomers, ΔV^0, the position of the spin equilibrium will be pressure dependent. The pressure dependence of the magnetic susceptibility has actually been measured by the Gouy method for some nickel(II) planar–tetrahedral equilibria (53), and for the spin equilibria of some iron(III) dithiocarbamate complexes (52). The Evans NMR method does not yet appear to have been used under high pressure to obtain ΔV^0 for spin equilibria, but with the increasing availability of high-pressure NMR techniques (47) this method will surely be used for this purpose.

The pressure dependence of other properties can be used to calculate ΔV^0. Because the volume differences between the spin states are relatively large, pressures of up to only 1000–3000 atm are sufficient to cause a significant shift in the spin equilibrium. Observation of the change in the electronic absorption spectrum, for example, enables calculation of ΔV^0, with the help of certain assumptions and ancillary experiments (19). The extinction coefficients for absorption by the two isomers must be obtained. In the simplest model they are assumed to be independent of pressure. In one approach (19) they were found by examination of the temperature dependence of the electronic absorption spectrum. This required knowledge of the temperature dependence of the spin-equilibrium constant, which was obtained from the temperature dependence of the susceptibility observed in the Evans NMR experiment. Clearly a more direct measurement is preferable.

The pressure dependence of the NMR spectrum of a nickel(II) complex which undergoes a coordination-spin equilibrium has been used to obtain the volume difference between the planar and octahedral isomers (118). In this case both the temperature and pressure dependence of the NMR spectra were analyzed simultaneously to yield five parameters, ΔH^0, ΔS^0, ΔV^0, and the chemical shifts of the two isomers. Subsequent determinations from the electronic spectra and ultrasonics relaxation are in good agreement with the NMR result (13).

The volume difference can be found without the use of high-pressure techniques in favorable cases from the amplitude of the sound absorption observed in ultrasonic relaxation of the spin equilibrium. This method will be described below in Section III,C.

Values of ΔV^0 obtained for a number of different spin equilibria by a variety of techniques are presented in Table II. Volumes of activation, ΔV^*, obtained from the pressure dependence of the rates of spin state interconversion, will be described in Section IV,A.

TABLE II

Volume Differences, ΔV^0, between Spin States

Complex	Method[a]	ΔV^0 (cm^3 mol^{-1})	Reference
Octahedral $\rightleftarrows$ Octahedral			
Fe^{II}(HB(pz)$_3$)$_2$	U	23.6	12
[Fe(pyim)$_3$]$^{2+}$	E	14.3[b]	115
	E	10.3[c]	115
	E	5.3[d]	115

pyim (structure: 2-(2-pyridyl)imidazole)

Complex	Method[a]	ΔV^0 (cm^3 mol^{-1})	Reference
[Fe(bzim)$_3$]$^{2+}$	E	12.4[b]	115
	E	9.6[c]	115
	E	4.3[d]	115
Fe^{III}(dtc)$_3$[e]	G	4–6	52
[Fe^{III}(acac)$_2$trien]$^+$	U	10.3	22
[Fe^{III}(sal)$_2$trien]$^+$	U	11.9	22
[Co^{II}(terpy)$_2$]$^{2+}$	E	10.1	18
Planar $\rightleftarrows$ Tetrahedral			
Ni(ethyl-ati)$_2$	G	7.5	53
Ni(n-propyl-ati)$_2$	G	8	53
Ni(β-naphthyl-ati)$_2$	G	8.5[f]	53
	G	5[g]	53
Ni(p-anisidyl-ati)$_2$	G	7.5[f]	53
	G	~4[g]	53

R-ati (structure)

R = Et, n-propyl, β-Naphthyl, p-Anisidyl

Complex	Method[a]	ΔV^0 (cm^3 mol^{-1})	Reference
Planar $\rightleftarrows$ Octahedral			
[Ni(Me$_4$cyclam)]$^{2+}$/(H$_2$O)$_2$	U	−10.1	13
	E	−8.6	13
	N	−10.0	118
[Ni(cyclam)]$^{2+}$/(H$_2$O)$_2$	E	−3.5	13
[Ni(2,3,2-tet)]$^{2+}$/(H$_2$O)$_2$	N	−3	123

[a] Methods: E, electronic spectra; G, Gouy; N, NMR; U, ultrasonic relaxation.
[b] CH$_3$CN.
[c] Acetone.
[d] CH$_3$OH plus 20% CH$_3$CN.
[e] A large number of complexes measured in different solvents.
[f] CHCl$_3$ solution.
[g] CH$_2$Cl$_2$ solution.

C. Electronic Spectra

The existence of spin states with significantly different metal–ligand bond lengths requires the isomers to exist with distinctly different electronic structures and properties. The spin-pairing energy for $3d$ electrons in a transition metal complex is of the order of 10,000–20,000 cm^{-1} (94). A useful illustration of this energy is the spectrum of chromium(III) complexes, in which the $^4A \rightleftarrows {}^2E$ transition corresponds to spin pairing from a quartet $(t_{2g})^3$ to a doublet $(t_{2g})^3$ d electron configuration. The transition occurs without any change in the $e_g{}^*$ antibonding population and produces accordingly a sharp spin-forbidden electronic band, exploited in the ruby laser. Its energy is about 16,000 cm^{-1}.

To compensate for this endothermic spin-pairing energy in the transition from the weak-field high-spin to the strong-field low-spin state, there is a substantial increase in the ligand-field stabilization energy. This arises from a significant increase in the ligand-field strength due to the shorter, stronger metal–ligand bonds. König and co-workers have evaluated Δ for the high-spin and low-spin states of [Fe(phen)$_2$(NCS)$_2$] to be 11,900 and 16,300 cm^{-1}, respectively (93). They subsequently measured an average metal–ligand bond length difference of 12 pm between the spin states, and observed that this is precisely what is predicted by ligand-field theory in which Δ depends inversely on the fifth power of the metal–ligand distance (96).

From these considerations it is clear that complexes in spin equilibrium do not exist at the crossover point between high-spin and low-spin configurations represented on a Tanabe–Sugano diagram. The two states are electronic isomers with geometric and electronic structures well separated on either side of the crossover point. The energy required to reach the crossover point represents at least part of the activation energy for the spin state interconversion.

Thus the electronic structure of each spin state leads to a distinctive electronic absorption spectrum for that state. The observed spectrum is the superposition of the spectra of the two isomers with different ligand fields. Interconversion is slow on the electronic time scale. There is no dynamic information in the spectrum but it can be used as a probe of the concentrations of the two spin states.

This description is most appropriate for those $\Delta S = 2$ spin equilibria of octahedral Fe(II) and Fe(III) in which spin-orbit coupling does not directly mix the two spin states. It is possibly less appropriate for $\Delta S = 1$ transitions for which sufficiently strong mixing could lead to a quantum mechanically mixed spin state as the ground state. That is, instead of two potential wells there is could be a single minimum (3, 6). There is little evidence for this phenomenon in octahedral complexes, but it has been described in five-coordinate complexes (*160*) and in tetragonally distorted complexes formed by porphryins and hemes. The latter will be discussed in Section VII,B.

There have been some measurements of the X-ray photoelectron spectra of spin-equilibrium complexes. Considerable difficulties have been encountered from X-ray-induced sample decomposition. Binding energy differences of a few tenths of an electron volt have been observed (*24, 103, 156*).

D. VIBRATIONAL SPECTRA

The description of spin states as electronic isomers with different metal–ligand distances requires that their vibrational spectra be a superposition of the spectra of the two separate spin states. The relative contribution of the two states to the observed spectrum will change with temperature as the population of the spin states changes. This has been observed (*76, 77, 122, 144, 145*). Difficulties occur with the assignment of the metal–ligand vibrational frequencies of particular interest for the analysis of the dynamics of spin state transitions. Some success has been achieved with the use of metal isotopic labeling (*82, 151*), but there are few reliable assignments.

The significance of these results is the demonstration that separate spin isomers do exist with distinctive vibrational spectra. Large differences have been observed, for example, in metal-sensitive infrared frequencies of $[Fe^{II}(pz)_3)_2]$. The low-spin isomer possesses metal–ligand bands in the region 450-400 cm^{-1} while the high-spin isomer has bands in the region 250–200 cm^{-1} (*82*). This is consistent with the 20-pm lengthening of the iron(II)–nitrogen bond which accompanies the spin state change (*125*).

There have been very few reports of the Raman spectra of spin-equilibrium complexes. In one experiment the presence of both high-spin and low-spin isomers of an iron(II) Schiff base complex was observed by the resonance Raman spectra of the imine region (*11*). The temperature dependence of the spectra was recorded for both solid and solution samples. Recently differences were described in the resonance Raman spectra of four- and six-coordinate nickel(II) porphyrin complexes which undergo coordination-spin equilibria. These studies are extensions of a considerable literature on spin state effects on the Raman spectra of iron porphyrins and hemes. There are apparently no reports of attempts to use time-resolved Raman spectra for dynamics experiments.

III. Techniques

The rapidity of spin state interconversions requires that relaxation methods be used to investigate spin equilibrium dynamics in solution. Only in the case of coordination-spin equilibria might the addition of low concentrations of ligands be observable by rapid mixing techniques with a high sensitivity detection method. In addition, the temperature coefficients of the rate constants are usually relatively small, so that lowering the temperature does not produce a large reduction in the rates. Consequently, the dynamics of spin state equilibria has generally been investigated with microsecond to nanosecond relaxation methods.

A. Magnetic Resonance Methods

In principle magnetic resonance methods are ideally suited to monitor the dynamics of spin equilibria. A spin crossover by definition produces a change in the number of unpaired electrons and hence in the magnetic moment of the complex. This necessarily affects the magnetic resonance spectra of the complex. Extensive use has been made of the resulting large paramagnetic chemical shifts for the purpose of determining the equilibrium constants and temperature dependence of spin equilibria (*69, 80, 86*).

NMR has not proved generally useful, however, for examining the dynamics of spin equilibria. Low-temperature proton NMR has been used successfully to obtain rates for some planar–tetrahedral equilibria in nickel(II) complexes (*99, 129, 130, 134*). Equation (1) illustrates the orbital occupancy and ground state terms for the d^8 equilibrium:

$$\begin{array}{cc}
\begin{array}{cc}
& \uparrow\downarrow \\
\uparrow\downarrow & \uparrow\downarrow \\
\uparrow\downarrow &
\end{array}
\rightleftarrows
&
\begin{array}{cc}
\uparrow\downarrow & \uparrow\ \ \uparrow \\
\uparrow\downarrow & \uparrow\downarrow
\end{array} \\
{}^1A(D_{4h}) & {}^3T(T_d)
\end{array} \qquad (1)$$

In this case the planar complex is diamagnetic and possesses the usual narrow line, high-resolution diamagnetic spectrum. The tetrahedral complex in T_d symmetry would possess a ^3T ground state. In approximately tetrahedral nickel(II) complexes the orbital angular momentum is incompletely quenched; the result is a very short electron spin relaxation time and an NMR spectrum with relatively narrow, paramagnetic shifted resonances.

The effect of interconversion between the planar and tetrahedral isomers on the NMR spectrum can be described by extension of the analysis for exchange between two diamagnetic sites (149). It should be noted that the well-known analysis of Swift and Connick (146, 148) is not directly applicable, because the paramagnetic site is not dilute, as it is in the case of solvent exchange on a paramagnetic metal ion.

Care must be taken to identify properly the exchange regime being observed. Only under slow exchange conditions can the exchange lifetime be extracted directly from the linewidth; otherwise the system must be analyzed to determine the chemical shift difference between the two sites and the relaxation time of the paramagnetic resonance. This identification is usually made from the temperature dependence of the observed linewidth. This procedure is complicated for spin equilibria by the temperature dependence of the mole fraction of the paramagnetic species as well as by the Curie temperature dependence of its paramagnetism. The dynamic properties of the spin equilibrium are superimposed on its anomalous magnetic properties. Despite these difficulties rate constants have been extracted.

For octahedral complexes only in the d^6 case does equilibrium occur between a diamagnetic state and a paramagnetic T state. In the other three cases one of the isomers is in a paramagnetic A or E ground state [Eqs. (2–5)].

$$\begin{array}{ccc} \overline{}\ \overline{}\ \uparrow\ \overline{} & & \\ \uparrow\downarrow\ \uparrow\ \uparrow & \rightleftarrows & \uparrow\ \uparrow\ \uparrow \\ {}^3T & & {}^5E \end{array} \quad (2)$$

$$\begin{array}{ccc} \overline{}\ \overline{}\ \uparrow\ \uparrow & & \\ \uparrow\downarrow\ \uparrow\downarrow\ \uparrow & \rightleftarrows & \uparrow\ \uparrow\ \uparrow \\ {}^2T & & {}^6A \end{array} \quad (3)$$

$$\begin{array}{ccc} \overline{}\ \overline{}\ \uparrow\ \uparrow & & \\ \uparrow\downarrow\ \uparrow\downarrow\ \uparrow\downarrow & \rightleftarrows & \uparrow\downarrow\ \uparrow\ \uparrow \\ {}^1A & & {}^5T \end{array} \quad (4)$$

$$\begin{array}{ccc} \uparrow\ \overline{} & & \uparrow\ \uparrow \\ \uparrow\downarrow\ \uparrow\downarrow\ \uparrow\downarrow & \rightleftarrows & \uparrow\downarrow\ \uparrow\downarrow\ \uparrow \\ {}^2E & & {}^4T \end{array} \quad (5)$$

These A and E states have long electron relaxation times which generally result in broad, unobservable NMR spectra. Hence it would be difficult to identify properly the exchange regime, even if dynamic effects were apparent in the NMR spectrum. In fact, where NMR spectra have been recorded, only exchange-averaged spectra are observed. This apparently arises because the spin equilibrium is established rapidly compared with the chemical shift difference between the spin states or the paramagnetic relaxation time, T_{2m}. Possibly even higher magnetic fields or more sophisticated pulse experiments could be used to address the problem.

Because NMR has in general been too slow a technique to observe the dynamics of octahedral spin equilibria, the higher frequency of observation of electron paramagnetic resonance (EPR) would seem well suited to this purpose. Despite the inevitable presence of unpaired electrons on complexes in spin equilibrium, there are serious limitations on the use of EPR, which are complementary to those of NMR. Diamagnetic states do not possess an EPR spectrum. Paramagnetic states with orbital angular momentum (T states) generally have a very short electron spin relaxation time, which leads to a broad EPR spectrum that is difficult to observe except at very low temperatures. In one novel attempt to circumvent this limitation, Mn(II) was doped into a lattice of an iron(II) spin-equilibrium complex and used as a probe of the magnetism (136). In general, however, only d^4 and d^7 octahedral complexes with paramagnetic E states are promising candidates for EPR experiments. While some signals have been observed for spin equilibrium complexes and can be used to place limits on the rates of spin state interconversion, the technique has not been widely applied.

B. Temperature-Jump Relaxation

All intramolecular spin equilibria have a nonzero enthalpy of reaction. This occurs because the high-spin isomer possesses greater entropy, both from its higher spin degeneracy and from its larger vibrational partition function, than does the low-spin isomer. Because ΔG is approximately zero, ΔH is therefore positive. Consequently, the equilibria are temperature dependent and can be perturbed by a rapid change in temperature.

The Joule heating temperature-jump experiment involves the discharge of a high voltage through a conducting solution to produce a temperature rise of a few degrees in several microseconds (59, 72). For nonconducting solutions microwave heating has been used, again with a heating rise time of some microseconds (25). Both of these techniques have been used to investigate coordination-spin equilibria, but have proved too slow to observe most intramolecular spin equilibria.

A major advance in the investigation of the intramolecular dynamics of spin equilibria was the development of the Raman laser temperature-jump technique (43). This uses the power of a laser to heat a solution within the time of the laser pulse width. If the relaxation time of the spin equilibrium is longer than this pulse width the dynamics of the equilibrium can be observed spectroscopically. At the time of its development only two lasers had sufficient power to cause an adequate temperature rise, the ruby laser at 694 nm and the neodymium laser at 1060 nm. Neither of these wavelengths is absorbed by solvents. Various methods were used in attempts to absorb the laser power, with partial success for microsecond relaxation times.

Nanosecond observations became possible with application of the stimulated Raman effect (16, 157). A schematic diagram of the experiment is shown in Fig. 2. Up to 10% of the power of the laser pulse can be scattered in the forward direction at a longer wavelength, which is the incident wavelength less the Raman energy of the scattering medium. With liquid N_2 as the Raman absorber, the energy of the Nd laser is shifted by the N–N Raman stretching frequency to 1410 nm (1.41 μm, 7090 cm^{-1}). This is the frequency of the first overtone of the O–H stretching frequency of water. Because the energy is absorbed in a vibrational mode of the solvent, and vibrational–translational energy transfer occurs within picoseconds, the temperature rise occurs within the time of the laser pulse width. The laser pulse triggers observation of the relaxation dynamics, which is typically performed spectrophotometrically with an interrogating beam focused on the heated volume of solvent and detected with single-shot transient recording. A few tenths of a millimeter thickness of water or other hydroxylic solvents can be heated several degrees within 20 nsec with this technique. High-pressure hydrogen can also be used as the Raman scatterer to shift the Nd laser energy to 1.89 μm (5290 cm^{-1}) which is absorbed by a number of nonaqueous solvents (2).

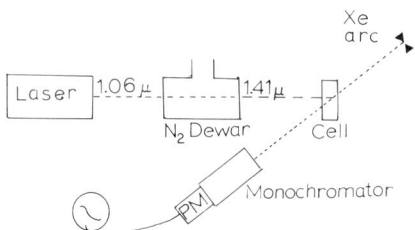

FIG. 2. Schematic diagram of the Raman laser temperature-jump experiment.

The Raman laser temperature-jump technique has been used in studies of a variety of spin-equilibrium processes. It was used in the first experiment to measure the relaxation time of an octahedral spin-equilibrium complex in solution (14). Its applications include investigations of cobalt(II), iron(II), iron(III), and nickel(II) equilibria.

C. Ultrasonic Relaxation

Spin equilibria are pressure dependent because the change in metal–ligand distances that accompanies the spin state transition produces a change in the volume of the complex. The single-step pressure-jump relaxation technique has an observation time of the order of 10^{-4} second, which is too slow to observe spin-equilibria dynamics. The periodic pressure variation of a sound wave, however, can be used for this purpose. Two different ultrasonic techniques have been used—the resonator method and the pulse method (50, 147). These span different frequency ranges. The resonator method operates from below 1 to above 15 MHz, or from 6×10^6 to 10^8 radians sec^{-1}, corresponding to relaxation times from longer than 10^{-7} to 10^{-8} second. The pulse method operates conveniently in the range 15–150 MHz, corresponding to relaxation times from 10^{-8} to 10^{-9} second.

Perturbation of a chemical equilibrium by ultrasound results in absorption of the sound. Ultrasonic methods determine the absorption coefficient, α (neper cm^{-1}), as a function of frequency. In the absence of chemical relaxation the background absorption, B, increases with the square of the frequency f (hertz); that is, α/f^2 is constant. For a single relaxation process the absorption increases with decreasing frequency, passing through an inflection point at the frequency ω (radians sec^{-1} = $2\pi f$) which is the inverse of the relaxation time, τ (seconds), of the chemical equilibrium [Eq. (6) and Fig. 3].

$$\alpha/f^2 = A(1 + \omega^2\tau^2)^{-1} + B \qquad (6)$$

The absorption due to the chemical relaxation, A, is obtained from the low-frequency absorption where $\omega \ll \tau$ [Eq. (7)].

$$A = \frac{2\pi^2 \rho v}{RT}\left[\Delta V^0 - \frac{\alpha_p}{\rho C_p}\Delta H^0\right]^2 \Gamma\tau \qquad (7)$$

In Eq. (7) ρ is the solution density, v is the sound velocity, α_p is the coefficient of thermal expansion, C_p is the specific heat, and Γ is the concentration dependence of the equilibrium. Neglecting activity

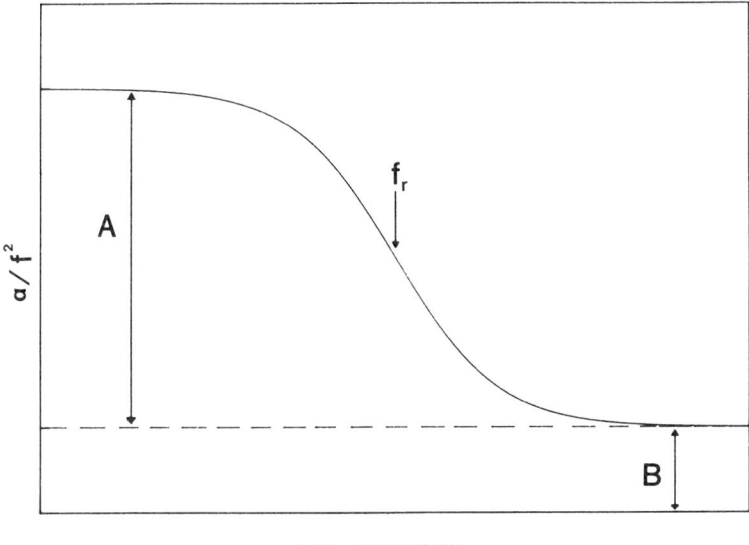

FIG. 3. Ultrasonic absorption curve for a single relaxation process.

coefficients, $\Gamma^{-1} = \sum_i v_i^2/c_i$, where v_i is the stoichiometric coefficient of the ith species and c_i its concentration. For a unimolecular spin equilibrium, $\Gamma^{-1} = [LS]^{-1} + [HS]^{-1}$, where [LS] and [HS] are the concentrations of the low-spin and high-spin species, respectively. Thus the ultrasonic relaxation absorption amplitude is a maximum when both spin isomers are present in equal concentrations.

Measurement of an ultrasonic relaxation curve enables evaluation of both the relaxation time, τ, and the relaxation amplitude, A. Interpretation of the relaxation time requires knowledge of the equilibrium constant. For a intramolecular isomerization such as a high-spin ⇌ low-spin equilibrium, the forward and reverse rate constants, k_1 and k_{-1}, respectively, can be evaluated from the relaxation time and the equilibrium constant from Eq. (8) (17).

$$\tau^{-1} = k_1 + k_{-1} = k_{-1}(K + 1) \qquad (8)$$

From the amplitude of the low-frequency excess sound absorption, A, the volume difference between the two isomers can be evaluated. This again requires knowledge of the equilibrium constant, in order to evaluate Γ, of the relaxation time τ, and of the standard enthalpy

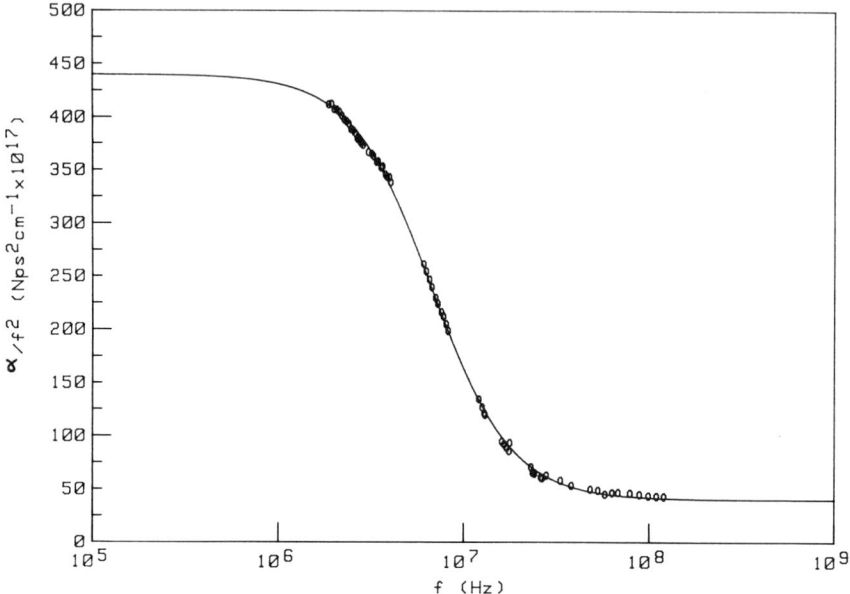

FIG. 4. Ultrasonic absorption data for an intramolecular spin-equilibrium relaxation.

change, ΔH^0. Consequently, a thorough investigation of the static equilibrium properties of the isomerization is required before the ultrasonics data can be fully evaluated. This is necessary in any case because it is necessary to have some independent evidence in order to assign confidently an observed ultrasonic absorption to a particular dynamic process. When this has been achieved, data of high precision can be obtained (Fig. 4).

D. PHOTOPERTURBATION

It is possible to perturb a spin equilibrium by photoexciting one of the isomers. Among the possible radiative and nonradiative fates of the excited state is intersystem crossing to the manifold of the other spin state. Internal conversion within this manifold ultimately results in the nonequilibrium population of the ground state. If these processes are rapid compared with the relaxation time of the spin equilibrium, then the dynamics of the ground state spin equilibrium can be observed. This experiment was first performed for spin equilibria with a coordination-spin equilibrium of a nickel(II) complex (85). More recently a similar phenomenon has been observed in the solid state at low temperatures (41). The nonequilibrium distribution can be trapped for long periods at sufficiently low temperatures; the effect has been termed "light-induced excited spin state trapping" (LIESST).

A question immediately arises whether the observed relaxation is that of the ground state or of some photochemical or photophysical process of the excited states. For the particular nickel(II) complex first studied in solution the relaxation was positively identified as the relaxation of the ground state equilibrium in an elegant experiment by Creutz and Sutin (37). Photoexcitation with the 1.06-μm radiation of a neodymium laser pulse results in depopulation of the octahedral complex. The subsequent reestablishment of the equilibrium is observed as a decrease in the absorbance of the planar complex, with a relaxation time of 0.30 ± 0.02 μsec. By Raman shifting the laser output to 1.41 μm the experiment is converted into a temperature-jump relaxation. Because ΔH^0 is negative for the planar-to-octahedral equilibrium, the increase in temperature results in an increase in the absorbance by the planar complex, again with the same relaxation time of 0.30 ± 0.02 μsec. This demonstrates that the photoexcitation does result in perturbation of the ground state equilibrium. An alternative means of establishing this is to scan the spectrum of the perturbed equilibrium to ascertain whether it corresponds to the ground state species.

This photoperturbation technique has been applied to a number of different spin-equilibrium complexes. Its success is apparently due to the fact that the relaxation times of the spin equilibria are longer in each case than the radiative and nonradiative processes in the excited states.

E. MÖSSBAUER SPECTROSCOPY

Mössbauer spectroscopy of the ^{57}Fe nucleus has been extensively used to investigate aspects of spin equilibria in the solid state and in frozen solutions. A rigid medium is of course required in order to achieve the Mössbauer effect. The dynamics of spin equilibria can be investigated by the Mössbauer experiment because the lifetime of the excited state of the ^{57}Fe nucleus which is involved in the emission and absorption of the γ radiation is 1×10^{-7} second. This is just of the order of the lifetimes of the spin states of iron complexes involved in spin equilibria. Furthermore, the Mössbauer spectra of high-spin and low-spin complexes are characterized by different isomer shifts and quadrupole coupling constants. Consequently, the Mössbauer spectrum can be used to classify the dynamic properties of a spin-equilibrium iron complex.

If the interconversion between the spin states is slow on the Mössbauer time scale, that is $\tau \gg 10^{-7}$ second, then the observed spectrum comprises the superposition of the spectra of the two separate spin states. This has been the most common observation. The areas of the two superimposed spectra can be used to estimate the relative populations of the two spin states. Their change with temperature can be used to ascertain whether the spin transition in the solid state is abrupt or gradual. These questions have been intensively investigated and reviewed (*35, 66–68*).

If the spin state interconversion is faster than the excited nuclear state lifetime, that is $\tau \ll 10^{-7}$ second, then the observed spectrum is an average of the spectra of the two spin states. Until recently this condition had been observed only for iron(III) complexes with thiocarbamate or selenocarbamate ligands—ferric dithiocarbamates (*119*), monothiocarbamates (*98*), or diselenocarbamates (*42*). Since 1982, however, there have been a number of reports of other iron(III) complexes which also display an averaged Mössbauer spectrum (*56, 57, 108–111, 124, 153, 155*).

If some iron(III) complexes undergo rapid spin interconversion on the Mössbauer time scale, and some undergo slow interconversion, then it is inevitable that a few will interconvert, at some accessible temperature, at a rate which produces dynamic effects on the Mössbauer spectrum. Such examples have now been found (*109, 111*). Rate constants have been extracted from these spectra and are necessarily of the order of 10^6–10^7 sec^{-1}. The interpretation of the spectral lineshapes is complex (*153, 154*), however, and further work will be needed to establish the reliability of the rate data obtained from such spectra.

IV. Solution Dynamics

A. Iron(II)

The dynamics of an octahedral spin equilibrium in solution was first reported in 1973 for an iron(II) complex with the Raman laser temperature-jump technique (*14*). A relaxation time of 32 ± 10 nsec was observed. Subsequently, further studies have been reported with the use of this technique, with ultrasonic relaxation, and with photoperturbation. Selected results are presented in Table III.

One striking feature of these results is the narrow range of relaxation times represented; they span only a range of four from 30 to 120 nsec. In part, this is an artifact of the relaxation techniques used, and in part, it

TABLE III

Dynamics of $^1A \rightleftarrows {}^5T$ Equilibria of Iron(II) Complexes in Solution at or near 298 K

Complex	Solvent	Method[a]	τ (nsec)	k_{15} (sec^{-1})	Reference
Fe(HB(pz)$_3$)$_2$	CH$_2$Cl$_2$/CH$_3$OH	T	32 ± 10	1 × 10^7	14
	THF	U	33.0 ± 0.7	4.9 × 10^6	12
Fe(6-MePy)(Py)$_2$tren^{2+}	CH$_3$OH; acetone/H$_2$O	T	120 ± 20	4 × 10^5	81
Fe(6-MePy)$_2$(Py)tren^{2+}	Acetone/H$_2$O; CH$_3$OH/H$_2$O	T	110 ± 30	4 × 10^6	81

RR'R''tren $N(CH_2CH_2N=\overset{H}{C}\overset{}{\underset{N}{\frown}}R,R',R'')_3$

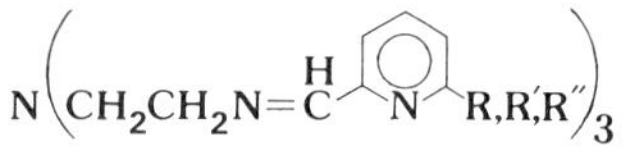

R,R' = 6-MePy; R'' = Py or
R = 6-MePy; R',R'' = Py

Fe(papth)$_2^{2+}$	H$_2$O	U	41.0 ± 0.2	1.7 × 10^7	12

papth

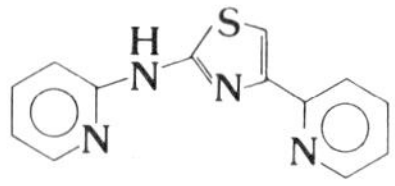

Fe(pyim)$_3^{2+}$	CH$_3$CN/CH$_3$OH	T	48 ± 8	1.1 × 10^7	137
	CH$_3$CN/CH$_3$OH	P	50 ± 3		114
	Acetone	P	45 ± 5		44
Fe(biz)$_3^{2+}$	CH$_3$CN	P	27		114

biz

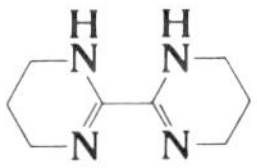

Fe(ppa)$_2^{2+}$	H$_2$O	P	102		114

ppa

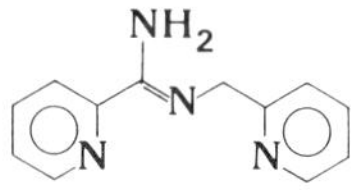

Fe(phenmethoxa)$_2^{2+}$	Acetone	P	110 ± 10	7.7 × 10^6	44

phenmethoxa

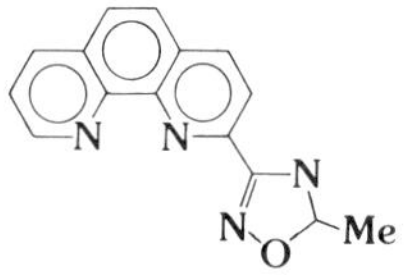

[a] Methods: T, Raman laser temperature-jump; U, ultrasonic relaxation; P, photoperturbation.

is a feature of the complexes that can be studied, as will be discussed below. The rate constants span a wider range, from 10^5 to 10^7 sec^{-1}, which reflects the range of the spin-equilibrium constants.

A second feature is the good agreement between the relaxation times observed using different techniques. After the first relatively imprecise study of $Fe(HB(pz)_3)_2$ with the Raman laser temperature-jump technique, the equilibrium was reinvestigated using the more precise ultrasonic relaxation method. Excellent agreement was found between the two methods. Similarly, the relaxation of the spin equilibrium in $Fe(pyim)_3^{2+}$ has been investigated in three different laboratories; two used the photoperturbation method and one the temperature-jump method. All three results were identical within experimental error. This is a particularly important observation, because of the uncertainty concerning the consequences of photoexcitation. For both the iron(II) complex described here and a nickel(II) complex described elsewhere, however, photoexcitation has been shown to result in perturbation of the spin equilibrium, which implies that the excited states decay rapidly and that at least some decay to the other isomer, perturbing the spin equilibrium.

Another feature of these results is that the relaxation time appears to be nearly solvent independent, in those cases in which studies have been performed in different solvents. This indicates that the dynamics of the spin equilibrium is predominately determined by intramolecular properties. There are some complexes for which hydrogen bonding to the solvent plays a significant role and both reaction volumes and activation volumes have been observed to be solvent dependent (*115*).

Consideration of the thermodynamics of a representative reaction coordinate reveals a number of interesting aspects of the equilibrium (Fig. 5). Because the complex is in spin equilibrium, $\Delta G^0 \approx 0$. Only complexes which fulfill this condition can be studied by the Raman laser temperature-jump or ultrasonic relaxation methods, because these methods require perturbation of an equilibrium with appreciable concentrations of both species present. The photoperturbation technique does not suffer from this limitation and can be used to examine complexes with a larger driving force, i.e., $\Delta G^0 \ll 0$. In such cases, however, ΔG^0 is difficult to measure and will generally be unknown.

For those systems with $\Delta G^0 \approx 0$, it follows that $\Delta H^0 \approx T\Delta S^0$. Conversion from the low-spin to the high-spin state involves lengthening and weakening the metal–ligand bonds. This results in a high-spin state with higher enthalpy and also greater entropy. Such a reaction profile is shown in Fig. 5 for the $Fe(HB(pz)_3)_2$ complex.

The activation parameters indicate that the quintet–singlet process,

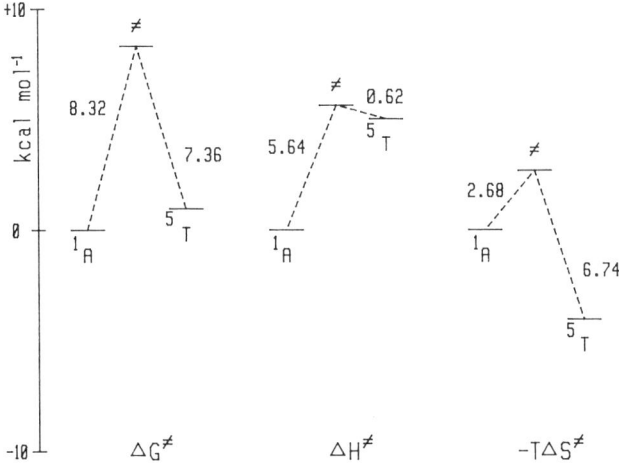

FIG. 5. Reaction coordinate profile for the octahedral spin equilibrium of [Fe(HB(pz)$_3$)$_2$].

represented by rate constant k_{51}, proceeds with a very small enthalpic barrier. This allows some inferences about the adiabaticity of the spin-forbidden transition. In conventional transition state theory the transmission coefficient, κ, is assumed to be unity.

$$k = \kappa(k_B T/h)\exp(-\Delta G^*/RT) \qquad (9)$$

For these spin-forbidden transitions, with $\Delta S = 2$, this assumption is very likely invalid and the process is nonadiabatic, i.e., $\kappa < 1$. If it is assumed that the entire entropic barrier of the quintet–singlet transition is due to nonadiabaticity, a minimum value of $\kappa \sim 10^{-4}$ is obtained.

The assumption that the entropic barrier to the spin state transition is due entirely to spin-forbidden nonadiabaticity is equivalent to assuming that the transition state has the structure of the high-spin state. This is unlikely to be the case, and insofar as the assumption is invalid, the minimum value of κ is increased. There is some evidence from the volume of activation for the spin state transition that the transition state lies well along the reaction coordinate between the low-spin and high-spin states. This novel experiment was accomplished independently by two groups with use of the photoperturbation technique with the sample subjected to variable pressure (44, 115). In the two cases reported, the volume of activation places the transition state about midway between the low-spin and high-spin states. If this is correct then there will be a considerable "chemical" contribution to the entropy of activation, due to the change in the structure and vibrational properties of the complex. The estimated minimum value of the transmission coefficient, κ, is correspondingly larger and the reaction process is correspondingly less "spin-forbidden." The available evidence would suggest that κ is no smaller than the order of 10^{-2}.

B. Iron(III)

The spin-equilibrium dynamics of iron(III) complexes in solution have been examined with the techniques of Raman laser temperature-jump, ultrasonic relaxation, and photoperturbation. The complexes investigated, the relaxation times observed, and one of the derived rate constants are presented in Table IV. Many of the relaxation times are quite short, and some of the original temperature-jump results (45) were found to be inconsistent with more accurate ultrasonic experiments (20) and later photoperturbation experiments (102). It has not been possible to repeat some of these laser temperature-jump observations. Instead, the expected absorbance changes and isosbestic points were found to occur within the heating rise time of the laser pulse, consistent with the ultrasonic and photoperturbation experiments (20). Consequently, none of the original Raman laser temperature-jump results is included in Table IV.

A number of observations indicate interconversion rates for some iron(III) complexes too fast to measure with existing techniques. Relaxation times less than the 30-nsec limit set by the heating rise time of the laser temperature-jump technique were observed for $[Fe(benzac_2trien)]^+$, $[Fe(Salmeen)_2]^+$, and $[Fe(Me_2dtc)_3]$ (45, 128). From ultrasonic observations a limit of less than 1 nsec was placed on the relaxation time for the first of these compounds,

TABLE IV

Dynamics of $^2T \rightleftarrows {}^6T$ Equilibria of Iron(III) Complexes in Solution at 298 K

Complex[a]	Solvent	Method[b]	τ (nsec)	k_{26} (sec^{-1})	Reference
$Fe(Sal_2trien)^+$	H_2O	U	5.33(5)	6.1×10^7	21, 20
	CH_3OH	P	5(1)	6.1×10^7	102
	Acetone	P	3(1)	6.1×10^7	102
$Fe(acac_2trien)^+$	H_2O	U	2.11(4)	1.6×10^8	20
$Fe(5\text{-}OCH_3Sal_2trien)^+$	Acetone/CH_3OH	P	5(1)	—	102

[a] For ligand abbreviations, see Table II.
[b] Methods: U, ultrasonic relaxation; P, photoperturbation.

[Fe(benzac$_2$trien)]$^+$ (20). These observations, together with the short relaxation times presented in Table IV, indicate that spin state interconversions in iron(III) complexes are slightly more rapid than those observed in iron(II) complexes. There are only a limited number of results available at present, however, and these may not be representative.

C. COBALT(II)

The cobalt(II) complexes which undergo spin equilibrium are of several different types. Octahedral high-spin complexes with a ^4T ground state are subject to Jahn–Teller distortion in the low-spin d^7 ^2E state. This effect is best documented in structures of the Co(terpy)$_2^{2+}$ spin-equilibrium complex. The high-spin isomer is nearly octahedral, with a difference in Co–N bond lengths between the central and distal nitrogens of only 6 pm. In the Jahn–Teller distorted low-spin state this difference has increased to 21 pm (58).

A second class of complexes which display spin equilibria includes those with a tetragonal ligand field (152). Complexes with a strong in-plane ligand field tend to give low-spin configurations. Addition of fifth and sixth axial ligands can lead to spin-equilibrium complexes (91, 164, 165). Such complexes could participate in coordination-spin equilibria as well as the intramolecular spin-equilibria observed in some five- and six-coordinate examples. The dynamics of the spin equilibria in these tetragonal complexes has not yet been examined.

The rate of the spin state change for the octahedral cobalt(II) complexes is expected to be faster than that observed for the iron(II) and iron(III) complexes. In the cobalt(II) case the spin state change involves only one electron, that is $\Delta S = 1$. The ^2E and ^4T states are directly mixed by spin-orbit coupling (10, 163). The spin state transition should be adiabatic, with $\kappa = 1$, without any "spin-forbidden" barrier. Furthermore, the coordination sphere reorganization involves a change in bond length of 21 pm along only two bonds, instead of all six bonds as in iron complexes. Both of these factors lead to the prediction of rapid spin state interconversion.

This prediction is confirmed by observation of a very rapid relaxation of the spin equilibrium in Co(terpy)$_2^{2+}$ in solution. A relaxation time of less than 15 nsec was observed in a Raman laser temperature-jump experiment (14). This is consistent with the absence of any relaxation of the small excess sound absorption found in ultrasonic experiments. An upper limit of 0.2 nsec for the relaxation time in water at 298 K can be calculated from the magnitude of the excess absorption, which is assumed to arise from a higher frequency relaxation process, and the values of ΔV^0 and ΔH^0 obtained from equilibrium measurements (18).

Such a short spin-equilibrium relaxation time raises the question of whether discrete spin state isomers exist. Their existence is affirmed by two observations. One is the persistence of electronic spectral bands typical of the low-spin ^2E state over a wide temperature range in solid samples (98). The other is the observation of EPR signals characteristic of the ^2E state in both solids and solutions between 4 and 293 K (98, 139). At very low temperatures EPR signals of both spin states can be observed simultaneously (98). At low temperatures hyperfine splitting into eight lines is observed from coupling with the $I = 7/2$ Co nucleus. As the temperature is raised the spectral features broaden and the hyperfine resolution is lost. This implies a relaxation process on the EPR time scale of 10^{10} sec^{-1}, or a relaxation time of the order 0.1 nsec, consistent with the upper limit set by the ultrasonic experiments.

These results make the observation of a "slower" relaxation time of 83 ± 23 nsec for the cobalt(II) complex $[\text{Co}(N\text{-NH}(\text{CH}_3)\text{-2,6-pyald})_2]^{2+}$ rather unusual (141). This experiment was performed with the Raman laser temperature-jump technique in acetonitrile–methanol solutions. The slow relaxation was tentatively ascribed to coordination and solvation sphere reorganization barriers (141). Complex and solvent-dependent phenomena were found with this complex in the ultrasonic experiment. Ultrasonic absorption also occurs, however, with a related high-spin t-butyl-substituted complex (18). No relaxation was observed in aqueous solutions, which implies a relaxation time of less than 1 nsec. In methanol solutions multiple absorption curves are found. The interpretation of all of these observations remains uncertain. There may be partial ligand dissociation processes occurring, which could be correlated with the spin equilibrium. Alternatively, the spin-equilibrium relaxation could be hindered by steric and solvent effects. It is generally agreed that electronic factors arising from the spin state transition are unlikely to inhibit the spin equilibrium.

D. NICKEL(II)

Nickel(II) complexes display a variety of equilibria which involve spin state changes. Planar four-coordinate complexes are invariably diamagnetic. These can undergo an intramolecular isomerization to paramagnetic tetrahedral four-coordinate species. Alternatively, the planar complexes can coordinate additional ligands to form five- and six-coordinate paramagnetic complexes. The additional ligand molecules can be Lewis bases in solution, or solvent molecules, or, in particular cases, uncoordinated donor atoms of the ligand coordinated in the planar complex.

In this section the dynamics of spin equilibria of nickel(II) will be described, beginning with intramolecular planar–tetrahedral equilibria and continuing with coordination-spin equilibria, in which bond formation and dissociation become involved.

1. Planar–Tetrahedral Equilibria

Planar–tetrahedral equilibria of nickel(II) complexes were the first spin-equilibria for which dynamics were measured in solution. It had been known that such complexes were in relatively rapid equilibrium in solution at room temperature, for their proton NMR spectra were exchange averaged, rather than a superposition of the spectra of the diamagnetic and paramagnetic species. At low temperatures, however, for certain dihalodiphosphine complexes, it is possible to slow the exchange and observe separate resonances for the two species. On warming the lines broaden and coalesce and kinetics parameters can be obtained. Two research groups reported such results almost simultaneously in 1970 (99, 129). Their results and others reported subsequently are summarized in Table V.

A second method used to examine the same equilibria is the laser photoperturbation technique. Irradiation with a Q-switched laser pulse at 1060 nm depletes the tetrahedral isomer; irradiation at the doubled frequency of 530 nm depletes the planar isomer. In both cases the same relaxation time of 0.93(4) μsec is observed for reestablishment of the equilibrium.

A remarkable feature of these results is that they have been obtained only for the series of diphosphinedihalonickel(II) complexes. Of the many planar–tetrahedral spin equilibrium complexes of nickel(II), no others show dynamic effects in low-temperature NMR spectra. All of the complexes which remain in fast exchange at low temperature are bis(chelate) complexes, including aminotroponeimines, salicylaldimines, and β-ketoimines. This difference in ligand structure could affect the temperature dependence of the interconversion rates. Inspection of the reaction coordinate profile for one of the diphosphinedihalonickel(II) complexes (Fig. 6) reveals that the free energy of activation is dominated by a large enthalpy of activation. This accounts for the success in slowing the rate significantly by lowering the temperature. If the enthalpy of activation were somewhat smaller for the bis(chelate) complexes, resolution of the spectra in the low-temperature NMR experiments would not be possible.

There is no reason, however, why the dynamics of these systems should not be observable with more rapid techniques. Exploratory

TABLE V
Dynamics of Planar–Tetrahedral Equilibria of Nickel(II)

Complex	Solvent	T (K)	k_{31} (sec^{-1})	Reference
Ni(PPh$_2$Me)$_2$Br$_2$	CHCl$_3$	223	1.1×10^3	129
	CHCl$_3$	298	8.5×10^5	129
Ni(P(p-ClC$_6$H$_4$)$_2$Me)$_2$Br$_2$	CHCl$_3$	223	5.2×10^2	129
	CHCl$_3$	298	3.8×10^5	129
Ni(P(p-MeOC$_6$H$_4$)$_2$Me)$_2$Br$_2$	CHCl$_3$	223	6.8×10^1	129
	CHCl$_3$	298	1.4×10^5	129
Ni(P(p-MeOC$_6$H$_4$)$_2$Me)$_2$Cl$_2$	CHCl$_3$	223	3.3×10^3	129
	CHCl$_3$	298	1.6×10^6	129
Ni(PPh$_2$Me)$_2$Cl$_2$	CHCl$_3$	298	2.6×10^5	99
Ni(PPh$_2$Me)$_2$Br$_2$	CHCl$_3$	298	4.5×10^4	99
Ni(PPh$_2$Me)$_2$I$_2$	CHCl$_3$	298	2.9×10^6	99
Ni(PPh$_2$Et)$_2$Cl$_2$	CHCl$_3$	298	9.2×10^5	99
Ni(PPh$_2$Et)$_2$Br$_2$	CHCl$_3$	298	1.6×10^5	99
Ni(PPh$_2$Pr)$_2$Cl$_2$	CHCl$_3$	298	6.6×10^5	99
Ni(PPh$_2$Pr)$_2$Br$_2$	CHCl$_3$	298	1.3×10^5	99
Ni(PPh$_2$Bu)$_2$Cl$_2$	CHCl$_3$	298	7.5×10^5	99
Ni(PPh$_2$Bu)$_2$Br$_2$	CHCl$_3$	298	1.4×10^5	99
Ni(PCyp$_3$)$_2$Cl$_2$	CH$_2$Cl$_2$	223	4.9×10^3	134
	CH$_2$Cl$_2$	298	1.2×10^6	134
Ni(PCyp$_3$)$_2$Br$_2$	CH$_2$Cl$_2$	223	1.1×10^3	134
Ni(PPhCyp$_2$)$_2$Br$_2$	CH$_2$Cl$_2$	223	3.2×10^2	134
	CH$_2$Cl$_2$	298	1.5×10^5	134
Ni(PPh$_2$Cyp)$_2$Br$_2$	CH$_2$Cl$_2$	223	4.5×10^2	134
	CH$_2$Cl$_2$	298	1.5×10^5	134
Ni(PPh$_2$Cyh)$_2$Br$_2$	CH$_2$Cl$_2$	223	1.0×10^3	134
	CH$_2$Cl$_2$	298	9.3×10^5	134
Ni(dpp)Cl$_2$	CH$_3$CN	296	6×10^5	116
Ni(dpp)Br$_2$	CH$_3$CN	281	$\tau = 0.5$ μsec	26
	CH$_3$CN	303	$\tau = 0.2$ μsec	26

dpp:

$$\begin{array}{c} \text{PPh}_2 \\ | \\ \text{PPh}_2 \end{array}$$

Complex	Solvent	T (K)	k_{31} (sec^{-1})	Reference
Ni(4-MeSal)$_2$	Cumene	298	$\tau \geq 0.2$ μsec	60

4-MeSal:

(2-hydroxy-5-methylbenzylidene)isopropylamine structure with Me, O, and N-iPr groups

Complex	Solvent	T (K)	k_{31} (sec^{-1})	Reference
Ni(anisydyl-ati)$_2$	Cumene	298	$\tau \geq 0.2$ μsec	60
Ni(benzyl-ati)$_2$	Cumene	298	$\tau \geq 0.2$ μsec	60

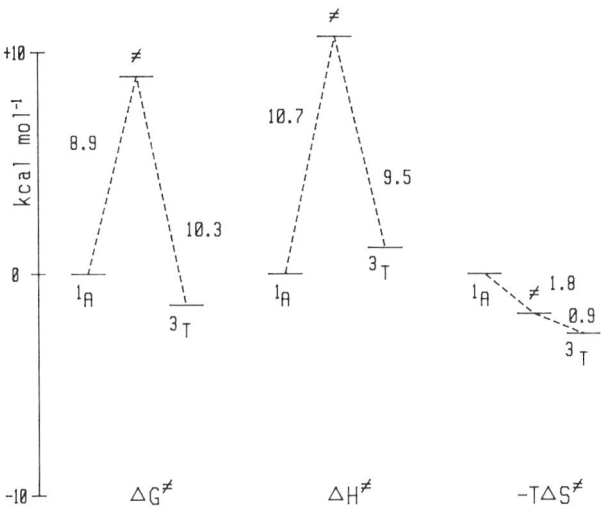

FIG. 6. Reaction coordinate profile for the planar–tetrahedral spin equilibrium of a diphosphinedihalonickel(II) complex.

experiments were undertaken with the ultrasonic resonance cell on cumene solutions of Ni(4-Mesal)$_2$, Ni(anisidyl-ati)$_2$, and Ni(benzyl-ati)$_2$. For the salicylaldimine complex a partial relaxation curve was observed, which indicates that relaxation occurs just to the low-frequency limit of this method, i.e., ≤1 MHz or 160 nsec. For the aminetroponeimine complexes no excess sound absorption was found. From the properties of the solvent and the known thermodynamics of the equilibrium this again indicates that the relaxation time is longer than 200 nsec, corresponding to rate constants less than about 5×10^6 sec^{-1}. Application of photoexcitation techniques should be possible with these complexes, as has been demonstrated by photoexcitation experiments in closely related systems (5, 105).

The available evidence thus suggests that relaxation times for planar–tetrahedral equilibria in nickel(II) complexes in solution at room temperature fall in the range 0.1–10 μsec, corresponding to rate constants of the order 10^5–10^7 sec^{-1}. These relaxation times are several orders of magnitude longer than those observed for octahedral spin equilibria. The reaction coordinate for the planar–tetrahedral equilibria is characterized by large enthalpies of activation for the reaction in both directions, in contrast with a relatively low enthalpy of activation for the high-spin to low-spin process in octahedral iron complexes.

There have been several theoretical analyses of the planar–tetrahedral interconversion process (*49, 106, 107*). The isomerization is not orbitally allowed without the inclusion of spin-orbit coupling, for the singlet ground state of the planar complex cannot correlate with the triplet ground state of the tetrahedral complex (*70, 161*). Inclusion of spin-orbit coupling mixes the ground states of both isomers sufficiently for the reaction coordinate to remain adiabatic; i.e., spin-orbit coupling for these $\Delta S = 1$ transitions mixes the two states so that there is an avoided crossing of the potential surfaces and the system remains on the lower energy surface from reactant to product.

An explanation for the slower rate and higher enthalpy of activation for the planar–tetrahedral isomerization compared with octahedral spin equilibria is apparent when the geometries of the processes are compared. For octahedral complexes the spin equilibrium produces a change in six metal–ligand bond lengths of not more than about 20 pm (0.2 Å). For the planar–tetrahedral isomerization the geometrical changes are much more extensive. Not only do the four bond lengths change by about 10 pm (0.1 Å), but there is a substantial change required as well along a bond bending or twisting coordinate. For a bis(bidentate) chelate complex the relative orientation of the two ligands must change by 90° between the planar and tetrahedral isomers. Motion of each donor atom with a metal–ligand bond length of 200 pm (2.00 Å) through an arc of 45° requires displacement by 157 pm (1.6 Å). This large geometry change undoubtedly contributes to the substantial activation enthalpy. There is no obvious explanation, however, for why the bis-bidentate chelate complexes undergo isomerization more rapidly than do the dihalophosphinenickel(II) complexes.

2. *Planar–Octahedral Equilibria*

Spin equilibria between diamagnetic planar four-coordinate complexes and paramagnetic five- or six-coordinate complexes clearly require bond formation and dissociation. The additional fifth and sixth ligands may be exogenous molecules or ions present in solution, solvent molecules themselves, or endogenous donors present in the planar complex as uncoordinated arms of a multidentate ligand. Although the thermodynamics of ligand addition to planar nickel(II) have been extensively studied, only recently have fast reaction techniques allowed examination of the dynamics (*162*).

Much of the focus of these studies has been on the relation between ligand substitution reaction mechanisms on octahedral nickel(II) and the dynamics of the planar–octahedral equilibria. For typical octa-

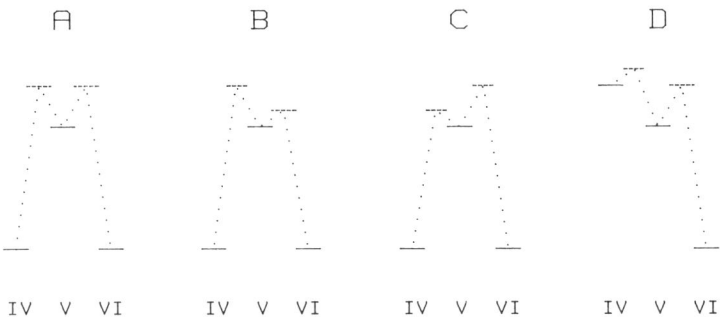

FIG. 7. Alternative reaction coordinate profiles for planar–octahedral equilibria of nickel(II).

hedral nickel(II) complexes, which undergo ligand substitution by a dissociative interchange mechanism, a five-coordinate species is a very unstable intermediate that closely resembles the transition state for substitution, and a four-coordinate species is of even higher energy (Fig. 7D). For complexes involved in a planar–octahedral equilibrium the four-coordinate species is stable, of energy comparable to the six-coordinate complex (Fig. 7A–C). Three possibilities exist for the reaction coordinate profile between the planar and octahedral complexes: in A (Fig. 7) the five-coordinate species closely resembles the transition state and is not an important intermediate; in B the interconversion between the planar and five-coordinate species is rate determining and the interconversion of five- and six-coordinate species is more rapid; in C the formation and dissociation of the six-coordinate, octahedral complex is rate determining and the interconversion of the planar and five-coordinate species is more rapid.

$$IV \underset{k_{54}}{\overset{k_{45}}{\rightleftarrows}} V \tag{10}$$

$$V \underset{k_{65}}{\overset{k_{56}}{\rightleftarrows}} VI \tag{11}$$

In some studies an assumption has been made about which of B or C (Fig. 7) is the rate-determining step, based more on prejudice about the role of the spin state change than on other evidence. At present it appears that only mechanism B can be distinguished from A and C. This is because mechanism B provides a pathway for ligand (or solvent) exchange from the six-coordinate complex, which is more rapid than the planar–octahedral interconversion and which can be observed by NMR independently. Thus if ligand (or solvent) exchange is more rapid than the planar–octahedral equilibrium, mechanism B obtains. In contrast, mechanisms A and C cannot be distinguished, unless the five-coordinate intermediate is particularly stable and accumulates. Otherwise, the ligand or solvent molecule does not have a sufficiently long residence time in the five-coordinate species to be relaxed, even if the five-coordinate species is paramagnetic.

The addition and dissociation of pyridine and substituted pyridine molecules to a planar nickel(II) complex with a quadridentate N_2O_2 ligand have been studied by the microwave temperature-jump technique in chlorobenzene solvent (38). The data were interpreted with the assumption of mechanism C (Fig. 7), i.e., that k_{65} is the smallest rate constant. Subsequently, however, ^{14}N NMR was used to measure the rate of pyridine exchange from the octahedral complex (138). The rates are the same for the two different experiments within a factor of two. This observation excludes mechanism B and is consistent with either mechanism A or C. The rate constants have consequently been presented in Table VI as k_{64}.

In contrast, solvent exchange on $Ni([12]aneN_4)(H_2O)_2^{2+}$ is greater than 100 times more rapid than the planar–octahedral equilibrium (33). Mechanism B obtains. This apparently arises because the ligand undergoes stereochemical reorganization between the planar complex and a cis configuration in the six-coordinate complex. In both $Ni(2,3,2\text{-tet})(H_2O)_2^{2+}$ and $Ni(cyclam)(H_2O)_2^{2+}$ water exchange is about a factor of 10 faster than the planar–octahedral equilibrium. In $Ni(trien)(H_2O)_2^{2+}$, however, water exchange proceeds at the same rate as the planar–octahedral interconversion.

These latter three tetraaminenickel(II) complexes are thought to possess a trans-octahedral geometry in the six-coordinate state. For these complexes little distinction can be inferred between mechanisms A, B, and C (Fig. 7). Both the four–five and five–six coordination number changes proceed at about comparable rates. A spin state change occurs in one of these steps. It is likely that the five-coordinate intermediate is paramagnetic, by analogy with the stable five-coordinate complexes described below. Hence the spin state change occurs in the four–five coordination step. This step is not substantially slower than the five–six coordination change, however, which does not involve a spin state change. This inference can be rationalized from the energetics of the reaction as follows. Dissociation of a ligand from the six-coordinate complex is endothermic, producing an unstable five-coordinate, but still high-spin, complex. The endothermic dissociation of the second ligand is compensated by the shortening of the in-plane

TABLE VI

Dynamics of Planar–Octahedral Equilibria at 298 K

Planar complex	Octahedral complex	Solvent	k_{64} (sec^{-1})	Reference
Ni(2,3,2-tet)$^{2+}$	Ni(2,3,2-tet)(H$_2$O)$_2$$^{2+}$	H$_2$O	4.9×10^5	85
		H$_2$O	$k_{65} = 8 \times 10^6$	127
	2,3,2-tet	H$_2$N–NH–NH–NH$_2$		
Ni(trien)$^{2+}$	Ni(trien)(H$_2$O)$_2$$^{2+}$	H$_2$O	1×10^6	85
		H$_2$O	$k_{65} = 1.1 \times 10^6$	135
Ni(bbh)	Ni(bbh)(py)$_2$	C$_6$H$_5$Cl	3.5×10^4	38
		py	5.6×10^4	138
	Ni(bbh)(2-Mepy)$_2$	C$_6$H$_5$Cl	3.8×10^5	38
		2-Mepy	8.5×10^5	138
	bbh	Ph$_2$C–O···O–CPh$_2$ / N–N···N–N / C(Me)=C(Me)		
Ni([12]aneN$_4$)	Ni([12]aneN$_4$)(H$_2$O)$_2$$^{2+}$	H$_2$O	1.5×10^4	33
		H$_2$O	$k_{65} = 4.2 \times 10^7$	33
	[12]aneN$_4$	(cyclic tetraamine)		
Ni(cyclam)$^{2+}$	Ni(cyclam)(H$_2$O)$_2$$^{2+}$	H$_2$O	3.6×10^6	60
		H$_2$O	$k_{65} = 4 \times 10^7$	127
Ni(Me$_4$cyclam)$^{2+}$	Ni(Me$_4$cyclam)(H$_2$O)$_2$$^{2+}$	H$_2$O	2.8×10^7	13
Ni(Salpip~)$_2$	Ni(Salpip)$_2$	Cumene	8.0×10^6	61
	Salpip	C$_6$H$_4$(OH)–CH=N–N(piperidine)		

metal–ligand bonds accompanying the spin state change. This confers the additional stability on the four-coordinate complex, which lowers its energy from that which obtains in the octahedral substitution mechanism (Fig. 7D). The spin state change affects the energetics and dynamics of the equilibria, but only as it affects the thermodynamics.

TABLE VII

Dynamics of Planar—Five-Coordinate Equilibria of Nickel(II)

Planar complex	Five-coordinate complex	Solvent	k_{54} (sec^{-1})	Reference
Ni(Et$_4$dien)Cl$^+$	Ni(Et$_4$dien)Cl$_2$	CH$_3$CN	7×10^5	78, 27
	Et$_4$dien = Et$_2$N–CH$_2$CH$_2$–NH–CH$_2$CH$_2$–NEt$_2$			
Ni(AEX)$^{2+}$	Ni(AEX)(H$_2$O)$^{2+}$	H$_2$O	1.5×10^7	28
	AEX = benzene-1,2-bis(SCH$_2$CH$_2$N)(=NMe)(NMe$_2$)	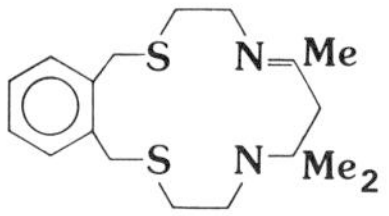		
Ni(Me$_4$cyclam)$^{2+}$	Ni(Me$_4$cyclam)(CH$_3$CN)$^{2+}$	CH$_3$CN	4.0×10^6	75a
Ni(Me$_4$cyclam)$^{2+}$	Ni(Me$_4$cyclam)(DMF)$^{2+}$	DMF	1.6×10^7	104

There is no evidence that the reactions are in any way nonadiabatic, i.e., that the electronic aspects of the spin state change inhibit the dynamics.

Some diamagnetic planar nickel(II) complexes add only one ligand to form paramagnetic five-coordinate species. The dynamics of several of these equilibria have been examined by photoperturbation or NMR methods. The rate constants present in Table VII are of the order 10^6 sec^{-1} for the dissociation of the ligand from the five-coordinate species. These rates are comparable with those of the planar–octahedral equilibria and are consistent with the mechanistic interpretation presented above.

V. Solid-State Dynamics

Only three aspects of the dynamics of spin equilibria in the solid state will be reviewed. One is the classification of the rates of spin interconversion based on spectroscopic properties observed in the solid. The second is the direct measurement of spin state lifetimes in crystals and powders. The third, the comparison of these with the dynamics observed in solutions, will be described in the course of the discussion.

The temperature dependence of the spin state populations in solids, as measured by magnetic susceptibility or any of the other properties which differ with the spin state, falls into two classes. In one class there is a gradual transition from one spin state to the other, more or less following a Boltzmann distribution, just as is observed in solutions. In the other class there is an abrupt change between the spin states at a particular transition temperature, often with hysteresis as the temperature is raised and lowered. In this second class the spin state transition is clearly a cooperative property of the lattice as well as of the discrete spin-equilibrium complex. The mechanisms of these cooperative spin state transitions have been extensively investigated and reviewed (*65, 95*) and will not be discussed further here. Attention will be focused on the gradual transitions which potentially reflect properties of the individual complexes. Even in these cases some degree of cooperativity can remain (*57, 74*).

A. LIFETIME LIMITS

The physical and spectroscopic properties of a spin-equilibrium complex can appear to be either the average or the superposition of the properties of the separate spin states. Which occurs is dependent on the time scale of the observation relative to the relaxation time of the equilibrium. Thus the electronic and vibrational spectra always appear as a superposition of the two isomers because each spin state possesses a distinctive potential energy surface with its characteristic electronic and vibrational properties. On the other hand, the NMR spectra appear as the average of the spectra of the two spin states, for all but the slowest interconversions, because the frequency of the interconversion is high compared with the frequency differences of the chemical shifts or the inverse of the spin relaxation times of the two isomers.

The two techniques which have been used effectively to set limits on the rates of spin state interconversions are Mössbauer and EPR spectroscopies. As described in Section III,E, the lifetime of the excited nuclear state involved in the Mössbauer effect is 10^{-7} second. Thus the observation of the Mössbauer spectrum can immediately classify the spin state lifetime as greater than or less than 10^{-7} second. Both conditions have been observed, as was described in Section III,E.

There is some evidence that the strength of intermolecular forces determines the degree of cooperativity and the rate of spin state interconversion in the lattice (*154, 155*). This is a reasonable hypothesis, for it assumes a continuum of behavior, from very weak interactions, which reflect intramolecular properties, to strong intermolecular forces, which cause cooperative phase transitions and abrupt spin state changes. Neutral complexes with a molecular lattice and little or no hydrogen bonding between the molecules, such as some iron(III) dithiocarbamate complexes, undergo rapid spin interconversion even in the solid state. Neutral complexes with larger volume changes between the isomers, such as the iron(II) pyrazolylborate complexes, which cause a greater reorganization of the crystal lattice, undergo slow interconversion on the Mössbauer time scale (67, 68). As hydrogen bonding causes the lattice to become more of a network, cooperativity increases and the rates of spin state interconversion decrease. Because the lattice properties and the intramolecular requirements of the spin state transition depend on a number of different factors, there will be no simple relationship between the intrinsic rate and the solid-state rate. Indeed, there is increasing evidence that solvation effects influence the properties of the spin equilibrium even in solution (115), as would be expected.

The EPR spectrum of a spin-equilibrium complex can be used to establish a lower limit to the spin state lifetimes of the order of 10^{-10} second. In an important paper in 1976, Hall and Hendrickson reported observation of EPR signals for both the high-spin and the low-spin isomers of iron(III) dithiocarbamate complexes at 4–12 K as powders, glasses, and doped solids (71). This resolved the question whether these complexes possess distinct high-spin and low-spin states. It also sets a lower limit on their interconversion lifetimes. Similarly, the observation of signals for both the high-spin and low-spin states of $[Co(terpy)_2^{2+}]$ (97) leads to the same conclusions about this complex. In both cases the interconversion rates in solution have proved too fast to measure, with lifetimes of less than 10^{-9} second indicated. The solution measurements were undertaken, of course, at room temperature and the EPR measurements at close to 4 K. Significant differences in the rates of solid and solutions at room temperature are still possible.

B. MEASURED RATES

Mössbauer spectroscopy can only be used to obtain rates of interconversion if the lifetimes are close to 10^{-7} second. As described in Section III,E a few examples satisfying this condition have been found. Some questions remain over the quantitative interpretation of the data. Nevertheless, spin-equilibrium relaxation lifetimes have been estimated from Mössbauer temperature-dependent linewidths for two salts of an iron(III) complex, $[Fe(acpa)_2]^+$. The lifetimes are of the order 10^{-5}–10^{-7} second over temperature ranges from 100 to 300 K (109, 111).

Interconversion rates have also been measured recently using a photoperturbation technique on solid samples. Irradiation of a sample at very low temperatures causes population of the higher energy spin state with an efficiency of 0.1–1% (*40, 41, 73, 77*). This "light-induced excited spin state trapping" occurs at temperatures too low for the thermally activated back intersystem crossing reaction to occur. On raising the temperature to ~60 K this process occurs within minutes (*39*). The relaxation is not first order, but the sample of the iron(II) complex undergoes a cooperative spin state transition. The effect has also been observed with the complex $[Fe(2-pic)_3]Cl_2 \cdot EtOH$, which undergoes a gradual spin state change as a crystal. In this case the relaxation of the photoperturbed equilibrium occurs at 25–30 K within an hour (*39*). No rates have yet been reported, but this appears to be a promising general method with convenient relaxation times for studying the interconversion kinetics in solid samples. The extrapolation to room temperature conditions is too long, however, for reliable, detailed comparisons to be made between solids and solutions. There is no reason, however, why increasingly sensitive nanosecond photometric detection will not enable the photoperturbation technique to be used on solids at room temperature as it has been used on solutions.

VI. Summary and Interpretation

The dynamics of spin equilibria in solution are rapid. The slowest rates are those for coordination-spin equilibria, in which bonds are made and broken; even these occur in a few microseconds. The fastest are the $\Delta S = 1$ transitions of octahedral cobalt(II) complexes, in which the population of the $e_g^* \, \sigma$ antibonding orbital changes by only one electron; these appear to occur in less than a nanosecond. For intramolecular interconversions without a coordination number change, the rates decrease as the coordination sphere reorganization increases. Thus the $\Delta S = 2$ transitions of octahedral iron(II) and iron(III) are slower than the $\Delta S = 1$ transitions of cobalt(II), and the planar-tetrahedral equilibria of nickel(II) are slower again, with lifetimes of about a microsecond.

A. Octahedral $\Delta S = 2$ Equilibria

The spin equilibria of octahedral iron(II) complexes are the best studied examples in both the solid state and in solution. Both the low-spin ^1A and the high-spin ^5T states are regular octahedra, so the transition can be considered a one-dimensional problem in the totally symmetric metal–ligand stretching coordinate. The bond length change is large (Table I), with an average difference of 20 pm per bond between the high-spin and low-spin states. Nevertheless, the rates of interconversion are rapid. The relaxation times observed for eight different complexes with the use of three different techniques span the narrow range from 27 to 120 nsec.

These relaxation times correspond to rates which are about 10^6 slower than the thermal vibrational frequency of $6 \times 10^{12} \sec^{-1}$ ($k_B T/h$) obtained from transition state theory. The question arises how much, if any, of this free energy of activation barrier is due to the spin-forbidden nature of the $\Delta S = 2$ transition. This question is equivalent to evaluating the transmission coefficient, κ, that is, to assess quantitatively whether the process is adiabatic or nonadiabatic.

It is not possible to evaluate κ directly, for it appears with the entropy of activation in the temperature-independent part of the rate constant. An estimate of κ requires an extrathermodynamic assumption. In two cases of iron(II) spin equilibria examined by ultrasonic relaxation the temperature dependence of the rates was precisely determined. If the assumption is made that all of the entropy of activation is due to a small value of κ, minimum values of 10^{-3} and 10^{-4} are obtained. Because there is an increase in entropy in the transition from the low-spin to the high-spin states, this assumption is equivalent to assuming that the transition state resembles the high-spin state. There is now evidence that this is not the case. Volumes of activation indicate that the transition state lies about midway between the two spin states. This is a more chemically reasonable and likely situation than the limiting assumption used to evaluate κ. In this case the observed entropy of activation includes some "chemical" contributions which arise from increased solvation and decreased vibrational partition functions as the high-spin state is compressed to the transition state. Consequently, the minimum value of κ is increased and is unlikely to be less than about 10^{-2}.

The singlet and quintet states are not mixed directly by spin-orbit coupling. According to this description the intersystem crossing should be completely forbidden. The two spin states are mixed, however, through second-order coupling with the excited triplet state. Thus the energy of that state is important in evaluating the extent of mixing of the singlet and quintet states and the magnitude of κ. Estimates of mixing of the order of 10–100 cm^{-1} have been made (23, 45). According to the Landau–Zener description of surface crossings, this corresponds to values of κ of the order 10^{-3} to 10^{-1}, in good agreement with the observations. Alternatively, a description of the reaction as a multiphonon radiationless transition has been given (23). Rate constants of 10^9 sec^{-1} are calculated, in reasonable agreement with the observed 10^7 sec^{-1}.

The description of the $^2T-{}^6A$ spin equilibrium of octahedral iron(III) complexes is similar to that for iron(II). Again the $\Delta S = 2$ transition occurs between octahedral states along a single metal–ligand coordinate. Again there is no direct mixing between the states, but coupling through the intermediate quartet state. There are fewer structural data for iron(III) complexes, but the difference in bonds lengths between the spin states appears to be somewhat smaller, with an average among three structures of 14 pm (Table I). If this is correct it would account for the somewhat faster rates observed for the relaxation of the spin equilibrium. For three complexes the relaxation times are 2–5 nsec, an order of magnitude shorter than those observed for iron(II). It should be noted that all three of these complexes are derived from a common N_4O_2 ligand structure and that, therefore, these results may not be representative. Minimum values of κ of 10^{-2} to 10^{-3} have again been made with the assumption that the transition state resembles the high-spin state. A more reasonable transition state intermediate between the doublet and sextet geometries will increase κ to 10^{-2} to 10^{-1}, again in agreement with estimates of the mixing between the spin states (94).

The conclusion is that these spin equilibria are only weakly "spin forbidden," with nonadiabatic transmission coefficients not smaller than 10^{-2}. The activation free energy is dominated by the inner-sphere reorganization energy which arises from the substantial bond length changes which accompany the transfer of two $e_g{}^*$ electrons. The dynamics can be properly described as an "intramolecular electron transfer reaction dominated by inner-sphere effects" (45). Solvent effects are small. In outer-sphere electron transfer reactions solvent reorganization is an important contributor to the activation free energy because of the transfer of charge. In spin equilibria the electrons are only redistributed on a single center. Solvent effects will be smaller and dependent on differences in solvation between the two spin states. These will include not only the change in volume between the two spin states, but also specific differences in hydrogen bonding in some cases.

B. OCTAHEDRAL $\Delta S = 1$ EQUILIBRIA

If the above analysis is correct, spin equilibria with $\Delta S = 1$ will relax significantly more rapidly than those with $\Delta S = 2$, for two reasons. One is that the transmission coefficient κ will be close to unity, because spin-orbit coupling directly mixes the two spin states. The second is that the coordination sphere reorganization barrier is lower, because the two spin states differ by only a single e_g^* electron.

The available evidence, which is limited to cobalt(II) complexes, is ambiguous. Crystallographic evidence on the $[\text{Co(terpy)}_2]^{2+}$ complexes indicates an average bond length difference of 12 pm between the high-spin and low-spin states. This is substantially less than the 20 pm found in iron(II) complexes, but not significantly different from the 14 pm seen in some iron(III) examples. Moreover, because the low-spin cobalt(II) ^2E state is subject to Jahn–Teller distortion, two of the six bond lengths differ by 21 pm, while the other four differ by only 7 pm (Table I). On a harmonic potential surface the energy required to distort the bonds to the transition state depends on the square of the distance, so the large difference for two of the bonds will contribute more than the average difference would imply. This suggests that the difference in spin-equilibria relaxation times between iron(III) and cobalt(II) might depend largely on the difference in κ.

The dynamics results are insufficient to resolve this question. For $[\text{Co(terpy)}_2]^{2+}$ the spin state relaxation appears to occur in only a few tenths of a nanosecond, consistent with the prediction that it could be an order of magnitude faster than that of iron(III). On the other hand, some relaxation process in another [Schiff base cobalt(II)] complex occurs with a slower relaxation time of 83 nsec. Until more data are acquired, this question must remain unresolved. It is a reasonable surmise, however, that these reactions are adiabatic.

C. Planar–Tetrahedral $\Delta S = 1$ Equilibria

If the electronic spin state change were the critical determinant of the dynamics of spin equilibria, then the $\Delta S = 1$ equilibration between planar and tetrahedral nickel(II) isomers would occur more rapidly than equilibration of the octahedral $\Delta S = 2$ spin states. This is not observed. Even though the $\Delta S = 1$ transition is most likely adiabatic, the large coordination sphere reorganization energy requirement causes these nickel(II) isomerizations to occur relatively slowly, with relaxation times of the order of a microsecond.

For those which have been studied by low-temperature NMR, in the dihalodiphosphinenickel(II) family, the activation parameters are characterized by a high enthalpy of activation. This is consistent with a large reorganization energy requirement. Although the bond length difference between the spin states is only 5 pm, there is a large change in the interligand bond angles between the planar and tetrahedral isomers. Because the energy required for distortion along a harmonic surface increases as the square of the displacement, a large bond angle change can produce a large activation energy barrier.

Only a few examples of these planar–tetrahedral equilibria have been successfully investigated and there remain unanswered questions about the dynamics. Nevertheless, it seems reasonable to regard these equilibria as intramolecular isomerizations in which the spin state change is not an important factor in the dynamics. Electron spin affects the bonding, geometry, and thermodynamics of the two isomers, but there is apparently sufficient mixing of the singlet and triplet states to allow their interconversion to be adiabatic.

D. Planar–Octahedral $\Delta S = 1$ Equilibria

Analysis of these equilibria is more complex than those described above because the planar–octahedral reactions are not intramolecular, but involve formation and dissociation of metal–ligand bonds. The processes are most likely to be adiabatic, a consequence of spin-orbit coupling between the singlet and triplet states. All of the dynamics have been observed with nickel(II). Consequently, the reactions observed for nickel(II) are usefully regarded in the context of the model for dissociative ligand substitution on that metal ion. The presence of the nearly equienergetic planar singlet state reduces the energy of the activated complex and increases the rate of the ligand dissociation reaction. The role of the five-coordinate intermediate is not clear, but for most of the reactions which have been examined it appears to be of high energy and not to accumulate. There is no evidence to indicate that electronic aspects of the spin state transition produce a rate-limiting process, except insofar as they contribute to the coordination sphere reorganization energy.

VII. Implications

The conclusions described in the previous section are inferred from a relatively small number of observations of spin-equilibrium dynamics. Nevertheless, they are internally self-consistent and also compatible with a much wider set of observations derived from studies of electron transfer reactions of metal complexes. For these reasons there is hope that they possess some generality and can be applied to other systems.

There are a few examples of spin equilibria with other metal ions which have not been mentioned above. In cobalt(III) chemistry there exist some paramagnetic planar complexes in equilibrium with the usual diamagnetic octahedral species (22). The equilibria are the converse of the diamagnetic-planar to paramagnetic-octahedral equilibria which occur with nickel(II). Their interconversions are also presumably adiabatic. Preliminary observations indicate relaxation times of tens of microseconds, consistent with slower ligand substitution on a metal ion in the higher (III) oxidation state (120).

Examples of spin equilibria among d^4 ions are rare. At least one Mn(III) example is known, which shows a cooperative spin state transition at low temperature (142a). No dynamics have been reported. The spin state transition has $\Delta S = 1$ and is therefore probably adiabatic. There is likely to be a substantial coordination sphere reorganization energy, however, for in this case the high-spin state is Jahn–Teller distorted, so there will be a large difference in some metal–ligand bond lengths between the two spin states.

A second example is the recently reported singlet–triplet equilibrium in a Mo(II) complex, cis-[Mo(bipy)$_2$(OPr$_i$)$_2$] (31). The interpretation of the magnetism is complicated by the possibility that the equilibrium is not entirely metal centered, but involves charge transfer to the bipyridyl ligands. Because a strong field is required to spin pair the d^4 configuration, however, it is not impossible that examples exist among Mo(II) chemistry. The possible dynamics are subject to considerations similar to those which apply to Mn(III).

Spin equilibria among organometallic complexes are also rare. Generally the ligand fields are too strong to allow population of the high-spin state. A well-documented exception occurs with bis(methylcyclopentadienyl)manganese(II), dimethylmanganocene. EPR and solution magnetic moments have established a $^2E\text{–}^6A$ equilibrium for this d^5 ion, analogous to that observed for iron(III) complexes (4, 150). An electron diffraction study gave a difference of 28 pm in the Mn–C distances in the two states (1). No dynamics have been reported; studies both in the solution and in the gas phase would be most interesting.

A. Reaction Mechanisms

1. Racemization and Isomerization

Spin state transitions have been invoked in explaining a number of reaction mechanisms. One is the intramolecular racemization of [Fe(phen)$_3$]$^{2+}$. The complex exists in the low-spin state, but expansion of the coordination sphere along a reaction coordinate for an intramolecular twisting racemization could reduce the ligand field enough to populate the high-spin state. Evidence for this hypothesis has been advanced from measurements of volumes of activation. These are large and positive ($+15$ cm^3 mol^{-1}) for both racemization and aquation, in contrast with values close to zero for the analogous reactions of [Ni(phen)$_3$]$^{2+}$ (*100*). The volume of activation for the iron(II) complex is comparable to those found for the spin equilibria of octahedral iron complexes (Table II). This lends credence to the hypothesis. A theoretical analysis of the mechanism has been presented, which includes an estimate of the transmission coefficient, κ, of the order of 10^{-1} (*132*).

Similar invocations of spin equilibria have been made for the racemization of a tris(dithiocarbamate)cobalt(III) complex (*101*) and a sexadentate iron(II) complex (*32*), and the isomerization of a cis-bis(phenanthroline)iron(II) complex (*133*). In the last case a semantic difficulty over the definition of rate constants obscures the argument. When the two spin states are not in thermal equilibrium, the forward and reverse rate constants are of course no longer nearly equal. The rate constant in the forward direction, k_{15}, from the ground state to the excited state, will be much less than in the reverse direction, from the excited state to the ground state, k_{51}. For a $\Delta S = 2$ transition in iron(II), k_{51} is likely to be of the order of 10^7 sec^{-1}. The value of k_{15} will be less than that of k_{51} by the energy difference between the ground and excited states. The reported rate for the isomerization reaction indicates that this energy difference between the ^1A and ^5T states must be less than 2000–2500 cm^{-1}. Only if the rate of the isomerization reaction in the excited state is greater than or competitive with k_{51} can the measured rate be identified with k_{15}. Since k_{51} is of the order of 10^7 s^{-1}, this condition is unlikely to be met. It is confusing, therefore, to identify the forward rate constant as that appropriate for a spin state change, because of the large thermal barrier to be surmounted.

2. *Electron Transfer*

Similar considerations apply to the role of spin equilibria in electron transfer reactions. For many years spin state restrictions were invoked to account for the slow electron exchange between diamagnetic, low-spin cobalt(III) and paramagnetic, high-spin cobalt(II) complexes. This explanation is now clearly incorrect. The rates of spin state interconversions are too rapid to be competitive with bimolecular encounters, except at the limit of diffusion-controlled reactions with molar concentrations of reagents. In other words, a spin equilibrium with a relaxation time of a few nanoseconds will always remain in equilibrium, even if one of the spin states participates in some bimolecular reaction. *Spin state changes are not rate determining in bimolecular reactions.* This conclusion could always have been inferred from the rate law for the cobalt(III)–cobalt(II) electron transfer reaction, for if the spin state change had been rate determining, the rate law would have been first order, not second order.

Spin restrictions may still affect the rate of electron transfer. If electron exchange between the ^1A and ^4T states is spin forbidden, then a spin-equilibrium transition prior to electron transfer is required. The rate of bimolecular electron exchange from this excited spin state will always be much lower than the thermal depopulation of this state to the other spin isomer, just as the rate of isomerization described above is less than the depopulation of the excited iron(II) spin state. For the cobalt(III)–cobalt(II) electron exchange, the spin state change is most likely to be in the cobalt(II) state from ^4T to ^2E. Note that this severely complicates the prediction of the rate because the ^2E state is subject to Jahn–Teller distortion. The usual one-dimensional analysis of the inner-sphere reorganization energy is not applicable. Both cobalt(II) and cobalt(III) will distort, requiring a two-dimensional analysis. In this sense spin restrictions affect the electron transfer by selecting a particular reaction coordinate. Alternatively, if the ^2E and ^1A surfaces were to cross above the ^4T and ^1A surfaces (Fig. 7B), spin restrictions would affect the rate by imposing a higher enthalpy of activation on the reaction. In neither case, however, would spin state changes be rate determining.

3. Substitution

Just as expansion of the coordination sphere by transition from the low-spin to the high-spin state enhances isomerization and racemization, so too should it enhance the rate of ligand substitution. Unpublished observations on the ligand substitution of $[Fe(pyim)_3]^{2+}$, which is coupled with its spin equilibrium, indicate that it is the high-spin state which preferentially undergoes the substitution reaction, as expected (*117*).

Less expected, perhaps, are results on substitution reactions of four-coordinate nickel(II) chelate complexes which occur in equilibrium between planar and tetrahedral isomers. Despite the longer bond lengths of the tetrahedral isomers, it is the planar isomers which undergo the substitution reactions (*140*).

B. Excited States

Spin equilibria are thermal intersystem crossing processes. The "ground state" and the "excited state" lie within a few hundred wavenumbers of each other and both are thermally populated. There are two photophysical processes in excited states related to the dynamics of thermal spin equilibria. One is the radiationless deactivation of an excited state to a ground state of different spin multiplicity. The other is intersystem crossing between excited states.

The results obtained from thermal spin equilibria indicate that $\Delta S = 1$ transitions are adiabatic. The rates, therefore, depend on the coordination sphere reorganization energy, or the Franck–Condon factors. Radiationless deactivation processes are exothermic. Consequently, they can proceed more rapidly than thermally activated spin-equilibria reactions, that is, in less than nanoseconds in solution at room temperature. Evidence for this includes the observation that few transition metal complexes luminesce under these conditions. Other evidence is the very success of the photoperturbation method for studying thermal spin equilibria; intersystem crossing to the ground state of the other spin isomer must be more rapid than the spin equilibrium relaxation in order for the spin equilibrium to be perturbed.

There are of course many examples of metal complexes with excited state lifetimes of longer than a few nanoseconds. These often involve vertical transitions, however, as in the well-studied case of the ^2E excited state of chromium(III) (51). The inferences to be drawn from spin-equilibrium dynamics apply only to nonvertical transitions along a metal–ligand reaction coordinate. These considerations imply that intersystem crossing between excited states is a closer analogy with thermal spin equilibria. The rates of such $\Delta S = 1$ processes should be determined by their thermodynamics and reorganization energy requirements, but not be limited by changes in spin multiplicity. This difference from singlet–triplet transitions in organic molecules arises from the larger spin-orbit coupling in transition metal complexes.

Even $\Delta S = 2$ transitions in excited states can be rapid. In ground state spin equilibria these transitions can be nonadiabatic due to the requirement of mixing through spin-orbit coupling with excited intermediate spin states. In photophysical processes these excited states can actually be populated. Again this conclusion follows from the success of photoperturbation methods applied to $\Delta S = 2$ spin equilibria. For example, excitation of the singlet state of iron(II) results in detectable population of the quintet spin-equilibrium state more rapidly than the nanosecond relaxation between the singlet and quintet spin-equilibrium states. The most likely cause of this is very rapid crossings from the excited singlet state to an intermediate triplet state to a quintet state. The efficiency of this process is high, with quantum yields at low temperatures in crystals of 0.1–1% (*73, 131*). This implies nearly adiabatic intersystem crossing.

C. Porphyrins and Heme Proteins

Spin equilibria are an important feature of the chemistry of porphyrins and heme proteins. The extensive literature on this subject cannot be reviewed here. Only some relevant implications from the studies of spin equilibria in metal complexes will be described.

The spin equilibria of iron(II) and iron(III) in a porphyrin coordination environment combine the features of $\Delta S = 2$ equilibria with the coordination-spin equilibria of nickel(II). There is often a coordination number change from five-coordinate high-spin to six-coordinate low-spin geometries. Consequently, there is substantial reorganization energy required for bond formation and dissociation. In addition, the $\Delta S = 2$ transitions are potentially nonadiabatic.

There is an important feature of porphyrin chemistry which increases the adiabaticity of these transitions. As the ligand field is distorted from octahedral to tetragonal, the energy of the intermediate spin states, triplet for iron(II) and quartet for iron(III), is lowered. In some cases these intermediate spin states become the ground states. These possess an intermediate magnetic moment which does not display the same temperature dependence as the Boltzmann equilibrium between high-spin and low-spin states. In other cases equilibria may exist between these intermediate spin states and either the high-spin or low-spin isomers. These possibilities obviously complicate the analysis and description of the magnetic and related properties of such complexes.

The presence of these low-lying intermediate spin states, however, also increases the mixing through spin-orbit coupling between the high-spin and low-spin states. By perturbation theory this mixing depends on the square of the energy difference between the ground states and the intermediate excited state. Since in octahedral iron complexes the $\Delta S = 2$ spin equilibria are estimated to be nonadiabatic only by a factor not less than 10^{-2}, the low-lying intermediate spin states should make spin equilibria in porphyrins and heme proteins adiabatic or nearly so.

If this interpretation is correct, then the rate of spin state transitions in these systems will depend on the reorganization energy requirements. Two examples will be described. Ferric myoglobin undergoes a spin equilibrium between a low-spin state in which a small molecule such as hydroxide or azide is coordinated in the sixth position and a high-spin state in which this ion is dissociated from the iron(III) but still located within the heme pocket. This equilibrium is estimated to have a relaxation time of about 20 ns (46). In contrast, ferricytochrome c undergoes a spin equilibrium in acidic solution in which the sixth endogenous ligand methioine dissociates. This equilibrium has a relaxation time of 100 μsec (48). The difference in rates arises from the difference in the reorganization energies required.

ACKNOWLEDGMENTS

Helpful comments on a draft of this review were made by Dr. R. A. Binstead and Professors H. A. Goodwin, S. F. Lincoln, and R. L. Martin. Special thanks are due to P. Del Favero for establishing the literature database and for preparation of the artwork. The research which led to this review was supported by the Australian Research Grants Scheme.

REFERENCES

1. Almenningen, A., Samdal, S., and Haaland, A., *J. Chem. Soc. Chem. Commun.* 14 (1977).
2. Ameen, S., *Rev. Sci. Instrum.* **46,** 1209 (1975).
3. Ammeter, J. H., *Nouv. J. Chim.* **4,** 631 (1980).
4. Ammeter, J. H., Bucher, R., and Oswald, N., *J. Am. Chem. Soc.* **96,** 7833 (1974).
5. Amir-Ebrahimi, V., and McGarvey, J. J., *Inorg. Chim. Acta* **89,** L39 (1984).
6. Bacci, M., *Inorg. Chem.* **25,** 2322 (1986).
7. Barefield, E. K., Bianchi, A., Billo, E. J., Connolly, P. J., Paoletti, P., Summers, J. S., and van Derveer, D. G., *Inorg. Chem.* **25,** 4197 (1986).
8. Barefield, E. K., Busch, D. H., and Nelson, S. M., *Q. Rev. Chem. Soc.* **22,** 457 (1968).
9. Barefield, E. K., Freeman, G. M., and Van Derveer, D. G., *Inorg. Chem.* **25,** 552 (1986).
10. Barraclough, C. G., *Trans. Faraday Soc.* **62,** 1033 (1966).
11. Batschelet, W. H., and Rose, N. J., *Inorg. Chem.* **22,** 2083 (1983).
12. Beattie, J. K., Binstead, R. A., and West, R. J., *J. Am. Chem. Soc.* **100,** 3044 (1978).
13. Beattie, J. K., Kelso, M. T., Moody, W. E., and Tregloan, P. A., *Inorg. Chem.* **24,** 415 (1985).
14. Beattie, J. K., Sutin, N., Turner, D. H., and Flynn, G. W., *J. Am. Chem. Soc.* **95,** 2052 (1973).
15. Becker, E. D., "High Resolution NMR," 2nd Ed., pp. 44–46. Academic Press, New York, 1980.
16. Beitz, J. V., Flynn, G. W., Turner, D. H., and Sutin, N., *J. Am. Chem. Soc.* **92,** 4130 (1970).
17. Bernasconi, C. F., "Relaxation Kinetics." Academic Press, New York, 1976.
18. Binstead, R. A., Ph.D. thesis, University of Sydney, 1979, and unpublished observations, 1986.

19. Binstead, R. A., and Beattie, J. K., *Inorg. Chem.* **25,** 1481 (1986).
20. Binstead, R. A., Beattie, J. K., Dewey, T. G., and Turner, D. H., *J. Am. Chem. Soc.* **102,** 6442 (1980).
21. Binstead, R. A., Beattie, J. K., Dose, E. V., Tweedle, M. F., and Wilson, L. J., *J. Am. Chem. Soc.* **100,** 5609 (1978).
22. Birker, P. J. M. W. L., Bour, J. J., and Steggerda, J. J., *Inorg. Chem.* **12,** 1254 (1973).
23. Buhks, E., Navon, G., Bixon, M., and Jortner, J. *J. Am. Chem. Soc.* **102,** 2918 (1980).
24. Burger, K., and Ebel, H., *Inorg. Chim. Acta,* **53,** L105 (1981).
25. Caldin, E. F., and Field, J. P., *J. Chem. Soc. Faraday Trans. I* **78,** 1923 (1982).
26. Campbell, L., and McGarvey, J. J., *J. Chem. Soc. Chem. Commun.* 749 (1976).
27. Campbell, L., and McGarvey, J. J., *J. Am. Chem. Soc.* **99,** 5809 (1977).
28. Campbell, L., McGarvey, J. J., and Samman, N. G., *Inorg. Chem.* **17,** 3378 (1978).
29. Cecconi, F., Di Vaira, M., Midollini, S., Orlandini, A., and Sacconi, L., *Inorg. Chem.* **20,** 3423 (1981).
30. Chandrasekhar, K., and Bürgi, H. B., *Acta Crystallogr.* **B40,** 387 (1984).
31. Chisholm, M. H., Kober, E. M., Ironmonger, D. J., and Thornton, P., *Polyhedron* **4,** 1869 (1985).
32. Christiansen, L., Hendrickson, D. N., Toftlund, H., Wilson, S. R., and Xie, C. L., *Inorg. Chem.* **25,** 2813 (1986).
33. Coates, J. H., Hadi, D. A., Lincoln, S. F., Dodgen, H. W., and Hunt, J. P., *Inorg. Chem.* **20,** 707 (1981).
34. Cotton, F. A., Diebold, M. P., O'Connor, C. J., and Powell, G. L., *J. Am. Chem. Soc.* **107,** 7438 (1985).
35. Cox, M., Darken, J., Fitzsimmons, B. W., Smith, A. W., Larkworthy, L. F., and Rogers, K. A., *J. Chem. Soc. Dalton Trans.* 1192 (1972).
36. Crawford, T. H., and Swanson, J., *J. Chem. Educ.* **48,** 382 (1971).
37. Creutz, C., and Sutin, N., *J. Am. Chem. Soc.* **95,** 7177 (1973).
38. Cusamano, M., *J. Chem. Soc. Dalton Trans.* 2133, 2137 (1976).
39. Decurtins, S., Gütlich, P., Hasselbach, K. M., Hauser, A., and Spiering, H., *Inorg. Chem.* **24,** 2174 (1985).
40. Decurtins, S., Gütlich, P., Köhler, C. P., and Spiering, H., *J. Chem. Soc. Chem. Commun.* 430 (1985).
41. Decurtins, S., Gütlich, P., Köhler, C. P., Spiering, H., and Hauser, A., *Chem. Phys. Lett.* **105,** 1 (1984).
42. De Filippo, D., Depalano, P., Diaz, A., Steffé, S., and Trogu, E. F., *J. Chem. Soc. Dalton Trans.* 1566 (1977).
43. Dewey, T. G., and Turner, D. H., *Adv. Mol. Relax. Interact. Process.* **13,** 331 (1978).
44. DiBenedetto, J., Arkle, V., Goodwin, H. A., and Ford, P. C., *Inorg. Chem.* **24,** 455 (1985).
45. Dose, E. V., Hoselton, M. A., Sutin, N., Tweedle, M. F., and Wilson, L. J., *J. Am. Chem. Soc.* **100,** 1141 (1978).
46. Dose, E. V., Tweedle, M. F., Wilson, L. J., and Sutin, N., *J. Am. Chem. Soc.* **99,** 3886 (1977).
47. Ducommun, Y., and Merbach, A., *In* "Inorganic High Pressure Chemistry" (R. van Eldik, ed.). Elsevier, New York, 1986.
48. Dyson, H. J., and Beattie, J. K., *J. Biol. Chem.* **257,** 2267 (1982).
49. Eaton, D. R., *J. Am. Chem. Soc.* **90,** 4272 (1968).
50. Eggers, F., and Kustin, K., *In* "Methods in Enzymology" (K. Kustin, ed.), Vol. XVI, pp. 55–80. Academic Press, New York, 1969.
51. Endicott, J. F., Lessard, R. B., Lei, Y., Ryu, C. K., and Tamilarasan, R., *ACS Symp. Ser.* **307,** 85 (1986).

52. Ewald, A. H., Martin, R. L., Sinn, E., and White, A. H., *Inorg. Chem.* **8,** 1837 (1969).
53. Ewald, A. H., and Sinn, E., *Inorg. Chem.* **6,** 40 (1967).
54. Evans, D. F., *J. Chem. Soc.* 2003 (1959).
55. Evans, D. F., and James, T. A., *J. Chem. Soc. Dalton Trans.* 723 (1979).
56. Federer, W. D., and Hendrickson, D. N., *Inorg. Chem.* **23,** 3861 (1984).
57. Federer, W. D., and Hendrickson, D. N., *Inorg. Chem.* **23,** 3870 (1984).
58. Figgis, B. N., Kucharski, E. S., and White, A. H., *Aust, J. Chem.* **36,** 1537 (1983).
59. French, T. C., and Hammes, G. G., *In* "Methods in Enzymology" (K. Kustin, ed.), Vol. XVI, pp. 3–30. Academic Press, New York, 1969.
60. Godfrey, A. F., Ph.D. thesis, University of Sydney, 1985.
61. Godfrey, A. F., and Beattie, J. K., *Inorg. Chem.* **22,** 3794 (1983).
62. Goodwin, H. A., *Coord. Chem. Rev.* **18,** 293 (1976).
63. Greenaway, A. M., O'Connor, C. J., Schrock, A., and Sinn, E., *Inorg. Chem.* **18,** 2692 (1979).
64. Greenaway, A. M., and Sinn, E., *J. Am. Chem. Soc.* **100,** 8080 (1978).
65. Gütlich, P., *Struct. Bond.* **44,** 83 (1981).
66. Gütlich, P., *Adv. Chem. Ser.* **194,** 405 (1981).
67. Gütlich, P., *In* "Mössbauer Spectroscopy Applied to Inorganic Chemistry" (G. J. Long, ed.), Vol. I, Chapter XI. Plenum, New York, 1984.
68. Gütlich, P., *In* "Chemical Mössbauer Spectroscopy" (R. H. Herber, ed.), Chap. II. Plenum, New York, 1984.
69. Gütlich, P., McGarvey, B. R., and Klaüi, W., *Inorg. Chem.* **19,** 3704 (1980).
70. Halevi, E. A., and Knorr, R., *Angew. Chem. Int. Ed. Engl.* **21,** 288 (1982).
71. Hall, G. R., and Hendrickson, D. N., *Inorg. Chem.* **15,** 607 (1976).
72. Hammes, G. G., *In* "Techniques of Chemistry" (G. G. Hammes, ed.), Vol. VI, Part II, 3rd Ed, pp. 147–185. Wiley (Interscience), New York, 1974.
73. Hauser, A., *Chem. Phys. Lett.* **124,** 543 (1986).
74. Hauser, A., Gütlich, P., and Spiering, H., *Inorg. Chem.* **25,** 4245 (1986).
75. Hay, R. W., Jeragh, B., Ferguson, G., Kaitner, B., and Ruhl, B. L., *J. Chem. Soc. Dalton Trans.* 1531 (1982).
75a. Helm, L., Meier, P., Merbach, A. E., and Tregloan, P. A., *Inorg. Chim. Acta* **73,** 1 (1983).
76. Herber, R. H., *Inorg. Chem.* **26,** 173 (1987).
77. Herber, R. H., and Casson, L. M., *Inorg. Chem.* **25,** 847 (1986).
78. Hirohari, H., Ivin, K. J., McGarvey, J. J., and Wilson, J., *J. Am. Chem. Soc.* **96,** 4435 (1974).
79. Holm, R. H., Everett, G. W., and Chakravorty, A., *Prog. Inorg. Chem.* **7,** 83 (1966).
80. Holm, R. H., and O'Connor, M. J., *Prog. Inorg. Chem.* **14,** 241 (1973).
81. Hoselton, M. A., Drago, R. S., Wilson, L. J., and Sutin, N., *J. Am. Chem. Soc.* **98,** 6967 (1976).
82. Hutchinson, B., Daniels, L., Henderson, E., Neill, P., Long, G. J., and Becker, L. W., *J. Chem. Soc. Chem. Commun.* 1003 (1979).
83. Imoto, H., and Simon, A., *Inorg. Chem.* **21,** 308 (1982).
84. Ito, T., Sugimoto, M., Ito, H., Toriumi, K., Nakayama, H., Mori, W., and Sekizaki, M., *Chem. Lett.* 121 (1983).
85. Ivin, K. J., Jamison, R., and McGarvey, J. J., *J. Am. Chem. Soc.* **94,** 1763 (1972).
86. Jesson, J. P., Trofimenko, S., and Eaton, D. R., *J. Am. Chem. Soc.* **89,** 3148 (1967).
87. Joedicke, I. B., Studer, H. V., and Yoke, J. T., *Inorg. Chem.* **15,** 1352 (1976).
88. Kaplan, M. L., Bovey, F. A., and Cheng, N. H., *Anal. Chem.* **47,** 1703 (1975).
89. Katz, B. A., and Strouse, C. E., *J. Am. Chem. Soc.* **101,** 6214 (1979).
90. Katz, B. A., and Strouse, C. E., *Inorg. Chem.* **19,** 658 (1980).

91. Kennedy, B. J., Fallon, G. D., Gatehouse, B. M. K. C., and Murray, K. S., *Inorg. Chem.* **23,** 580 (1984).
92. Kilbourn, B. T., and Powell, H. M., *J. Chem. Soc. (A)* 1688 (1970).
93. König, E., and Madeja, K., *Inorg. Chem.* **6,** 48 (1967).
94. König, E., and Kremer, S., *Theor. Chim. Acta* **23,** 12 (1971).
95. König, E., Ritter, G., and Kulshreshtha, S. K., *Chem. Rev.* **85,** 219 (1985).
96. König, E., and Watson, K. J., *Chem. Phys. Lett.* **6,** 457 (1970).
97. Kremer, S., Henke, W., and Reinen, D., *Inorg. Chem.* **21,** 3013 (1982).
98. Kunze, K. R., Perry, D. L., and Wilson, L. J., *Inorg. Chem.* **16,** 594 (1977).
99. La Mar, G. N., and Sherman, E. O., *J. Am. Chem. Soc.* **92,** 2691 (1970).
100. Lawrance, G. A., and Stranks, D. R., *Inorg. Chem.* **17,** 1804 (1978).
101. Lawrance, G. A., Suvachittanont, S., Stranks, D. R., Tregloan, P. A., Gahan, L. R., and O'Connor, M. J., *J. Chem. Soc. Chem. Commun.* 757 (1979).
102. Lawthers, I., and McGarvey, J. J., *J. Am. Chem. Soc.* **106,** 4280 (1984).
103. Lazarus, M. S., Hoselton, M. A., and Chou, T. S., *Inorg. Chem.* **16,** 2549 (1977).
104. Lincoln, S. F., Hambley, T. W., Pisaniello, D. L., and Coates, J. H., *Aust. J. Chem.* **37,** 713 (1984).
105. Lockwood, G., McGarvey, J. J., and Devonshire, R., *Chem. Phys. Lett.* **86,** 127 (1982).
106. Lohr, L. L., Jr., *J. Am. Chem. Soc.* **100,** 1093 (1978).
107. Lohr, L. L., Jr., and Grimmelmann, E. K., *J. Am. Chem. Soc.* **100,** 1100 (1978).
108. Maeda, Y., Ohshio, H., and Takashima Y., *Chem. Lett.* 943 (1982).
109. Maeda, Y., Ohshio, H., Takashima Y., Mikuriya, M., and Hidaka, M., *Inorg. Chem.* **25,** 2958 (1986).
110. Maeda, Y., Tsutsumi, N., and Takashima, Y., *Chem. Phys. Lett.* **88,** 248 (1982).
111. Maeda, Y., Tsutsumi, N., and Takashima, Y., *Inorg. Chem.* **23,** 2440 (1984).
112. Martin, R. L., and White, A. H., *Trans. Metal. Chem.* **5,** 113 (1969).
113. Matsumoto, N., Ohta, S., Yoshimura, C., Ohyoshi, A., Kohata, S., Okawa, H., and Maeda, Y., *J. Chem. Soc. Dalton Trans.* 2575 (1985).
114. McGarvey, J. J., and Lawthers, I., *J. Chem. Soc. Chem. Commun.* 906 (1982).
115. McGarvey, J. J., Lawthers, I., Heremans, K., and Toftlund, H., *J. Chem. Soc. Chem. Commun.* 1575 (1984).
116. McGarvey, J. J., and Wilson, J., *J. Am. Chem. Soc.* **97,** 2531 (1975).
117. McMahon, K. J., and Beattie, J. K., unpublished observations.
118. Merbach, A. E., Moore, P., and Newman, K. E., *J. Magn. Reson.* **41,** 30 (1980).
119. Merrithew, P. B., and Rasmussen, P. G., *Inorg. Chem.* **11,** 325 (1972).
120. Moody, W. F., and Beattie, J. K., unpublished observations.
121. Mulay, L. N., and Mulay, I. L., *Anal. Chem.* **56,** 293R (1984).
122. Müller, E. W., Ensling, J., Spiering, H., and Gütlich, P. *Inorg. Chem.* **22,** 2074 (1983).
123. Nielson, R. M., Dodgen, H. W., Hunt, J. P., and Wherland, S. E., *Inorg. Chem.* **25,** 582 (1986).
124. Ohshio, H., Maeda, Y., and Takashima, Y., *Inorg. Chem.* **22,** 2684 (1983).
125. Oliver, J. D., Mullica, D. F., Hutchinson, B. B., and Milligan, W. O., *Inorg. Chem.* **19,** 165 (1980).
126. Ostfeld, D., and Cohen, I. A., *J. Chem. Educ.* **49,** 829 (1972).
127. Pell, R. J., Dodgen, H. W., and Hunt, J. P., *Inorg. Chem.* **22,** 529 (1983).
128. Petty, R. H., Dose, E. V., Tweedle, M. F., and Wilson, L. J., *Inorg. Chem.* **17,** 1064 (1978).
129. Pignolet, L. H., Horrocks, W. DeW., and Holm, R. H., *J. Am. Chem. Soc.* **92,** 1855 (1970).
130. Pignolet, L. H., and La Mar, G. N. *In* "NMR of Paramagnetic Molecules"

(G. N. LaMar, W. DeW. Horrocks, and R. H. Holm, eds.), pp. 333–369. Academic Press, New York, 1973.
131. Poganiuch, P., and Gütlich, P., *Inorg. Chem.* **26,** 455 (1987).
132. Purcell, K. F., *J. Am. Chem. Soc.* **101,** 5147 (1979).
133. Purcell, K. F., and Zapata, J. P., *J. Chem. Soc. Chem. Commun.* 497 (1978).
134. Que, L., and Pignolet, L. H., *Inorg. Chem.* **12,** 156 (1972).
135. Rablen, D. P., Dodgen, H. W., and Hunt, J. P., *Inorg. Chem.* **15,** 931 (1976).
136. Rao, P. S., Reuveni, A., McGarvey, B. R., Ganguli, P., and Gütlich, P., *Inorg. Chem.* **20,** 204 (1981).
137. Reeder, K. A., Dose, E. V., and Wilson, L. J., *Inorg. Chem.* **17,** 1071 (1978).
138. Sachinidis, J., and Grant, M. W., *J. Chem. Soc. Chem. Commun.* 157 (1978).
139. Schmidt, J. G., Brey, W. S., and Stoufer, R. C., *Inorg. Chem.* **6,** 268 (1967).
140. Schumann, M., and Elias, H., *Inorg. Chem.* **24,** 3187 (1985).
141. Simmons, M. G., and Wilson, L. J., *Inorg. Chem.* **16,** 126 (1977).
142. Simon, A., Schnering, H. G., and Schäfer, H., *Z. Anorg. Allg. Chem.* **355,** 295 (1967).
142a. Sim, P. G., and Sinn, E., *J. Am. Chem. Soc.* **103,** 241 (1981).
143. Sinn, E., Sim, G., Dose, E. V., Tweedle, M. F., and Wilson, L. J., *J. Am. Chem. Soc.* **100,** 3375 (1978).
144. Sorai, M., *J. Inorg. Nucl. Chem.* **40,** 1031 (1978).
145. Sorai, M., and Seki, S., *J. Phys. Chem. Solids* **35,** 555 (1974).
146. Stengle, T. R., and Langford, C. H., *Coord. Chem. Rev.* **2,** 349 (1967).
147. Stuehr, J., *In* "Techniques of Chemistry" (G.G. Hammes, ed.), Vol. VI, Part II, 3rd Ed, pp. 237–283. Wiley (Interscience), New York, 1974.
148. Swift, T. J., and Connick, R. E., *J. Chem. Phys.* **37,** 307 (1962).
149. Swift, T. J., *In* "NMR of Paramagnetic Molecules" (G. N. LaMar, G. N., W. DeW. Horrocks, and R. H. Holm, eds.), pp. 53–83. Academic Press, New York, 1973.
150. Switzer, M. E., Wang, R., Rettig, M. F., and Maki, A. H., *J. Am. Chem. Soc.* **96,** 7669 (1974).
151. Takemoto, J. H., and Hutchinson, B., *Inorg. Chem.* **12,** 705 (1973).
152. Thuéry, P. and Zarembowitch, J., *Inorg. Chem.* **25,** 2001 (1986).
153. Timken, M. D., Abdel–Mawgoud, A. M., and Hendrickson, D. N., *Inorg. Chem.* **25,** 160 (1986).
154. Timken, M. D., Hendrickson, D. N., and Sinn, E., *Inorg. Chem.* **24,** 3947 (1985).
155. Timken, M. D., Strouse, C. E., Soltis, S. M., Daverio, S. A., Hendrickson, D. N., Abdel–Mawgoud, A. M., and Wilson, S. R., *J. Am. Chem. Soc.* **108,** 395 (1986).
156. Tricker, M. J., *J. Inorg. Nucl. Chem.* **36,** 1543 (1974).
157. Turner, D. H., Flynn, G. W., Sutin, N., and Beitz, J. V., *J. Am. Chem. Soc.* **94,** 1554 (1972).
158. van Geet, A. L., *Anal. Chem.* **42,** 679 (1970).
159. Vos, G., De Graaff, R. A. G., Haasnoot, J. G., Van Der Kraan, A. M., De Vaal, P., and Reedijk, J., *Inorg. Chem.* **23,** 2905 (1984).
160. Wells, F. V., McCann, S. W., Wickman, H. H., Kessel, S. L., Hendrickson, D. N., and Feltham, R. D., *Inorg. Chem.* **21,** 2306 (1982).
161. Whitesides, T. H., *J. Am. Chem. Soc.* **91,** 2395 (1969).
162. Wilkins, R. G., Yelin, R., Margerum, D. W., and Weatherburn, D. C., *J. Am. Chem. Soc.* **91,** 4326 (1969).
163. Williams, D. L., Smith, D. W., and Stoufer, R. C., *Inorg. Chem.* **6,** 590 (1967).
164. Zarembowitch, J., Claude, R., and Thuéry, P., *Nouv. J. Chim.* **9,** 467 (1985).
165. Zarembowitch, J., and Kahn, O., *Inorg. Chem.* **23,** 589 (1984).

HYDROXO-BRIDGED COMPLEXES OF CHROMIUM(III), COBALT(III), RHODIUM(III), AND IRIDIUM(III)

JOHAN SPRINGBORG

Chemistry Department, Royal Veterinary and Agricultural University,
DK-1871 Frederiksberg C, Denmark

I. Introduction
II. Structural Considerations
    A. Determination of Structures
    B. Crystallographic Data
III. Spectroscopic and Magnetic Properties
IV. Formation of Polynuclear Complexes
    A. Hydrolysis Reactions
    B. Solid-State Reactions
    C. Formation by Oxidation Reactions
    D. Formation from Other Polynuclear Species
V. Stability Constants
    A. Determination of Stability Constants
    B. Monohydroxo-Bridged Species
    C. Dihydroxo-Bridged Species
    D. Trihydroxo-Bridged Species
    E. The Aqua Chromium(III) System
VI. Acid–Base Equilibria
    A. Bridging Hydroxide
    B. Bridging Water
    C. Terminally Coordinated Water
VII. Kinetics for the Condensation Reaction of Mononuclear Species to Give Dinuclear Species
VIII. Cleavage of Polynuclear into Mononuclear Species
    A. Cleavage in Strong Acids
    B. Kinetics of the Cleavage of Dinuclear Monohydroxo-Bridged Complexes
IX. Equilibria between Mono- and Dihydroxo-Bridged Dinuclear Complexes of Chromium(III), Rhodium(III), and Iridium(III)
    A. General Remarks
    B. Uncatalyzed Cleavage
    C. Acid-Catalyzed Cleavage
    D. Bridge Formation
    E. Base-Catalyzed Bridge Cleavage
X. Kinetics of the Hydrolysis of Dihydroxo-Bridged Cobalt(III) Complexes

XI. Equilibria between Tri- and Dihydroxo-Bridged Complexes
   A. Cobalt(III)
   B. Chromium(III)
   C. Rhodium(III)
XII. Bridge Cleavage of Mixed Bridge Complexes
   A. Hydroxo Bridge Cleavage of Amido-Bridged Dicobalt(III) Complexes
   B. Other Complexes
XIII. Concluding Remarks
XIV. Abbreviations for Ligands and Solvents
   A. General
   B. Specific
   References

I. Introduction

Polynuclear complexes are known for most metal ions and they are of current interest because of their rather special chemical and physical properties resulting from the mutual interaction of two or more metal centers. Most of the common ligands in inorganic chemistry have been shown to function as bridging groups. The bridging ligands are often monoatomic ions, such as Cl^-, O^{2-}, and S^{2-}, or simple polyatomic ions, such as OH^- or NH_2^-, in which cases a single ligating atom is bound to two (or more) metal centers. Polydentate ligands such as $RCOO^-$, SO_4^{2-}, PO_4^{3-}, etc., usually act as bidentate ligands when they form a bridge.

Polynuclear complexes with hydroxide (or oxide) as bridging ligands constitute an important class of complexes. They are formed by hydrolysis of mononuclear aqua complexes of most metal ions and they therefore constitute an important aspect of the hydrolytic chemistry of metal ions. They display a chemistry which is interesting in itself, but which is also relevant in relation to applied chemistry and to biochemistry, as mentioned in Section XIII.

The present review is concerned mainly with the chemistry of hydroxo-bridged chromium(III) and cobalt(III) complexes, but the comparatively few studies of the related rhodium(III) and iridium(III) systems which have been reported have also been included. The review deals primarily with the chemical properties of these complexes, and special emphasis has been made on reactions which involve the cleavage or formation of a hydroxo bridge. Reactions of bridging groups other than hydroxide and reactions of nonbridging ligands are also discussed, but only to the extent that they have been considered relevant to the main issue. A large number of X-ray crystal-structure determinations have been made during the last two decades and references to all the published structures are given. The hydroxo-bridged chromium(III) oligomers constitute a very interesting class of complexes from a spectroscopic and magnetochemical point of view. This interesting field, however, lies beyond the main scope of the present article and only a brief review of the latest progress is given.

The literature has, in principle, been covered completely, but since the intention has not been to describe the historical development of the subject, many of the oldest references have been left out in the following discussions. It therefore seems appropriate to round off this section by giving credit to the pioneering work done at the beginning of the century by S. M., Jørgensen (1–4), A. Werner (5–13), and N. Bjerrum (14–15). As early as 1882 Jørgensen ($1, 3$) made a singly bridged dichromium(III) complex, the so-called rhodo complex, $(NH_3)_5Cr(OH)Cr(NH_3)_5^{5+}$, and later he prepared a tetranuclear complex, the so-called rhodoso complex (4), $Cr_4(NH_3)_{12}(OH)_6^{6+}$. Jørgensen also prepared a tetranuclear cobalt(III) complex, $Co_4(NH_3)_{12}(OH)_6^{6+}$ (2). This complex was subsequently the first inorganic salt to be resolved into its optical isomers, by Werner (9). Werner, as well as Pfeiffer ($16, 17$) and Dubsky (18), synthesized a very impressive number of hydroxo-bridged chromium(III) and cobalt(III) complexes during a comparatively short period. Using classical chemical methods, they also established the structures of many of these polynuclear species correctly. The study of these complexes clearly contributed significantly to the acceptance of Werner's proposal of octahedral coordination geometry rather than the chain theory of C. W. Blomstrand and S. M. Jørgensen. At the same time that Werner studied the ammine and amine complexes, Bjerrum started investigating the hydrolytic behavior of chromium(III) aqua ions and established the formation of hydroxo-bridged oligomers ($14, 15$). Bjerrum's pioneer work was taken up during the 1950s by L. G. Sillén and co-workers (19–21), who have made a systematic study of the hydrolytic behavior of a large number of metal aqua ions and have introduced the use of modern computer programs for the numerical treatment of the data.

II. Structural Considerations

A. Determination of Structures

In this section the methods which have been used to gain structural information are briefly summarized. The term structure is used in this context in its broadest sense, including more qualitative observations concerning the skeleton of the bridging atoms. As a general rule, the hydroxo-bridged polynuclear complexes of chromium(III) and cobalt(III) can be isolated as well-defined crystalline salts and it is therefore quite natural that single-crystal X-ray structure analysis has become the method of choice for obtaining unambiguous and detailed structural information.

The X-ray crystal data are often used in relation to solution chemistry, and the general question of the identity of the species in solution and those present in the crystals has to be considered, since rapid interconversion between different bridged species is frequently seen for both chromium(III) and cobalt(III). Supplementary solution measurements are therefore generally required in order to be reasonably certain that the solid-state structure corresponds to the structure in solution.

The most important methods for obtaining structural information about solutions are probably potentiometric pH measurements and visible/ultraviolet spectroscopy, but many other techniques have been applied, including bridge-cleavage experiments, magnetic measurements, and electron spin resonance (ESR) spectroscopy ($22, 23$).

A knowledge of the nuclearity, i.e., the number of metal ions per oligomer, and of the charge on the species can normally be obtained with good accuracy by classical methods. The molecular weight of the polynuclear species has been determined by measurement of the freezing point depression of a eutectic solution of a strong electrolyte in water ($24-28$). This method has been used by Ardon and Linnenberg (25) to determine the molecular weight of the blue dinuclear species $(H_2O)_4Cr(OH)_2Cr(H_2O)_4^{4+}$ and by Schwarzenbach and Magyar (24) to determine the molecular weights of the sulfate salts of the cations $(NH_3)_5Cr(OH)Cr(NH_3)_5^{5+}$ and cis-$(NH_3)_5Cr(OH)Cr(NH_3)_4X^{n+}$ (X = H_2O, OH^-, NCS^-, and F^-). The ratio charge/metal center can be determined accurately by ion-exchange chromatography, which from elution rates may also provide an estimate of the charge per species ($27, 29-31$).

It will therefore usually be possible to determine the values of x and y in the formula $M_x(OH)_y$, but the number of hydroxo bridges may remain undetermined, e.g., $M_2(OH)_2^{n+}$ may correspond to the monohydroxo-, dihydroxo-, or oxo-bridged structures: $M(OH)M(OH)^{n+}$, $M(OH)_2M^{n+}$, or $M-O-M^{n+}$. Classical methods which can often distinguish between these structures are potentiometric acid–base titration and measurements of the $d-d$ absorption spectra as a function of pH. In combination, such measurements may provide information about the number of terminally coordinated water ligands, the number of hydroxo bridges, and even about the nuclearity. The dinuclear ammine and amine chromium(III) complexes $L_4Cr(OH)_2CrL_4^{4+}$ and $(H_2O)L_4Cr(OH)L_4(OH)^{4+}$ [$L_4 = (NH_3)_4$ or $(en)_2$] can easily be distinguished from such measurements: the dihydroxo-bridged species exhibits no acid–base properties in the pH region 0–10, whereas the monohydroxobridged species does, and the spectral changes (positions of the first ligand-field bands (see Table VI) could be interpreted according to the proposed structures (*32–36*). Potentiometric and spectrophotometric acid–base studies of $\Delta,\Delta/\Lambda,\Lambda$-(phen)$_2$Cr(OH)$_2$Cr(phen)$_2^{4+}$ and its basic oxo-bridged forms showed that the nuclearity was $2n$ (n integer) and that a dinuclear ion should be dihydroxo-bridged, as later confirmed by crystal-structure analysis (*37–39*).

Hydrolysis is strong acid in which all hydroxo bridges are cleaved, followed by identification of the various mononuclear species and a determination of their molar ratios, may provide extremely valuable information. A straightforward example is the cleavage of the tetranuclear species Cr$_4$(NH$_3$)$_{12}$(OH)$_6^{6+}$, which yields Cr(H$_2$O)$_6^{3+}$ and *cis*-Cr(NH$_3$)$_4$(H$_2$O)$_2^{3+}$ in a ratio of 1:3 (*40*). Since it could be demonstrated at the same time that the polynuclear cation does not exhibit acid–base properties in the pH region for terminally coordinated water, it was concluded that the only possible structure was **6** in Fig. 1, as later confirmed by a crystal-structure analysis (*41*).

Magnetic measurements on the paramagnetic chromium(III) complexes are useful diagnostically and may distinguish a dihydroxo-bridged binuclear structure from an oxo-bridged structure (see Section III). Similarly, electron spin resonance studies on chromium(III) complexes, frozen glass mixture, or solids could be a powerful technique for distinguishing different bridged structures. Until now, however, they have been used mostly to characterize complexes of known structure (*40, 42*).

Vibrational (infrared and Raman) spectroscopy, nuclear magnetic resonance (NMR) spectroscopy, and X-ray techniques are frequently used to study aqueous solutions of the hydrolysis products of metal aqua ions, but have not been applied to any great extent to the present metal complexes; few infrared studies have been reported (*43–46*). ^{59}Co NMR studies of a series of cobalt(III) oligomers suggest that NMR is a good method for establishing the structures of such complexes in solution (*47*). Similarly, ^{2}H NMR spectroscopy has recently been used to study the solution chemistry of chromium(III) oligomers (*48*).

B. Crystallographic Data

Figure 1 shows schematically the skeletal structures of hydroxo-bridged oligomers which have been observed in X-ray crystal structures, including a few speculative but relevant structures (**4b**, **7b**, and **7c**).

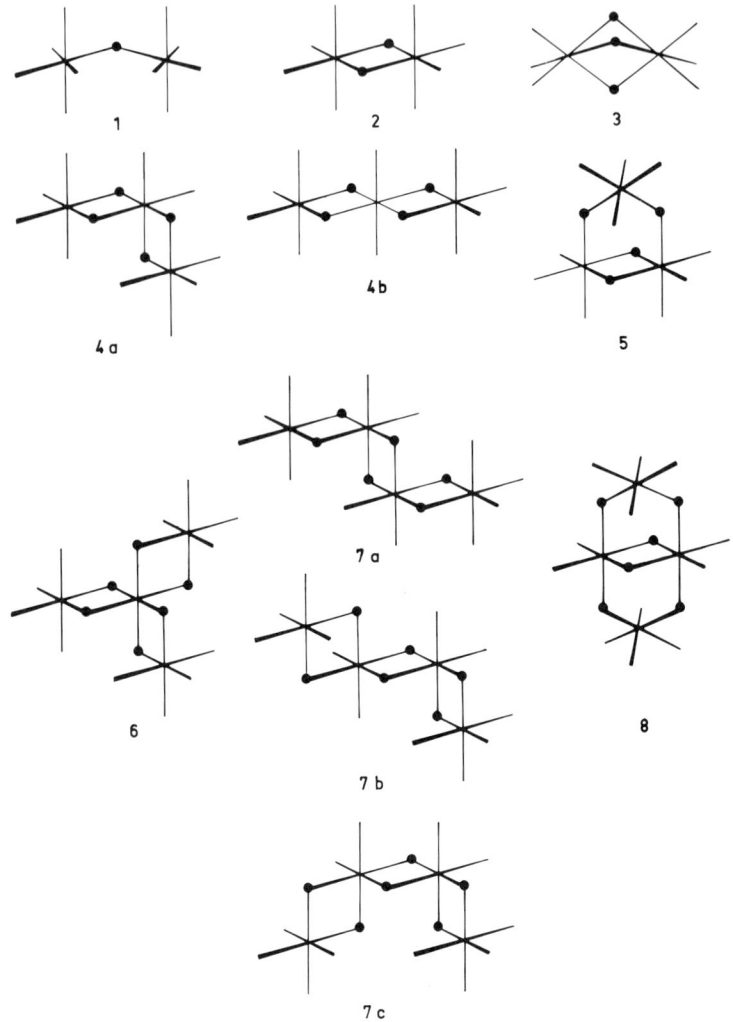

FIG. 1. The di-, tri-, and tetranuclear structures observed in X-ray crystal structures of hydroxo-bridged oligomers of cobalt(III), rhodium(III), iridium(III), or chromium(III); structures **4b**, **7b**, and **7c** have never been observed, but the last two have been mentioned as possible structures for two of the known isomers of $Cr_4(OH)_6^{6+}$

In the dinuclear complexes two octahedrally coordinated metal ions are bound together by one, two, or three hydroxide ions by sharing a corner, an edge, or a face, as shown in structures **1–3**. These mono-, di-, and trihydroxo-bridged binuclear structures are known for each of the metal ions chromium(III), cobalt(III), rhodium(III), and iridium(III), and many of these have been structurally characterized (cf. Tables I–III).

Further condensation of the dinuclear species may occur either by continuing the "edge-to-edge condensation," which leads to "linear" trinuclear structures such as **4a** and **4b**, or by condensation of a third chromium ion to each of the two chromium(III) centers in structure **2**, yielding the cyclic structure **5**. The $(NH_3)_4Co(OH)_2Co(CN)_2(OH)_2$-$Co(NH_3)_4^{3+}$ cation (49) possesses structure **4a**, and there is good evidence that the trinuclear amine complex $Cr_3(en)_5(OH)_4^{5+}$ has the same structure (42). The linear trinuclear structure **4b** has not been observed in crystal structures. The cyclic, trinuclear structure **5** has been observed in crystal structures of chromium(III) complexes with ammonia, tacn, and bispicam (40, 50, 51).

The three different tetranuclear structures which have been observed in the crystalline state are the two compact structures **6** and **8** and the chain structure **7a**. Structure **6** is found in $[Co_4(NH_3)_{12}(OH)_6]Cl_6 \cdot 8H_2O$ and its amine analogs (52–59). The analogous ammonia and ethylenediamine chromium(III) complexes $Cr_4(NH_3)_{12}(OH)_6^{6+}$ and $Cr_4(en)_6(OH)_6^{6+}$ have been characterized quite recently (40, 41, 42, 60). Structure **7a** has so far been observed (42) only in a chromium(III) amine complex, $Cr_4(en)_6(OH)_6^{6+}$, but, as discussed in Section IV, both structures **7b** and **7c** are possible structures for the tetranuclear aqua chromium(III) species. Structure **8** is known from the so-called rhodoso complex, $Cr_4N_{12}(OH)_6^{6+}$ [$N_{12} = (NH_3)_{12}$ or (en)$_6$] (61, 62).

In Tables I, II, and III selected bond distances and angles for mono-, di-, and trihydroxo-bridged complexes are compared. Most of the singly bridged species are chromium(III) complexes. The flexible hydroxo bridge in these complexes does not impose strain and the first coordination spheres of the two metal centers are generally close to being undistorted octahedra. The M–O–M angles vary greatly within the range 132–166°. Linear coordination is observed when oxide is the bridging ligand, as found in the so-called basic rhodo, $(NH_3)_5CrOCr(NH_3)_5^{5+}$, which has an M–O–M angle of exactly 180°.

When more than one bridge connects two metal ions, strain is imposed as reflected by the deviation from 90° for the O–M–O angles and by decreased M–O–M angles. The latter angles are typically 99–101° in dihydroxo-bridged complexes and 85–90° in trihydroxo-bridged complexes. These figures show that the triply bridged structures are, as anticipated, more strained than the doubly bridged structures. As discussed later, this correlates with the order of decreasing kinetic stability $OH > (OH)_2 > (OH)_3$.

TABLE I

CRYSTALLOGRAPHIC DATA FOR MONOHYDROXO-BRIDGED COMPLEXES[a]

Complex	M–O–M (deg)	M–M (Å)	⟨M–O⟩ (Å)	Reference
[(NH$_3$)$_5$Cr(OH)Cr(NH$_3$)$_5$]$^{5+}$	165.6(9)	3.852(9)	1.97	63[b]
trans-[(NH$_3$)$_5$Cr(OH)Cr(NH$_3$)$_4$(H$_2$O)]$^{5+}$	155.1(3)	3.873(2)	1.98	64[c]
cis-[(NH$_3$)$_5$Cr(OH)Cr(NH$_3$)$_4$(OH)]$^{4+}$	142.8(5)	3.744(3)	1.97	65[d]
Δ,Λ-[(OH)(en)$_2$Cr(OH)Cr(en)$_2$(OH)]$^{3+}$	135.4(2)	3.677(2)	1.99	66[e]
Δ,Λ-[(H$_2$O)(en)$_2$Ir(OH)Ir(en)$_2$(OH)]$^{4+}$	131.9(3)	3.797(1)	2.08	67[f]

[a] In this and the following tables one standard deviation is given in parentheses.
[b] Chloride monohydrate. The structure of the dihydrate has also been reported and gave a Cr–O–Cr angle of 158.4(7) deg (68).
[c] Chloride trihydrate.
[d] Dithionate trihydrate.
[e] Perchlorate monohydrate.
[f] Dithionate perchlorate hydrate.

TABLE II

Crystallographic Data for Dihydroxo-Bridged Complexes[a]

Compound	M–O–M (deg)	M–M (Å)	⟨M–O⟩ (Å)	Reference
(NH$_3$)$_4$Cr(OH)$_2$Cr(NH$_3$)$_4^{4+}$	101.54(5)	3.045(1)	1.97	69[b]
ΔΛ-(en)$_2$Cr(OH)$_2$Cr(en)$_2^{4+}$	103.42(8)	3.059(2)	1.95	70[c]
ΔΔ/ΛΛ-(phen)$_2$Cr(OH)$_2$Cr(phen)$_2^{4+}$	102.1(3)	2.986(4)	1.92	39[d]
ΔΛ-(gly)$_2$Cr(OH)$_2$Cr(gly)$_2$	98.2(2)	2.974(2)	1.97	71, 72
ΔΛ-(mal)$_2$Cr(OH)$_2$Cr(mal)$_2^{4-}$	99.34(7)	3.031(2)	1.99	82[e]
(NH$_3$)$_4$Co(OH)$_2$Co(NH$_3$)$_4^{4+}$	101.1(5)	2.932(5)	1.91	73, 74[f]
ΔΛ-(en)$_2$Co(OH)$_2$Co(en)$_2^{4+}$	101.1(1)	2.951(1)	1.93	75[g]
trans-(H$_2$O)(tacn)Rh(OH)$_2$Rh(tacn)(H$_2$O)$^{4+}$	100.5(3)	3.138(1)	2.04	76[h]

[a] The crystal structures of the uncharged complexes [CrL$_4$(OH)]$_2$ with L$_4$ = (S-pro)$_2$ (77), (S-ala)$_2$ (78), (dipic)(H$_2$O) (79), (hydroxodipic)(H$_2$O) (80), (chlorodipic)(H$_2$O) (80), (edda) (81), and (en)(mal) (82); of [Rh(η^3-C$_3$H$_5$)$_2$(OH)]$_2$ (83); and of the salt of the ions [Cr(mepic)$_2$(OH)]$_2^{4+}$ (84), [Cr(bispicen)(OH)]$_2^{4+}$ (85), [Cr(bispictn)(OH)]$_2^{4+}$ (86), [Cr(NH$_3$)$_3$(H$_2$O)(OH)]$_2^{4+}$ (87), [Co(NH$_3$)$_3$(H$_2$O)(OH)]$_2^{4+}$ (88), [Rh(η^5-C$_5$Me$_5$)(dmpz)(OH)]$_2^{2+}$ (89), and [Cr(ox)$_2$(OH)]$_2^{4-}$ (90) have also been reported.
[b] Dithionate tetrahydrate. Data for the chloride tetrahydrate have been reported both for monoclinic (69) and for triclinic crystals (91).
[c] Diperchlorate dichloride dihydrate. Data for the dithionate (92), chloride (93), and bromide (93) salts have also been reported.
[d] Iodide tetrahydrate. The structure of the chloride hexahydrate has also been reported (38).
[e] Sodium salt pentahydrate.
[f] Chloride tetrahydrate.
[g] Nitrate.
[h] Perchlorate tetrahydrate.

TABLE III

CRYSTAL DATA FOR TRIHYDROXO-BRIDGED COMPLEXES

Complex	⟨M–O–M⟩ (deg)	M–M (Å)	⟨M–O⟩ (Å)	Reference
$(NH_3)_3Co(OH)_3Co(NH_3)_3^{3+}$	83.3(6)	2.565	1.93	94[a]
$(dpt)Co(OH)_3Co(dpt)^{3+}$	85	2.579(1)	1.91	95[b]
$(C_5Me_5)Rh(OH)_3Rh(C_5Me_5)^+$	89.6(1)	2.9738(8)	2.109	96[c]
$(C_5Me_5)Ir(OH)_3Ir(C_5Me_5)^+$	92.8(3)	3.0709(7)	2.120	96[d]
$(metacn)Cr(OH)_3Cr(metacn)^{3+}$	84.1(3)	2.642(2)	1.97	97[e]
$(tacd)Cr(OH)_3Cr(tacd)^{3+}$	83	2.666(3)	1.99	98[f]

[a] Dithionate. The bromide and iodide salts have also been described (99).
[b] Perchlorate, racemic isomer. The three M–O–M angles are 85.2, 84.1, and 85.0°.
[c] Hydroxide undecahydrate. $C_5Me_5 = \eta^5$-pentamethylcyclopentadienyl.
[d] Acetate tetradecahydrate.
[e] Iodide trihydrate.
[f] Bromide dihydrate. The M–O–M angles vary from 81 to 84°.

The metal–oxygen (bridge) distance is close to that for terminally coordinated water or hydroxide. There seems to be no significant change in metal–oxygen (bridge) bond distances on going from singly to triply bridged complexes.

Crystal data for the trinuclear (structure **5**) and tetranuclear (structures **6, 7a,** and **8**) complexes reveal bond distances and angles which are comparable to those observed in the mono- and dihydroxo-bridged binuclear complexes (40, 42, 49, 50, 53, 54, 62, 87).

Mixed bridge systems have been reported for chromium(III), cobalt(III), and rhodium(III), and the most common types are dinuclear doubly or triply bridged species: M(OH)(X)M, M(OH)$_2$(X)M, and M(OH)(X)$_2$M. Table IV summarizes crystallographic data for M(OH)(X)M species. For X = O_2^{2-} the M–O–M angle approximates that in dihydroxo-bridged species, whereas for larger bidentate bridging groups such as sulfate, phosphate, and carboxylates the M–O–M angle is similar to that found in singly bridged complexes. In μ-oxo–μ-hydroxo chromium(III) complexes it is the oxo bridge which has the larger Cr–O–Cr angle; cf. the linearly coordinated oxo bridge in the so-called basic rhodo, $(NH_3)_5CrOCr(NH_3)_5^{4+}$ (Table V). This tendency of an oxo bridge toward linear coordination is rationalized by invoking π bonding stabilization, and it is clear that linear coordination gives maximum overlap between the lone pairs of the O^{2-} ligand and the d orbitals (see also Section III).

Crystallographic data for M(OH)$_2$(X)M species (X = NO_2^-, CO_3^{2-}, SO_4^{2-}, and CH_3COO^-) and for M(OH)(X)$_2$M species [X$_2$ = $(CH_3COO^-)(NH_2^-)$ and $(SO_4^{2-})_2$] have also been reported (49, 51, 100–104).

TABLE IV

Variations of the M–O–M Geometry with the Ligand X^{n-} in Hetero-Bridged Complexes

Complex	M–O(H)–M (deg)	M–O–M (deg)	M–M (Å)	⟨M–O⟩ (Å)	Reference
Δ,Λ-[(en)$_2$Cr(OH)(HPO$_4$)Cr(en)$_2$]$^{3+}$		138.2(1)	3.662(1)	1.96	105[a]
Δ,Λ-[(en)$_2$Cr(OH)(SO$_4$)Cr(en)$_2$]$^{3+}$		137.4(2)	3.706(2)	1.99	106[b]
Δ,Λ-[(en)$_2$Cr(OH)(CF$_3$COO)Cr(en)$_2$]$^{4+}$		134.9(6)	3.616(2)	1.96	107[c]
Δ,Λ-[(en)$_2$Co(OH)(O$_2$)Co(en)$_2$]$^{3+}$		117.2(4)	3.289	1.93	108[d]
$\Delta,\Delta/\Lambda,\Lambda$-[(en)$_2$Co(OH)(O$_2$)Co(en)$_2$]$^{3+}$		114.4	3.272	1.95	109[e]
[(NH$_3$)$_4$Co(OH)(NH$_2$)Co(CO$_3$)$_2$]		96.4(1)	2.862(1)	1.92	110[f]
$\Delta,\Delta/\Lambda,\Lambda$-[(mepic)$_2$Cr(OH)(O)Cr(mepic)$_2$]$^{3+}$	95.0(2)	95.0(2)	2.883(2)	1.96	111[g]

[a] Perchlorate monohydrate.
[b] Dithionate monohydrate.
[c] Bromide triperchlorate hydrate.
[d] Perchlorate.
[e] Nitrate dithionate dihydrate. The corresponding superoxo-bridged complex has similar structural data (109).
[f] Pentahydrate: Co–N–Co angle is 97.2(1)°.
[g] Bromide pentahydrate; see also Table V.

TABLE V

Crystallographic Data for Oxo-Bridged Complexes

Complex	M–O(H)–M (deg)	M–O–M (deg)	M–M (Å)	⟨M–O(H)⟩ (Å)	⟨M–O⟩ (Å)	Reference
[(NH$_3$)$_5$Cr(O)Cr(NH$_3$)$_5$]$^{5+}$	—	180	3.642(1)	—	1.82	112, 113[a]
[(SCN)(tpyea)Cr(O)Cr(tpyea)(NCS)]$^{2+}$	—	176.5(6)	3.628(12)	—	1.82	114[b]
$\Delta,\Delta/\Lambda,\Lambda$-[(mepic)$_2$Cr(OH)(O)Cr(mepic)$_2$]$^{3+}$	95.0(2)	100.6(2)	2.883(2)	1.96	1.87	111[c]

[a] Chloride monohydrate.   [b] Tetraphenyl borate.   [c] Bromide pentahydrate.

Each of the polynuclear systems (structures 1–8) may exist in a number of isomers, depending on the nature of the nonbridging ligands. As seen from Tables I–V, the nonbridging ligands are in most cases polydentate amine ligands, and, particularly when these are unsymmetrical, a considerable number of geometric or optical isomers are possible. A few very common examples are discussed in the following paragraphs and other examples are given in subsequent sections.

The monohydroxo-bridged complexes of the type $L_5M(OH)ML_4X^{n+}$ are known for M = Cr(III) and may exist as cis and trans isomers; cis and trans refer to the position of X relative to the bridge. Both isomers of the "aqua erythro" cation $(NH_3)_5Cr(OH)Cr(NH_3)_4(H_2O)^{5+}$ are known and crystal structures of their salts have been reported (cf. Table I).

The dihydroxo-bridged binuclear complexes are among the most common, structurally well-characterized polynuclear complexes. For symmetrical bidentate nonbridging ligands, LL, the cation $(LL)_2M(OH)_2M(LL)_2^{n+}$ may exist in three isomeric forms: a meso isomer (Δ,Λ) with configuration Δ and Λ, respectively, at the two metal centers, and a racemic pair of isomers (Δ,Δ and Λ,Λ) with configuration Δ or Λ at each of the two metal centers (cf. Fig. 2). If the bidentate

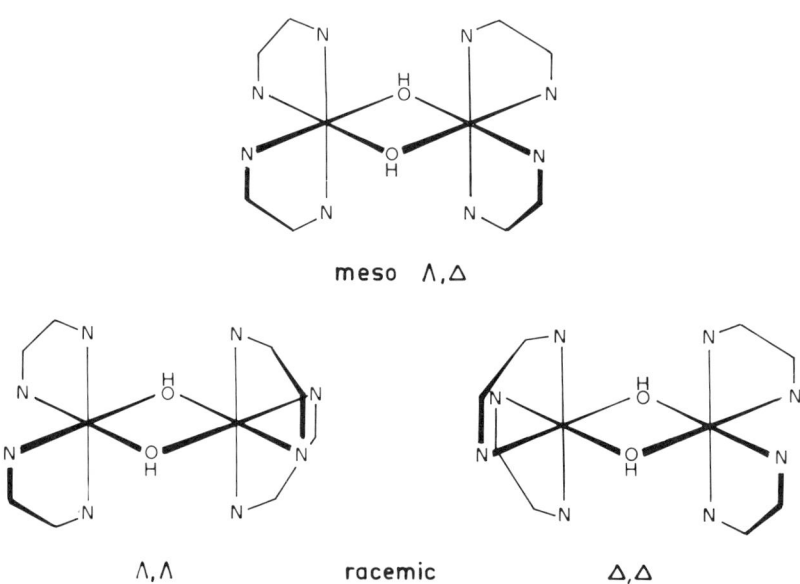

Fig. 2. Both the meso (Δ,Λ) and the racemic (Λ,Λ/Δ,Δ) isomers of the $(en)_2Cr(OH)_2Cr(en)_2^{4+}$ cation are known.

ligand is unsymmetrical or asymmetrical, the number of isomers increases considerably, and as an example it may be mentioned that unsymmetrical ligands, such as 1-(2-pyridyl)methylamine, give rise to 24 possible isomers (115). With chiral ligands such as 1-(2-pyridyl)-ethylamine, the number of possible isomers increases considerably (84, 116).

When higher polynuclears are taken into account, the number of isomers becomes enormous. Figure 1 shows five bridged skeletons for tetranuclear species containing six hydroxo bridges, but it is easily seen that there are many more possibilities. Each of these possible structures may further exist as a number of isomers, depending on the nature of the nonbridging ligand. The ethylenediamine complex $Co\{(OH)_2Co(en)_2\}_3^{6+}$ — "Werner's brown salt" — can be taken as an example. This compound has structure **6**, and all eight possible configurational isomers, $\Delta(\Delta\Delta\Delta)$, $\Delta(\Delta\Delta\Lambda)$, $\Delta(\Delta\Lambda\Lambda)$, and $\Delta(\Lambda\Lambda\Lambda)$ and their enantiomers, have been characterized (57) (the first chirality symbol refers to the configuration about the CoO_6 chromophore and those in brackets to the remaining three CoN_4O_2 chromophores). If the conformations of the chelate ring systems are also considered, then these eight *configurational* isomers embrace a total of 208 *conformational* isomers (117).

Intramolecular hydrogen bonds play a significant structural role; the most important interactions involve hydroxide, either as a terminally coordinated ligand (α-type interactions) or as a bridging ligand (β-type interactions). In α-type interactions the hydrogen donor group may be a terminally coordinated water or ammonia ligand or another terminal hydroxide ligand (Fig. 3A). Interaction of this kind was first proposed to account for the unusual acid properties of monohydroxo-bridged

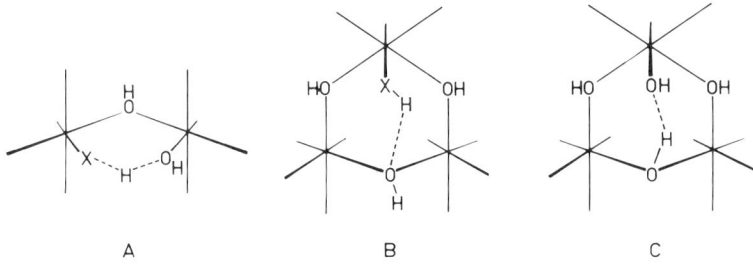

FIG. 3. Intramolecular hydrogen bonds involving two terminally coordinated ligands (α type) as shown in A (X = OH_2, NH_3, or OH^-) or a terminally coordinated ligand and bridging hydroxide (β type) as shown in B (X = OH_2 or NH_3) or in C.

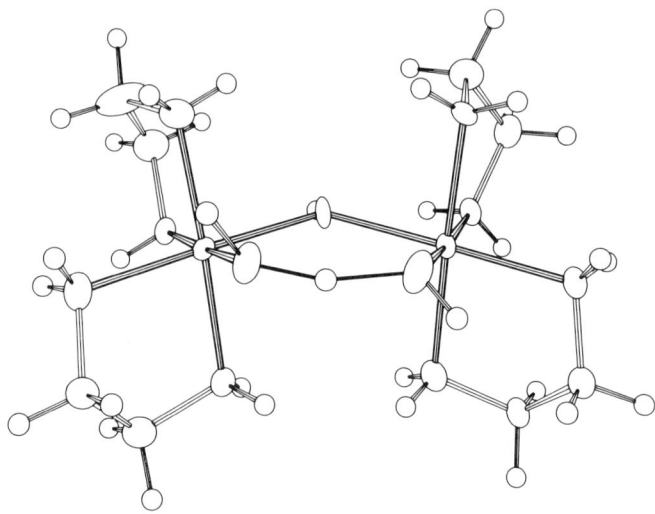

FIG. 4. The Δ,Λ-$(H_2O)(en)_2Ir(OH)Ir(en)_2(OH)^{4+}$ cation is forced into a fixed conformation by a strong and symmetrical intramolecular hydrogen bond interaction of the α type. The O_1-O_2 distance is 2.429(9) Å. Reprinted with permission from Galsbøl, F., Larsen, S., Rasmussen, B., and Springborg, J., *Inorg. Chem.* **25**, 290 (1986). Copyright (1986) American Chemical Society.

complexes of the cis type, $(H_2O)L_4M(OH)ML_4(H_2O)^{5+}$, as further discussed in Section VI. Evidence for such α-type interactions in the solid state has recently been obtained from an X-ray crystal-structure analysis (67) of the perchlorate dithionate salt of the Δ,Λ-$(H_2O)(en)_2Ir(OH)Ir(en)_2(OH)^{4+}$ cation (Fig. 4). The terminally coordinated hydroxo and aqua groups in this cation interact by a very short symmetrical hydrogen bond with an O–O distance of 2.429(9) Å. A similar, but weaker, α-type interaction has been observed in the structure (66) of Δ,Λ-$[(OH)(en)_2Cr(OH)Cr(en)_2(OH)](ClO_4)_3$, as shown in Fig. 5. The M–O–M angle in the latter complex (135.4°) is larger than that in the aqua hydroxo iridium(III) complex (131.9°), but both angles are much smaller than the Cr–O–Cr angle in, e.g., the $(NH_3)_5Cr(OH)Cr(NH_3)_5^{5+}$ cation (165.6°). The fact that the cis hydroxo "erythro" ion has a significantly smaller M–O–M angle than that of the trans aqua isomer is also explicable in terms of a weak α-type interaction in the former cation. These variations demonstrate how hydrogen bond interactions force the monohydroxo-bridged cation into a fixed conformation, with the result that the M–O–M angle

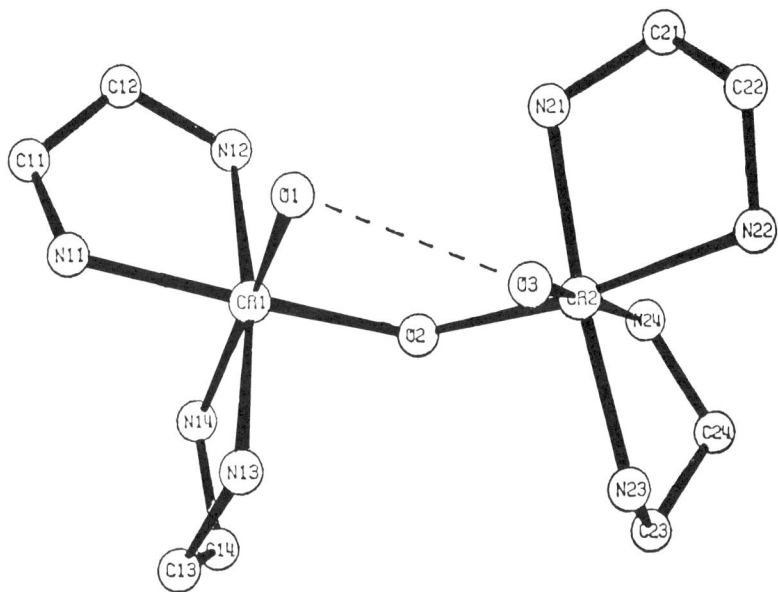

FIG. 5. The Δ,Λ-(HO)(en)$_2$Cr(OH)Cr(en)$_2$(OH)$^{3+}$ cation, like the iridium(III) analog in Fig. 4, is forced into a fixed conformation by a hydrogen bond interaction of the α type. The O O distance is 3.004(6) Å. Redrawn with permission from Kaas, K., *Acta Crystallogr.* B **35**, 1603 (1979).

approaches that in M(OH)(X)M complexes, where X is a bidentate-bridging ligand (cf. Tables I and IV.)

In β-type interactions the hydroxo bridge may act as hydrogen acceptor when the other group is a stronger acid, such as a terminally coordinated ammonia or water ligand (Fig. 3B); it may also act as hydrogen donor with a basic group such as a terminally coordinated hydroxide ligand (Fig. 3C). Interaction of the first kind has been observed in the crystal structures of the trinuclear complex (*40*) Cr$_3$(NH$_3$)$_{10}$(OH)$_4^{5+}$ (structure **5**, Fig. 1) and the tetranuclear rhodoso complexes (*61*, *62*) Cr$_4$L$_{12}$(OH)$_6^{6+}$ [L$_{12}$ = (NH$_3$)$_{12}$ or (en)$_6$; structure **8**, Fig. 1]. Figure 6 shows how the CrN$_4$O$_2$ units in the tetranuclear rhodoso complexes are tilted as a result of two β-type interactions. Similar interactions (X = H$_2$O in Fig. 3B) have been proposed to play an important role in the stability differences observed for various isomers of the aqua chromium(III) oligomers (*118*). The crystal structure (*50*) of the trinuclear complex Cr$_3$(tacn)$_3$(OH)(μ-OH)$_4^{5+}$ indicates a β-type interaction of the kind shown in Fig. 3C.

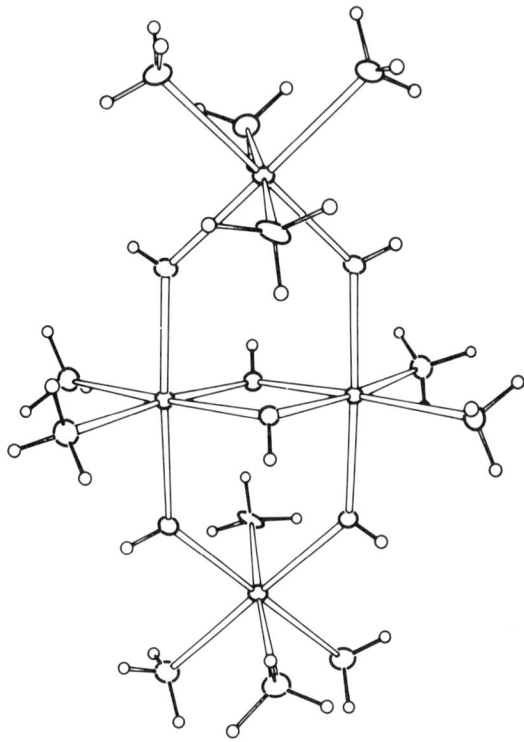

FIG. 6. The cation $Cr_4(NH_3)_{12}(OH)_6^{6+}$. The two CrN_4O_2 units are tilted, owing to hydrogen bond interaction of the β type (between μ-OH and NH_3). Reprinted with permission from Bang, E., *Acta Chem. Scand.* **A38**, 419 (1984).

III. Spectroscopic and Magnetic Properties

Cobalt(III), rhodium(III), and iridium(III) complexes are d^6 systems with diamagnetic ground states. The absorption spectra of the hydroxo-bridged cobalt(III) complexes usually show two broad bands in the visible region assigned as spin-allowed transitions from $^1A_{1g}$ to $^1T_{1g}$ and $^1T_{2g}$ (O_h symmetry), respectively. In, e.g., the dihydroxo-bridged complexes $(NH_3)_4Co(OH)_2Co(NH_3)_4^{4+}$ and Δ,Λ-$(en)_2Co(OH)_2$-$Co(en)_2^{4+}$, the second ligand-field bands are partially obscured by overlap with an intense charge transfer band. The visible absorption spectra of the rhodium(III) and iridium(III) complexes are interpreted in the same way, but for these complexes the second ligand field is normally completely obscured by an intense charge transfer band.

TABLE VI

Absorption Spectra of Some Hydroxo-Bridged Ammine and Amine Complexes in Aqueous Solution

Complex	$(\varepsilon, \lambda)_{max}$ [a]			Reference
$(NH_3)_5Cr(OH)Cr(NH_3)_5^{5+}$	(95, 498)	(76, 372)		129
trans-$(NH_3)_5Cr(OH)Cr(NH_3)_4(H_2O)^{5+}$	(76, 502)	(78, 373)		129
cis-$(NH_3)_5Cr(OH)Cr(NH_3)_4(H_2O)^{5+}$	(96, 505)	(69, 376)		129
cis,cis-$(H_2O)(NH_3)_4Cr(OH)Cr(NH_3)_4(H_2O)^{5+}$	(98, 514)	(63, 380)		36
cis,cis-$(H_2O)(NH_3)_4Cr(OH)Cr(NH_3)_4(OH)^{4+}$	(98, 520)	(77, 386)		36
cis,cis-$(HO)(NH_3)_4Cr(OH)Cr(NH_3)_4(OH)^{3+}$	(100, 534)	(91, 402)		36
$(NH_3)_4Cr(OH)_2Cr(NH_3)_4^{4+}$	(125, 536)	(67, 387)		36
Δ,Λ-$(en)_2Cr(OH)_2Cr(en)_2^{4+}$	(199, 540)	(107, 386)		33
$\Delta,\Delta/\Lambda,\Lambda$-$(en)_2Cr(OH)_2Cr(en)_2^{4+}$	(190, 539)	(102, 384)		35
(metacn)Cr(OH)$_3$Cr(metacn)$^{3+}$	(103, 505)	(56, 378)$_{sh}$	(90, 342)	97
$(NH_3)_5Co(OH)Co(NH_3)_5^{5+}$	(174, 529)[b]	(282, 370)$_{sh}^b$		130
$(NH_3)_4Co(OH)_2Co(NH_3)_4^{4+}$	(195, 528)	(187, 370)$_{sh}$	(2650, 287)	131
$(NH_3)_3Co(OH)_3Co(NH_3)_3^{3+}$	(135, 526)	(263, 364)	(1780, 296)	132
cis,cis-$(H_2O)(NH_3)_4Rh(OH)Rh(NH_3)_4(H_2O)^{5+}$	(464, 338)	(426, 274)		133
$(NH_3)_4Rh(OH)_2Rh(NH_3)_4^{4+}$	(305, 336)			133
(tacn)Rh(OH)$_3$Rh(tacn)$^{3+}$	(437, 340)			76
Δ,Λ-$(H_2O)(en)_2Ir(OH)Ir(en)_2(H_2O)^{5+}$	(520, 271)			67
Δ,Λ-$(en)_2Ir(OH)_2Ir(en)_2^{4+}$	(393, 271)			67

[a] ε in liters cm^{-1} (mol dinuclear)$^{-1}$ and λ in nanometers.
[b] Medium is dimethyl sulfoxide.

Some representative spectral data are given in Table VI. The spectroscopic properties of μ-hydroxo–μ-peroxo and μ-hydroxo–μ-superoxo dicobalt(III) complexes have recently been reviewed by Fallab and Mitchell (*119*) and seem reasonably well rationalized in terms of the molecular orbital (MO) models formulated by Lever and Gray (120). The spectra of μ-hydroxo–μ-peroxo and μ-hydroxo–μ-superoxo complexes of rhodium(III) show features similar to those of the cobalt(III) species (*121–124*).

The spectral and magnetic properties of hydroxo-bridged polynuclear chromium(III) complexes have been studied intensely during the last two decades and two slightly different theoretical models have been presented by Güdel and co-workers and by Glerup, Hodgson, and Pedersen (*125–128*). The contributions from each of these two groups of authors suffer from a complete lack of discussion of the work done by the other group. For a nonspecialist within this field, such as the present author, it has therefore been difficult to decide to what extent the two groups employ different physical models rather than merely different formalisms. The following presentation is based essentially on the work presented by Glerup, Hodgson, and Pedersen.

The visible absorption spectra typically show (Table VI) two broad and intense bands which correspond to the spin-allowed transitions from $^4A_{2g}$ to $^4T_{2g}$ and $^4T_{1g}$ (O_h symmetry), respectively, observed in mononuclear species. In addition, weak but sharp absorption bands are often observed (*131*), and this is particularly pronounced in the spectrum of the classical compound "basic rhodo," $(NH_3)_5CrOCr(NH_3)_5{}^{4+}$, which also shows three very intense and sharp bands in the near UV region, as shown in Fig. 7. The oxo-bridged basic rhodo has a magnetic moment of about 1 BM per chromium at room temperature, which is very much lower than the theoretical "spin-only" value of 3.87 BM and the values found experimentally for monomeric chromium(III) complexes (in the range 3.5–4.0 BM) (*134*). This is explained in terms of coupling between the two paramagnetic ($S = 3/2$) chromium(III) centers. This mixing of the two quartet ground states ($^4B_1$ in C_{4v} or $^4A_{2g}$ in O_h symmetry) gives rise to $^1A_{1g}$, $^3A_{2u}$, $^5A_{1g}$, and $^7A_{2u}$ ground levels (D_{4h} symmetry). The Heisenberg Hamiltonian $\mathbf{H} = J\mathbf{S_1} \cdot \mathbf{S_2}$ describes the splitting of these levels. As shown by Pedersen, the magnetic properties of basic rhodo are in accordance with such a model, a separation of $J = 450$ cm^{-1} between the singlet ground state and the triplet state being found (*134*). Within the framework of the Angular Overlap Model, Glerup has presented a model to calculate the antiferromagnetic coupling and absorption spectra of basic rhodo (*125*). The model assumes that two p orbitals of the bridging

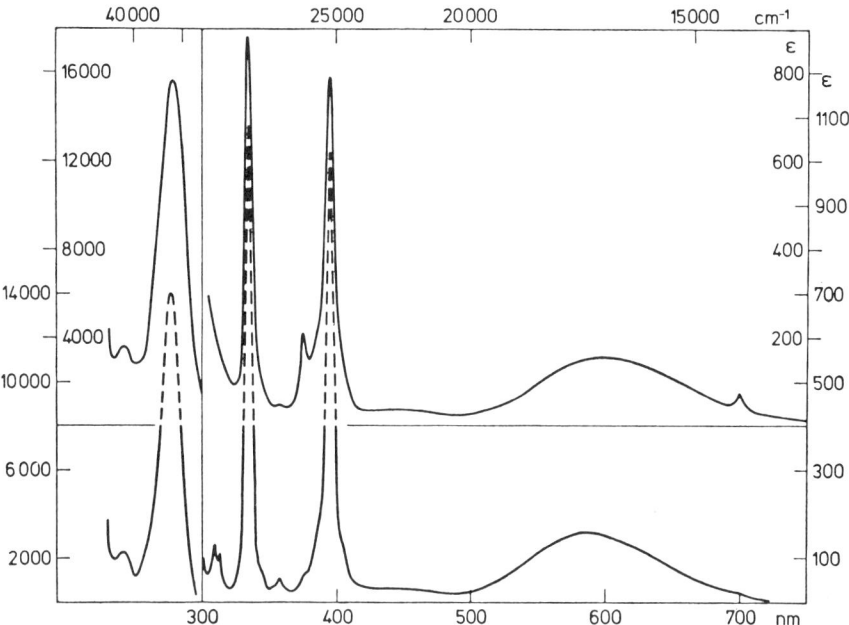

FIG. 7. Antiferromagnetic coupling. Absorption spectra of $(NH_3)_5CrOCr(NH_3)_5^{4+}$ in 4:1 ethanol:methanol mixture. The upper curve is measured at 230 K and the lower curve at 121 K. Reprinted with permission from Glerup, J., *Acta Chem. Scand.* **26**, 3775 (1972).

oxygen participate in π bonds with d orbitals at each of the chromium centers. The antiferromagnetic coupling in the dinuclear system is then explained by a one-electron interaction between atomic orbitals situated on each of the chromium atoms. Through configuration interaction between the lowest levels and the charge transfer levels this one-electron interaction gives rise to the Landé interval rule for an antiferromagnetic coupling with the singlet as the ground state. Furthermore, one-electron interaction mixes doubly excited levels with allowed one-electron charge transfer levels, whereby transitions to doubly excited levels become allowed. The sharp bands mentioned above (Fig. 7) are thus assigned as spin-allowed transitions (double excitations) between the singlet and triplet ground states of the coupled chromophores ($^1A_{1g}$ and $^3A_{2u}$) to the excited singlet and triplet states.

Glerup's model can be extended to many other systems (*135, 136*). For example, an extension of the model has been used to interpret the magnetic properties of di-μ-hydroxo dichromium(III) complexes. It was shown (*126*) that the model semiquantitatively relates the magnitude of the antiferromagnetic coupling to the Cr–O–Cr bridging angle, the Cr–O bond length, and the angle Θ between the bridging plane and the OH vector of the bridging group. The model uses these parameters in the Angular Overlap Model to account for the influence of geometry on competing ferro- and antiferromagnetic contributions. These have previously been expressed either by the van Vleck Hamiltonian $\mathbf{H} = J\mathbf{S}_1 \cdot \mathbf{S}_2$, the expanded Hamiltonian including a biquadratic exchange term j $(\mathbf{S}_1 \cdot \mathbf{S}_2)^2$ or an entirely generalized Hamiltonian. A very satisfactory contribution made by the Angular Overlap Model is a dominating relationship between the angle Θ and the magnitude of the exchange coupling. This correlation demonstrates clearly the π interaction between d orbitals at the two chromium(III) centers (d_{zx} and d_{yz}) through the p_z orbitals on the bridging oxygens. If the hydrogen atom of the bridging OH lies in the Cr_2O_2 ring plane (Θ = 0°), then the lone pair is a pure p_z orbital perpendicular to the plane, corresponding to a maximum value of the π-overlap integral. This value decreases as the hydrogen atom moves out of the plane. The model therefore predicts, in agreement with experiment, that an increasing Θ angle is associated with a decreasing π interaction and therefore also with a decreasing J value (*69, 85, 86, 91, 126*). The model has been applied equally successfully to di-μ-oxo and di-μ-alkoxo complexes (*111, 126*). Table VII illustrates how J varies with Θ for some dibridged binuclear complexes.

Only a few trihydroxo-bridged chromium(III) complexes have been studied. The cation $(metacn)Cr(OH)_3Cr(metacn)^{3+}$ has $J = 128$ cm^{-1} (137, 138), and the corresponding tacd complex has $J = 96$ cm^{-1} (*98*).

Inelastic neutron scattering experiments have been used to study the ground state splitting of singly bridged dichromium(III) complexes of

TABLE VII

Variation of the Coupling Constant J with the Angle Θ between the OH Vector and the Bridging Plane[a]

Compound	Θ (deg)	J (cm^{-1})
Δ,Λ-[(en)$_2$Cr(OH)$_2$Cr(en)$_2$](S$_2$O$_6$)$_2$	57(3)	3.4(1)
[(NH$_3$)$_4$Cr(OH)$_2$Cr(NH$_3$)$_4$]Cl$_4$·4H$_2$O	41(3)	5.23(1)
[(NH$_3$)$_4$Cr(OH)$_2$Cr(NH$_3$)$_4$](S$_2$O$_6$)$_2$4H$_2$O	24(3)	9.12(1)
Δ,Λ-[(en)$_2$Cr(OH)$_2$Cr(en)$_2$]Cl$_4$·2H$_2$O	5(3)	29.4(5)
Δ,Δ/Λ,Λ-[(mepic)$_2$Cr(OH)$_2$Cr(mepic)$_2$](S$_2$O$_6$)$_2$·2H$_2$O	0(5)	32.9(1)
Δ,Δ/Λ,Λ-[(phen)$_2$Cr(OH)$_2$Cr(phen)$_2$]Cl$_4$·6H$_2$O	0(10)	43.0(5)

[a] From Ref. *126*.

the type mentioned above, and the calculated exchange parameters were found to agree with those obtained from the optical and magnetochemical studies (*139–141*).

Only a very few polynuclear complexes containing more than two chromium(III) centers have been studied so far. However, magnetochemical and inelastic neutron scattering studies, heat capacity measurements, and emission spectroscopy have been reported for various tetranuclear species (*40, 142–151*). Two review articles dealing with the spectroscopic and magnetic properties of chromium(III) oligomers have recently appeared (*127, 128*).

The circular dichroism (CD) spectra of optically active di-, tri-, and tetranuclear complexes of chromium(III) and cobalt(III) have been reported and used to establish the complexes' absolute configurations (*55–59, 111, 115, 116, 152–157*). The changes in circular dichroism resulting from ion pairing have been studied for the tetranuclear "hexol" $Co\{(OH)_2Co(NH_3)_4\}_3^{6+}$ and have been shown to be attributable to the vicinal effect of the chiral oxygen centers produced stereospecifically by the ion-pair formation (*56*). For a series of trinuclear cobalt(III) amine complexes, $cis\text{-}Co(CN)_2\{(OH)_2Co(N_4)_2\}_2^{3+}$, it was shown that the main CD contributions due to the two chiral $Co(OH)_4(CN)_2$ and $Co(N_4)(OH)_2$ centers are additive (*155*). In the case of the related tetranuclear complex $Co\{(OH)_2Co(en)_2\}_3^{6+}$ this postulate of additivity of CD spectra proved unsatisfactory (*57*).

IV. Formation of Polynuclear Complexes

A. Hydrolysis Reactions

Formation of hydroxo-bridged complexes by hydrolysis in aqueous solution is, not surprisingly, the most common preparative method. As a rule, such reactions give quite complex product mixtures containing species with different nuclearities, each of which may be present in many isomeric forms. The fact that most of the preparative procedures employed lead to the isolation of one single and pure isomer probably more often reflects favorable solubility properties rather than stereospecificity. In some cases ion-exchange chromatography has been used to isolate the polynuclear species, but systematic analysis of hydrolysis mixtures by this technique has been reported for only a few systems.

1. Dinuclear Complexes

The majority of the oligomers which have been prepared are dinuclear species containing one, two, or three hydroxo bridges. Although the hydrolysis process must always lead to initial formation of monohydroxo-bridged species, these are practically never synthesized by this route, essentially because of their unfavorable thermodynamic properties. For example, single bridged chromium(III) and cobalt(III) species of the type $L_5M(OH)ML_5^{(2n-1)+}$ (L is an inert ligand such as ammonia) are well known but have never been prepared by hydrolysis of $ML_5(H_2O)^{n+}$, probably as a consequence of small formation constants (Q_{21}; see Section V). The brown monohydroxo-bridged cobalt(III) complex $[(PBu_3)(salen)Co(OH)Co(salen)(PBu_3)]^+$ has been isolated as a perchlorate by adding aqueous NaOH to an ethanolic solution of the green monomer $[Co(PBu_3)(salen)]ClO_4$. The binuclear complex is stable only as a solid or in a nonpolar solvent (158). A report to the effect that the trans isomers of $Co(en)_2(Cl)(H_2O)^{2+}$ and $Co(en)_2(Cl)(OH)^+$ react together in cold aqueous solution to give a monohydroxo-bridged species would seem to require further corroboration (159). Singly bridged species of the type $(H_2O)L_4M(OH)ML_4(OH)^{4+}$ are in general easily formed from the corresponding mononuclear species, but they are often (although not always; see Table XXVIII) unstable with respect to further condensation to, e.g., dihydroxo-bridged species, and they are therefore more conveniently prepared via the dihydroxo-bridged species, as described in Section IV,D.

A very large number of dihydroxo-bridged chromium(III) and cobalt(III) complexes have been synthesized from the parent mononuclear species by aqueous hydrolysis, as shown in Eq. (1) for a cationic species, but also neutral and

$$2\textit{cis-}ML_4(H_2O)(OH)^{n+} \rightleftharpoons L_4M(OH)_2ML_4^{2n+} + 2H_2O \qquad (1)$$

anionic species have been made by this reaction. Cationic chromium(III) complexes which have been prepared by reaction Eq. (1) have been reported for $L_4 = (H_2O)_4$ (26, 28, 29), $(NH_3)_4$ (131), $(NH_3)_3(H_2O)$ (87), ibn (160), $(en)_2$ (131), $(tacn)(H_2O)$ (50), $(tame)(H_2O)$ (50), $(phen)_2$ (37, 161), $(bipy)_2$ (37, 161), and (phen)(bipy) (153), and for a series of pyridyl-substituted bi-, tri-, and tetradentate N ligands (51, 55, 85, 86, 111, 115, 116, 152, 154, 157, 162). Neutral complexes have been reported for $L_4 =$ edda (81, 163), (en)(mal) (82), $(H_2O)(dipic)$ (79), $(H_2O)(chlorodipic)$ (80), $(H_2O)(hydroxodipic)$ (80), and a very large number of amino acids, $L_4 = [RCH(NH_2)COO^-]_2$ (*77, 78, 164–169*). Anionic complexes are known for $L_4 = (ox)_2$ (*8*), $(mal)_2$ (*170*), and nta (*48, 171*). The ligand squarate leads to formation of a quadruply bridged complex, $(H_2O)_2Cr(C_4O_4)_2(OH)_2Cr(H_2O)_2$, which has been characterized by a crystal-structure analysis (*172, 173*). One of the few cationic cobalt(III) complexes which can be prepared by reaction Eq. (1) is the tren complex (*174*). Neutral cobalt(III) complexes made by reaction Eq. (1) have been reported for $L_4 = (NO_2)_2(NH_3)_2$ (*174*), $(RCH(NH_2)COO^-)_2$ (*169*), and for $(H_2O)(bh)$ (*175*) (bh is a series of benzoyl hydrazones). Anionic cobalt(III) complexes made by reaction Eq. (1) have been reported for $L_4 = (ox)_2$ (*176*), nta (*177–179*), and for a series of analogs of nta (*180*).

The yields of the dihydroxo-bridged complexes vary and are often relatively low, probably owing to the formation of different isomers and higher polynuclear complexes and to hydrolysis of the ligand L. For many of the complexes listed above, the preparative procedures involved the preparation of the mononuclear species *in situ*.

The chromium(III) complexes with the ligands 1,10-phen and 2,2'-bipy have been obtained by boiling aqueous chromium(III) solutions in the presence of the ligand and slowly adding the appropriate amount of strong base (*37, 161, 181*). Similarly, the chromium(III) complexes with pyridyl-substituted bidentate and tetradentate amines (*86, 111, 115, 116, 152, 157*) have been prepared by the reaction of $Cr(H_2O)_6^{3+}$ with the amine in the presence of zinc dust [or chromium(II)]. The use of organic solvents such as 2-methoxyethanol seems to facilitate the formation of dinuclear hydroxo-bridged complexes to a greater extent than the formation of mononuclear complexes. Chromium(III) complexes with amino acid anions as nonbridging bidentate ligands have been prepared by allowing an aqueous solution of $Cr(NH_3)_6^{3+}$ to react with amino acid at the appropriate pH (*77, 164–166*). Strictly speaking, therefore, it is by no means certain that these reactions proceed via Eq. (1), although such a route seems very feasible.

It should be stressed that some of the complexes mentioned above are more easily made by other routes. The complex $(H_2O)_4Cr(OH)_2(H_2O)_4^{4+}$ is produced quantatively by the Tl^{3+} oxidation of aqueous chromium(II), but only in low yield by reaction Eq. (1) (*30, 31*). The complex Δ,Λ-$(en)_2Cr(OH)_2Cr(en)_2^{4+}$ is obtained (*16*) in good yield by reaction Eq. (1), and the yield may be increased by working in methanolic solution (*182*). However, the solid-state reaction (*131*) described below (Section IV,B) gives this complex in essentially quantitative yield.

With a few exceptions, all of the dihydroxo-bridged complexes listed above may exist in several, and often many, isomeric forms. The topic is discussed briefly in Section II, and a few additional comments seem relevant in the present context.

Hydrolysis of cis-$Cr(en)_2(H_2O)(OH)^{2+}$ probably gives a mixture of racemic (Δ,Δ and Λ,Λ) and meso (Δ,Λ) isomers (Fig. 2). Both forms are known but only the meso isomer has been isolated from the reaction mixture as the sparingly soluble bromide salt (33, 35). The bromide salt of the racemic form is very soluble, and it has been prepared as described in Section IV,D.

Hydrolysis of the analogous mononuclear 1,10-phen and 2,2'-bipy chromium(III) complexes yields the racemic (Δ,Δ and Λ,Λ) dinuclear isomers in high yield, the formation of the meso (Δ,Λ) isomer not being observed. The apparent stereospecificity in these reactions has been explained in terms of steric hindrance in the meso isomer arising from repulsion between the α-hydrogen atoms of ligand molecules attached to each metal ion, as illustrated in Fig. 8 (37). Direct resolution of both the 1,10-phen and the 2,2'-bipy complex into the enantiomers has been reported (161, 183), thereby providing evidence for the racemic configuration; this has later been confirmed by a single-crystal X-ray structure analysis of the 1,10-phen complex (38, 39). The $(-)_{589}$ isomers of these 1,10-phen and 2,2'-bipy complexes have been tentatively assigned (153) the absolute configuration Δ,Δ on the basis of the

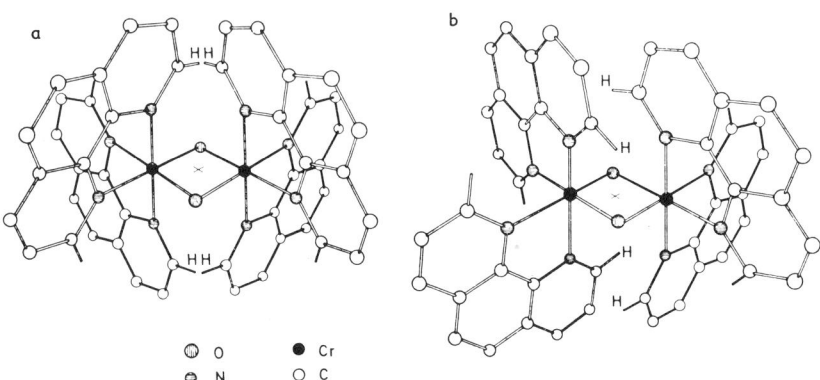

FIG. 8. Schematic drawing of the Δ,Λ and Λ,Λ forms of $(phen)_2Cr(OH)_2Cr(phen)_2^{4+}$. (a) Δ,Λ Isomer—the hydrogen atoms on different ligand molecules, HH, near the plane of symmetry, are seen to be very close; (b) Λ,Λ isomer—nonbonded hydrogen interactions are unimportant. Reprinted with permission from Josephsen, J., and Schäffer, C. E., Acta Chem. Scand. 24, 2929 (1970).

empirical rule (derived originally for mononuclear species) that a negative major CD component in the first ligand absorption band is associated with configuration Δ (*184*).

A series of pyridyl-substituted bi-, tri-, and tetradentate amine complexes of chromium(III) has been studied by Michelsen and co-workers, and like the ethylenediamine complex, for example, these complexes may in principle occur in racemic (Δ,Δ and Λ,Λ) and meso (Δ,Λ) forms (*51, 85, 86, 111, 115, 116, 152, 157, 162*). However, the total number of possible isomers is considerably larger as a result of diastereoisomerism. Despite the large number of possible isomers, the hydrolysis reaction Eq. (1) apparently occurs with a high degree of stereoselectivity, and for each ligand only a few isomers are formed in reasonably high yield, and they are all of the types Δ,Δ and Λ,Λ. These selectivities are presumably a consequence of interactions between two pyridyl moieties attached either to one chromium(III) or to two different chromium(III) centers. The structures of some of these complexes have been established by single-crystal X-ray structure analyses (*84–86, 111*), but strong evidence for the proposed structures has also been provided by steric considerations, by qualitative analysis of the CD spectra, and by analysis of the cleavage products of hydrolysis in concentrated strong acids.

As mentioned above, several chromium(III) complexes containing amino acids as nonbridging ligands have been reported. In general only a very limited investigation of the chemistry of these complexes has been made, but several of the complexes have been characterized by single-crystal X-ray structure analysis. Despite the fact that the number of possible isomers (e.g., 24 with glycine) is very large, the isolated products appear in most cases to be pure, single isomers. This would seem to indicate a high degree of stereoselectivity, as substantiated by the observation that the reaction with glycine yields a Δ,Λ isomer which has the same structure as the complex formed with S-alanine (*71, 72, 78*). However, differences in the solubilities of the various isomers are likely to play an equally important role in this apparent stereospecificity—e.g., with L-proline an isomer with configuration Δ,Δ can be isolated (*77*).

Hydrolysis of $ML_3(H_2O)_3^{3+}$ by reaction Eq. (1) gives dihydroxo-bridged species which may exist in several isomeric forms. For $(L)_3 = (NH_3)_3$ there are five possible isomers; if only facial coordination is considered then the number of isomers is reduced to two, the so-called cis and trans isomers (see Fig. 9). The ammonia, tame, and tacn chromium(III) complexes mentioned previously have all been isolated as their trans isomers. Similarly it has been found that

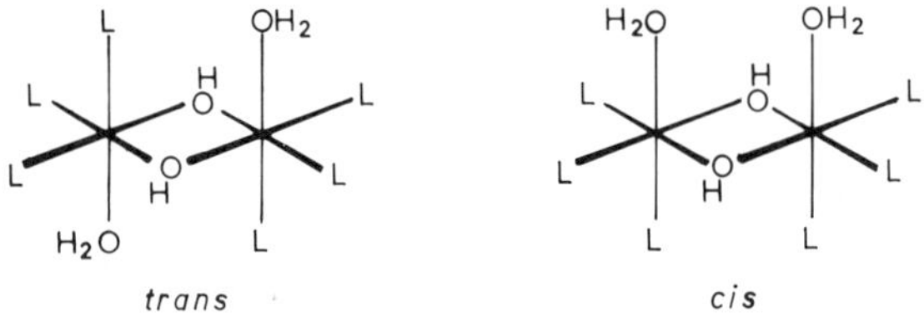

FIG. 9. The so-called trans and cis isomers obtained by condensation of two facial $ML_3(H_2O)_3^{3+}$ cations.

$Rh(tacn)(H_2O)_3^{3+}$ reacts by Eq. (1) forming a trans isomer (76). Likewise, the analogous cobalt(III) complexes with $L_3 = (NH_3)_3$, dien, tach, and tacn are also exhaustively trans isomers (88, 132, 174, 185–187).

As well as yielding dihydroxo-bridged species, hydrolysis of fac-$ML_3(H_2O)_3^{3+}$ may give trihydroxy-bridged species by reaction Eq. (2). This behavior has been reported for chromium(III), cobalt(III), and rhodium(III), but not for iridium(III).

$$fac\text{-}ML_3(H_2O)(OH)_2^+ + fac\text{-}ML_3(H_2O)_2(OH)^{2+} \rightleftharpoons L_3M(OH)_3ML_3^{3+} + 3H_2O \quad (2)$$

Cobalt(III) complexes made by reaction Eq. (2) have been reported for $L_3 = (NH_3)_3$, dien, $(py)_3$, tacn, and tach (99, 132, 174, 185, 186, 188). For $L_3 =$ dpt and metacn, trihydroxo-bridged cobalt(III) complexes have been prepared by oxidation (O_2 or H_2O_2) of aqueous cobalt(III) solutions in the presence of the amine (95, 97). The dpt complex has also been made by hydrolysis of $Co(dpt)_2^{3+}$ in aqueous solution (95).

Trihydroxo-bridged complexes of chromium(III) appear to be much more difficult to synthesize than their cobalt(III) analogs. As mentioned above, hydrolysis of fac-$Cr(NH_3)_3(H_2O)_3^{3+}$ yields predominantly dihydroxo-bridged species, and if any hydrolysis does occur by reaction Eq. (2), then the content of the trihydroxo-bridged species is at the very least considerably smaller (40) than clamined in an early report (189). Similarly, for $L_3 =$ tame and tacn, the facial isomers of $CrL_3(H_2O)_3^{3+}$ do not appear to hydrolyze by reaction Eq. (2), but give dihydroxo-bridged species. To date, the only well-documented trihydroxo-bridged complexes of chromium(III) are $(metacn)Cr(OH)_3Cr(metacn)^{3+}$ and the corresponding tacd complex, which have both been obtained by reaction Eq. (2) (97, 98). The tacn and metacn rhodium(III) analogs have been obtained by reaction Eq. (2) (76, 97).

Finally it should be mentioned that several dinuclear trihydroxo-bridged η^5-pentamethylcyclopentadienylrhodium(III) and iridium(III) complexes have been characterized, as has also a dinuclear dihydroxo-bridged η^3-allylrhodium(III) species (83, 89, 96, 190–192).

2. Higher Oligomers

There are several examples of well-characterized tri- and tetranuclear hydroxo-bridged complexes of chromium(III) and cobalt(III). Penta- and hexanuclear aqua chromium(III) complexes have been prepared in solution, but their structure and properties are unknown. Oligomers of nuclearity higher than four have not been reported for cobalt(III), with the exception of some hetero-bridged heteronuclear species (193, 194). There appear to be no reports of rhodium(III) or iridium(III) complexes of nuclearity higher than two.

Most of the higher chromium(III) oligomers are made by hydrolytic condensation in aqueous solution. Condensation of mono- and dinuclear species gives trinuclear species and, as shown for the aqua chromium(III) system, tetranuclear species may be formed either by condensation of two dinuclear species or by condensation of mononuclear species with trinuclear species (31). As expected, these higher oligomers are obtained under more vigorous conditions than the dinuclear species described above, either by prolonged aging or by aging at higher temperature. As an example it can be mentioned that aging of an aqueous solution of fac-Cr(tacn)(H$_2$O)$_3{}^{3+}$ under mild conditions (25°C, 1 day) leads to formation of the dinuclear species $trans$-(H$_2$O)(tacn)Cr(OH)$_2$Cr(tacn)(H$_2$O)$^{4+}$ in high yield, but under more vigorous conditions (100°C, 3 days) such a solution forms the trinuclear complex Cr$_3$(tacn)$_3$(OH)$_5{}^{4+}$ (50). Optimal conditions may, as in this example, lead to a high yield of a specific isomer, but in general such aged solutions contain a large number of higher oligomers, and the isolation of the individual species usually involves separation by cation-exchange chromatography, possibly in combination with fractional crystallization procedures.

a. Chromium Ammine and Amine Oligomers. Andersen et al. have studied the equilibria between chromium(III) and nitrogen ligands such as ammonia and ethylenediamine in aqueous solution (195, 196). It was found that solutions in which equilibrium between the mononuclear species has been established can be prepared by employing the combined catalytic effects of chromium(II) and charcoal. Depending on the conditions, a variable content of polynuclear hydroxo-bridged complexes was also obtained, but equilibrium with respect to these species was not attained (40, 42, 60, 87).

Chromium(II)–charcoal catalysis has been used to synthesize several new tri- and tetranuclear chromium(III) ammine species (*40*). Anaerobic charcoal-catalyzed oxidation of Cr(II) in aqueous ammonia/ammonium buffer solutions leads to evolution of hydrogen and the formation of a mixture of mononuclear and oligomeric chromium(III) ammine complexes. Under appropriate conditions a high content of polynuclear species is obtained and it is also possible to optimize the yield of certain species by using specific conditions. After quenching by removal of the catalysts, the desired polynuclear species was isolated either by direct precipitation from the product solution or by separation of the mixtures by cation-exchange chromatography in combination with fractional crystallization. Among the complexes which have been isolated by direct precipitation from such solutions is the so-called rhodoso ion, $Cr_4(NH_3)_{12}(OH)_6^{6+}$, which was first prepared by Jørgensen (*4*). The Cr(II)–charcoal catalytic method is faster and more reliable than the original method, and gives a good yield in the form of the bromide salt, which can be crystallized directly from the hydrolysis mixture.

The trinuclear cation $Cr_3(NH_3)_{10}(OH)_4^{5+}$ and the tetranuclear cation $Cr\{(OH)_2Cr(NH_3)_4\}_3^{6+}$ have been prepared similarly by the Cr(II)–charcoal catalytic method and were separated by cation-exchange chromatography (*40*). The trinuclear species was isolated as a bromide salt and has structure **5** in Fig. 1. The tetranuclear species $Cr\{(OH)_2Cr(NH_3)_4\}_3^{6+}$ has been shown to be a chromium ammonia analog of the so-called Werner's brown salt, $Co\{(OH)_2Co(en)_2\}_3^{6+}$ (structure **6** in Fig. 1) (*41*).

Polynuclear chromium(III) ethylenediamine complexes have been synthesized by methods similar to those applied for the ammine systems by using the combined catalytic effect of chromium(II) and charcoal on aqueous ethylenediamine buffer solutions (pH ~ 8) of chromium(III) (*40, 42, 60*). As mentioned above, the use of catalysts is important when equilibration between the mononuclear species is required, but it is unnecessary when the aim is to produce polynuclear species. In fact, identical polynuclear species are formed in approximately the same ratio when buffered chromium(III) solutions ([Cr] = 0.1 M, [en] = 0.3 M) without catalyst are kept for a few days at 40–50°C (*40*).

A trinuclear complex, $Cr_3(en)_5(OH)_4^{5+}$, and three tetranuclear complexes with the common formula $Cr_4(en)_6(OH)_6^{6+}$ have been isolated as salts and characterized. The structures of the trinuclear and one of the tetranuclear complexes have, however, not been established, although cleavage experiments and spectral properties suggest that they both have a linear structure; e.g., the most probable structure for the trinuclear species is structure **4a** in Fig. 1. X-Ray studies have shown that the remaining two tetranuclear species are the racemic cation $\Delta(\Delta\Lambda\Lambda)/\Lambda(\Lambda\Delta\Delta)$-$Cr\{(OH)_2Cr(en)_2\}_3^{6+}$ and the centrosymmetric cation $\Delta,\Lambda,\Delta,\Lambda$-$lel_6$-$(en)_2Cr(OH)_2Cr(en)(OH)_2Cr(en)(OH)_2Cr(en)_2^{6+}$, i.e., with structures **6** and **7a**, respectively, in Fig. 1. In addition to the trinuclear ammonia and tacn complexes mentioned above, it should be mentioned that analogous complexes (structure **5** in Fig. 1) have been reported for pyridyl-substituted amines (*51, 162*).

b. Oligomers of the Hexaaquachromium(III) Ion. Studies of the hydrolytic behavior of the hexaaqua chromium(III) ion were first reported by Niels Bjerrum in 1908 in his thesis "Studies on Basic Chromic Compounds. A Contribution to the Theory of Hydrolysis" (*14*). Bjerrum showed that aging of partially neutralized chromium(III) solutions leads to formation of hydroxo-bridged oligomers, and he was able to determine the composition and stability constants for some of these. The conclusions originally arrived at by Bjerrum have essentially been confirmed by several other groups (*25–31, 197–202*).

Hydrolysis of $Cr(H_2O)_6^{3+}$ yields a large number of polynuclear species. The most well-characterized species are the green monohydroxo-bridged species, $(H_2O)_5Cr(OH)Cr(H_2O)_5^{5+}$, and the blue dihydroxo-bridged species, $(H_2O)_4Cr(OH)_2Cr(H_2O)_4^{4+}$. Although neither of these cations has been isolated as crystalline salts, their respective mono- and dihydroxo-bridged binuclear structures have been established with great certainty by a variety of methods, including cation-exchange experiments (*29, 30*), ^{18}O-exchange experiments (*197*), magnetic measurements (*26*), cryoscopic molecular weight determination (*25*), and kinetic and thermodynamic studies (*26, 28, 31, 199, 200*). A report (*203*) to the effect that the blue dihydroxo-bridged species can be crystallized as a *meta-vanadate* $[VO_2(OH)_2^-]$ would seem to require reexamination before it can be concluded that the solid does in fact contain the dinuclear cation. In some recent studies of these species (*204–209*), structures such as $(H_2O)_4Cr(ClO_4)(OH)Cr(H_2O)_4^{4+}$ and $(H_2O)_5CrOCr(H_2O)_5^{4+}$ have been claimed, but these proposals appear to be in glaring conflict with the large body of evidence indicating mono- and dihydroxo-bridged structures, respectively.

In addition to binuclear species, trinuclear $Cr_3(OH)_4^{5+}$ and tetranuclear $Cr_4(OH)_6^{6+}$ cations (water ligands omitted) have been prepared in solution by separation of the components in the aged solution by cation-exchange chromatography (*31*). The trinuclear species is believed to occur as only one isomer, but the tetranuclear species occur as two forms which interconvert within ≤1 second at room temperature (*200*).

The two tetranuclear forms, which are not necessarily isomers and which might differ in the number of coordinated water molecules, are referred to in the following discussion as α-$Cr_4(OH)_6^{6+}$ and β-$Cr_4(OH)_6^{6+}$, corresponding to the stable and unstable forms, respectively, and $Cr_4(OH)_6^{6+}$ refers to an equilibrium mixture of both forms.

Hydrolysis of $Cr(H_2O)_6^{3+}$ gives tetranuclear species in only low yield, and a more convenient method for preparing these species is to partially neutralize solutions of the dinuclear species, yielding the tetranuclear species in high yield (21%) together with other species. Similarly a 1:1 mixture of $Cr(H_2O)_6^{3+}$ and $Cr_3(OH)_4^{5+}$ gives rise to a good yield of tetranuclear species, and it was shown that the yield of tetranuclear species decreases drastically (from 25 to 2%) if $Cr(H_2O)_6^{3+}$ is absent from the initial solution (31).

Bridge cleavage of the tetranuclear species in acidic solutions occurs, at least predominantly, with stepwise release of $Cr(H_2O)_6^{3+}$, as indicated in Eqs. (3)–(5), where the half-lives for $[H^+] = 1.0\ M$ at 25°C are (31)

$$Cr_4(OH)_6^{6+} \xrightarrow{t_{1/2}\ =\ 3\ \text{hours}} Cr^{3+} + Cr_3(OH)_4^{5+} \qquad (3)$$

$$Cr_3(OH)_4^{5+} \xrightarrow{t_{1/2}\ =\ 21\ \text{days}} Cr^{3+} + Cr_2(OH)^{5+} \qquad (4)$$

$$Cr_2(OH)^{5+} \xrightarrow{t_{1/2}\ =\ 7\ \text{days}} 2Cr^{3+} \qquad (5)$$

It is interesting that the first step in the hydrolysis of the tetranuclear species gives a pure trinuclear species almost without interference from side reactions, such as, e.g., cleavage into two dinuclear species or into other isomers of the trinuclear species. This observation is in keeping with the observation that a mixture of trinuclear species and $Cr(H_2O)_6^{3+}$ at pH ~4 yields tetranuclear species in high yield. These results are further supported by Cr labeling experiments (31).

As mentioned, the tetranuclear species occurs in two forms which interconvert rapidly. These interconversion reactions have been studied spectrophotometrically in the pH range 0.3–4.0 (25°C) and the data have been interpreted in terms of the equilibria shown in Scheme 1.

The structures of the trinuclear species $Cr_3(OH)_4^{5+}$ and the two tetranuclear species α- and β-$Cr_4(OH)_6^{6+}$ have not been established. Linear as well as cyclic trimeric structures (structures **4a**, **4b**, and **5** in Fig. 1) have been considered (27, 31), as has another cyclic structure such as that shown in Fig. 10. However, the kinetic, thermodynamic, spectroscopic, and magnetic properties appear to provide no evidence to favor the one rather than the other (27). The number of possible

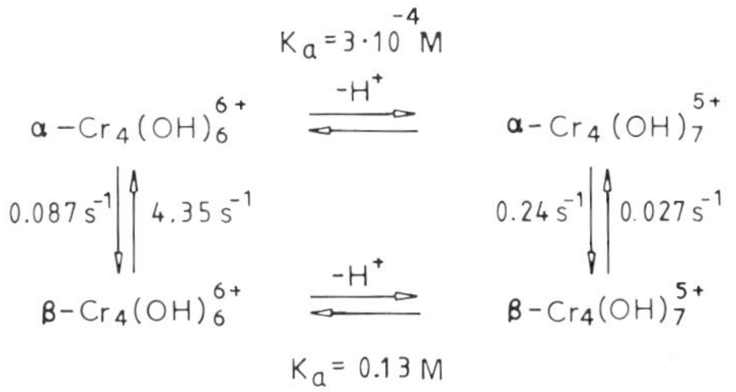

Scheme 1

$$\alpha\text{-Cr}_4(\text{OH})_6^{6+} \underset{-\text{H}^+}{\overset{K_a = 3 \cdot 10^{-4} \text{ M}}{\rightleftharpoons}} \alpha\text{-Cr}_4(\text{OH})_7^{5+}$$

$$0.087 \text{ s}^{-1} \updownarrow 4.35 \text{ s}^{-1} \qquad 0.24 \text{ s}^{-1} \updownarrow 0.027 \text{ s}^{-1}$$

$$\beta\text{-Cr}_4(\text{OH})_6^{6+} \underset{K_a = 0.13 \text{ M}}{\overset{-\text{H}^+}{\rightleftharpoons}} \beta\text{-Cr}_4(\text{OH})_7^{5+}$$

SCHEME 1. The α and β isomers of the two known tetranuclear aqua chromium(III) species interconvert rapidly. The values are for 25°C in 1 M (Na,H)ClO$_4$.

structures for the $\text{Cr}_4(\text{OH})_6^{6+}$ species is quite large. Stüntzi and Marty have proposed cyclic structures which for the β isomer involve a hydroxide ligand coordinated to four chromium(III) ions (31). This proposal has been criticized in a recent note by L. Mønsted, O. Mønsted, and the present author (118). It was pointed out that the interpretation of the kinetic and thermodynamic properties does not require the postulation of unorthodox structures, and as an example it was shown that the trinuclear structure **4a** and linear tetranuclear structures such as structures **7b** and **7c** (Fig. 1) account for the properties of these species in a much more satisfactory way.

c. Cobalt(III) Oligomers. Only very few higher oligomers of cobalt(III) have been made by hydrolysis of mononuclear cobalt(III) aqua complexes. One of the few examples is provided by the hydrolysis of *cis*-Co(NH$_3$)$_4$(H$_2$O)$_2^{3+}$ in dilute ammonia solution, forming the

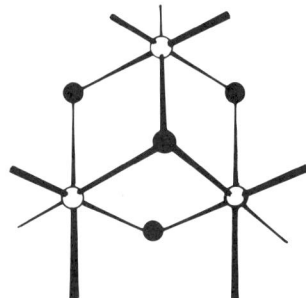

FIG. 10. A possible structure of the trinuclear aqua chromium(III) species $\text{Cr}_3(\text{OH})_4^{5+}$. The metal ions and hydroxo bridges are indicated by ○ and ●, respectively. Another very plausible structure is **4a** in Fig. 1.

tetranuclear complex $Co\{(OH)_2Co(NH_3)_4\}_3{}^{6+}$ (*2, 9, 56, 156, 210*):

$$4\text{cis-}Co(NH_3)_4(H_2O)_2{}^{3+} + 2OH^- \longrightarrow$$
$$Co\{(OH)_2Co(NH_3)_4\}_3{}^{6+} + 4NH_4{}^+ + 4H_2O \qquad (6)$$

This tetranuclear complex was first prepared by Jørgensen (*2*), and was resolved into optical isomers by Werner (*9*), who thereby disposed of the vitalist contention that organic carbon is an essential concomitant of optical activity. Werner named the complex "hexol" after the bridging ligands. The structure of this tetranuclear species has now been established by an X-ray crystal-structure analysis of the racemic salt (*52*). The complex racemizes fairly readily and rates of racemization depend heavily on pH; the first-order rate constants at pH 2.0, 7.0, and 8.1, are respectively, $k = 2 \times 10^{-6}, 2 \times 10^{-3},$ and 4×10^{-2} sec^{-1} at 22°C. Tartrate and selenate decrease the rate of racemization, probably as a result of ion pairing between the hydroxo bridges and the anions (*56*).

Reaction Eq. (6) has been studied in the pH range 8–10 by the pH-stat method (*211*). The reaction was found to be first order in the mononuclear species, and the rate-determining step was proposed to involve a dissociation reaction of the $trans\text{-}Co(NH_3)_4(OH)_2{}^+$ ion.

Corresponding hexols containing aliphatic diamines have also been prepared, although not by the hydrolysis reaction Eq. (6) (see Section IV,C).

Heteronuclear complexes of the types $M(II)\{(OH)_2Co(tren)\}_2{}^{4+}$ and $M(II)\{(OH)_2Co(tren)\}_3{}^{5+}$ have been prepared by mixing aqueous solutions of M(II) and $Co(tren)(OH)_2{}^+$ (M = Co, Ni, Cu, Zn, and Cd). The stability constants were determined and some of the oligomers were isolated as crystalline perchlorate salts (*212*). The $cis\text{-}Co(en)_2(OH)_2{}^+$ ion reacts similarly (*213*).

B. Solid-State Reactions

Dinuclear dihydroxo-bridged complexes can often be obtained from the parent mononuclear complexes by the solid-state reaction Eq. (7). This was first reported by Werner (*7, 11*) and Dubsky (*18*), and it is generally the most convenient method for the preparation of dihydroxo-bridged complexes of Cr(III), Co(III), Rh(III), and Ir(III) with $L_4 = (NH_3)_4$ or $(en)_2$ [and $(tn)_2$ in the case of chromium(III)] (*67, 131, 133, 214–219*). With the exception of the ammonia chromium(III) complex, these reactions are essentially quantitative and the rate of reaction follows the order chromium(III) > cobalt(III) > rhodium(III) ≳

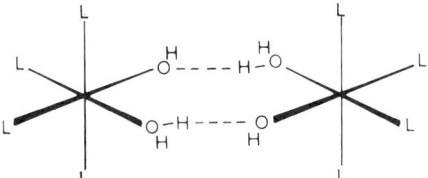

FIG. 11. Intermolecular hydrogen-bonded pairs of aqua hydroxo complexes have been found in several crystal structures and are probably always present in those salts of the type cis-[ML$_4$(H$_2$O)(OH)]X$_2$, which yield dinuclear dihydroxo-bridged complexes upon heating in the solid state.

iridium(III). With L$_4$ = (en)$_2$, the reaction for the chromium(III) complex (*220*) is complete within 15 minutes at 120°C, but at the same temperature the iridium(III) complex (*133*) requires about 14 hours (see Fig. 11).

$$2cis\text{-}[ML_4(H_2O)(OH)]X_2 \xrightarrow{\text{heat}} [L_4M(OH)_2ML_4]X_4 + 2H_2O \quad (7)$$

The solid-state reactions of cis-[Co(NH$_3$)$_4$(H$_2$O)(OH)]X$_2$ (X = Cl$^-$, Br$^-$, NO$_3^-$, and $\frac{1}{2}$S$_2$O$_6^{2-}$) have been studied using differential thermal analysis and thermogravimetry (*221–223*). The reactions, which are all endothermic, are very sensitive to the anions and ΔH is in the region 8–40 kJ mol^{-1}. The activation energies also show a strong dependence on the anions, and for X = Cl$^-$, $\frac{1}{2}$SO$_4^{2-}$, Br$^-$, and $\frac{1}{2}$S$_2$O$_6^{2-}$, the values E_a = 84, 105, 180, and 205 kJ mol^{-1} have been reported. Thermal data have also been reported for a series of cobalt(III) complexes containing coordinated amino acid anions (*224*). ΔH values in the region 30–160 kJ mol^{-1} were found for the reactions of cationic and neutral aqua hydroxo mononuclear species, and anionic mononuclear

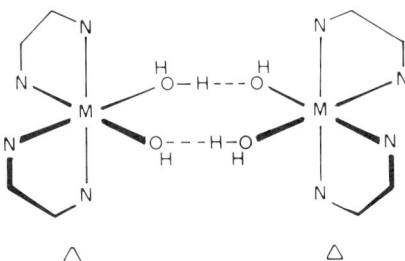

FIG. 12. Crystal packing of the cations in cis-[M(en)$_2$(H$_2$O)(OH)]S$_2$O$_6$ in the manner shown explains why heating of the solid salts for M = Co(III), Rh(III), Ir(III), or Cr(III) yields stereospecifically the meso (Δ,Λ) isomers of (en)$_2$M(OH)$_2$M(en)$_2^{4+}$.

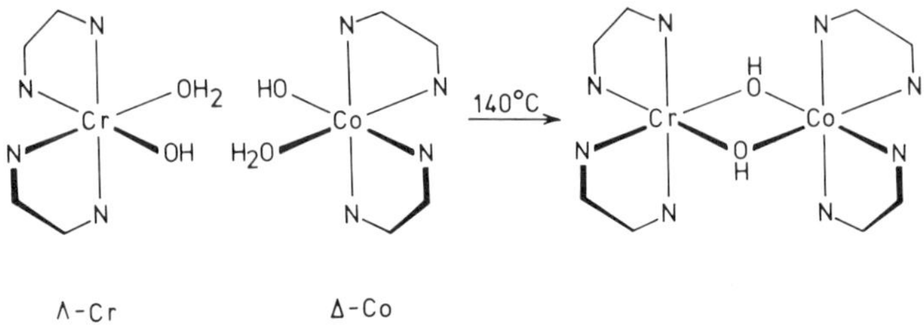

Λ-Cr   Δ-Co

FIG. 13. In a syncrystallized mixture of Λ-cis-Cr(en)$_2$(H$_2$O)(OH)$^{2+}$ and Δ-cis-Co(en)$_2$-(H$_2$O)(OH)$^{2+}$ (as the dithionate), the cations in the crystals are positioned with Λ opposite to Δ, as shown on the left, resulting in stereospecific formation of the heterodinuclear species $(-)_D$-Λ,Δ-(en)$_2$Cr(OH)$_2$Co(en)$_2^{4+}$ upon heating of the solid salt.

complexes yielded values between -20 and -80 kJ mol^{-1}. The activation energies varied from 200 to 400 kJ mol^{-1} with 330–380 kJ mol^{-1} as typical values. It was found that the effect of cations on the thermal condensation of anionic mononuclear species is smaller than the effect of anions on the condensation of cationic species.

It is probably a prerequisite for reaction Eq. (7) that the cations in the crystals of cis-[ML$_4$(H$_2$O)(OH)]X$_2$ pack in the fashion shown schematically in Fig. 11. This structure has been observed in a number of crystal structures and, as discussed at greater length in Section V, the pairs of cations in these structures are held together in the crystal by a strong hydrogen bond interaction, as indicated in the figure. With L$_4$ = (en)$_2$ reaction Eq. (7) is stereospecific and yields the meso (Δ,Λ) isomer quantitatively. This stereospecificity probably results from crystal packing with the Δ enantiomer opposite to the Λ enantiomer, as shown in Fig. 12. Indirect but strong evidence for crystal packing in this manner is provided by the fact that a cocrystallized 1:1 mixture of the dithionates of $(+)_D$-Λ-cis-Cr(en)$_2$(H$_2$O)(OH)$^{2+}$ and $(-)_D$-Δ-cis-Co(en)$_2$(OH)(H$_2$O)$^{2+}$ does not yield the dichromium or dicobalt complex upon heating, but instead yields stereospecifically the heterobinuclear diastereoisomer $(-)_D$-Λ,Δ-[(en)$_2$Cr(OH)$_2$Co(en)$_2$](S$_2$O$_6$)$_2$, as shown in Fig. 13 (214, 215). The fact that many species cannot be made by reaction Eq. (7) presumably reflects unfavorable packing in the crystals of the corresponding mononuclear aqua hydroxo complexes.

C. FORMATION BY OXIDATION REACTIONS

A variety of chromium(III) and cobalt(III) oligomers can be prepared by oxidation of an aqueous solution of the appropriate divalent metal-ion complex with oxidants such as O$_2$ and Tl^{3+}. The chromium(II) reactions with dioxygen are believed to involve the intermediate formation of chromium(IV). It is now well documented that the cobalt(II) reactions occur via the formation of peroxo-bridged dicobalt(III) complexes.

1. Chromium(III) Complexes

The so-called rhodo ion, $(NH_3)_5Cr(OH)Cr(NH_3)_5^{5+}$, is obtained by the oxidation of chromium(II) in an ammonia buffer (*3, 225–227*). The reaction is not quantitative and other unidentified products are formed, but the rhodo ion is the major component under optimal conditions.

Other oxidants such as ClO^-, BrO^-, and H_2O_2 have been investigated but do not lead to formation of the rhodo ion. H_2O_2 is a common product in the reduction of molecular oxygen, and the observation that this oxidant does not give the rhodo ion restricts the choice of possible mechanisms. Joyner and Wilmarth have studied the reaction (*226*) by using ^{18}O labeling followed by analysis of the percentage of ^{18}O among the bridging oxygen atoms in the rhodo ion. The proposed reaction sequence, Eqs. (8)–(12), explains the major results of the labeling experiments [Cr = $Cr(NH_3)_5$]:

$$2Cr(II) + {}^{18}O_2 \longrightarrow Cr(III) \ {}^{18}O_2 \ Cr(III) \tag{8}$$

$$Cr(III) \ {}^{18}O_2 \ Cr(III) + Cr(II) \xrightarrow{2H^+} Cr(III) \ {}^{18}OH \ Cr(III) + Cr(IV)(OH) \tag{9}$$

$$Cr(IV)({}^{18}OH) + H_2O \longrightarrow Cr(IV)(OH) + H_2{}^{18}O \tag{10}$$

$$Cr(IV)(OH) + Cr(II) \longrightarrow Cr(III) \ OH \ Cr(III) \tag{11}$$

$$Cr(IV)({}^{18}OH) + Cr(II) \longrightarrow Cr(III) \ {}^{18}OH \ Cr(III) \tag{12}$$

The oxidation of ammoniacal solutions of Cr(II) may be compared with reaction Eq. (13), for which there are no appreciable side reactions

$$2Cr(aq)^{2+} + \tfrac{1}{2}O_2 \longrightarrow (H_2O)_4Cr(OH)_2Cr(H_2O)_4^{4+} \tag{13}$$

(*30, 198*). The overall reaction rate is proportional to $[Cr(II)]^2[O_2]$, but at the same time it was found that the rate was proportional to the oxygen flow rate, and the results are not easily interpreted. The ^{18}O labeling experiments show that both of the oxygen atoms from O_2 are incorporated into the dinuclear species without exchanging with solvent during the oxidation (*197*). These observations have been interpreted in terms of the mechanism of Eqs. (14)–(16).

The blue dinuclear species in Eq. (13) can be obtained using oxidants other than O_2. It is obtained pure and in quantitative yield when the

$$2\text{Cr(II)} + {}^{18}\text{O}_2 \xrightarrow{\text{slow}} \text{Cr(III)}-{}^{18}\text{O}_2-\text{Cr(III)} \quad (14)$$

$$\text{Cr(III)}-{}^{18}\text{O}_2-\text{Cr(III)} \xrightarrow{\text{fast}} 2\text{Cr(IV)}({}^{18}\text{O}) \quad (15)$$

$$\text{Cr(IV)}({}^{18}\text{O}) + \text{Cr(II)} \xrightarrow{\text{fast}} \text{Cr(III)}-{}^{18}\text{O}-\text{Cr(III)} \quad (16)$$

oxidation is performed with Tl^{3+}, and in lower yields when oxidation is carried out with ClO_3^- (60%), $HClO$ (30%), $Cr_2O_7^{2-}$ (50%), or H_2O_2 (14%). One-electron oxidants such as Fe^{3+}, Cu^{2+}, Cl_2, or Br_2 do not give rise to polynuclear species (30).

A reaction sequence similar to that of Eqs. (14)–(16) has been observed for the reaction of chromium(II) porphyrins with oxygen (228, 229). Similarly, the green oxo-bridged complex [{Cr(NCS)-(tpyea)}$_2$O](BPh$_4$)$_2$ has been prepared by the O$_2$ oxidation of [Cr(NCS)-(tpyea)](BPh$_4$) and its structure has been determined by single-crystal X-ray diffraction (114) [tpyea is the tetradentate ligand tris(2-pyrazal-1-ylethyl)amine]. An oxo-bridged salen dichromium(III) complex has been reported (230).

2. Cobalt(III) Complexes

The oxidation of aqueous solutions of cobalt(II) salts in the presence of ligands may give a variety of products depending on the conditions. Air oxidation of $Co(en)_2(H_2O)_2^{2+}$ in the presence of charcoal as catalyst yields $Co(en)_3^{3+}$, and without charcoal catalyst the reaction gives either the peroxo-bridged dicobalt(III) complex $(en)_2Co(OH)(O_2)Co(en)_2^{3+}$ or the tetranuclear complex $Co\{(OH)_2Co(en)_2\}_3^{6+}$, depending on the conditions.

The above μ-hydroxo–μ-peroxo complex represents an important class of complexes, and a series of $L_4Co(OH)(O_2)CoL_4^{3+}$ complexes has been prepared with other amines such as NH_3, tn, tren, and trien (108, 231–244); the structures of these have been established by X-ray crystal-structure analysis (108, 109, 242, 243). The kinetics of formation and the chemical properties of these dinuclear species have been studied in detail, as discussed in a recent review by Fallab and Mitchell (119).

The tetranuclear complex $Co\{(OH)_2Co(en)_2\}_3^{6+}$ was first prepared by Werner (6) and has subsequently been the subject of several studies (53, 54, 57, 156, 245–247). Its ammonia analog has already been mentioned in Section IV,A. As mentioned in Section II,B, the en tetranuclear species may exist in eight optically isomeric forms. All eight isomers are formed during the oxidation reaction and have been separated by column chromatography (57). The yields of the isomers decrease in the order $\Delta(\Lambda\Lambda\Lambda) > \Delta(\Delta\Lambda\Lambda) > \Delta(\Delta\Delta\Lambda) > \Delta(\Delta\Delta\Delta)$. The X-ray crystal structures of salts of the racemates $\Delta(\Lambda\Lambda\Lambda)/\Lambda(\Delta\Delta\Delta)$ and $\Delta(\Lambda\Lambda\Delta)/\Lambda(\Delta\Delta\Lambda)$ have been reported (53, 54, 245). It is the latter of these racemates which was originally isolated as a nitrate by Werner, and which is now often referred to as Werner's brown salt.

Optical isomers of analogous tetranuclear species (hexols) containing other diamines have been reported for *meso*-2,3-bn (all eight possible optical isomers), R-pn (six isomers), and R,R-chxn (four isomers) (58, 59). As observed for the ammonia hexol, some diamine hexol isomers have been reported to mutarotate (56, 58).

Oxidation of cobalt(II) with O_2 or H_2O_2 may also yield di- or trihydroxo-bridged dinuclear complexes, some of which have already been mentioned in Section IV,A. Other examples may be found in the literature (248–250).

3. Iridium(III) Complexes

Electrochemical reduction of iridium(IV) aqua ions has been reported to give $(H_2O)_5Ir(OH)Ir(H_2O)_5^{5+}$ and $(H_2O)_4Ir(OH)_2Ir(H_2O)_4^{4+}$ (251).

D. Formation from Other Polynuclear Species

1. Condensation of Smaller Oligomers

Although it would be logical to assume that larger oligomers could be made in improved yield by starting with suitable smaller oligomers, this strategy has seldom been used preparatively. One of the few and successful examples is the preparation of the tetranuclear aqua chromium(III) species either by condensation of two dinuclear species or by condensation of Cr^{3+} with the trinuclear species (31).

2. Substitution of the Nonbridging Ligands

Hydrolysis of ammonia or amines is often observed, but only in a few cases have such reactions proved to be useful synthetically. Base hydrolysis (aqueous NH_3) of the so-called rhodo ion, $(NH_3)_5Cr(OH)Cr(NH_3)_5^{5+}$, yields the so-called cis hydroxo erythro ion, *cis*-$(NH_3)_5Cr(OH)Cr(NH_3)_4(OH)^{4+}$, and both this ion and its corresponding acid form, cis aqua erythro have been isolated as salts (227, 252, 253). The hydrolysis is complete within minutes, and unlike the hydrolysis of many other ammine chromium(III) complexes, is quite a clean reaction, at least in solutions of moderate alkalinity (225). The corresponding trans aqua isomer has been prepared by heating the solid

TABLE VIII

Rate Constants for the Hydrolysis of
"Rhodo" and "Erythro" Ions in Base at
$I = 0.1\ M$ and $20°C^a$

Reactant	$k\ (\mathrm{sec}^{-1})$
$(NH_3)_5Cr-O-Cr(NH_3)_5^{4+}$	0.016
$cis\text{-}(NH_3)_5Cr-O-Cr(NH_3)_4(en)^{4+}$	~0.01
$cis\text{-}(NH_3)_5Cr-O-Cr(NH_3)_4(NCS)^{3+}$	0.30
$trans\text{-}(NH_3)_5Cr-O-Cr(NH_3)_4Cl^{3+}$	0.54
$cis\text{-}(NH_3)_5Cr-O-Cr(NH_3)_4F^{3+}$	0.011^b

^{a} Reactions as shown in Eq. (17). From Refs. 24 and 256.
^{b} $(HO)(NH_3)_4Cr(OH)Cr(NH_3)_4F^{3+}$ is also formed.

chloride salt of cis aqua erythro, giving trans chloro erythro (227, 252), followed by Hg(II)-assisted hydrolysis of the latter to yield the trans aqua erythro ion, isolated as a perchlorate (64, 252).

Substitution reactions of the aqua erythro ion with other nucleophiles, usually under acidic conditions, have led to $cis\text{-}(NH_3)_5Cr(OH)Cr(NH_3)_4X^{(5-n)+}$ species for $X^{n-} = NO_2^-$, F^-, SCN^-, and Cl^- (the last of these being isolated only in solution) (24, 227, 254, 255). Reaction of $cis\text{-}(NH_3)_5Cr(OH)Cr(NH_3)_4(SCN)^{4+}$ in ethylenediamine gives $(NH_3)_5Cr(OH)Cr(NH_3)_4(NH_2(CH_2)_2NH_3)^{6+}$, which presumably has the cis configuration (256). The kinetics for the base hydrolysis reactions, Eq. (17), have been studied for $X^{n-} = SCN^-$ (cis), Cl^- (trans), F^- (cis), NH_3, and en (cis) (24, 256, 257) (Table VIII).

$$(NH_3)_5Cr(OH)Cr(NH_3)_4X^{(5-n)+} + OH^- \xrightarrow{fast}$$
$$(NH_3)_5Cr-O-Cr(NH_3)_4X^{(4-n)+} + H_2O \xrightarrow{slow}$$
$$(NH_3)_5Cr(OH)Cr(NH_3)_4(OH)^{4+} + X^{n-} \qquad (17)$$

A kinetic study of the acid hydrolysis of the cis nitro erythro cation has been reported (258).

Replacement of the nonbridging ammine ligands in $(NH_3)_4Co(OH)(NH_2)Co(NH_3)_4^{4+}$ has been reported to take place during acid hydrolysis (259), leading to $(H_2O)(NH_3)_4Co(NH_2)Co(NH_3)_3(H_2O)_2^{5+}$, and likewise the reaction with aqueous bicarbonate (110) gives $(NH_3)_4Co(OH)_2(NH_2)Co(CO_3)_2$. The latter complex reacts with acid, forming $(NH_3)_4Co(OH)(NH_2)Co(H_2O)_4^{4+}$.

3. Hydroxo Bridge Cleavage or Formation

Bridge cleavage of di- and trihydroxo-bridged complexes in acidic or basic solution provides a convenient method for the synthesis of the corresponding monohydroxo-bridged (*32–36, 133, 214–218*) and dihydroxo-bridged (*50, 76, 100, 132, 174, 185, 186, 188*) species, as exemplified by reaction Eqs. (18)–(20).

$$(NH_3)_4Rh(OH)_2Rh(NH_3)_4^{4+} + H_3O^+ \longrightarrow$$
$$cis,cis\text{-}(H_2O)(NH_3)_4Rh(OH)Rh(NH_3)_4(H_2O)^{5+} \quad (18)$$

$$\Delta,\Lambda\text{-}(en)_2Cr(OH)_2Cr(en)_2^{4+} + OH^- \longrightarrow$$
$$\Delta,\Lambda\text{-}(OH)(en)_2Cr(OH)Cr(en)_2(OH)^{3+} \quad (19)$$

$$(NH_3)_3Co(OH)_3Co(NH_3)_3^{3+} + H_3O^+ \longrightarrow$$
$$trans\text{-}(H_2O)(NH_3)_3Co(OH)_2Co(NH_3)_3(H_2O)^{4+} \quad (20)$$

The heterobinuclear complex $\Lambda,\Delta\text{-}(H_2O)(en)_2Cr(OH)Co(en)_2(H_2O)^{5+}$ can be obtained similarly but has been characterized only in solution (*215*). $(H_2O)_5Cr(OH)Cr(H_2O)_5^{5+}$ has likewise been characterized only in solution, and can be made by acid hydrolysis of the dihydroxo-bridged complex (*26*).

Reaction with nucleophiles other than water (or hydroxide) may lead to products of the types **I, II,** and **III**.

```
        H                    H                   H
        O                    O                   O
      /   \                /   \               /   \
   L₄M    ML₄          L₄M    ML₄          L₄M    ML₄
   |      |            |      |             \    /
   OH₂    X            X      X               X

     I                   II                  III
```

The products obtained in the bridge-cleavage reactions of the chromium(III), cobalt(III), rhodium(III), and iridium(III) complexes $\Delta,\Lambda\text{-}(en)_2M(OH)_2M(en)_2^{4+}$ with different nucleophiles X^{n-} are summarized in Table IX. These reactions have been carried out under acidic conditions, except for $X^{n-} = OH^-$ (neutral or basic) and for $X = O_2^{2-}$ (neutral). These reactions are generally quantitative, and most of the products have been characterized by X-ray crystal-structure analysis. Two different pathways are possible. As illustrated in Scheme 2, the hydroxo-bridge cleavage and the entering of the nucleophile could occur simultaneously; alternatively, the initial reaction could be water-assisted hydroxo-bridge cleavage with subsequent substitution of OH_2 by X^{n-}. A pathway similar to that of reaction A (Scheme 2) has been demonstrated for the reaction of $(NH_3)_4Co(NH_2)(OH)Co(NH_3)_4^{4+}$ with a series of nucleophiles (see Section XII,A). For the dihydroxo-

TABLE IX

Products Formed by Bridge Cleavage of $\Delta,\Lambda\text{-(en)}_2\text{M(OH)}_2\text{M(en)}_2^{4+}$ by Reaction with Various Nucleophiles X^{n-} [a]

M	Structure I X^{n-}	Structure II X^{n-}	Structure III X^{n-}
Cr(III)	H_2O, OH^- (33), CF_3COO^- (261)	OH^- (33), CF_3COO^- (107), Cl^- (260)	SO_4^{2-} (262); PO_4^{3-} (105); CF_3COO^- (107); $HCOO^-$, CH_3COO^-, and $RCH(NH_2)COO^-$ (263)
Co(III)	—	—	O_2^{2-} (121)
Rh(III)	H_2O, OH^- (217)	OH^- (217)	O_2^{2-} (121), PO_4^{3-} (220)
Ir(III)	H_2O, OH^- (67)	OH^- (67), Br^- (264)	—

[a] References are given in parentheses.

bridged complexes the reaction with H_2O is fast compared to the anation reactions studied, and pathway B cannot be ruled out. Cleavage of trihydroxo-bridged complexes in the presence of mono- or bidentate nucleophiles (other than water) usually gives dihydroxo-bridged complexes of types **IV** or **V**.

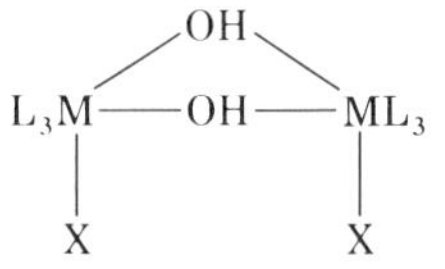

$$L_3M\underset{X}{\overset{OH}{\diagup\diagdown}}\underset{X}{\overset{}{\text{—OH—}}}ML_3 \qquad L_3M\underset{X}{\overset{OH}{\diagup\diagdown}}\text{—OH—}ML_3$$

IV 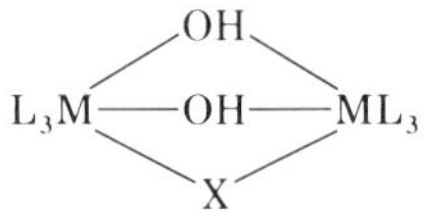   V

Starting from $(NH_3)_3Co(OH)_3Co(NH_3)_3^{3+}$, derivatives of type **IV** have been obtained for $X^{n-} = F^-$, N_3^-, and SCN^- (*132*) and of type **V** for $X^{n-} = SO_3^{2-}$ (*265*), NO_2^- (*266*), and a very large number of carbox-

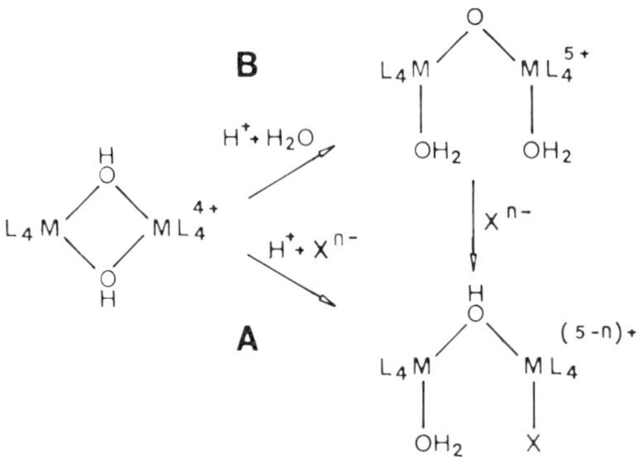

Scheme 2. The reactions of dinuclear dihydroxo-bridged species with a nucleophile X^{n-} in acid solution may occur by pathway A or B.

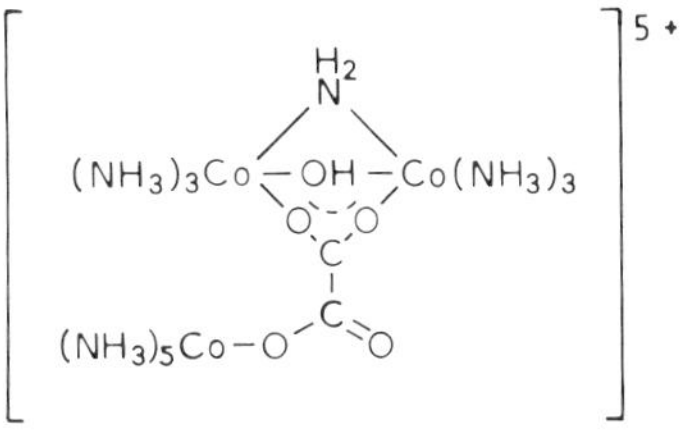

ylates *(193, 194, 265–282)*. The bidentate coordination of both carboxylate functions in a dicarboxylate anion can lead to the formation of tri-, tetra-, and pentanuclear species, as exemplified by structures **VI**, **VII**, and **VIII**. The amido-bridged complex $(NH_3)_3Co(NH_2)(OH)_2$-$Co(NH_3)_3^{3+}$ reacts similarly, i.e., with hydroxo-bridge cleavage rather than with amido-bridge cleavage, as discussed in Section XII,A. The kinetics of reduction of $(NH_3)_3Co(OH)_2(X)Co(NH_3)_3^{(4-n)+}$ and $(NH_3)_3Co(NH_2)(OH)(X)Co(NH_3)_3^{(4-n)+}$, containing a variety of ligands X^{n-}, and of other related species have been studied thoroughly *(193, 194, 267–275, 283–299)*.

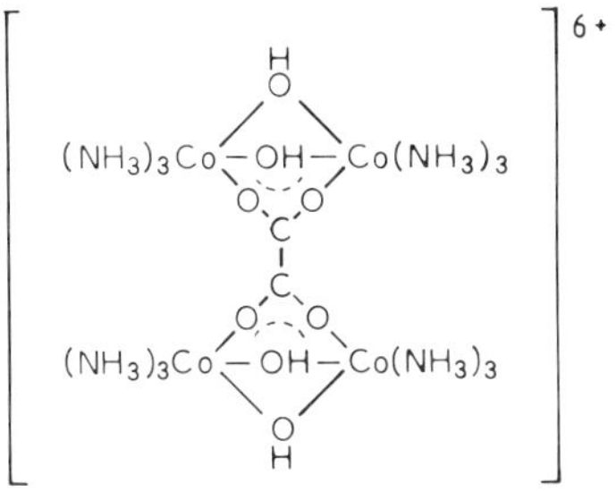

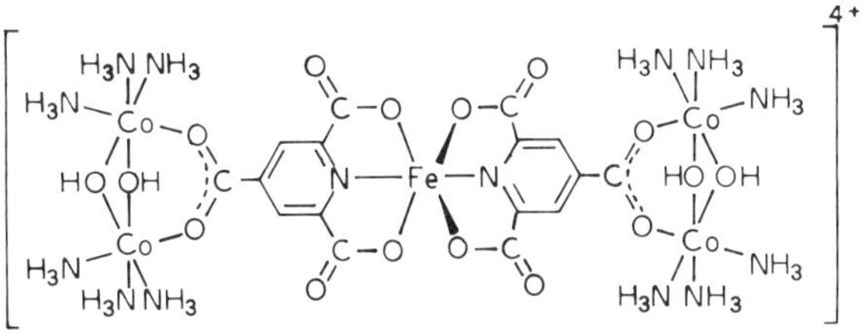

The reactions of hydroxo-bridged aqua chromium(III) oligomers with a variety of anions have been studied intensively because of their relevance to the use of chromium(III) in tanning *(300–317)*. Partially neutralized chromic sulfate solutions are among the must widely used inorganic tanning agents. Such solutions contain a complex mixture of oligomeric chromium(III) species, which in the tanning process are believed to cross-link the polypeptide chains of collagen together by reactions with the carboxylate groups of acidic amino acids in the polypeptide chains. The composition of the active chromium(III) species is not known, but the size of the oligomer is believed to play an important role. Mononuclear chromium(III) species are too small to effect cross-linking, and the large oligomers (tetranuclear or larger) are not effective, perhaps because they are too large to penetrate the cavities in the collagen *(303)*. Addition of inorganic or organic anions is often used to modify the tanning process, and these so-called masking (and demasking) agents are bound to the chromium(III) oligomers in either a terminal or a bridging fashion, thereby strongly influencing the nuclearity, the structure, and the reactivity of the oligomers (*305, 308, 311, 314*).

Studies of such reactions are very difficult and are seriously complicated by the fact that very few of the oligomers have been isolated or characterized. The most studied complexes are those containing sulfate, and of these the more well-characterized species (*315, 317*) appear to be $(H_2O)_4Cr(OH)(SO_4)Cr(H_2O)_4^{3+}$, $(H_2O)_4Cr(OH)(SO_4)Cr(H_2O)_3SO_4^+$, and $(H_2O)_3Cr(OH)(SO_4)_2Cr(H_2O)_3^+$, the last of which (*317*) has been isolated as a green chloride salt. Analogous amine complexes are well known, and X-ray crystal structures of Δ,Λ-[(en)$_2$Cr(OH)(SO$_4$)-Cr(en)$_2$]$^{3+}$ and of [(bispicam)Cr(OH)$_2$(SO$_4$)Cr(bispicam)]$^{2+}$ have been reported (*51, 106, 262*).

Hydroxo-bridged complexes may be synthesized from hetero-bridged complexes, e.g., of the types M(OH)(X)M or M(OH)$_2$(X)M by reactions involving cleavage of the X^{n-} bridge. Reactions shown in Eqs. (21) and (22) are examples (*100, 130*) of this type of process.

$$\text{Cl(NH}_3)_3\text{Co} \overset{\text{NO}_2}{\underset{\text{O}\atop\text{H}}{\diamond}} \text{Co(NH}_3)_3\text{Cl}^{2+} \xrightarrow{\text{NH}_3\text{(liq)}} (\text{NH}_3)_5\text{Co}\overset{\text{H}}{-}\text{O}-\text{Co(NH}_3)_5^{5+} \quad (21)$$

$$(\text{tacn})\text{Rh}\overset{\overset{\text{H}}{\text{O}}}{\underset{\overset{\text{O}}{\underset{\text{C}\atop\|\atop\text{O}}{\diamond}}}{-}}\text{OH}\overset{}{-}\text{Rh(tacn)}^{2+} \xrightarrow[-\text{CO}_2]{2\text{H}_3\text{O}^+} (\text{H}_2\text{O})(\text{tacn})\text{Rh}\overset{\overset{\text{H}}{\text{O}}}{\underset{\overset{\text{O}}{\text{H}}}{\diamond}}\text{Rh(tacn)(H}_2\text{O})^{4+} \quad (22)$$

The tetranuclear cobalt(III) hexol reacts with cyanide (*318*) as shown in Eq. (23). The red trinuclear complex has been shown (*49*) by an X-ray

$$\text{Co}\{(\text{OH})_2\text{Co(NH}_3)_4\}_3^{6+} + 2\text{CN}^- \longrightarrow$$

$$(\text{NH}_3)_4\text{Co}\overset{\overset{\text{H}}{\text{O}}}{\underset{\overset{\text{O}}{\text{H}}}{\diamond}}\text{Co(CN)}_2\overset{\overset{\text{H}}{\text{O}}}{\underset{\overset{\text{O}}{\text{H}}}{\diamond}}\text{Co(NH}_3)_4^{3+} + \text{Co(NH}_3)_4(\text{OH})_2^+ \quad (23)$$

crystal-structure analysis of the racemic salt to contain cis-coordinated cyanide groups (structure **4a** in Fig. 1). Analogous reactions with hexols containing en, *R*-pn, and *S,S*-chxn have been reported. The optical isomers of the resulting trinuclear complexes have been separated and their absolute configuration determined on the basis of their circular dichroism and ^{13}C NMR spectra, and by characterization of the mononuclear complexes formed by cleavage in strong acid (155).

4. Isomerization Reactions

The cis/trans isomerization reaction, Eq. (24), has been applied in the preparation of salts of the cis isomers of the chromium(III) complexes with $L_3 = (NH_3)_3$ or tacn (319). For these species Eq. (24) equilibrium is shifted to the right, while the corresponding equilibria with the diaqua or dihydroxo species, respectively, are shifted to the left (Table X). The increased stability of the cis aqua hydroxo species can be explained in terms of intramolecular hydrogen bond formations (Section VI,C). As mentioned above, the corresponding cobalt(III) and rhodium(III) complexes have been isolated as salts only in the case of the trans-$(H_2O)L_3M(OH)_2ML_3(H_2O)^{4+}$ cations, but it seems very probable that their cis isomers could be prepared by reaction Eq. (24).

$$trans\text{-}(H_2O)L_3M(OH)_2ML_3(OH)^{3+} \rightleftharpoons cis\text{-}(H_2O)L_3M(OH)_2ML_3(OH)^{3+} \quad (24)$$

The isomerization between meso and racemic isomers of $(OH)(en)_2Cr(OH)Cr(en)_2(OH)^{3+}$ has been used to synthesize the racemic isomer (34, 35). The equilibrium constant for Eq. (25) has been estimated to be $K = 4$ (45°C, 8 M ethylenediamine), which is significantly larger than the statistical value of unity.

$$\Delta,\Lambda\text{-}(OH)(en)_2Cr(OH)Cr(en)_2(OH)^{3+} \xrightarrow{K}$$
$$\Delta,\Delta/\Lambda,\Lambda\text{-}(OH)(en)_2Cr(OH)Cr(en)_2(OH)^{3+} \quad (25)$$

Isomerization in the solid state has also been reported. One example is the preparation of trans-$[(NH_3)_5Cr(OH)Cr(NH_3)_4Cl]Cl_4$ from cis-$[(NH_3)_5Cr(OH)Cr(NH_3)_4(H_2O)]Cl_5$ (252). Solid-state isomerization

TABLE X

Equilibrium Constants for the Reaction[a]

$$trans\text{-}(X)L_3Cr(OH)_2CrL_3(Y)^{n+} \xrightleftharpoons{K(D^{n+})} cis\text{-}(X)L_3Cr(OH)_2CrL_3(Y)^{n+}$$

L_3	$X = Y = H_2O$ $K(D^{4+})$	$X = H_2O, Y = OH^-$ $K(D^{3+})$	$X = Y = OH^-$ $K(D^{2+})$
$(NH_3)_3$	0.09(2)	8(2)	0.21(6)
tacn	0.037(3)	6.8(5)	0.19(2)

[a] At 25°C in 1 M NaClO$_4$ (319).

has also been reported for salts of the racemic isomers $\Delta,\Delta/\Lambda,\Lambda$-$(en)_2Cr(OH)_2Cr(en)_2^{4+}$ and $\Delta,\Delta/\Lambda,\Lambda$-$(H_2O)(en)_2Cr(OH)Cr(en)_2$-$(OH)^{4+}$. Heating of the solid salt $\Delta,\Delta/\Lambda,\Lambda$-$[(en)_2Cr(OH)_2Cr(en)_2]$-$Br_4 \cdot 4H_2O$ yields the corresponding meso complex, although in low yield (20%). When heated to 115°C, the solid salt $\Delta,\Delta/\Lambda,\Lambda$-$[(H_2O)(en)_2$-$Cr(OH)Cr(en)_2(OH)]Br_4 \cdot H_2O$ loses the water of crystallization, but neither formation of a dihydroxo-bridged complex, nor isomerization, occurs at this temperature. Heating at 140°C for 1 hour gave an almost quantitative (95%) conversion to the meso isomer of the dihydroxo-bridged complex. The very different yields in the two processes are somewhat surprising and might imply that two different mechanisms are operating (*34, 35*).

5. Photochemical Reactions

Photochemical reactions have not been studied to any great extent (*320–322*). Photochemical trans → cis isomerization of $(NH_3)_5Cr(OH)$-$Cr(NH_3)_4X^{(5-n)+}$ ($X^{n-} = Cl^-$ or H_2O) has been reported (*320*). The partial photoresolution of $(ox)_2Cr(OH)_2Cr(ox)_2^{4-}$ has been reported (*322*).

6. Redox Reactions

Redox equilibria between μ-peroxo–μ-hydroxo dicobalt(III) complexes and their oxidized superoxo-bridged form have been discussed in a recent review article by Fallab and Mitchell (*119*). Corresponding dirhodium(III) complexes have recently been reported (*121–124*). Reduction of peroxo- (or superoxo)-bridged dicobalt(III) complexes provides a method for inserting a hydroxo bridge, as shown in reaction Eq. (26) (*323*).

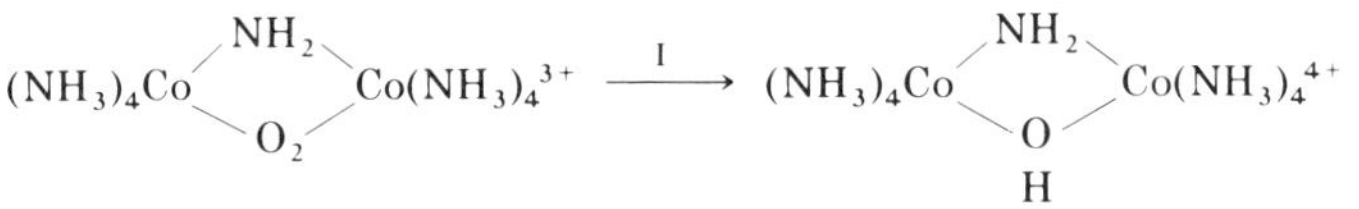

$$(NH_3)_4Co\underset{O_2}{\overset{NH_2}{\diagdown\diagup}}Co(NH_3)_4^{3+} \xrightarrow{\text{I}} (NH_3)_4Co\underset{\underset{H}{O}}{\overset{NH_2}{\diagdown\diagup}}Co(NH_3)_4^{4+} \quad (26)$$

V. Stability Constants

A. Determination of Stability Constants

The simplest and probably the most fundamental reaction, since it initiates all oligomerization reactions, is the condensation of two appropriate metal ions to give a dinuclear monohydroxo-bridged species, M(OH)M. However, the singly bridged species is seldom very stable and further condensation generally takes place, either through intramolecular bridge formation to give, e.g., $M(OH)_2M$, or through further oligomerization. Formation constants are often given in terms of concentration equilibrium constants Q_{xy} defined by Eq. (27).

$$xM(OH_2)_n^{z+} \underset{}{\overset{Q_{xy}}{\rightleftharpoons}} M_x(OH)_y(OH_2)_{nx-my}^{(xz-y)+} + yH^+ + (m-1)yH_2O \quad (27)$$

This equation shows that not only a high metal-ion concentration, but also a high pH, often favors the formation of higher polynuclear species, since y generally increases more rapidly than x. For many aqua metal ions, however, the precipitation of insoluble hydroxides sets an upper pH limit, so that in practice it is possible to study the oligomerization reactions only within a narrow pH region defined by the magnitude of the first acid dissociation constant of the monomeric aqua ion and the pH at which insoluble hydroxide formation occurs.

A wide variety of methods has been used in studies of oligomerization reactions. The most important quantitative method is potentiometric measurement of pH as a function of the total metal concentration and of the concentration of the analytical excess of acid or base. Other quantitative methods which are often used are potentiometric determination of metal ion concentration, calorimetry, spectrophotometry, and ion exchange. These, together with a number of other techniques, have recently been discussed thoroughly by Baes (22).

In the case of inert systems such as cobalt(III) and chromium(III) it is possible, at least in principle, to separate the different species by means of ion-exchange chromatography. It is thus possible to characterize the polynuclear species much better than in the case of labile systems, and their equilibrium concentrations can be determined by direct measurement. The majority of chromium(III) and cobalt(III) oligomers reported are ammine and amine complexes. The N ligators reduce the number of possible skeletons by blocking ligand positions, which in certain respects facilitates the study of such amine complexes. Also, they reduce the number of water ligands, thereby greatly simplifying studies of the acid–base properties of such complexes. The drawback in connection with the study of amine complexes is that the number of isomers which can be formed for a given bridge skeleton may be quite large, as discussed in the previous sections. Another problem is that hydrolysis of the amine ligands may take place on the same time scale as the oligomerization reaction studies, and this complication appears to be more pronounced for chromium(III) than for cobalt(III), rhodium(III), and iridium(III) ammine systems. The hydrolysis of cis-$Cr(en)_2(H_2O)_2^{3+}$ is a typical example. The reaction gives dinuclear mono- and dihydroxo-bridged species of the types $(H_2O)(en)_2Cr(OH)$-$Cr(en)_2(OH)^{4+}$ and $(en)_2Cr(OH)_2Cr(en)_2^{4+}$, of which both meso (Δ,Λ) and racemic (Δ,Δ-Λ,Λ) isomers are known (33, 35). Quantitative studies, however, are complicated by the fact that hydrolysis of the ethylenediamine ligands takes place at the same time as the condensation processes, leading to the formation of mono- and dinuclear species with less than two ethylenediamine ligands per chromium, together with polynuclear complexes of nuclearity higher than two (42, 324). The resulting solution thus contains a complex mixture of known and unidentified species, and attempts to achieve equilibrium conditions have so far failed.

One way to overcome the above problem would be to suppress hydrolysis of the amine ligands by working with an appropriate amine buffer medium. This strategy has been used with great success by Andersen et al. to obtain quantitative equilibrium data for the formation of mononuclear amine complexes (195, 196). Andersen et al. have also studied the formation of polynuclear complexes under similar conditions, but equilibrium was not attained with respect to these species (40, 42, 60, 87). The fact, however, that both thermal hydrolysis and charcoal/chromium(II)-catalyzed hydrolysis in such an amine buffer medium give the same polynuclear species in almost identical ratios would seem to indicate that some degree of equilibration had been achieved. It therefore seems likely that these methods could, in principle, be modified so as to also be applicable for equilibrium studies. Quite a different approach would be to study complexes with macrocyclic amines such as cyclam, which are known to have a reduced tendency to hydrolysis. However, such systems have not as yet been studied in detail.

B. MONOHYDROXO-BRIDGED SPECIES

Condensation to monohydroxo-bridged complexes is often described by Eq. (28), for which the equilibrium constants K_d are related to those defined by Eq. (27) by $K_d = Q_{21}/Q_{11}$, where Q_{11} is the first acid dissociation constant of the mononuclear aqua ion.

$$ML_5(H_2O)^{n+} + ML_5(OH)^{(n-1)+} \xrightleftharpoons{K_d} L_5M(OH)ML_5^{(2n-1)+} + H_2O \quad (28)$$

The few constants K_d which have been reported for chromium(III) and rhodium(III) are listed in Table XI and are compared with those

TABLE XI

Equilibrium Constants $K_d = Q_{21}/Q_{11}$ for Eq. (28) at 25°C

Dinuclear ion	K_d (M^{-1})	Ionic strength (M)	Reference
$(NH_3)_5Co(OH)Co(NH_3)_5^{5+}$	$\ll 1$	—	174
$(NH_3)_5Rh(OH)Rh(NH_3)_5^{5+}$	$\ll 1$	—	220
$(H_2O)_5Cr(OH)Cr(H_2O)_5^{5+}$	$\sim 10^{-0.8}$	1.0–2.0	—[a]
$(H_2O)(en)_2Rh(OH)Rh(en)_2(H_2O)^{5+}$	$\sim 10^{-2.5}$	1.0	325[b]
$(H_2O)_5Cr(OH)Cr(H_2O)_4(OH)^{4+}$	$\sim 10^{1.9}$	1.0–2.0	—[a]
$(H_2O)(en)_2Rh(OH)Rh(en)_2(OH)^{4+}$	$\sim 10^{1.4}$	1.0	325[b]
$(OH)(en)_2Rh(OH)Rh(en)_2(OH)^{3+}$	$\sim 10^{0.5}$	1.0	325[b]
$Zn_2(OH)^{3+}$	$10^{1.8}$	3.0	22
$Cd_2(OH)^{3+}$	$10^{1.1}$	3.0	22
$Hg_2(OH)^{3+}$	$10^{0.9}$	3.0	22
$Pb_2(OH)^{3+}$	$10^{1.7}$	2.0	22

[a] Calculated from data given in Tables XIII, XIX, and XXVIII and from $Q_{11} = 10^{-4.29} M$ (1 M NaClO$_4$) from Ref. 31.
[b] Approximate values for Δ,Λ isomer (see text).

published for aqua ions of other metals. The values in Table XI demonstrate that K_d increases as the charge on the mononuclear cations decreases, and this is in agreement with the trends that one would expect on the basis of purely electrostatic considerations. However, other factors such as intra- and intermolecular interactions may also influence the equilibrium, as discussed below.

Dinuclear complexes of the type cis,cis-$(H_2O)L_4M(OH)ML_4(OH)^{4+}$ may be stabilized by intramolecular hydrogen bond formation of the α type, as mentioned briefly in Section II,B and as further corroborated in the next section. The same kind of stabilization is not possible in dinuclear complexes of the type $L_5M(OH)ML_5^{5+}$ (L = H_2O or NH_3). The significantly greater stability of the $(H_2O)L_4M(OH)ML_4(OH)^{4+}$ species relative to the $L_5M(OH)ML_5^{5+}$ species (a factor of 10^3–10^4) is therefore most likely due, in addition to charge effects, to a strong intramolecular hydrogen bond stabilization of the former species. Similarly, the greater stability of the $(H_2O)(en)_2Rh(OH)Rh(en)_2$-$(OH)^{4+}$ ion relative to its deprotonated form is the reverse of that expected on the basis of electrostatic effects and is explicable by hydrogen bond stabilization of the former species.

Intermolecular hydrogen bond formation between the mononuclear species in aqueous solution has been proposed by Ardon and Bino,

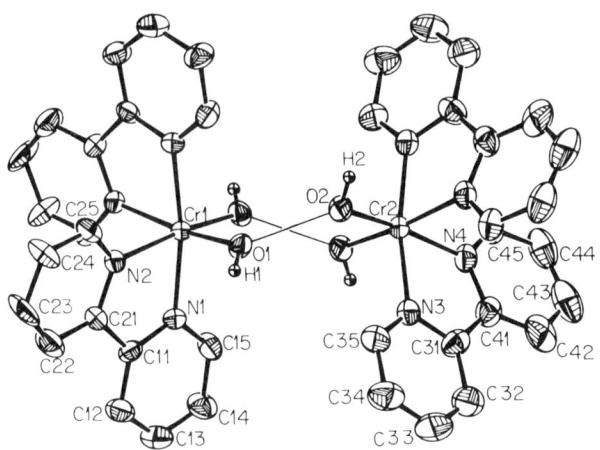

FIG. 14. Structure of cis-[Cr(bipy)$_2$(H$_2$O)(OH)]$^{2+}$. Reprinted with permission from Ardon, M., and Bino, A., *Inorg. Chem.* **24**, 1343 (1985). Copyright (1986) American Chemical Society.

essentially on the basis of crystal-structure studies (*326–330*). Formation of cation pairs of the type shown in Figs. 11 and 14 has been observed in a large number of crystal structures of salts of the type cis-[ML$_4$(H$_2$O)(OH)]X$_2$. A similar kind of interaction has been observed in, e.g., the crystal structure of the salt trans-[Co(en)$_2$(H$_2$O)(OH)](ClO$_4$)$_2$, which forms an infinite chain of Co–OH---H---HO–Co units. The intermolecular hydrogen bond distances quoted in Table XII show that these bonds are strong with an O–O distance varying from 2.41 to 2.59 Å. It should be noted that Ardon and Bino use a different

TABLE XII

STRUCTURAL DATA FOR INTERMOLECULAR HYDROGEN-BONDED CATION PAIRS

Compound	O$_1$·····O$_2$ (Å)	M$_1$·····M$_2$ (Å)	Reference
cis-[Cr(pic)$_2$(H$_2$O)(OH)]S$_2$O$_6$	2.586(6)	6.18	*333*
[Co(tren)(H$_2$O)(OH)](NO$_3$)$_2$ · H$_2$O	2.450(4)	4.82	*326*
cis-[Cr(bipy)$_2$(H$_2$O)(OH)]I$_2$ · H$_2$O	2.446(5)	5.03	*331*
cis-[Cr(bipy)$_2$(H$_2$O)(OH)](NO$_3$)$_2$	2.442(4)	4.96	*326*
trans-[Co(en)$_2$(H$_2$O)(OH)](ClO$_4$)$_2$	2.441(2)	5.72	*331*
trans-[Co(en)$_2$(NO$_2$)(H$_2$O)], trans-[Co(en)$_2$(NO$_2$)(OH)](ClO$_4$)$_3$ · 2H$_2$O	2.412(9)	5.67	*326*
trans-[Co(en)$_2$(NCS)(H$_2$O)], trans-[Co(en)$_2$(NCS)(OH)](CF$_3$SO$_3$)$_3$ · H$_2$O	2.415(6)	5.77	*326*

terminology for these cation pairs: they refer to the hydrogen-bonded cation pair as a dimer and describe the H–O–H⋯OH⁻ entity as an $H_3O_2^-$ ligand. However, the fact that the O–O distance of the $H_3O_2^-$ entity varies more than is usually observed for ligands has made the present author reluctant to adopt this terminology.

Evidence for the formation of such hydrogen-bonded cation pairs in solution has been obtained by three-phase vapor tensiometry studies on solutions of cis-$Cr(bipy)_2(H_2O)(OH)^{2+}$ in a saturated solution of barium nitrate (the total molality of chromium species was 0.1–0.2 m) (326, 330). It was shown that the equilibrium of Eq. (29) in concentrated solutions lies considerably to the right, and it was estimated that $K_{ip} \sim 1\ M^{-1}$.

$$2cis\text{-}ML_4(H_2O)(OH)^{n+} \underset{}{\overset{K_{ip}}{\rightleftharpoons}} L_4M \underset{OH\cdots H_2O}{\overset{OH_2\cdots HO}{\diagup\diagdown}} ML_4^{2n+} \qquad (29)$$

It is noted that the formation of such hydrogen-bonded cation pairs corresponds well to results obtained in a study of the behavior of freshly precipitated chromic hydroxide (199).

It is readily seen that a large value of K_{ip} would imply that the ratio [dinuclear]:[mononuclear] becomes constant and independent of the total metal-ion concentration. The observation that K_d for Eq. (28) is a true constant, which does not vary within a reasonably large metal-ion concentration range, is therefore indirect evidence that K_{ip} is small, and this seems to be the case for the chromium(III) and rhodium(III) complexes listed in Table XI.

It may thus be concluded that variations in K_d can to a large extent be rationalized in terms of charge effects, but that intramolecular hydrogen bond interactions (α type) may contribute significantly to stabilization of the dinuclear species. Intermolecular hydrogen bond interactions may stabilize the mononuclear species, but in the case of +2 charged cations such interactions are measurable only in very concentrated solutions. The possible kinetic consequences of these inter- and intramolecular hydrogen bond interactions are discussed in Sections VII and VIII.

C. Dihydroxo-Bridged Species

The formation of dihydroxo-bridged species may be considered to take place via stepwise formation of a monohydroxo-bridged species with subsequent intramolecular bridge formation. Thermodynamic

TABLE XIII

Equilibrium Constants $K_d = Q_{22}/(Q_{11})^2$ for Eq. (31)

Mononuclear species	K_d (M^{-1})	Temperature (°C)	Ionic strength (M)	Reference
$Sc(OH)^{2+}$	$10^{4.2}$	25	1.0	22
$Y(OH)^{2+}$	$10^{4.0}$	25	3.0	22
$Al(OH)^{2+}$	$10^{3.0}$	25	1.0	22
$Cr(OH)^{2+}$	$10^{3.6}$	25	1.0	22, 31[a]
$Cr(nta)(H_2O)(OH)^-$	$10^{3.28}$	25	1.0	171
$Co(tach)(H_2O)_2(OH)^{2+}$	$10^{3.5}$	20	0.1	174
$cis\text{-}Co(NH_3)_2(NO_2)_2(H_2O)(OH)$	$10^{2.2}$	20	0.1	174
$Co(tren)(H_2O)(OH)^{2+}$	<10	20	0.1	174
$fac\text{-}Co(NH_3)_3(H_2O)_2(OH)^{2+}$	>1	20	0.1	174
$fac\text{-}Co(bamp)(H_2O)_2(OH)^{2+}$	$\ll 1$	20	0.1	174
$cis\text{-}Co(en)_2(H_2O)(OH)^{2+}$	$\ll 1$	20	0.1	332

[a] Calculated from $Q_{22}(av) = 10^{-5.0}$ M (Table XIV) and $Q_{11} = 10^{-4.29}$ M (31).

data for reaction Eq. (30) are known for a number of systems and are discussed in Section IX.

$$(H_2O)L_4M(OH)ML_4(OH)^{n+} \rightleftharpoons L_4M(OH)_2ML_4^{n+} + H_2O \quad (30)$$

Condensation to give dihydroxo-bridged species is often described by Eq. (31), which is related to Eq. (27) by $K_d = Q_{22}/(Q_{11})^2$. The cation $(H_2O)_4Cr(OH)_2Cr(H_2O)_4^{4+}$ and the anion $(nta)Cr(OH)_2Cr(nta)^{2-}$ are so far the only chromium(III) complexes for which K_d has been determined. A number of amine Co(III) complexes have been shown to hydrolyze according to Eq. (31), but only in a few instances has it been possible to determine the equilibrium constant K_d.

$$2cis\text{-}ML_4(H_2O)(OH)^{(n-1)+} \overset{K_d}{\rightleftharpoons} L_4M(OH)_2ML_4^{(2n-2)+} + 2H_2O \quad (31)$$

From the values given in Table XIII it is noted that K_d is apparently much more sensitive to variation of the nonbridging ligands than to variation of the metal ion. It is seen that K_d for all the aqua metal ions lies within a relatively narrow range of about $10^{3.5}$ M^{-1}. In contrast, chromium(III) and cobalt(III) amine complexes have K_d values which vary by at least five orders of magnitude.

D. Trihydroxo-Bridged Species

Although condensation of $ML_3(H_2O)_2(OH)^{2+}$ and $ML_3(H_2O)(OH)_2^+$ to trihydroxo-bridged species is known for chromium(III), cobalt(III), and rhodium(III), the equilibrium constants have been determined only for the tach cobalt(III) system, which has $K_d = 10^{9.4}\ M^{-1}$ [Eq. (2)] (*174*).

E. THE AQUA CHROMIUM(III) SYSTEM

Stability constants for higher oligomers have, until now, been reported only for the aqua chromium(III) system. In addition to Bjerrum's classical study of this system (*14*), stability data for this very complicated system have been reported by several independent groups (*26, 27, 28, 200*). The Q_{xy} values given in Table XIV show some deviation between the values reported by different authors, and it should further be noted that Q_{46} applies to a mixture of at least two tetranuclear species (see Section IV). The stability constants K_n for the stepwise polymerization process, Eq. (32), have been determined as $K_2 \approx 1 \times 10^5\ M^{-1}$, $K_3 \approx 6 \times 10^6\ M^{-1}$, and $K_4 \approx 2 \times 10^5\ M^{-1}$. The change in free

$$\mathrm{Cr}_{n-1}(\mathrm{OH})_{2n-4}^{(n+1)+} + \mathrm{Cr}(\mathrm{OH})_2^+ \xrightleftharpoons{K_n} \mathrm{Cr}_n(\mathrm{OH})_{2n-2}^{(n+2)+} \quad (32)$$

energy is therefore approximately the same when $\mathrm{Cr}(\mathrm{H}_2\mathrm{O})_4(\mathrm{OH})_2^+$ is added successively to form di-, tri-, and tetranuclear species, respectively, and this regular trend may indicate structural similarity between the different polynuclear species, as discussed in Section IV. It should be noted, however, that the trinuclear species is significantly more stable than both the di- and tetranuclear species. A similar

TABLE XIV

EQUILIBRIUM DATA FOR THE REACTION[a]

$$x\mathrm{Cr}^{3+} \xrightleftharpoons{Q_{xy}} \mathrm{Cr}_x(\mathrm{OH})_y^{(3x-y)+} + y\mathrm{H}^+$$

$\mathrm{Cr}_x(\mathrm{OH})_y^{(3x-y)+}$	$-\log Q_{xy}$[c]	ΔH_{xy}[b] (kJ mol^{-1})	ΔS_{xy}[b] (J mol^{-1} K^{-1})
$\mathrm{Cr}_2(\mathrm{OH})_2^{4+}$	5.34,[d] 5.1,[e] 4.9[b]	53[f]	84[f]
$\mathrm{Cr}_3(\mathrm{OH})_4^{5+}$	8.3,[e] 7.95[b]	96[g]	314[g]
$\mathrm{Cr}_4(\mathrm{OH})_6^{6+}$	≤14.1[d]	—	—

[a] At $I = 1.0\ M$ and 25°C.
[b] From Ref. 22, extrapolated on the basis of data given in Refs. 28 and 334.
[c] $Q_{xy} = [\mathrm{H}^+]^y[\mathrm{Cr}_x(\mathrm{OH})_y^{(3x-y)+}]/[\mathrm{Cr}^{3+}]^x$.
[d] From Ref. 31.
[e] From Ref. 31, extrapolated on the basis of data given in Refs. 26 and 27.
[f] From Ref. 26.
[g] From Ref. 27.

increased stability of trinuclear relative to dinuclear species has been observed for the tacn system, where heating of an aqueous solution of $trans$-$(H_2O)(tacn)Cr(OH)_2Cr(tacn)(H_2O)^{4+}$ under reflux yields a trinuclear species in high yield (50).

VI. Acid–Base Equilibria

The acid–base properties of the polynuclear complexes are closely connected with the reactivity of these compounds, this being primarily a question of the acid–base properties of terminally coordinated water or hydroxide and bridging hydroxide. The protolytic properties of water and hydroxide ligand are influenced by the nature of the other ligands, by the overall charge, by the medium, and by the central ion in the same way as is known for mononuclear species. However, for both ligands the acid–base properties can be predicted only in a few cases on the basis of the properties of the corresponding mononuclear species. In most cases, other properties specific to the polynuclear compounds are of importance, particularly the formation of intramolecular hydrogen bonds, and the effect of these properties on the acid strength of terminally coordinated water in particular can be very pronounced.

A. Bridging Hydroxide

The bridging hydroxo ligand is a weak acid as well as an extremely weak base. The acid properties are by far the best investigated, and acid dissociation constants have been reported for singly and doubly bridged dinuclear complexes containing one or two hydroxo bridges (cf. Tables XV and XVI).

TABLE XV

Acid Strength of Bridging Hydroxide in Singly Bridged Chromium(III) Complexes at 20°C

Complex	pK_a	Medium	Reference
cis-$(NH_3)_5Cr(OH)Cr(NH_3)_4(enH)^{6+}$	6.36[a]	0.1 M KCl	256
$(NH_3)_5Cr(OH)Cr(NH_3)_5^{5+}$	7.63	0.12 M NaClO$_4$	24
$trans$-$(NH_3)_5Cr(OH)Cr(NH_3)_4(OH)^{4+}$	~9	1 M NaClO$_4$	129
cis-$(NH_3)_5Cr(OH)Cr(NH_3)_4NCS^{4+}$	10.62	0.11 M NaClO$_4$	24
$trans$-$(NH_3)_5Cr(OH)Cr(NH_3)_4Cl^{4+}$	11.37	0.2 M NaCl	24
cis-$(NH_3)_5Cr(OH)Cr(NH_3)_4F^{4+}$	13.4	1.0 M (KCl, NaOH)	24
cis-$(NH_3)_5Cr(OH)Cr(NH_3)_4(OH)^{4+}$	>16	0.14 M NaClO$_4$	24

[a] The coordinated $NH_2CH_2CH_2NH_3^+$ ligand has a $pK_a = 8.42$ (256).

TABLE XVI

Acid Strength of Bridging Hydroxide in $L_4Cr(OH)_2CrL_4{}^{4+}$ Species at 25°C and $I = 1.0\ M$

$L_4{}^a$	Configuration	pK_{a1}	pK_{a2}	Reference
$(NH_3)_4$	—	~12	—	220
$(en)_2$	Δ,Λ	~12	>14	33
$(tn)_2$	—	~12	—	216
$(mepic)_2$	Δ,Δ(α,α)(RR,RR)	10.7	—	111[a]
bispicen	Δ,Δ(α,α)(RR,RR)	9.3	—	152[a]
bispicpn	Δ,Δ(α,α)	~9	>14	152
$(bipy)_2$	Δ,Δ	7.60	11.9	37[a]
$(phen)_2$	Δ,Δ	7.40	11.8	37[a]
nta[b]	—	8.7	9.8	171

[a] pK_a measured for the racemic salt.
[b] $(nta)Cr(OH)_2Cr(nta)^{2-}$

1. Monohydroxo-Bridged Species

The species must studied in this context are the so-called rhodo ion, $(NH_3)_5Cr(OH)Cr(NH_3)_5{}^{5+}$, and its derivatives cis- and trans-$(NH_3)_5Cr(OH)Cr(NH_3)_4X^{(5-n)+}$, the so-called aniono erythro ions. In strong base these ions deprotonate, giving oxo-bridged blue cations, as shown in Eq. (33) and (34).

$$(NH_3)_5Cr(OH)Cr(NH_3)_5{}^{5+} \rightleftharpoons (NH_3)_5CrOCr(NH_3)_5{}^{4+} + H^+ \quad (33)$$

$$(NH_3)_5Cr(OH)Cr(NH_3)_4X^{(5-n)+} \rightleftharpoons (NH_3)_5CrOCr(NH_3)_4X^{(4-n)+} + H^+ \quad (34)$$

In Eq. (34) the configuration may be either cis or trans and X can be, for example, OH^-, Cl^-, F^-, SCN^-, or $-NH_2CH_2CH_2NH_3{}^+$ (cf. Table XV). The blue species are unstable, and one cis ammonia ligand in the basic rhodo ion and the X groups in the basic erythro ions are hydrolyzed rapidly in the basic solutions. The K_a values have therefore been determined by a rapid-flow technique in combination with potentiometric or spectrophotometric measurements (24).

The basic rhodo ion has been isolated as a salt and its crystal structure has been reported (Table V), but the instability and lability of the other oxo-bridged species have prevented their isolation as salts.

The rhodo ion is the only +5 charged species in this series and it has a significantly higher acidity than expected. From the pK_a values of H_3O^+ ($pK_a \sim -1.7$) and of $Cr(NH_3)_5(H_2O)^{3+}$ ($pK_a \sim 5.2$) it is anticipated that $(NH_3)_5Cr(OH)Cr(NH_3)_5{}^{5+}$ should have $pK_a \sim 12$. This is very much higher than the observed value of ~ 8. The increased acidity is a consequence of π bond formation between chromium(III) and the oxide ligand in the linear oxo-bridged form, i.e., stabilization of the base form by donation of electrons from oxygen to chromium(III).

The cation cis-$(NH_3)_5Cr(OH)Cr(NH_3)_4(NH_2CH_2CH_2NH_3)^{6+}$ is formally a $+6$ charged species, but since the charge of the ammonium group is remote from the hydroxo bridge, it is reasonable that the acidity of the latter is comparable to that for the $+5$ charged rhodo ion.

The lower acidities of the $+4$ charged erythro ions relative to that of the $+5$ charge rhodo ion are to some extent explicable on the basis of charge effects. However, if one considers the variation of pK_a with X^- for the erythro cations of the same charge, $(NH_3)_5Cr(OH)Cr(NH_3)_4X^{4+}$, it is clear that the very large variation in pK_a cannot be explained in terms of simple inductive effects. The substitution of Cl^- by OH^- in $trans$-$(NH_3)_5Cr(OH)Cr(NH_3)_4Cl^{4+}$ reduces the acid strength by more than 2 pK units, while a similar substitution in $Cr(H_2O)_5Cl^{2+}$ causes a change in pK_a of only 0.4 units (24). Differences in π bond stabilization resulting from differences in the properties of the nonbridging ligands could be responsible for the large variations observed, but no quantitative considerations of this issue have been reported. In any case, it is not easy to find an explanation for the very large difference (>7 pK units) between the pK_a values of the two isomers of the $(NH_3)_5Cr(OH)Cr(NH_3)_4(OH)^{4+}$ ion.

A number of $+3$ charged singly bridged species of the type $(HO)L_4M(OH)ML_4(OH)^{3+}$ have been reported for chromium(III), rhodium(III), and iridium(III), all of them having $pK_a > 14$ (33, 35, 67, 133).

2. *Dihydroxo-Bridged Species*

Dihydroxo-bridged complexes may deprotonate to form μ-hydroxo–μ-oxo and di-μ-oxo species, as shown in Eqs. (35) and (36).

$$L_4M(OH)_2ML_4{}^{n+} \xrightleftharpoons{K_{a1}} L_4M(OH)(O)ML_4{}^{(n-1)+} + H^+ \quad (35)$$

$$L_4M(OH)(O)ML_4{}^{(n-1)+} \xrightleftharpoons{K_{a2}} L_4M(O)_2ML_4{}^{(n-2)+} + H^+ \quad (36)$$

Nearly all the reported studies have been on chromium(III) complexes (cf. Table XVI). The chromium(III) complexes $L_4Cr(OH)_2CrL_4{}^{4+}$ with ammonia or aliphatic diamines (en and tn) deprotonate in strongly basic solutions to form blue μ-hydroxo–μ-oxo species, which in some cases have been isolated as stable and crystalline salts, e.g., Δ,Λ-

[(en)$_2$Cr(OH)(O)Cr(en)$_2$](ClO$_4$)$_3$·2H$_2$O (*33*). Deprotonation to yield a di-μ-oxo bridged species has not been observed. Studies of these basic forms are made difficult by the fact that they rapidly undergo bridge cleavage to form the singly bridged species, e.g., Δ,Λ-(OH)(en)$_2$Cr(OH)Cr(en)$_2$(OH)$^{3+}$. In the case of the ammonia complex, hydrolysis of the ammonia ligands further complicates the issue.

A number of dinuclear species with bi- or tetradentate ligands containing heteroaromatic nitrogen ligators show acid properties similar to those of the ammine and diamine analogs, and they appear to be much more robust (or more stable) in basic media with respect to bridge cleavage than the latter. The chromium(III) complexes with 1,10-phen and 2,2'-bipy deprotonate in basic solution, forming yellow μ-hydroxo–μ-oxo species and brown di-μ-oxo species (*37, 336*), and these basic forms have been isolated as crystalline salts (*37, 161*) and their magnetic properties studied (*335*). Similar acid properties have also been reported for the chromium(III) complexes with pyridyl-substituted amines (mepic, bispicen, or bispicpn) (*111, 152*).

From the pK_a values listed in Table XVI it can be seen that pK_{a1} decreases as the number of heteroaromatic N donors increases: values are p$K_{a1} \sim 12$ when all N donors are aliphatic, p$K_{a1} \sim 10$ when there is an equal number of aliphatic and heteroaromatic N donors, and p$K_{a1} \sim 7.5$ when all N donors are heteroaromatic. A similar trend has been observed for the mononuclear species *cis*-CrL$_4$(H$_2$O)$_2^{3+}$, which for L$_4$ = (en)$_2$ and (phen)$_2$ show pK_{a1} = 4.75 (*337*) and 3.4 (*181*), respectively.

The difference between the first and the second pK_a values for the dihydroxo-bridged complexes is more than 4 pK units, which is significantly larger than the differences observed in the corresponding mononuclear species, which is about 2 pK units. The large difference for the dinuclear species is contrary to the effect expected on the basis of electrostatic consideration, since the charge is distributed among two metal centers in the binuclear species. The larger separation of the pK_a values may reflect that a large bridging angle and thereby good π bonding can be achieved more easily for an oxo bridge in a mono-deprotonated complex than in a doubly deprotonated complex.

It is interesting that the acid strength of the (nta)Cr(OH)$_2$Cr(nta)$^{2-}$ ion is comparable to that of the tetrapositive 1, 10-phen complex, for example, a fact which is in keeping with the above suggestion that factors other than the overall charge strongly influence the acid strength of these complexes.

The few and rather preliminary pK_a values for dihydroxo-bridged rhodium(III) and iridium(III) complexes which have been measured indicate that these are weaker acids than the chromium(III) species

(67, 133, 217). This is consistent with trends observed for the mononuclear species and with the reduced tendency toward π bonding of the metal ions with filled d_π orbitals.

A number of hetero-bridged complexes of the type $(en)_2Cr(OH)(X)Cr(en)_2^{(5-n)+}$ for $X^{n-} = SO_4^{2-}$ or $RCOO^-$ have been studied and they all exhibit $pK_{a1} \sim 12$. The observation that the $+3$ charged sulfato complex is more acidic than the $+3$ charged μ-hydroxo–μ-oxo analog ($pK_a > 14$) reflects the greater flexibility of the former ring system, implying that the Cr–O–Cr angle in the deprotonated sulfato complex can attain a larger value than is possible in the di-μ-oxo complex (33, 262, 263).

3. Trihydroxo-Bridged Species

Only a few trihydroxo-bridged complexes have been studied: $(tacd)Cr(OH)_3Cr(tacd)^{3+}$ has $pK_a = 12.9$ (25°C, $I = 0.4\ M$) (98) and $(tach)Co(OH)_3Co(tach)^{3+}$ has $pK_a \sim 14$ (186).

B. Bridging Water

The hydroxo bridge possesses very weak basic properties, but there has been no report of any direct evidence for the formation of aqua-bridged complexes. From an extrapolation of the pK_a values of water ($pK_a \sim 15.7$) and of $Cr(NH_3)_5(H_2O)^{3+}$ ($pK_a \sim 5.2$) it is anticipated that an aqua-bridged complex such as $(NH_3)_5Cr(OH_2)Cr(NH_3)_5^{6+}$ should have a pK_a in the region of -5. Spectroscopic studies support this proposal, but from the available data only an upper limit of about -1 can be estimated (33, 35, 36, 67, 133, 217). From the kinetic data presented in the following sections it is concluded that the aqua-bridged species are unstable and labile with respect to bridge cleavage, which explains why such species have not been isolated as salts. In this context, however, it is of interest to mention that X-ray crystal structures of aqua-bridged species with copper(II), nickel(II), cadmium(II), cobalt(II), and ruthenium(II) have been reported (338–345).

C. Terminally Coordinated Water

The acid strength of the terminally coordinated water ligands in many polynuclear complexes is strongly influenced by intramolecular hydrogen bond interaction. The main issue of the last part of this section will be a discussion of such interactions, although other

TABLE XVII

Acid Strength of Water Ligands in Dinuclear Complexes with Negligible Intramolecular Hydrogen Bond Interaction at 25°C and $I = 1.0\ M$

Complex	pK_{a1}	pK_{a2}	ΔpK_a	Reference
cis-$(NH_3)_5Cr(OH)Cr(NH_3)_4(H_2O)^{5+}$	3.5^a	—	—	347
trans-$(NH_3)_5Cr(OH)Cr(NH_3)_4(H_2O)^{5+}$	≈4.5	—	—	129
trans-$(H_2O)(NH_3)_3Cr(OH)_2Cr(NH_3)_3(H_2O)^{4+}$	6.15	7.48	1.33	319
trans-$(H_2O)(tacn)Cr(OH)_2Cr(tacn)(H_2O)^{4+}$	5.08	7.25	2.17	319

$^a\ pK_a = 2.8$ at 20°C in 0.14 M NaClO$_4$ (24).

factors such as the charge on the complex ion and the nature of the other ligands need also to be considered. The influence of the latter is best illustrated by first considering some of the few polynuclear complexes in which intramolecular hydrogen bond interactions must be expected to play a minor role.

1. Comparison with Mononuclear Species

The acid strength of terminally coordinated water in polynuclear species is influenced by the charge, and the effect may be compared with that in mononuclear species (Tables XVII and XVIII). Since the charge

TABLE XVIII

Acid Dissociation Constants for Mononuclear Aqua Complexes at 25°C in 1.0 M (Na,H)ClO$_4$

Complex	pK_{a1}	pK_{a2}	ΔpK(monomer) $= pK_{a2} - pK_{a1}$	Reference
trans-$Cr(NH_3)_4(H_2O)_2^{3+}$	4.38	7.78	3.40	337
cis-$Cr(NH_3)_4(H_2O)_2^{3+}$	4.96	7.53	2.57	337
cis-$Cr(en)_2(H_2O)_2^{3+}$	4.75	7.35	2.60	337
fac-$Cr(NH_3)_3(H_2O)_3^{3+}$	5.00	7.27	2.27	347
$Cr(tacn)(H_2O)_3^{3+}$	4.47	6.64	2.17	347
cis-$Co(NH_3)_4(H_2O)_2^{3+}$	5.69^a	7.99^a	2.30	174
cis-$Co(en)_2(H_2O)_2^{3+}$	6.06^b	8.19^b	2.13	332
cis-$Rh(NH_3)_4(H_2O)_2^{3+}$	6.40	8.32	1.92	346
cis-$Rh(en)_2(H_2O)_2^{3+}$	6.34	8.24	1.90	217
cis-$Ir(en)_2(H_2O)_2^{3+}$	6.29	8.10	1.81	348
$Ir(H_2O)_6^{3+}$	4.37	5.2	0.8	349

^{a} 0.1 M NaClO$_4$, 20°C.
^{b} 1 M NaNO$_3$.

on the polynuclear species is distributed among several metal centers, it would seem reasonable when making such comparisons to consider the average charge per metal center rather than the total charge. The $trans$-$(H_2O)(NH_3)_3Cr(OH)_2Cr(NH_3)_3(H_2O)^{4+}$ cation is a significantly stronger acid than cis-$Cr(NH_3)_4(OH)(H_2O)^{2+}$: $pK_{a1} = 6.2$ and 7.5, respectively [here, and in the following discussion, the minor statistical corrections which ought to be made due to the presence of different numbers of water (hydroxide) ligands have been ignored]. Similarly, cis-$(NH_3)_5Cr(OH)Cr(NH_3)_4(H_2O)^{5+}$ has $pK_a = 3.5$, which is significantly lower than the average value $\frac{1}{2}(pK_{a1} + pK_{a2}) = 6.25$ for the cis-$Cr(NH_3)_4(H_2O)_2^{3+}$ ion. The same is found if $trans$-$(NH_3)_5Cr(OH)Cr(NH_3)_4(H_2O)^{5+}$ is compared with the trans diaqua mononuclear species. These few examples indicate that the acid strength of a water ligand in a polynuclear species is greater than that in the corresponding mononuclear species, of charge equal to the average charge per metal center in the polynuclear species.

The difference between the two pK_a values for $trans$-$(H_2O)(NH_3)_3$-$Cr(OH)_2Cr(NH_3)_3(H_2O)^{4+}$ is 1.3, which is significantly less than the difference of 2.6 observed for cis-$Cr(NH_3)_4(H_2O)_2^{3+}$. This is in accordance with the fact that the water ligands are separated more in the dinuclear cation than in the mononuclear cation.

The influence of the other ligands, as long as they do not participate in intramolecular hydrogen bonds, is expected to be the same as that observed in the corresponding mononuclear species. The number of complexes for which this can be examined is small, but the pK_a values for the trans isomers of $(H_2O)L_3Cr(OH)_3CrL_3(H_2O)^{4+}$ with $L_3 = (NH_3)_3$ or tacn do indeed show that the complex with the more bulky amine is the stronger acid, as observed for the mononuclear complexes (Tables XVII and XVIII).

The difference in acid strength of the cis and trans isomers of $(NH_3)_5Cr(OH)Cr(NH_3)_4(H_2O)^{5+}$ may be explained as an effect of the ligand trans to water. A comparison with the effect of trans ligand on the water ligand acidity observed for mononuclear ammine and amine complexes of chromium(III) or rhodium(III) (346) leads to the sequence $\mu(OH) < OH \sim NH_3 < H_2O$ for increasing acidity due to the trans ligand.

2. Hydrogen Bond Interactions

a. Monohydroxo-Bridged Species. Chromium(III), rhodium(III), and iridium(III) complexes of the type cis,cis-$(H_2O)L_4M(OH)ML_4$-$(H_2O)^{5+}$ are known for $L_4 = (NH_3)_4$, $(en)_2$, $(NH_3)_3(H_2O)$, or $(H_2O)_4$.

TABLE XIX

ACID STRENGTH OF WATER LIGANDS IN MONOHYDROXO-BRIDGED COMPLEXES AFFECTED BY
INTRAMOLECULAR HYDROGEN BONDS AT 25°C IN 1.0 M (Na,H)ClO$_4$

Complex ion	pK_{a1}	pK_{a2}	ΔpK(dinuclear) = pK_{a2} − pK_{a1}	Reference
$(H_2O)_5Cr(OH)Cr(H_2O)_5^{5+}$	1.6	—	—	28[a]
cis,cis-$(H_2O)(NH_3)_4Cr(OH)Cr(NH_3)_4(H_2O)^{5+}$	1.75	7.50	5.75	36
Δ,Λ-$(H_2O)(en)_2Cr(OH)Cr(en)_2(H_2O)^{5+}$	0.48	7.94[b]	7.27[b]	33, 36
$\Delta,\Delta/\Lambda,\Lambda$-$(H_2O)(en)_2Cr(OH)Cr(en)_2(H_2O)^{5+}$	0.54	6.87[c]	6.30[c]	35
Δ,Λ-$(H_2O)(en)_2Cr(OH)Co(en)_2(H_2O)^{5+}$	1.31	~8	~6.7	36, 214, 215
fac-$(H_2O)_2(NH_3)_3Cr(OH)Cr(NH_3)_3(H_2O)_2^{5+}$	1.5	5.52	4.0	319
cis,cis-$(H_2O)(NH_3)_4Rh(OH)Rh(NH_3)_4(H_2O)^{5+}$	3.41	8.80	5.39	133
Δ,Λ-$(H_2O)(en)_2Rh(OH)Rh(en)_2(H_2O)^{5+}$	2.37	9.13	6.76	217
Δ,Λ-$(H_2O)(en)_2Ir(OH)Ir(en)_2(H_2O)^{5+}$	1.91	9.04	7.13	67
$(H_2O)_5Ir(OH)Ir(H_2O)_5^{5+}$	0.8[d]	—	—	251

[a] Determined from kinetic data, 2 M (Li,H)ClO$_4$. The p$K_a \approx$ 1.3–1.6 has recently been reported in an independent study (350).
[b] At 0.8°C; pK_{a1} = 0.67 at this temperature.
[c] At 20°C; pK_{a1} = 0.57 at this temperature.
[d] 2 M (Li,H)ClO$_4$.

A common and very important feature for all these species is that the first and second acid dissociation constants are significantly larger and smaller, respectively, than predicted from the behavior of the corresponding mononuclear species, following the lines of reasoning given above (Tables XVIII and XIX). This may be illustrated by comparing the difference ΔpK(dinuclear) = pK_{a2} − pK_{a1} = 7.3 for Δ,Λ-$(H_2O)(en)_2Cr(OH)Cr(en)_2(H_2O)^{5+}$ with the difference ΔpK(mononuclear) = pK_{a2} − pK_{a1} = 2.6 for the corresponding mononuclear species cis-$Cr(en)_2(H_2O)_2^{3+}$. The observation that ΔpK(dinuclear) $\gg$ ΔpK(mononuclear) is the opposite of what would be expected from the fact that the protolytic groups are further apart in the dinuclear ion than in the mononuclear ion. The large value for ΔpK(dinuclear) reflects an unexpectedly large stabilization of the aquahydroxo cation relative to its corresponding acid or base forms. Similar strong stabilization of the aqua hydroxo species has been observed for a series of analogous chromium(III), rhodium(III), and iridium(III) complexes (cf. Table XIX). This increased stability of the aqua hydroxo cation has been explained as arising from intramolecular hydrogen bond formation between the hydroxide ligand bound at one metal center and the water ligand bound at the other, as

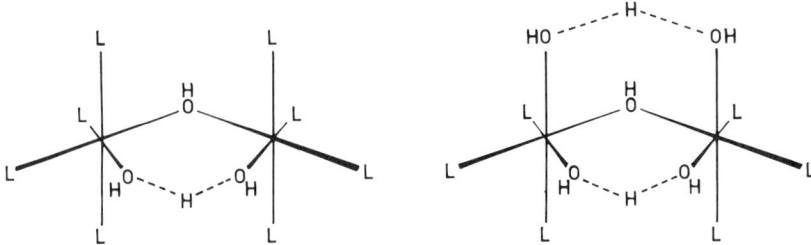

FIG. 15. Intramolecular hydrogen bond stabilization of monohydroxo-bridged complexes.

shown in Fig. 15 (*33, 35, 36, 67, 133, 217*). Crystallographic evidence for this kind of hydrogen bond formation (α type) has been discussed in Section II,B.

For the complex $fac,fac\text{-}(H_2O)_2(NH_3)_3Cr(OH)Cr(NH_3)_3(H_2O)_2{}^{5+}$ the first acid dissociation constant clearly shows that the singly deprotonated species must be hydrogen bond stabilized (Table XIX). However, the fact that the difference between the first and the second acid dissociation constants for this system is relatively small is consistent with stabilization also of the doubly deprotonated species (by two intramolecular hydrogen bonds) as shown in Fig. 15. A similar effect is expected for the cations $(H_2O)_5Cr(OH)Cr(H_2O)_5{}^{5+}$ and $(H_2O)_5Ir(OH)Ir(H_2O)_5{}^{5+}$.

b. Dihydroxo-Bridged Species. Intramolecular hydrogen-bond stabilization also explains the significant difference between the cis and trans isomers of the ions $(H_2O)L_3Cr(OH)_2CrL_3(OH_2)^{4+}$ (see Fig. 16).

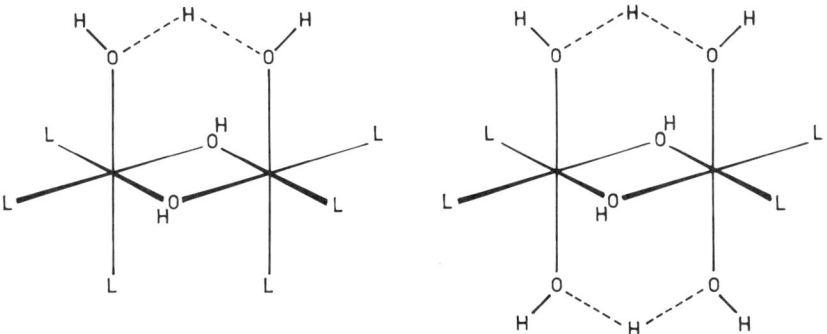

FIG. 16. Intramolecular hydrogen bond stabilization of dihydroxo-bridged complexes.

TABLE XX

ACID STRENGTH OF WATER LIGANDS IN CATIONS OF THE TYPE
$(H_2O)L_3Cr(OH)_2CrL_3(H_2O)^{4+}$ AFFECTED BY INTRAMOLECULAR
HYDROGEN BONDS AT 25°C IN 1.0 M (Na,H)ClO$_4$

L$_3$	pK_{a1} (cis isomer)	pK_{a2} (cis isomer)	ΔpK^a (cis isomer)	ΔpK^b (trans isomer)	Reference
(NH$_3$)$_3$	4.19	9.05	4.86	1.33	319
(tacn)	2.82	8.8	5.98	2.17	319
	pK_{a1}	pK_{a2}	ΔpK	—	—
(H$_2$O)$_3$	3.68	6.04	2.36	—	31

^{a} ΔpK = pK_{a2} - pK_{a1}.
^{b} See also Table XVII.

From the data in Tables XVII and XX it is seen that for both the ligands NH$_3$ and tacn ΔpK(cis) $\gg$ ΔpK(trans) and that this is due to an increase in K_{a1} as well as a decrease in K_{a2} for the cis isomer relative to the trans isomer.

A similar effect may explain the rather high strength of $(H_2O)_4Cr(OH)_2Cr(H_2O)_4^{4+}$, but in this system the doubly deprotonated cation may also be stabilized by two identical intramolecular hydrogen bonds (Fig. 16), which explains the significantly smaller ΔpK value (cf. Table XX) for this system than for cis-$(H_2O)L_3Cr(OH)_2CrL_3(H_2O)^{4+}$.

c. Quantitative Considerations. For the monohydroxo-bridged complexes of the type cis,cis-$(H_2O)L_4M(OH)ML_4(H_2O)^{5+}$ and the dihydroxo-bridged complexes of the type cis-$(H_2O)L_3M(OH)_2ML_3(H_2O)^{4+}$ (L$_3$ and L$_4$ refer to ammine or amine ligands) it is possible to obtain an estimate for the magnitude of the hydrogen bond stabilization, i.e., the equilibrium constant K_H for Eq. (37).

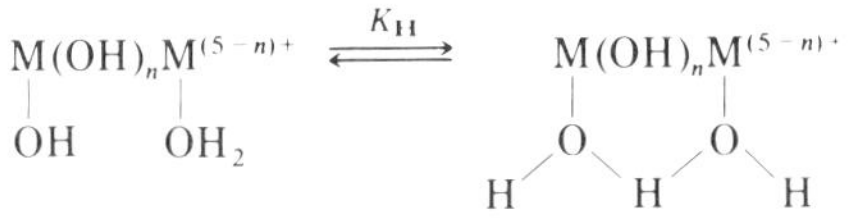

$$M(OH)_nM^{(5-n)+} \underset{}{\overset{K_H}{\rightleftharpoons}} M(OH)_nM^{(5-n)+} \quad (37)$$
$$\begin{array}{cc} | & | \\ OH & OH_2 \end{array} \qquad \begin{array}{cc} O & O \\ \diagdown & \diagdown \\ H \quad H \quad H \end{array}$$

The observed acid dissociation constants K_{a1} and K_{a2} may then be expressed by Eqs. (38) and (39), where K'_{a1} and K'_{a2} are the microscopic

$$K_{a1} = K'_{a1}(1 + K_H) \quad (38)$$

$$K_{a2} = K'_{a2}/(1 + K_H) \quad (39)$$

equilibrium constants for the equilibria, Eqs. (40) and (41), between the non-hydrogen-bonded species.

$$\underset{\underset{OH_2}{|}\underset{OH_2}{|}}{M(OH)_nM^{(6-n)+}} \overset{K'_{a1}}{\rightleftharpoons} \underset{\underset{OH}{|}\underset{OH_2}{|}}{M(OH)_nM^{(5-n)+}} + H^+ \qquad (40)$$

$$\underset{\underset{OH}{|}\underset{OH_2}{|}}{M(OH)_nM^{(5-n)+}} \overset{K'_{a2}}{\rightleftharpoons} \underset{\underset{OH}{|}\underset{OH}{|}}{M(OH)_nM^{(4-n)+}} + H^+ \qquad (41)$$

Since $K_H \gg 1$, Eqs. (38) and (39) may be combined to give Eq. (42) and

$$\log K_H = \tfrac{1}{2}[\Delta pK_a(\text{dinuclear}) - \Delta pK'_a(\text{dinuclear})] \qquad (42)$$

K_H can be calculated if $\Delta pK'_a(\text{dinuclear})$ is known. The microscopic equilibrium constants K'_a cannot be determined directly, but a good estimate for $\Delta pK'_a(\text{dinuclear}) = pK'_{a2} - pK'_{a1}$ can be obtained. Following the discussion above, the difference between the pK_a values for trans-$(H_2O)L_3Cr(OH)_2CrL_3(H_2O)^{4+}$ ions can be taken as a reliable estimate for $\Delta pK'_a(\text{dinuclear})$. This gives $\Delta pK'_a = 2.2$ for the tacn chromium(III) complex and $\Delta pK'_a = 1.3$ for the ammine and ethylenediamine complexes of chromium(III) (Table XVII). The corresponding values for the ammine and ethylenediamine complexes of rhodium(III) and iridium(III) are probably slightly smaller, as suggested by the data for the mononuclear species; $\Delta pK_a(\text{mononuclear})$ for chromium(III) is about 0.7 pK units greater than for rhodium(III) and iridium(III) (TableXVIII). If this difference (0.7) is assumed to apply to the dinuclear complexes, the value $\Delta pK'_a(\text{dinuclear}) = 0.6$ is then obtained for rhodium(III) and iridium(III). These estimated $\Delta pK'_a(\text{dinuclear})$ values yield the K_H values quoted in Table XXI.

TABLE XXI

Hydrogen Bond Stabilization of $(H_2O)L_{5-n}M(OH)_nML_{5-n}(OH)^{(5-n)+}$ Species at 25°C in 1.0 M (Na,H)ClO$_4$ [a]

Complex ion	$K_H[Cr(III)]$	$K_H[Rh(III)]$	$K_H[Ir(III)]$
cis,cis-$(H_2O)(NH_3)_4M(OH)M(NH_3)_4(OH)^{4+}$	$10^{2.2}$	$10^{2.4}$	—
Δ,Λ-$(H_2O)(en)_2M(OH)M(en)_2(OH)^{4+}$	$10^{3.0}$ [b]	$10^{3.1}$	$10^{3.3}$
cis-$(H_2O)(NH_3)_3M(OH)_2M(NH_3)_3(OH)^{3+}$	$10^{1.8}$	—	—
cis-$(H_2O)(tacn)M(OH)_2M(tacn)(OH)^{3+}$	$10^{1.9}$	—	—

[a] K_H is defined in Eq. (37); see the text.
[b] Estimated on the basis of ΔpK_a at 0°C and $\Delta pK'_a$ at 25°C.

TABLE XXII

Thermodynamic Parameters for the Equilibrium[a]

$$(H_2O)L_4M(OH)ML_4(H_2O)^{5+} \underset{}{\overset{K_{a1}}{\rightleftharpoons}} (H_2O)L_4M(OH)ML_4(OH)^{4+} + H^+$$

	$L_4 = (NH_3)_4$		$L_4 = (en)_2$ [b]		
M	ΔH^0 (kJ mol^{-1})	ΔS^0 (J mol^{-1} K^{-1})	ΔH^0 (kJ mol^{-1})	ΔS^0 (J mol^{-1} K^{-1})	Reference
Cr(III)	49(6)	129(20)	12(3)	30(9)	*33, 36*
Rh(III)	53(2)	111(5)	28(4)	49(13)	*133, 217*
Ir(III)	—	—	11(7)	−1(22)	*67*

[a] At 1 M (Na,H)ClO$_4$.
[b] Δ,Λ isomers.

The K_H values in Table XXI show some obvious and very reasonable trends. For the monohydroxo-bridged complexes it is seen that K_H is very sensitive to changes in the first coordination sphere, but is practically insensitive to the nature of the metal centers, i.e., log K_H(NH$_3$) ∼ 2.3 and log K_H(en) ∼ 3.1, independent of the metal ion.

The observation that K_H(en) > K_H(NH$_3$) is qualitatively supported by the enthalpy and entropy changes associated with the first acid dissociation equilibrium (Table XXII). Increased hydrogen bond stabilization should contribute negatively to both $\Delta H^0(K_{a1})$ and $\Delta S^0(K_{a1})$, which is consistent with the observation for both chromium(III) and rhodium(III) that it is the ammine complexes which have the highest $\Delta H^0(K_{a1})$ and $\Delta S^0(K_{a1})$ values. The greater acid strength of the ethylenediamine systems is then due to a decrease of ΔH^0, which is greater than the decrease of ΔS^0. The $\Delta H^0(K_{a1})$ and $\Delta S^0(K_{a1})$ parameters for the iridium(III) complex are, within the uncertainty limits, similar to those for the ethylenediamine complexes of chromium(III) and rhodium(III), which is consistent with the finding that K_H is almost identical for all three complexes.

The dihydroxo-bridged chromium(III) complexes are, not unexpectedly, significantly less hydrogen bond stabilized than the more flexible monohydroxo-bridged species. The variation of K_H with the nature of the nonbridging ligands which is observed for the monohydroxo-bridged species is not apparent for the dihydroxo-bridged chromium(III) species, and this may be due to the inherently rigid conformation in the latter.

From Eqs. (38) and (39) together with the K_H values in Table XXI the microscopic equilibrium constants K'_{a1} and K'_{a2} can be calculated.

For the Δ,Λ-$(H_2O)(en)_2M(OH)M(en)_2(H_2O)^{5+}$ ions this gives the values $pK'_{a1} = 3.5$, 5.5, and 5.2 for chromium(III), rhodium(III), and iridium(III), respectively. It is noted that the K'_{a1} varies in the order Cr(III) $\gg$ Ir(III) $\sim$ Rh(III), as observed for K_{a1} for the mononuclear diaqua complexes. It is also quite reasonable that the value $pK'_{a1} = 3.5$ for the chromium(III) complex is equal to $pK_a = 3.5$ for the equally charged cis aqua erythro ion.

3. Higher Polynuclear Species

The few data available for tri- and tetranuclear species can be rationalized along the lines given above with the additional prospect of β-type hydrogen bonds being involved, i.e., interaction between the hydroxo bridges and the nonbridging ligands (see also Section II,B). The latter type of interaction has been proposed to be responsible for the enhanced acid strength ($pK_a \leq 1$) of the trinuclear chromium(III) complex $Cr_3(tacn)_3(OH)_4(H_2O)^{5+}$ (50).

The acid dissociation constants for tri- and tetranuclear aqua chromium(III) species are summarized in Table XXIII. The structures of these species are not known. If, however, it is assumed that they have linear structures, such as structures **4a**, **7b**, and **7c** shown in Fig. 1, then the observed acid strengths can be rationalized in terms of α- and β-type hydrogen bond interactions, as discussed recently (118).

TABLE XXIII

ACID DISSOCIATION CONSTANTS FOR AQUA CHROMIUM(III) OLIGOMERS AT 25°C IN 1.0 M (Na,H)ClO$_4$

Complex	pK_{a1}	pK_{a2}	pK_{a3}	Reference
Cr_2OH^{5+}	1.6[a]	—	—	28
$Cr_2(OH)_2^{4+}$	3.68	6.04	—	31
$Cr_3(OH)_4^{5+}$	4.35	5.63	6.0	31
$Cr_4(OH)_6^{6+}$	2.55[b]	5.08[b]	—	31[b]
α-$Cr_4(OH)_6^{6+}$	3.53	—	—	200
β-$Cr_4(OH)_6^{6+}$	0.89	—	—	200

[a] Determined from kinetic data; in 2 M (Li,H)ClO$_4$ a $pK_a \approx$ 1.3–1.6 has recently been reported in an independent study (350).

[b] Apparent dissociation constants for an equilibrium mixture of the α and β isomer; see also Section IV,A.

VII. Kinetics for the Condensation Reaction of Mononuclear Species to Give Dinuclear Species

The condensation of two mononuclear aqua species to form a monohydroxo-bridged dinuclear species is of fundamental interest, since it is the first step in polymerization reactions [Eq. (43)]. However,

$$ML_5(H_2O)^{n+} + ML_5(OH)^{(n-1)+} \underset{k_r}{\overset{k_f}{\rightleftharpoons}} L_5M(OH)ML_5^{(2n-1)+} + H_2O \quad (43)$$

kinetic studies of the forward reaction in Eq. (43) are generally complicated by subsequent (often faster) bridge-formation reactions to give di- or trihydroxy-bridged species, or by further condensation to higher polynuclear species, and this probably explains the scarcity of kinetic data for reactions of this type.

From simple electrostatic considerations it is expected that the rate of condensation will depend strongly on the charge of the mononuclear species. This is qualitatively confirmed by the observation that condensation of the neutral species $Co(NH_3)_2(NO_2)_2(OH_2)(OH)$ is much faster than condensation of +2 charged species such as $Co(tach)(H_2O)_2(OH)^{2+}$ or $Co(tren)(H_2O)(OH)^{2+}$ (*174*).

The kinetics of the condensation of the $Cr(H_2O)_6^{3+}$ ion and its corresponding deprotonated species have been studied in the pH region 3.5–5.0 [25°C, $I = 1.0\ M$ (NaClO$_4$)] (*201*). The study of this reaction is complicated by the formation of higher oligomers. Chromatographic analysis of the products as a function of time established the dinuclear species to be the main product for the first 5% of reaction, and the initial-rate kinetics of condensation were studied by a pH-stat technique. The observed pH dependence of the rate was interpreted in terms of the second-order rate constants defined by Eq. (44), and values for

$$Cr(OH)_p^{(3-p)+} + Cr(OH)_q^{(3-q)+} \xrightarrow{k_{p,q}} Cr(OH)Cr(OH)_s^{(5-s)+} \quad (44)$$

k_{11}, k_{12}, and k_{22} were determined (Table XXIV). In the pH range studied the k_{00} and k_{10} processes do not contribute significantly to the rate, but k_{10} can be estimated from equilibrium data together with the rate constant for the cleavage of $(H_2O)_5Cr(OH)Cr(H_2O)_5^{5+}$. The condensation constants seem to be unexpectedly large. Thus the smallest rate constant, k_{10}, is 140 times larger than the bimolecular rate constant, k_{H_2O}, for water exchange of Cr^{3+} ($k_{H_2O} = k_{ex}/[H_2O] = 4.3 \times 10^{-8}\ M^{-1}\ s^{-1}$) (*351*). A rather regular increase by a factor of ~ 30–200 is found for each deprotonation step in the series k_{10}, k_{11}, k_{12},

TABLE XXIV

KINETIC DATA FOR SOME DIMERIZATION REACTIONS AT 25°C

Reactants	$k_{p,q}$	k (M^{-1} sec^{-1})		
		M = Cr(H$_2$O)$_4$[a]	M = Rh(en)$_2$[b,c]	M = Co(trans[14]diene)[d]
M(OH$_2$)$_2$$^{3+}$ + M(OH$_2$)$_2$$^{3+}$	k_{00}	—	≈5 × 10^{-14}	—
M(OH$_2$)(OH)$^{2+}$ + M(OH$_2$)$_2$$^{3+}$	k_{10}	6 × 10^{-6}	≈1 × 10^{-8}	—
M(OH$_2$)(OH)$^{2+}$ + M(OH$_2$)(OH)$^{2+}$	k_{11}	2.0(4) × 10^{-4}	≈6 × 10^{-5}	2.4 × 10^{-2}
M(OH)$_2$$^+$ + M(OH$_2$)(OH)$^{2+}$	k_{12}	3.8(10) × 10^{-2}	—	—
M(OH)$_2$$^+$ + M(OH)$_2$$^+$	k_{22}	1.8(2)	—	—

[a] From Ref. 201.
[b] From Ref. 325.
[c] k_{00}, k_{10}, and k_{11} correspond to k_{-5}, k_{-4}, and k_{-3}, respectively, in Scheme 5, and refer to the formation of Δ,Λ isomers of the dinuclear species.
[d] From Ref. 352.

and k_{22}. These increments correspond well to the increase in reactivity for water exchange found on going from Cr^{3+} to $Cr(OH)^{2+}$, $k_{ex}(Cr(OH)^{2+})/k_{ex}(Cr^{3+}) = 75$ (351).

Kinetic data for the formation of dinuclear species from cis-$Rh(en)_2$-$(H_2O)_2^{3+}$ and the corresponding deprotonated species are also given in Table XXIV. These values have been calculated from the kinetic data for the cleavage reaction, k_r, in Eq. (43) (see Section VIII), together with the relevant equilibrium constants. The bimolecular rate constants for these rhodium(III) species show trends similar to those for the aqua chromium(III) ions. The significantly reduced activity of two cis-$Rh(en)_2(H_2O)_2^{3+}$ ions with respect to condensation probably reflects the fact that the nucleophile in this case is a coordinated water ligand, whereas in the other reactions it can be a coordinated hydroxide ligand. Alternatively, the enhanced reactivity of the latter species may be the result of the labilizing effect of coordinated hydroxide.

The kinetics for the dimerization of trans-$Co(trans$-[14]diene)(H_2O)-$(OH)^{2+}$ have been reported (352). The product was not isolated, nor identified, but a structure such as trans,trans-$(H_2O)L_4Co(OH)$-$CoL_4(OH)^{4+}$ would seem likely. The rate constant for the dimerization reaction is given in Table XXIV.

Grant and Hamm have studied the kinetics of the reaction Eq. (45) (353). In dilute solution the reaction is second order, as expected from Eq. (45), but with increasing concentration the order decreases until it

$$2cis\text{-}Cr(C_2O_4)_2(H_2O)(OH)^{2-} \longrightarrow (C_2O_4)_2Cr(OH)_2Cr(C_2O_4)_2^{4-} + 2H_2O \quad (45)$$

becomes first order for chromium(III) concentrations above ~ 0.01 M. As pointed out by Ardon and Bino, this behavior can be explained in terms of cation-pair formation [Eq. (29)] if K_{ip} is large (326).

VIII. Cleavage of Polynuclear into Mononuclear Species

A. Cleavage in Strong Acids

Hydrolysis of polynuclear hydroxo-bridged chromium(III) complexes in concentrated solutions of strong acid yields the corresponding mononuclear species. Such cleavage reactions are fast in comparison with the hydrolysis in dilute acid and proceed with retention of configuration of the mononuclear entities. A few representative examples are shown in Eqs. (46)–(49) (40, 42, 161, 252).

$$(NH_3)_5Cr(OH)Cr(NH_3)_5{}^{5+} + H_3O^+ \longrightarrow 2Cr(NH_3)_5(H_2O)^{3+} \quad (46)$$

$$\Delta,\Lambda\text{-}(en)_2Cr(OH)_2Cr(en)_2{}^{4+} + 2H_3O^+ \longrightarrow 2cis\text{-}Cr(en)_2(H_2O)_2{}^{3+} \quad (47)$$

$$Cr_3(NH_3)_{10}(OH)_4{}^{5+} + 4H_3O^+ \longrightarrow$$

$$cis\text{-}Cr(NH_3)_4(H_2O)_2{}^{3+} + 2fac\text{-}Cr(NH_3)_3(H_2O)_3{}^{3+} \quad (48)$$

$$(-)_{589}\text{-}\Delta,\Delta\text{-}(phen)_2Cr(OH)_2Cr(phen)_2{}^{4+} + 4HBr \longrightarrow$$

$$2(-)_{589}\text{-}\Delta\text{-}cis\text{-}Cr(phen)_2Br_2{}^+ + 2H_3O^+ \quad (49)$$

Analysis of the products of these cleavage reactions has often served as proof of the structures of the polynuclear species. Cleavage of hydroxo-bridged complexes of nuclearity higher than two will in most cases yield at least two different mononuclear species. Identification of these species and determination of the relative ratio in which they are formed reduce the number of possible bridged skeletons greatly, and the studies of polynuclear ammine and amine chromium(III) made by Andersen et al. (mentioned in Section IV) provide many examples of this, one of which is shown in Eq. (48) above (see also Section II,A).

Cleavage reactions have also been utilized as a convenient method for the synthesis of "parent" mononuclear species. An example would be the acid cleavage of optically active dinuclear dihydroxo-bridged 1,10-phen and 2,2'-bipy chromium(III) species, providing a facile synthetic route to optically pure $(-)_{589}$ enantiomers of $cis\text{-}Cr(phen)_2X_2^{(3-2n)+}$ or $cis\text{-}Cr(bipy)_2X_2^{(3-2n)+}$ ($X^{n-} = H_2O$, Cl^-, and Br^-) (161).

Similar cleavage reactions have been reported for hydroxo-bridged Co(III) complexes, and the product analyses have been taken as evidence for the proposed structures. However, the greater tendency of cobalt(III) than of chromium(III) to undergo isomerization and redox reactions limits to some extent the applicability of such reactions in terms of structural analyses and synthetic methods. These difficulties can be illustrated by the results obtained for the acid hydrolysis of $(py)_3Co(OH)_3Co(py)_3{}^{3+}$. Cleavage in 12 M HCl yields the mer isomer and not, as would have been expected, the fac isomer of $Co(py)_3Cl_3$. Cleavage in 12 M HClO$_4$ probably yields $Co(py)_3(H_2O)_3{}^{3+}$, but the formation of other species [including Co(II) species] hampered a further study of this reaction (188).

The cleavage of polynuclear hydroxo-bridged rhodium(III) and iridium(III) complexes into the corresponding mononuclear fragments has been reported in only a few instances, but the well-established tendency of mononuclear complexes of these metal ions to undergo substitution reactions with retention of configuration indicates the possibility of analytical and synthetic applications such as described above for chromium(III).

B. KINETICS OF THE CLEAVAGE OF DINUCLEAR MONOHYDROXO-BRIDGED COMPLEXES

1. Acid Hydrolysis

In the majority of the kinetic studies reported $HClO_4$–$NaClO_4$ media have been employed. The acid cleavage process can be described by Eq. (50), but in the presence of additional nucleophiles other than water, i.e., either anionic ($X^{n-} = Cl^-$, SCN^-, SO_4^{2-}, etc.) or neutral (X = DMSO, CH_3CN, etc.), there is competition between Eq. (50) and the reaction shown in Eq. (51). The presence of anionic nucleophiles may either increase or decrease (relative to ClO_4^-) the rate of cleavage by the formation of ion doublet and triplets.

$$L_5M(OH)ML_5^{5+} + H^+ + H_2O \longrightarrow 2ML_5(H_2O)^{3+} \tag{50}$$

$$L_5M(OH)ML_5^{5+} + H^+ + X^{n-} \longrightarrow ML_5(H_2O)^{3+} + ML_5X^{(3-n)+} \tag{51}$$

The cleavage reaction, Eq. (50), may proceed by an uncatalyzed path (k_0) and/or by an acid-catalyzed path ($k_a = k'_a/K_a$), as shown in Scheme 3.

The acid-catalyzed path is proposed to involve a labile aqua-bridged intermediate which is a strong acid with $K_a \gg 1$, and the observed rate constant for the acid-catalyzed path is thereby a composite term: $k_a = k'_a/K_a$.

It seems to be a general feature that the acid-catalyzed path plays a more dominant role in the reactions of cobalt(III) and rhodium(III) than in those of chromium(III). A rationalization for this could be that the latter species do not readily protonate, in keeping with the observation that chromium(III) aqua complexes are significantly more acidic than cobalt(III) or rhodium(III) aqua complexes (Table XVIII).

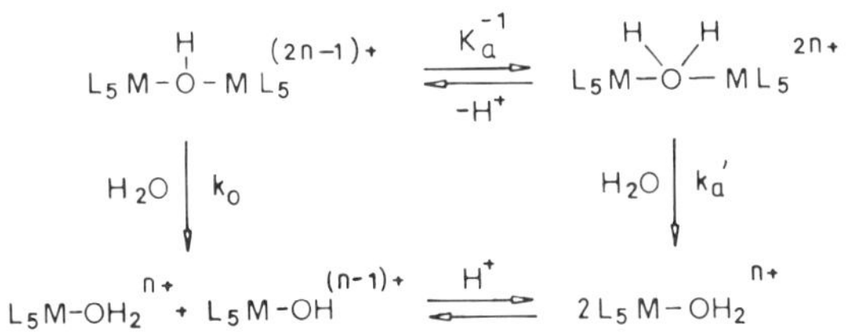

$$\begin{array}{ccc}
L_5M\overset{H}{\overset{|}{-}}O-ML_5^{(2n-1)+} & \underset{-H^+}{\overset{K_a^{-1}}{\rightleftharpoons}} & L_5M\overset{H\diagdown\diagup H}{-O-}ML_5^{2n+} \\
H_2O \downarrow k_0 & & H_2O \downarrow k'_a \\
L_5M-OH_2^{n+} + L_5M-OH^{(n-1)+} & \underset{}{\overset{H^+}{\rightleftharpoons}} & 2L_5M-OH_2^{n+}
\end{array}$$

SCHEME 3. Acid cleavage of dinuclear monohydroxo-bridged species may occur uncatalyzed (k_0) and acid-catalyzed ($k_a = k'_a/K_a$).

a. *Chromium(III) Complexes.* The cleavage of the rhodo ion, $(NH_3)_5Cr(OH)Cr(NH_3)_5^{5+}$, and of the cis or trans isomers of the erythro ions, $(NH_3)_5Cr(OH)Cr(NH_3)_4X^{(5-n)+}$, $(X^{n-} = H_2O$, Cl^-, SCN^-, or F^-) has been studied using acidic perchlorate media. The rhodo ion yields $Cr(NH_3)_5(H_2O)^{3+}$, whereas the erythro ions yield both $Cr(NH_3)_5(H_2O)^{3+}$ and the respective cis or trans isomers of $Cr(NH_3)_4(H_2O)X^{(3-n)+}$. The kinetic data are consistent with a contribution from the uncatalyzed path (k_0 in Scheme 3) alone, and yield the parameter values listed in Table XXV. Bond rupture in the unsymmetric species is assumed to occur at the $Cr(NH_3)_4X$ fragment, since, when trans chloroerythro is cleaved in the presence of Cl^- or Br^-, the halide is incorporated into the tetraammine fragment while the pentaammine fragment remains as $Cr(NH_3)_5(H_2O)^{3+}$ *(252)*. The cleavage of the cis aquaerythro ion in the presence of Cl^- and Br^-, respectively, has been studied *(254)*, and both systems follow the rate law $-d[\text{dinuclear}]/dt = (k_1 + k_2[X])[\text{dinuclear}]$, where $X = Cl^-$ or Br^-. In the reaction with chloride the products are $Cr(NH_3)_5(H_2O)^{3+}$ and *cis*-$Cr(NH_3)_4Cl(H_2O)^{2+}$, and the formation of cis chloroerythro as an active intermediate was proposed. In contrast to this, the products in the bromide reaction are $Cr(NH_3)_5(H_2O)^{3+}$ and *cis*-$Cr(NH_3)_4(H_2O)_2^{3+}$. The enhancement of the cleavage rate by bromide has been explained in terms of ion-pair formation.

TABLE XXV

Kinetic Data for the Cleavage Reactions of Monohydroxo-Bridged Chromium(III) Complexes at 25°C in 1.0 M (Na,H)ClO$_4$

Complex	$10^5 \times k_0$ (sec^{-1})	$\Delta H^{\ddagger}$ (kJ mol^{-1})	$\Delta S^{\ddagger}$ (J mol^{-1} K^{-1})	Reference
$(NH_3)_5Cr(OH)Cr(NH_3)_5^{5+}$	1.03	113(1)	40(3)	*252, 254*[a]
trans-$(NH_3)_5Cr(OH)Cr(NH_3)_4(H_2O)^{5+}$	0.83	113(2)	37(6)	*354*
cis-$(NH_3)_5Cr(OH)Cr(NH_3)_4(H_2O)^{5+}$	0.24	105(3)	−1(8)	*252, 254*[a]
$(H_2O)_5Cr(OH)Cr(H_2O)_5^{5+}$	0.1	107(10)	0(30)	*28*[b]
cis-$(NH_3)_5Cr(OH)Cr(NH_3)_4Cl^{4+}$		Faster than *trans*-Cl		*252*
trans-$(NH_3)_5Cr(OH)Cr(NH_3)_4Cl^{4+}$	1.47	115(1)	50(4)	*252, 254*[a]
cis-$(NH_3)_5Cr(OH)Cr(NH_3)_4F^{4+}$	1.21	97(3)	−14(3)	*254*
cis-$(NH_3)_5Cr(OH)Cr(NH_3)_4(NCS)^{4+}$	0.21	71(6)	−116(18)	*254*

[a] There are some discrepancies between the calculated activation parameters originally published in Ref. *252* and some parameters later recalculated in Ref. *254*. The data given here are from Ref. *254*.

[b] These data are for approximately 2 M HClO$_4$. Values for $\Delta H^{\ddagger}$ and $\Delta S^{\ddagger}$ have been calculated from the rate constants at 25, 35.5, and 45°C given in Ref. *28*, and $\Delta H^{\ddagger}$ deviates by a factor of ~ 2 from the value quoted in this reference.

The mechanistic aspects of the variation of the activation parameters with the nature of the substitutent X in $(NH_3)_5Cr(OH)Cr(NH_3)_4X^{n+}$ have been discussed previously (254, 354), but the paucity of data for ions of the same geometry and the same charge renders discussion difficult.

Cleavage of $(H_2O)_5Cr(OH)Cr(H_2O)_5{}^{5+}$ in various media has been investigated by Connick and co-workers, and only an uncatalyzed path (k_0) was observed (Table XXV). The activation parameters are very similar to those for cis-$(NH_3)_5Cr(OH)Cr(NH_3)_4(H_2O)^{5+}$, which is a very reasonable result. It should be noted that there is no agreement between Connick's results quoted in Table XXV and data from some recent studies by Holwerda and co-workers (204–208). According to Holwerda the cleavage of the dinuclear aqua species in 1 M HClO$_4$ is approximately 10^3 times faster than was found by Connick, and it is also claimed (204) that $(H_2O)_5Cr(OH)Cr(H_2O)_5{}^{5+}$ deprotonates to form an oxo-bridged ion even in 1 M HClO$_4$. This would require that the acid strength of the hydroxo bridge in $(H_2O)_5Cr(OH)Cr(H_2O)_5{}^{5+}$ should be at least 10^8 times greater than that observed in the related rhodo ion, $(NH_3)_5Cr(OH)Cr(NH_3)_5{}^{5+}$ ($pK_a = 7.6$), and would seem to be a very surprising result.

Cleavage of Δ,Λ-(en)$_2$Cr(OH)$_2$Cr(en)$_2{}^{4+}$ in concentrated strong acids (e.g., 12 M HClO$_4$) proceeds in two kinetically well-separated steps. The first and fast step ($t_{1/2} \sim 40$ seconds at 25°C) yields Δ,Λ-(H$_2$O)(en)$_2$-Cr(OH)Cr(en)$_2$(H$_2$O)$^{5+}$, which then undergoes bridge cleavage much more slowly ($t_{1/2} \sim 3$ hours at 25°C) to yield cis-Cr(en)$_2$(H$_2$O)$_2{}^{3+}$ quantitatively (33, 42). The kinetics for the first bridge cleavage have been studied for ionic strength 1.0 M, as described in Section IX. Kinetic studies of the second bridge cleavage are seriously complicated by the occurrence of side reactions at lower acidities. A spectrophotometric study of the second bridge-cleavage reaction at lower acidities ($[H^+] = 0.02$–1.0 M at $I = 1.0$ M) has been reported (324), and the observed rate constants in this study were obtained on the basis of the assumption that the only side reaction occurring is subsequent hydrolysis of the product cis-Cr(en)$_2$(H$_2$O)$_2{}^{3+}$ to Cr(en)(H$_2$O)$_4{}^{3+}$. This is, however, in conflict with a more recent report that the hydrolysis of Δ,Λ-(en)$_2$Cr(OH)$_2$Cr(en)$_2{}^{4+}$ (and of its parent monohydroxo-bridged species) in 1 M HClO$_4$ yields a complex mixture of new polynuclear species, cis-Cr(en)$_2$(H$_2$O)$_2{}^{3+}$ and mononuclear species with fewer than four nitrogens coordinated per chromium (42). It therefore seems uncertain to which reaction(s) the observed rate constant should be related, and a more detailed analysis of the products seems to be called for.

The cleavage is accelerated in the presence of nitrate and a rate law of the form $k_{obs} = a + b[NO_3^-] + c[H^+][NO_3^-]$ has been reported (*324*). The acceleration is probably due to rapid formation of a labile nitrate complex, e.g., $(H_2O)(en)_2Cr(OH)Cr(en)_2(NO_3)^{4+}$, but it does not seem possible on the basis of the present data to establish an unambiguous reaction scheme.

b. Cobalt(III) Complexes. The cleavage of $(NH_3)_5Co(OH)Co(NH_3)_5^{5+}$ in perchlorate media has been studied several times (*130, 355–358*). In neutral or alkaline medium the reaction is complicated by the formation of products other than $Co(NH_3)_5(H_2O)^{3+}$ and $Co(NH_3)_5(OH)^{2+}$; these products apparently result from loss of ammonia in the binuclear species since the aqua or hydroxo products are known to be comparatively stable under these conditions (*130, 356*). In acidic media and in the absence of complexing ligands the only cleavage product is $Co(NH_3)_5(H_2O)^{3+}$. Wharton and Sykes (*355*) have studied the kinetics in 2 M LiClO$_4$ in the [H$^+$] region $10^{-4} \leq$ [H$^+$] $\leq 2\,M$ and the results were interpreted in terms of uncatalyzed (k_0) and acid-catalyzed (k_a) bridge-cleavage reactions (Scheme 3). Kinetic data for lower ionic strength support this reaction scheme (Table XXVI). However, the observation that k_0 shows a greater ionic strength dependence than does k_a is somewhat surprising. The aqua-bridged intermediate is a strong acid, and an estimated value of $K_a > 50\,M$ was

TABLE XXVI

Kinetic Parameters for Cleavage of $(NH_3)_5Co(OH)Co(NH_3)_5^{5+}$ in Various Media at 25°C

Medium	k_0 (sec^{-1})	k_a (M^{-1} sec^{-1})	Reference
2 M LiClO$_4$	7.58(10) × 10^{-3} [a,b]	5.37(7) × 10^{-3} [a]	355
1 M NaClO$_4$	4.0 × 10^{-3}	6.2 × 10^{-3}	356
0.2 M NaClO$_4$	≈2.2 × 10^{-3}		130[c]
CH$_3$CN–water (X$_{CH_3CN}$ = 0.5)	2.3 × 10^{-3}	4.05 × 10^{-4}	356[d]
DMSO–water (X$_{DMSO}$ = 0.25)	≤3 × 10^{-5}	1.11 × 10^{-4}	356[d]
DMSO–water (X$_{DMSO}$ = 0.20)	1.8 × 10^{-4}	1.35 × 10^{-3}	356[d]

[a] $\Delta H^{\ddagger}(k_0) = 84(3)$ kJ mol^{-1}, $\Delta S^{\ddagger}(k_0) = -4(9)$ J mol^{-1} K^{-1}; $\Delta H^{\ddagger}(k_a) = 50(3)$ kJ mol^{-1}, $\Delta S^{\ddagger}(k_a) = -120(1)$ J mol^{-1} K^{-1}.

[b] $k_0 = 6.0 \times 10^{-3}$ sec^{-1} in 2 M NaClO$_4$ (*355*).

[c] Calculated from the value $k_0 = 0.65 \times 10^{-3}$ sec^{-1} at 15°C from Ref. *130*, using $E_a = 86.6$ kJ mol^{-1} for 2 M LiClO$_4$ taken from Ref. *355*.

[d] k_a in sec^{-1} kg mol^{-1}

obtained (355). Since k_a is a composite term, the corresponding activation energies are also composite: $\Delta H^{\ddagger}(k_a) = \Delta H^{\ddagger}(k'_a) - \Delta H^0(K_a)$ and $\Delta S^{\ddagger}(k_a) = \Delta S^{\ddagger}(k'_a) - \Delta S^0(K_a)$. Since the aqua bridged species is a strong acid, $\Delta H^0(K_a)$ is probably negative. This implies that $\Delta H^{\ddagger}(k'_a) <$ 50 kJ mol^{-1}, in keeping with the presumed poor bridging-ligand properties of a water molecule.

Cleavage in aqueous solution in the presence of other anions, Y^-, has been studied for $Y^- = NO_3^-$, SCN^-, Cl^-, and $CH_3SO_3^-$ (355–358). In all these reactions variable amount of $Co(NH_3)_5Y^{2+}$ are produced together with the aqua complex, and extensive competition studies have been reported for the uncatalyzed as well as for the acid-catalyzed cleavage reaction. The results of these competition experiments are in keeping with an essentially dissociative mechanism for bridge cleavage.

The kinetic data reveal a complex dependence on the anion concentration and the hydrogen-ion concentration and have been interpreted on the basis of ion-pair and ion-triplet formation. The uncatalyzed path (k_0) has been shown to involve $(NH_3)_5Co(OH)Co(NH_3)_5^{5+}$ ($= M^{5+}$) and the ion pair $M \times Y^{4+}$, and it was proposed that the ion pair $M \times Y^{4+}$ scavenges Y^- from solution and not from the second coordination sphere (357). It was shown that the reactive intermediates are quite selective for anions (as well as being selective for the N terminus of NCS^-, the ratio for N-bound:S-bound being approximately 4), and this has been interpreted as arising from a genuine, coordinately unsaturated intermediate. The acid-catalyzed path has been interpreted in terms of the formation of protonated unaggregated reactant, MH^{6+}, and small concentrations of the protonated ion pairs and ion triplets $MH \times Y^{5+}$ and $MH \times Y_2^{4+}$ (355, 356).

c. *Rhodium(III) Complexes.* Acid cleavage of the monohydroxo-bridged species $\Delta,\Lambda\text{-}(H_2O)(en)_2Rh(OH)Rh(en)_2X^{n+}$ (X = H_2O or OH^-) has been studied in the [H$^+$] region 10^{-5}–1.0 M (325). Since equilibration of the monohydroxo-bridged species with the parent dihydroxo-bridged species takes place much faster than cleavage into mononuclear species, it was necessary to take account of these equilibria in the calculations, as shown in Scheme 4. The kinetic experiments were conducted at low complex concentrations [[Rh(III)] = 2 × [dinuclear] + [mononuclear] ~ 10^{-4} M] so that the influence of the condensation reactions (k_{-3}, etc.) could be disregarded ($K_d = k_{-3}/k_3 \sim 25$, as discussed in Section V,B). The observed rate constants were found to obey Eq. (52) derived from Scheme 4. Since the equilibrium constants K_{a1} and

$$k_{obs} = \frac{k_3 K_{a1} + k_4[H^+] + (k_5/K_{a5})[H^+]^2}{K_{a1}/K_1 + K_{a1} + [H^+]} \quad (52)$$

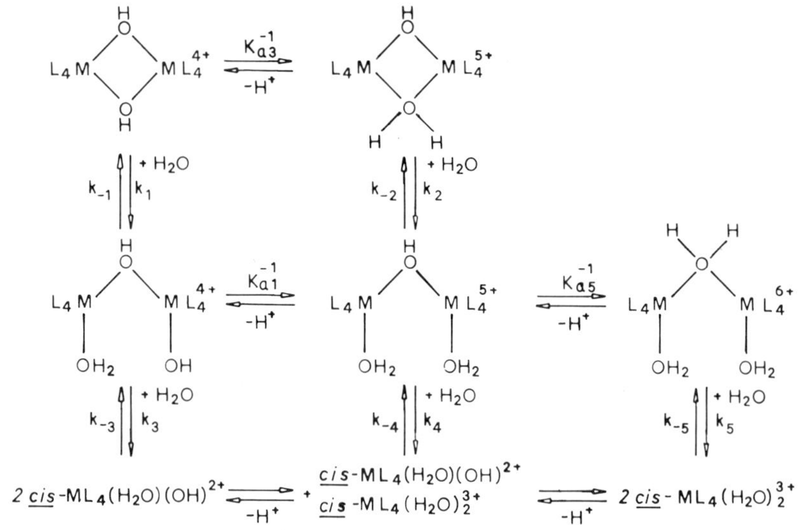

SCHEME 4. Complete reaction scheme for the formation and cleavage of dinuclear dihydroxo-bridged species in acidic solution.

K_1 have been determined in an independent study (see Section IX), it was possible to determine the rate constants k_3 and k_4 (k_0 in Scheme 3) and the ratio k_5/K_{a5} ($k_a = k'_a/K_a$ in Scheme 3), given in Table XXVII. The fact that $\Delta H^{\ddagger}(k_0)$ is much larger than $\Delta H^{\ddagger}(k_a)$ for the cleavage of the diaqua complex may be rationalized as discussed above for the cobalt(III) species. The rate constant for the uncatalyzed bridged cleavage of the aqua hydroxo species, $\Delta,\Lambda\text{-}(H_2O)(en)_2Rh(OH)Rh(en)_2\text{-}(OH)^{4+}$, k_0, is about 20 times smaller than that for cleavage of the diaqua species, $\Delta,\Lambda\text{-}(H_2O)(en)_2Rh(OH)Rh(en)_2(H_2O)^{5+}$, which is in keeping with their respective thermodynamic stabilities (Table XI). As discussed in Sections V and VI, the increased thermodynamic stability of the aqua hydroxo species can be explained on the basis of intramolecular hydrogen bond formation (Fig. 15; Table XXI). The kinetic consequences of this hydrogen bond interaction are strongly dependent on the structure of the transition state, i.e., on whether the hydrogen bond persists or is broken in the transition state, as shown in Scheme 5. Cleavage of the hydrogen bond prior to the formation of the transition state would provide a good explanation for the increased kinetic stability. The experimental results thus indicate the non-hydrogen-bonded transition state structure II in Scheme 5. It should be noted that this is not consistent with the proposal (*326*) that the reactive species for the reverse reaction is a cation pair with intermolecular

TABLE XXVII

KINETIC DATA FOR UNCATALYZED AND ACID-CATALYZED CLEAVAGE OF MONOHYDROXY-BRIDGED RHODIUM(III) COMPLEXES AT 25°C IN 1.0 M (Na,H)ClO$_4$ [a,b]

Compound	Uncatalyzed path			Acid-catalyzed path			Reference
	$10^6 \times k_0$ (sec^{-1})	$\Delta H^{\ddagger}$ (kJ mol^{-1})	$\Delta S^{\ddagger}$ (J mol^{-1} K^{-1})	$10^5 \times k_a$ (M^{-1} sec^{-1})	$\Delta H^{\ddagger}$ (kJ mol^{-1})	$\Delta S^{\ddagger}$ (J mol^{-1} K^{-1})	
(H$_2$O)(en)$_2$Rh(OH)Rh(en)$_2$(H$_2$O)$^{5+}$	3.82	116(3)	40(10)	3.48	78(4)	−68(14)	325
(H$_2$O)(en)$_2$Rh(OH)Rh(en)$_2$(OH)$^{4+}$	0.23	107(2)	−12(6)	—	—	—	325
(H$_2$O)$_2$(tacn)Rh(OH)Rh(tacn)(H$_2$O)$_2^{5+}$	1.4	81(15)	−86(48)	3.4	68(2)	−102(8)	100

[a] k_0 equal to k_3 or k_4 and k_a equal to k_5/K_{a5} in Scheme 4.
[b] As discussed in the text, the data for the tacn system are based on a tentative interpretation (325).

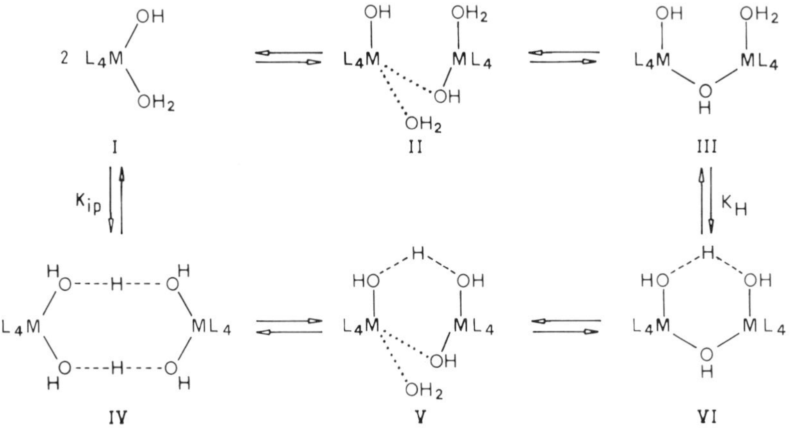

SCHEME 5. Stabilization of dinuclear species in aqueous solution by intramolecular hydrogen-bond formation (VI) is well established. Stabilization of the mononuclear species in aqueous solution by intermolecular hydrogen bonds (IV) may be important in some systems. Interconversion between mononuclear and dinuclear species may occur via non-hydrogen-bonded and hydrogen-bonded transition states, respectively, as schematically shown in II and V. Dashed lines denote hydrogen bonds; dotted lines denote bond making and bond breaking.

hydrogen bonding (Fig. 11), since such a species would lead to a hydrogen-bonded transition state structure (structure V in Scheme 5).

Acid hydrolysis of the dihydroxo-bridged complex trans-$(H_2O)(tacn)$-$Rh(OH)_2Rh(tacn)(H_2O)^{4+}$ yields mononuclear triaqua complex (100). Only one reaction step was observed for $[H^+] = 0.2$–1.0 M, the rate expression being of the form $k_{obs} = a + b[H^+]$. If it is assumed that the first bridge cleavage is faster than the second, as has been found for the corresponding tetraammine and bis(ethylenediamine) complexes, the observed $[H^+]$ dependence can be interpreted in terms of Scheme 4, which for large $[H^+]$ leads to the approximate expression $k_{obs} = k_4 + (k_5/K_{a5})[H^+]$. Values for k_4 and k_5/K_{a5} are listed in Table XXVII.

2. Base Hydrolysis

Studies of the hydroxo-bridge cleavage in basic solution are often complicated by hydrolysis of the nonbridging ligands, and few kinetic studies have been made. The kinetics of hydrolysis of the $(en)_2Co(OH)_2Co(en)_2^{4+}$ ion in basic solution indicate two consecutive steps, and different mechanisms involving cleavage of the presumed intermediate $(H_2O)(en)_2Co(OH)Co(en)_2(OH)^{4+}$ (or its deprotonated form) have been considered (*359*). Base hydrolysis in the presence of carbonate has been proposed to involve cleavage of the presumed intermediate $(H_2O)(en)_2Co(OH)Co(en)_2(HCO_3)^{4+}$ (*360*).

3. Chromium(II)-Catalyzed Hydrolysis

The Cr^{2+}-catalyzed hydrolysis reaction, Eq. (53), has been studied for 1 M $HClO_4$ medium (*361*). The reaction is followed by a much slower reaction in which Cr^{2+}-catalyzed hydrolysis of $(NH_3)_5Cr(H_2O)^{3+}$ takes place. The kinetic data for Eq. (53) are consistent with a rate law of the

$$(NH_3)_5Cr(OH)Cr(NH_3)_4Cl^{4+} + 5H_3O^+ \xrightarrow{Cr^{2+}}$$
$$(NH_3)_5Cr(H_2O)^{3+} + 4NH_4^+ + (H_2O)_5CrCl^{2+} \qquad (53)$$

form $k_{obs} = k[Cr^{2+}][\text{dinuclear}]$, with $k = 3.0$ M^{-1} sec^{-1} at 25°C and $E_a = 39$ kJ mol^{-1}. The form of the rate law and the occurrence of $Cr(H_2O)_5Cl^{2+}$ as a reaction product suggest that the rate-determining step in the chromium(II)-catalyzed reaction is an electron-transfer process in which the chloride ion serves as a bridging ligand.

4. Photochemical Hydrolysis

The photochemical (254 nm) hydrolysis of $(NH_3)_5Cr(OH)Cr(NH_3)_5^{5+}$ to $Cr(NH_3)_5(H_2O)^{3+}$ has been reported (*321*).

IX. Equilibria between Mono- and Dihydroxo-Bridged Dinuclear Complexes of Chromium(III), Rhodium(III), and Iridium(III)

A. General Remarks

The equilibrium Eq. (54) has been studied for a series of chromium(III), rhodium(III), and iridium(III) complexes. Nearly all

$$L_4M(OH)_2ML_4^{4+} + H_2O \underset{}{\overset{K_1}{\rightleftharpoons}} (H_2O)L_4M(OH)ML_4OH^{4+} \qquad (54)$$

the species studied contain ammonia or ethylenediamine ligands and mono- and dihydroxo-bridged cations have been isolated and characterized as crystalline salts, as described in Section IV. For these ammine and ethylenediamine complexes the equilibration reaction, Eq. (54), is relatively fast, so it has been possible to study the kinetics and thermodynamics of the process without interference from other

TABLE XXVIII

Equilibrium between Mono- and Dihydroxo-Bridged Species at 25°C in 1.0 M (Na,H)ClO$_4$ [a]

M(III)	L$_4$	K_1	ΔH^0 (kJ mol^{-1})	ΔS^0 (J mol^{-1} K^{-1})	Reference
Cr(III)	Δ,Λ-(en)$_2$	0.75(2)	−3.1(9)	−13(3)	33, 36
Cr(III)	Δ,Δ/Λ,Λ-(en)$_2$	0.22(3)	−2(4)	−19(13)	35, 36
Cr(III)/Co(III)	Δ,Λ-(en)$_2$	4	—	—	36, 215
Cr(III)	(NH$_3$)$_4$	0.318(11)	6(4)	11(14)	36
Cr(III)	(H$_2$O)$_4$	0.02	—	—	28
Cr(III)	Δ,Δ/Λ,Λ-(phen)$_2$	≪1	—	—	336
Rh(III)	Δ,Λ-(en)$_2$	11.2(5)	−14(3)	−28(8)	217
Rh(III)	(NH$_3$)$_4$	3.03(4)	2(1)	15(4)	133
Ir(III)	Δ,Λ-(en)$_2$	5.7(8)	−2(5)	10(15)	67

[a] See Scheme 6.

reactions such as hydrolysis of the N-donor ligands or cleavage of the monohydroxo-bridged species to give mononuclear species.

Thermodynamic data for the equilibrium Eq. (54) have been obtained by spectrophotometric and potentiometric (pH) measurements (cf. Table XXVIII). The K_1 values lie in the region 1–10, the only significant exceptions being the small values for the aqua and 1,10-phenanthroline chromium(III) systems. The enthalpy changes are small, as anticipated for reactions which involve bond breaking and bond making of similar kinds of bonds.

Most of the kinetic studies made have been spectrophotometric and have employed acidic or neutral solutions. For the ammine and ethylenediamine complexes pseudo first-order rate constants were obtained for solutions initially containing either monohydroxo-bridged species [k_{obs}(M)] or dihydroxo-bridged species [k_{obs}(D)]. It was found that $k_{obs}(M) = k_{obs}(D)$, which is consistent with reversible first-order kinetics.

The [H$^+$] dependence of k_{obs} has been interpreted in terms of an acid-catalyzed and an uncatalyzed reaction path, as shown in Scheme 6. The acid-catalyzed path has been proposed to involve protonation of one hydroxo bridge to give a labile aqua-bridged intermediate. The aqua-bridged complexes have in no case been identified, but spectroscopic results indicate that such species are very strong acids with $K_{a3} \gg 1$. The kinetic parameters calculated from the rate expression, Eq. (55), are

$$k_{calc} = k_1 + \frac{[H^+]K_1 k_{-2}}{K_{a1}} + \frac{K_{a1} K_1^{-1} k_1 + [H^+]k_{-2}}{K_{a1} + [H^+]} \quad (55)$$

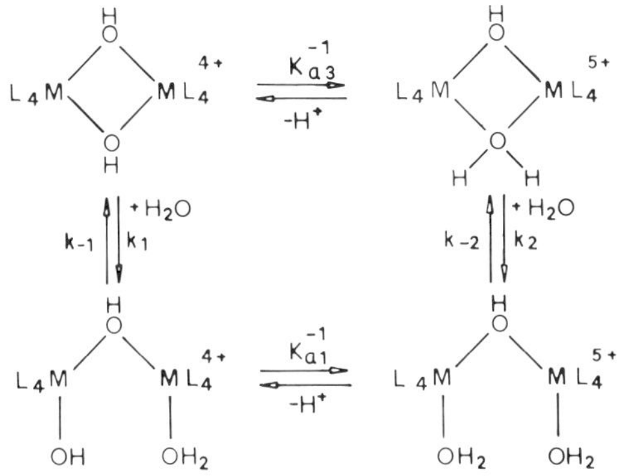

SCHEME 6. Reaction scheme for the equilibria between dinuclear mono- and dihydroxo-bridged species of chromium(III), rhodium(III), and iridium(III).

given in Tables XXIX–XXXII. As mentioned previously, the parameters K_{a1} and K_1 have been determined independently of the kinetic measurements by spectrophotometry or potentiometry (Tables XIX and XXVIII). The kinetic parameters for uncatalyzed bridge cleavage and formation (k_1 and k_{-1}) have been determined for each of the ammine and ethylenediamine complexes of chromium(III), rhodium(III), and iridium(III), whereas the parameters for the acid-catalyzed path (k_2/K_{a3} and k_{-2}) have been determined only for some of these systems. In the meso and racemic Cr(III) ethylenediamine systems and in the mixed CrCo ethylenediamine system the contribution from the acid-catalyzed path is insignificant even in 1 M HClO$_4$, and only an upper limit value for the rate constants has been determined (see Tables XXXI and XXXII).

A kinetic study of the aqua chromium(III) system has also been interpreted in terms of Scheme 6. In this system the parameters K_{a1} and K_1 are not known with great precision and consequently the rate constants are somewhat uncertain and have not been taken into account in the discussion below. The following values have been reported for 25°C [2 M (H$^+$,Li)ClO$_4$]: $k_1 = 1.5 \times 10^{-5}$ sec^{-1}, $k_{-1} = 8 \times 10^{-4}$ sec^{-1}, $k_2/K_{a3} = 9 \times 10^{-5}$ M^{-1} sec^{-1}, and $k_{-2} = 1.2 \times 10^{-4}$ sec^{-1} (28). It should be noted that these values are similar to those observed for the chromium(III) ammonia system (Tables XXIX–XXXII).

The kinetics for the acid hydrolysis of Δ,Δ/Λ,Λ-(phen)$_2$Cr(OH)$_2$-Cr(phen)$_2^{4+}$ revealed that the formation of cis diaqua mononuclear species proceeds in one step without significant buildup of monohydroxo-bridged intermediates (336). The rate law observed is $k_{obs} = a + b[H^+]$ at low acidities ([H$^+$] = 0.01–0.05 M) and $k_{obs} = b[H^+]$ at

TABLE XXIX

Uncatalyzed Cleavage of $L_4M(OH)_2ML_4^{4+}$ Species at 25°C in 1.0 M (Na,H)ClO$_4$

M(III)	L$_4$	$10^4 \times k_1$ (sec^{-1})	$\Delta H^\ddagger$ (kJ mol^{-1})	$\Delta S^\ddagger$ (J mol^{-1} K^{-1})	Reference
Cr(III)/Co(III)	Δ,Λ-(en)$_2$	69(9) [138(17)]	78(3)	−25(12) [−20(12)]	36, 215[a]
Cr(III)	Δ,Λ-(en)$_2$	69(5)	80(2)	−18(7)	33, 36
Cr(III)	$\Delta,\Delta/\Lambda,\Lambda$-(en)$_2$	46.1(19)	85(1)	−4(5)	35, 36
Cr(III)	(NH$_3$)$_4$	1.21(2)	87(2)	−29(5)	36
Cr(III)	$\Delta,\Delta/\Lambda,\Lambda$-(phen)$_2$	0.69(3)	99(3)	7(11)	336[b]
Rh(III)	Δ,Λ-(en)$_2$	4.56(11)	86(3)	−19(8)	217
Rh(III)	(NH$_3$)$_4$	0.426(18)	91(5)	−24(16)	133
Ir(III)	Δ,Λ-(en)$_2$	0.0148(11)	113(2)	22(6)	67

[a] Δ,Λ-(en)$_2$Cr(OH)$_2$Co(en)$_2^{4+}$. If it is assumed that the bridge cleavage in the mixed system occurs essentially with Cr−O bond breaking, the k_1 value should be corrected with a statistical factor of 2 when compared with the remaining complexes. This then gives the values in square brackets.

[b] 0.1 M (Na,H)NO$_3$.

TABLE XXX

ACID-CATALYZED BRIDGE CLEAVAGE OF $L_4M(OH)_2ML_4{}^{4+}$ SPECIES AT 25°C IN 1.0 M (Na,H)ClO$_4$

M(III)	L$_4$	k_2/K_{a3} (M^{-1} sec^{-1})	$\Delta H^{\ddagger}(k_2) - \Delta H^0(K_{a3})$ (kJ mol^{-1})	$\Delta S^{\ddagger}(k_2) - \Delta S^0(K_{a3})$ (J mol^{-1} K^{-1})	Reference
Cr(III)	Δ,Λ-(en)$_2$	$<3.5 \times 10^{-4}$	—	—	33[a]
Cr(III)	(NH$_3$)$_4$	$4.9(5) \times 10^{-5}$	82(11)	$-53(35)$	36
Rh(III)	Δ,Λ-(en)$_2$	$1.34(3) \times 10^{-1}$	58(2)	$-68(7)$	217
Rh(III)	(NH$_3$)$_4$	$7.4(2) \times 10^{-2}$	58(3)	$-72(9)$	133
Ir(III)	Δ,Λ-(en)$_2$	$2.40(5) \times 10^{-4}$	69(2)	$-84(6)$	67

[a] Estimated from data in Ref. 33.

high acidities ($[H^+] = 0.1$–1.0 M). The rate at which the dinuclear species exchange ^{18}O was found to be independent of $[H^+]$ over the range $[H^+] = 0.01$–0.1 M [0.1 M (Na,H)NO$_3$] and the exchange is fast compared with the cleavage reaction, viz. $k_{\text{exchange}}/k_{\text{cleavage}} = 38$ at 60°C and $[H^+] = 0.1$ M. Experiments involving cleavage in D$_2$O indicated that there is a rapid prequilibrium entailing protonation prior to the rate-determining step. These observations suggest that rapid equilibration between mono- and dihydroxo-bridged species takes place prior to the rate-determining bridge cleavage of the monohydroxo-bridged species (Scheme 4). The observations that the monohydroxo-bridged species never attains a high concentrations and that the dihydroxo-bridged complex exchanges ^{18}O rapidly provide strong evidence that $K_1 \ll 1$, from which it then follows that $k_1 = 4k_{\text{ex}}$ (Table XXIX). The fact that the ^{18}O exchange rate is independent of $[H^+]$ rules out the possibility that bridge cleavage proceeds by an acid-catalyzed pathway to any significant extent, i.e., $[H^+]k_2/K_{a3} \ll k_1$. Then $[H^+]$ dependence of k_{obs} can be interpreted in terms of Scheme 4, but it is not possible to determine any values for the individual parameters as long as neither K_1 nor K_{a1} is known.

Preliminary kinetic data for the cleavage of $(tn)_2Cr(OH)_2Cr(tn)_2{}^{4+}$ have been reported: $k_1 + k_{-1} = 6(1) \times 10^{-4}$ sec^{-1} at 25°C (216). Cleavage of the trans-$(H_2O)(NH_3)_3Cr(OH)_2Cr(NH_3)_3(H_2O)^{4+}$ ion (Fig. 9) has been studied qualitatively. In 4 M HClO$_4$, cleavage of the first bridge is much faster than that of the second bridge and the product fac,fac-$(H_2O)_2(NH_3)_3Cr(OH)Cr(NH_3)_3(H_2O)_2{}^{5+}$ has been isolated. It was shown that the monohydroxo-bridged complex at pH 2 reforms the dihydroxo-bridged species (87).

The kinetic parameters for the bridge cleavage and formation reactions show a number of trends which in some cases parallel those

TABLE XXXI

Bridge Formation of $(H_2O)L_4M(OH)ML_4(OH)^{4+}$ Species at 25°C in 1.0 M (Na,H)ClO$_4$

M(III)	L$_4$	k_{-1} (sec^{-1})	$\Delta H^{\ddagger}$ (kJ mol^{-1})	$\Delta S^{\ddagger}$ (J mol^{-1} K^{-1})	K_{a1} ^{a} (M)	Reference
Cr(III)	$\Delta,\Delta/\Lambda,\Lambda$-(en)$_2$	2.07(8) × 10^{-2}	87(1)	15(5)	0.29	35, 36
Cr(III)	Δ,Λ-(en)$_2$	0.93(7) × 10^{-2}	84(2)	−1(7)	0.33	33, 36
Cr(III)/Co(III)	Δ,Λ-(en)$_2$	0.18(2) × 10^{-2}	78(3)	−37(12)	5 × 10$^{-2\,b}$	36, 215
Cr(III)	(NH$_3$)$_4$	3.80(4) × 10^{-4}	81(2)	−40(5)	1.8 × 10^{-2}	36
Rh(III)	Δ,Λ-(en)$_2$	4.07(10) × 10^{-5}	101(3)	9(9)	4.2 × 10^{-3}	217
Rh(III)	(NH$_3$)$_4$	1.41(6) × 10^{-5}	89(5)	−39(16)	3.9 × 10^{-4}	133
Ir(III)	Δ,Λ-(en)$_2$	2.58(35) × 10^{-7}	114(4)	13(14)	1.2 × 10^{-2}	67

^{a} K_{a1} is the acid dissociation constant for the $(H_2O)L_4M(OH)ML_4(H_2O)^{5+}$ ion.
^{b} Measured at 0.8°C.

TABLE XXXII

Bridge Formation of $(H_2O)L_4M(OH)ML_4(H_2O)^{5+}$ Species at 25°C in 1.0 M (Na,H)ClO$_4$

M(III)	L$_4$	k_{-2} (sec^{-1})	$\Delta H^{\ddagger}$ (kJ mol^{-1})	$\Delta S^{\ddagger}$ (J mol^{-1} K^{-1})	Reference
Cr(III)	Δ,Λ-(en)$_2$	<2 × 10^{-4}	—	—	*33*[a]
Cr(III)	(NH$_3$)$_4$	2.7(3) × 10^{-6}	123(11)	63(35)	*36*
Rh(III)	Δ,Λ-(en)$_2$	5.1(1) × 10^{-5}	100(2)	9(7)	*217*
Rh(III)	(NH$_3$)$_4$	9.6(3) × 10^{-6}	109(3)	24(11)	*133*
Ir(III)	Δ,Λ (en)$_2$	5.1(8) × 10^{-7}	81(8)	−94(27)	*67*

[a] Estimated from data given in Ref. *33*.

observed for the corresponding mononuclear species. Other trends are due to effects specific for the polynuclear species, and in particular intramolecular hydrogen bond interactions are important factors in relation to the bridge-formation reactions. The most important of these trends are summarized in the following discussion.

B. Uncatalyzed Cleavage

From the data in Table XXIX it can be seen that the rate constant, k_1, for the bridge-cleavage reaction of different dihydroxo-bridged chromium(III) complexes decreases in the ligand order *meso*-en > *racemic*-en > NH$_3$ > 1,10-phen, and that this order is accompanied by an increase in $\Delta H^{\ddagger}$. The data for the heteronuclear complex, Λ,Δ-(en)$_2$Cr(OH)$_2$Co(en)$_2^{4+}$, place this complex in the above sequence close to the corresponding dichromium complex with respect to both k_1 and $\Delta H^{\ddagger}$, which might suggest that the k_1 path in the CrCo complex essentially involves Cr–O bond cleavage.

A comparison of the activation parameters $\Delta H^{\ddagger}(k_1)$ for the ammonia and ethylenediamine (meso isomer) complexes shows that $\Delta H^{\ddagger}(\text{NH}_3) > \Delta H^{\ddagger}(\text{en})$ for both chromium(III) and rhodium(III), and that $\Delta H^{\ddagger}(\text{Ir}) > \Delta H^{\ddagger}(\text{Rh}) > \Delta H^{\ddagger}(\text{Cr})$ for both NH$_3$ and en. These trends are in keeping with the data for substitution reactions of the corresponding mononuclear complexes (*364*).

Although it is tempting to suggest that the variations in k_1 could be the result of differing degrees of strain in the M$_2$O$_2$ ring system, it is not possible to establish such a relationship on the basis of available crystal data since the observed variations in the M–O–M angles for different cations are equal to the variations found for salts of the same cation with different anions. As an example, it could be mentioned that the Cr–O–Cr angles found in the chloride salts of the NH_3, en, and 1,10-phen chromium(III) complexes, respectively, are 99.9, 102.4, and 102.7°, whereas the Cr–O–Cr angles in different salts of the $\Delta,\Lambda\text{-(en)}_2Cr(OH)_2Cr(en)_2^{4+}$ cation vary from 100.0 to 103.4° (38,69, 70, 91–93).

C. Acid-Catalyzed Cleavage

Only values for the ratio k_2/K_{a3} have been determined and the activation parameters are therefore composite terms, $\Delta H^\ddagger(k_2) - \Delta H^0(K_{a3})$ and $\Delta S^\ddagger(k_2) - \Delta S^0(K_{a3})$. For all three metal ions it is found that the aqua-bridged complexes are very strong acids, implying that the $\Delta H^0(K_{a3})$ values are undoubtedly all negative. It therefore follows that the values for $\Delta H^\ddagger(k_2) - \Delta H^0(K_{a3})$ represent upper limits to the values of $\Delta H^\ddagger(k_2)$. The small $\Delta H^\ddagger(k_2)$ reflect the fact that water is a poor bridging group, and similar low $\Delta H^\ddagger$ values are found for the acid-catalyzed bridge cleavage of trihydroxo-bridged complexes (see Section XI).

The quantity $(k_2/K_{a3})/k_1$ gives the ratio between the rate of acid-catalyzed bridge cleavage at $[H^+] = 1\ M$ and the rate of uncatalyzed bridge cleavage. For the ammonia systems this ratio is 1700 and 0.4, respectively, for rhodium(III) and chromium(III). For the meso ethylenediamine systems the ratio is 300, 160, and less than 0.05, respectively, for rhodium(III), iridium(III), and chromium(III). The considerably enhanced relative efficiency of the acid-catalyzed path in iridium(III) and rhodium(III) compared with chromium(III) could be due to the differences in K_{a3}, which would be qualitatively in keeping with the greater acidity of mononuclear chromium(III) complexes compared to complexes of iridium(III) and rhodium(III) (Table XVIII). Also, the observation that for both chromium(III) and rhodium(III) the ratio $(k_2/K_{a3})/k_1$ is greater for NH_3 than for ethylenediamine can be explained qualitatively along the same lines.

D. Bridge Formation

The bridge-formation reactions k_{-1} are strongly influenced by the intramolecular hydrogen bond interactions which have been discussed in Section VI (Figs. 15 and 16, and Table XXI). For the chromium(III) complexes (and the mixed CrCo complex) it is found that k_{-1} decreases in the order *racemic*-(en)$_2$ > *meso*-(en)$_2$ > *meso*-(en)$_2$(CrCo) > (NH$_3$)$_4$, which is also the order of decreasing acid strength (K_{a1}) of the corresponding diaqua monohydroxo-bridged species (Table XXXI). For the rhodium(III) systems the order of decreasing k_{-1} and K_{a1} is (en)$_2$ > (NH$_3$)$_4$. This correlation has been explained in terms of intramolecular hydrogen bond stabilization of the aquahydroxo species. The hydrogen-bond-stabilized conformation resembles the transition state structure more than other conformations and should therefore contribute to k_{-1} by an enhancement of $\Delta S^{\ddagger}$. On the other hand, the intramolecular hydrogen bond has to be broken and this bond breaking will contribute to k_{-1} in terms of increasing $\Delta H^{\ddagger}$. It is therefore anticipated that increasing hydrogen bond stabilization, i.e., increasing K_{a1}, should be associated with increasing $\Delta S^{\ddagger}$ and increasing $\Delta H^{\ddagger}$. This is seen to be in good agreement with the experimental results for both chromium(III) and rhodium(III). The fact that it is found that increasing hydrogen bond stabilization is associated with increasing k_{-1} values is seen to be due to an increase larger in $T\Delta S^{\ddagger}$ than in $\Delta H^{\ddagger}$.

The values for k_{-1} may be compared with those for k_{-2} (Tables XXXI and XXXII). For the chromium(III) species $k_{-2} \ll k_{-1}$ for each of the systems studied. Since coordinated water is undoubtedly a much poorer nucleophile than coordinated hydroxide, this is in keeping with an essentially associative mechanism for both the k_{-1} and the k_{-2} pathways. For the rhodium(III) and iridium(III) species it is found that k_{-1} is roughly equal to k_{-2}, which is in keeping with an essentially dissociative mechanism. This may be illustrated further by comparing the activation parameters. Comparison of the activation parameters for bridge cleavage of the ammine complexes of chromium(III) and rhodium(III) with the kinetic data for water exchange (k_{ex}) of the corresponding mononuclear diaqua complexes shows the sequences $\Delta H^{\ddagger}(k_{-1}) < \Delta H^{\ddagger}(k_{ex}) < \Delta H^{\ddagger}(k_{-2})$ and $\Delta S^{\ddagger}(k_{-1}) < \Delta S^{\ddagger}(k_{ex}) < \Delta S^{\ddagger}(k_{-2})$ for both of the metal centers (Table XXXIII). The large separation in the $\Delta H^{\ddagger}$ values for the chromium(III) reactions is evidence for a high degree of associative behavior and this is in keeping with the properties of mononuclear tetraamminechromium(III) complexes. The sequence further confirms the expected order, Cr–OH > H$_2$O > Cr–OH$_2$, of decreasing nucleophilic character. It is seen that the nucleophilicity of water decreases tremendously on coordination, as reflected by a very large $\Delta H^{\ddagger}(k_{-2})$ value. However, despite this, k_{-2} is a factor of only 20 less than the rate of water exchange, this being due to a proximity effect, i.e., a very large $\Delta S^{\ddagger}(k_{-2})$ value. The observation that the $\Delta H^{\ddagger}$ values for rhodium(III) are much less separated indicates that these

TABLE XXXIII

KINETIC PARAMETERS FOR BRIDGE-FORMATION AND WATER-EXCHANGE REACTIONS OF AMMINE COMPLEXES AT 25°C IN 1.0 M (Na,H)ClO$_4$ [a]

	Rhodium(III)			Chromium(III)		
	$10^6 \times k$ (sec^{-1})	$\Delta H^\ddagger$ (kJ mol^{-1})	$\Delta S^\ddagger$ (J mol^{-1} K^{-1})	$10^6 \times k$ (sec^{-1})	$\Delta H^\ddagger$ (kJ mol^{-1})	$\Delta S^\ddagger$ (J mol^{-1} K^{-1})
k_{-1}	14.1	89	−39	380	81	−40
k_{ex}	7.5	105	11	59	95	−7
k_{-2}	9.6	109	24	2.7	123	63

[a] k_{-1} and k_{-2} are the rate constants for the formation of $(NH_3)_4M(OH)_2M(NH_3)_4^{4+}$ as defined in Scheme 6 (36, 133). Values of k_{ex} represent the statistically corrected rate constants for exchange of one water in cis-$M(NH_3)_4(H_2O)_2^{3+}$ for M = Rh(III) or Cr(III) (362, 363).

reactions are less associative than those of chromium(III), a conclusion which is in keeping with results for mononuclear ammine systems (364).

The activation parameters for the ethylenediamine complexes of rhodium(III) and iridium(III) are also in keeping with an essentially dissociative mechanism. The observation that $\Delta H^\ddagger(k_{-1})$ is *larger* than $\Delta H^\ddagger(k_{-2})$ for iridium(III) has been rationalized in terms of stabilization of the aquahydroxo species by intramolecular hydrogen bond formation. Similarly, the observation for the rhodium(III) system that $\Delta H^\ddagger(k_{-1}) < \Delta H^\ddagger(k_{-2})$ for ammonia, whereas $\Delta H^\ddagger(k_{-1}) \sim \Delta H^\ddagger(k_{-2})$ for ethylenediamine may, in part, by rationalized in terms of the observed differences in the degree of intramolecular hydrogen bond stabilization of the aqua hydroxo species in the two systems [$K_H(en) > K_H(NH_3)$; see Table XXI].

E. Base-Catalyzed Bridge Cleavage

Detailed studies of the base hydrolysis of the dihydroxo-bridged species have not been reported. The hydrolysis of Δ,Λ-(en)$_2$-Rh(OH)$_2$Rh(en)$_2^{4+}$ to give Δ,Λ-(HO)(en)$_2$Rh(OH)Rh(en)$_2$(OH)$^{3+}$ has been reported to be five times faster at pH 12 than the uncatalyzed (k_1) bridge-cleavage reaction. The dihydroxo-bridged species is a weak acid (p$K_a > 12$) and the mechanism could either involve cleavage via formation of an oxo-bridged species or via direct attack by hydroxide (217).

X. Kinetics of the Hydrolysis of Dihydroxo-Bridged Cobalt(III) Complexes

The kinetics of the acid hydrolysis of dihydroxo-bridged cobalt(III) complexes have been studied for both cationic and anionic species. The stoichiometry of the hydrolysis reaction for cationic complexes can be expressed by Eq. (56). The equilibrium lies completely to the right at low pH (typically less than 3) and the reverse process in Eq. (56) can normally be disregarded. For all the systems studied to date the observed rate laws can be interpreted in terms of Scheme 4.

$$L_4Co(OH)_2CoL_4^{2n+} + 2H_2O + 2H^+ \underset{k_r}{\overset{k_f}{\rightleftarrows}} 2cis\text{-}CoL_4(H_2O)_2^{(n+1)+} \tag{56}$$

A major problem in the chemistry of the cobalt(III) species has been the fact that, although reaction Eq. (56) has to proceed in two steps, only one reaction stage is normally observed. The species which apparently follow first-order kinetics are $L_4Co(OH)_2CoL_4^{4+}$ for $L_4 = (NH_3)_4$, $(NH_3)_3(H_2O)$, $(en)_2$, and $(dien)(H_2O)$ (*132, 186, 187, 365, 366*; a so-called induction period has usually been observed for these reactions. In a few other cases, however, hydrolysis has been reported to follow consecutive first-order kinetics (*177, 359*). In none of the systems studied so far has it been possible to isolate the monohydroxo-bridged intermediate, and in the absence of additional information on the properties of this intermediate, unambiguous interpretation of the kinetic data is not possible. The following discussion should be read with this in mind.

Hoffman and Taube have studied the acid hydrolysis of $(NH_3)_4\text{-}Co(OH)_2Co(NH_3)_4^{4+}$ spectrophotometrically in the $[H^+]$ range 0.05–1.0 M (*365*). The cleavage was found to follow first-order kinetics and k_{obs} showed an $[H^+]$ dependence of the form

$$k_{obs} = a[H^+]/(1 + b[H^+]) \tag{57}$$

Additional studies of the reduction of the dinuclear species by Eu^{2+} and Cr^{2+} revealed the rate law, Eq. (58). The c term corresponds to direct

$$-d(\ln \text{dimer})/dt = a[H^+] + c[M^{2+}] \tag{58}$$

attack by the reducing agents on the complex ion, and, as expected, the values of c for the two reducing agents are different. The parameter a has the same value for the two reducing agents and is identical with the parameter in Eq. (57). The latter connection between the two processes is explained if Cr^{2+} and Eu^{2+} are assumed to scavenge the monohydroxo-bridged intermediate efficiently, which in terms of Scheme 4 implies that $a = k_2/K_{a3}$. A rate expression for the hydrolysis reaction which is consistent with the reduction kinetics was then proposed by applying the steady-state approximation for the monohydroxobridged intermediate. Assuming that $k_3 K_{a1} > [H^+]k_4 + [H^+]^2 k_5/K_{a5}$ and $k_3 > k_{-1}$, this gives an expression, Eq. (59), which for large $[H^+]$

$$k_{obs} = \frac{k_1 + (k_2/K_{a3})[H^+]}{1 + k_{-2}[H^+]/(k_3 K_{a1})} \qquad (59)$$

may be reduced to Eq. (60), in which the $[H^+]$ dependence is the same as for the observed rate law, Eq. (57). This leads to the values $k_2/K_{a3} = 1.2 \times 10^{-3} \ M^{-1} \ \text{sec}^{-1}$ and $k_{-2}/(k_3 K_{a1}) = 0.57 \ M^{-1}$ (25°C, $I = 1.0 \ M$).

$$k_{obs} = \frac{(k_2/K_{a3})[H^+]}{1 + k_{-2}[H^+]/(k_3 K_{a1})} \qquad (60)$$

It has been reported, however, that the length of the induction period has an $[H^+]$ dependence which does not agree with this interpretation, although the full mechanistic implications have not yet been evaluated (369).

The acid hydrolysis of $(en)_2Co(OH)_2Co(en)_2^{4+}$ has been studied several times, but with very conflicting results (359, 366–368). Rasmussen and Bjerrum studied the reaction at low acidities ($[H^+] = 0.004$–$0.1 \ M$) and found a pseudo first-order rate expression of the form given in Eq. (61) (367).

$$k_{obs} = a + b[H^+] \qquad (61)$$

DeMaine and Hunt studied the reaction by spectrophotometry and by pH-stat measurements, but covering a wider $[H^+]$ range (10^{-4}–$1.0 \ M$), and found a pseudo first-order rate expression of the form in Eq. (62)

$$k_{obs} = k_0 + \frac{a[H^+]}{1 + b[H^+]} \approx \frac{k_0 + a[H^+]}{1 + b[H^+]} \qquad (62)$$

(366). It should be noted that although Eq. (62) approaches Eq. (61) at low acidities, the observed rate constants found in the two studies are not in agreement.

A very different result has been reported by El-Awady and Hugus, who studied the hydrolysis in acidic and basic solutions by spectrophotometry and by pH-stat measurements, respectively. According to these authors, the spectral changes occurring during the acid hydrolysis ($[H^+] = 10^{-3}-10^{-1}$ M) are best interpreted in terms of two consecutive first-order reactions (*359*).

An unambiguous interpretation of these conflicting results is quite impossible, not only because there is disagreement as to whether the kinetics are simple first order or consecutive first order, but also because it appears to be possible to interpret each of the rate laws in terms of different mechanisms. However, with the kinetic results for the ammine system in mind it seems reasonable to use the approximate expression Eq. (59) to interpret DeMaine and Hunt's data (*366*), and this approach leads to the values $k_1 = 1.4 \times 10^{-5}$ sec^{-1}, $k_2/K_{a3} = 2.7 \times 10^{-3}$ M^{-1} sec^{-1}, and $k_{-2}/(k_3 K_{a1}) = 0.50$ M (25°C, $I = 1.0$ M). An alternative and quite different interpretation would be that a fast preequilibrium between mono- and dihydroxo-bridged species precedes a much slower bridge cleavage of the singly bridged species. This leads to Eq. (52), which for $k_4 > k_5[H^+]/K_{a5}$ reduces to an expression which has an $[H^+]$ dependence of the observed form.

The hydrolysis has also been studied in chloride media. After an induction period, the reaction was found to follow pseudo first-order kinetics with $k_{obs} = k_{Cl}[H^+][Cl^-]$, where $k_{Cl} = 2.8 \times 10^{-2}$ M^{-2} sec^{-1} at 25°C and $I = 1.0$ M ($E_a = 61$ kJ mol^{-1}). Evidence for a chloride-containing intermediate was obtained and its structure could be either Cl(en)$_2$Co(OH)Co(en)$_2$(H$_2$O)$^{4+}$ or (en)$_2$Co(OH)(Cl)Co(en)$_2$$^{4+}$ (*366*).

The base hydrolysis of (en)$_2$Co(OH)$_2$Co(en)$_2$$^{4+}$ studied by the pH-stat method (pH 9.3–10.4) revealed two consecutive first-order reactions which could, however, be interpreted in terms of quite different reaction schemes (*359*).

Schwarzenbach and colleagues have studied the reaction, Eq. (56), for L$_4$ = (NH$_3$)$_3$(H$_2$O) and (dien)(H$_2$O) (*186*). Only one reaction stage was observed and the observed rate constants showed an $[H^+]$ dependence as in Eq. (57). The cleavage of the ammine complex has also been studied by Lindhard and Siebert (*132*) and by Jentsch *et al.* (*187*), and in contrast to the results of Schwarzenbach, a rate expression of the form $k_{obs} = a + b[H^+]$ was observed in both studies. Jentsch *et al.* proposed slow bridge cleavage of the dibridged species followed by fast cleavage of the second bridge. This is contradictory to the proposal given by Schwarzenbach, who proposed the second bridge cleavage to be the slower.

The anionic oxalato complex $(ox)_2Co(OH)_2Co(ox)_2^{4-}$ hydrolyzes completely in acidic aqueous solution to cis-$Co(ox)_2(H_2O)_2^-$, but kinetic studies are complicated by decomposition of the mononuclear species to give cobalt(II) and CO_2 (370). The hydrolysis has been studied in a variety of buffer solutions (formate, acetate, and chloroacetate), and an increase in rate with increasing buffer concentration was observed for each buffer system. The limiting rates were identical for the different buffers at a given pH, and have been interpreted as being equal to $k_2/K_{a3} = 4.1\ M^{-1}\ sec^{-1}$ (25°C). The observation that k_2/K_{a3} for the anionic oxalato complex is 10^3 times larger than the value observed for the cationic ammine complex is qualitatively in agreement with the expected variation in K_{a3} due to the difference in charge.

The kinetics of the acid hydrolysis of $(nta)Co(OH)_2Co(nta)^{2-}$ have been reported in two independent studies (177, 178). At $[H^+] \geq 10^{-2}\ M$, the hydrolysis yields mononuclear $Co(nta)(H_2O)_2$ quantitatively and proceeds in two kinetically well-separated steps, a fast step ($t_{1/2} \sim$ 0.04 second at pH 0) followed by a much slower step (about 10^2 times slower under similar conditions). The occurrence of two steps has, however, been interpreted quite differently in the two studies. Thacker and Higginson assumed that their samples contained one of the two possible isomers contaminated with small amounts of the other isomer and it was suggested that the "rapid" and "slow" reactions correspond to hydrolysis of the two isomeric forms (178).

The proposal of Thacker and Higginson has been questioned by Meloon and Harris (177), who found that a pure potassium salt can be made and that this compound exhibits the same kinetic properties as originally reported for the "impure" compound. Meloon and Harris therefore assumed that the kinetic data should be interpreted in terms of consecutive cleavage of the two bridges in one of the two possible isomers. A comparison of the kinetic data obtained in the two studies is complicated by different choices of media. Meloon and Harris interpreted the occurrence of $[H^+]^2$ terms in the rate expressions by assuming that the rate-determining step in both paths involves protonation of the hydroxo bridge to give a μ-aqua complex which then undergoes *acid-catalyzed* bridge cleavage. This interpretation would imply [Eq. (4) in Ref. 177] that the μ-aqua complex, $(nta)Co(OH)(OH_2)Co(nta)^-$, has $K_a = 0.02\ M$ (25°C, 2 M $NaNO_3$), which would appear to be an unusually low value (see Section VI,B). Furthermore, this value for K_a seems to be in conflict with the observation made in the same study that the dihydroxo-bridged complex does not exhibit basic properties when titrated with dilute acid.

XI. Equilibria between Tri- and Dihydroxo-Bridged Complexes

A. COBALT(III)

The kinetics for acid hydrolysis of $L_3Co(OH)_3CoL_3^{3+}$ ions have been reported for $L_3 = (NH_3)_3$, dien, tach, and tacn (*132, 185–187, 371*). The ammonia system has been studied in detail (*132, 186, 187, 371*), and it is now generally accepted that the kinetic data should be interpreted in terms of Scheme 7.

The main features of this reaction scheme are a fast, acid-catalyzed bridge-cleavage and -formation reaction, followed by a slower, but still fast, isomerization reaction between the cis and trans diaqua isomers of the dihydroxo-bridged species. The thermodynamic and kinetic data are listed in Table XXXIV. The aqua-bridged intermediate is assumed to be a very strong acid ($K_a \gg 1$) and it has therefore been possible only to determine the ratio k_1/K_a. Since $\Delta H^0(K_a)$ is probably negative, the low value for $\Delta H^{\ddagger}(k_1) - \Delta H^0(K_a)$ shows that $\Delta H^{\ddagger}(k_1) < 47$ kJ mol^{-1} and emphasizes the fact that water acts as a poor bridging ligand.

The kinetics of the acid cleavage of the trihydroxobridged complexes with the tridentate amines dien and tach have also been studied (*186*). These systems are very similar to the ammonia system, and the kinetic and thermodynamic data are therefore likely to be interpretable in the same way (Table XXXV). The cleavage of the NH_3, dien, and tach complexes in chloride media has also been studied (*186*). These reactions were found to give dihydroxo-bridged species in two

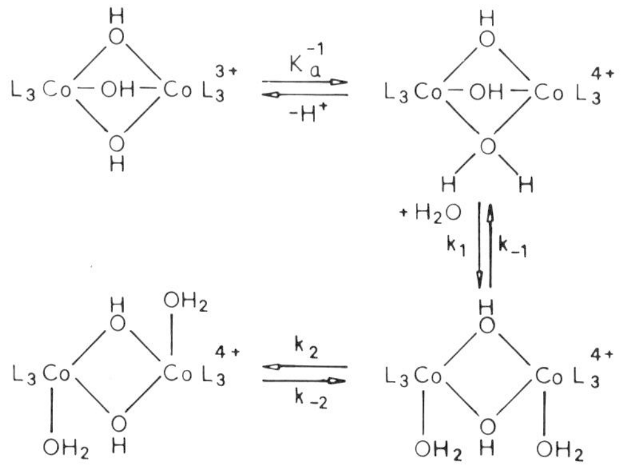

SCHEME 7. The acid cleavage of dinuclear trihydroxo-bridged species is normally followed by a fast cis–trans isomerization process.

TABLE XXXIV

CLEAVAGE OF $(NH_3)_3Co(OH)_3Co(NH_3)_3^{3+}$ AT
25°C IN 1.0 M (Li,H)ClO$_4$ ^{a}

Constant	Values	$\Delta H^\ddagger$ or ΔH^0 (kJ mol^{-1})	$\Delta S^\ddagger$ or ΔS^0 (J mol^{-1} K^{-1})
k_1/K_a	4.22 M^{-1} sec^{-1}	47(8)	−77(28)
k_{-1}	1.80 sec^{-1}	73(7)	3(21)
K_1/K_a	2.34 M^{-1}	−26(11)	−80(35)
k_2	0.186 sec^{-1}	98(11)	69(37)
k_{-2}	0.00662 sec^{-1}	95(6)	32(18)
K_2	27.9	2.5(13)	38(41)

^{a} See Scheme 7. From Refs. *187* and *371*.

kinetically well-separated steps, but substitution by chloride in the dihydroxo-bridged species complicated the study.

Acid hydrolysis of $(tacn)Co(OH)_3Co(tacn)^{3+}$ yields *trans*-(H_2O)-$(tacn)Co(OH)_2Co(tacn)(H_2O)^{4+}$, as confirmed by an X-ray crystal-structure analysis of the perchlorate salt (*185*). The kinetics of the equilibration reactions in Scheme 7 showed only one step for which the observed rate law $k_{obs} = k_r + k_f[H^+]$ was interpreted in terms of a rapid bridge-cleavage equilibration reaction followed by a rate-determining isomerization reaction (*185*). This interpretation, which was preferred to the alternative mechanism involving a rate-determining bridge-cleavage reaction, gave the values for $k_2K_1/K_a = k_f$ and $k_{-2} = k_r$ shown in Table XXXV. Deuterium isotope effects on the kinetic parameters have also been measured, and they support the proposed mechanism. The reduction of the tacn complex by Cr(II) was found to follow a rate law of the form $k_{obs} = a + b[H^+] + c[H^+]\cdot[Cr(II)]$ (*97*).

The data in Table XXXV show that common features for these ammonia and amine complexes are very fast isomerization between the cis and trans isomers of the diaqua species and the fact that the trans diaqua isomers are generally more stable than the cis isomers. In the ammine system the activation parameters for k_2 and k_{-2} are consistent with an isomerization process at cobalt(III), but it is at present not clear how this occurs. It need not be a simple cis–trans isomerization occurring at one of the Co(III) centers, but might involve the participation of both metal centers. The isomerization reaction may proceed via intramolecular proton transfer between a water ligand and one of the two hydroxo bridges with simultaneous bridge cleavage and formation

TABLE XXXV

Bridge Cleavage of $L_3Co(OH)_3CoL_3^{3+}$ Species at 20°C in 1.0 M (Li,H)ClO$_4$

L_3	k_1/K_a (M^{-1} sec^{-1})	k_{-1} (sec^{-1})	K_1/K_a (M^{-1})	k_2 (sec^{-1})	k_{-2} (sec^{-1})	K_2	K_1K_2/K_a [a] (M^{-1})	Reference
(NH$_3$)$_3$	2.65	0.94	2.8	0.09	0.0027	33	93	187
dien	11.4	10.7	1.07	0.119	0.00043	277	296	186
tach	260	12.5	21	0.14	~0.04	~3.5	~74	186
tacn	—	—	—	—[b]	0.0024[c]	—	9	185

[a] $K_1K_2/K_a \ll 1$ for L_3 = metacn and (py)$_3$ (97, 188).
[b] $k_2K_1/K_a = 0.0225(7) \, M^{-1}$ sec^{-1}, $\Delta H^\ddagger - \Delta H^0 = 63(2)$ kJ mol^{-1}, $\Delta S^\ddagger - \Delta S^0 = -63(5)$ J mol^{-1} K^{-1}.
[c] $\Delta H^\ddagger = 100(8)$ kJ mol^{-1}, $\Delta S^\ddagger = 50(25)$ J mol^{-1} K^{-1}.

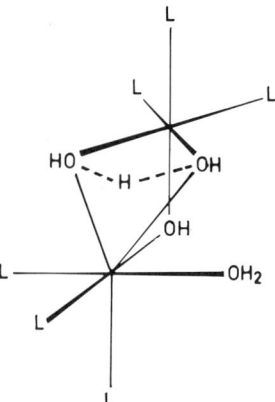

FIG. 17. A possible transition state structure for the "bridge-shift" mechanism for cis–trans equilibria of $XL_3M(OH)_2ML_3Y^{n+}$ species shown for $X = Y = H_2O$.

of the new hydroxo bridge. The transition state structure of this intramolecular "bridge shift" mechanism is shown in Fig. 17. This bridge shift mechanism has first been discussed in relation to isomerization reactions of several polynuclear chromium(III) complexes (118) (see also Table X).

Trihydroxo-bridged complexes with $L_3 = (py)_3$, dpt, and metacn have also been reported and are unusually stable with respect to bridge cleavage (95, 97, 188). The pyridine complex does not react measurably in $HClO_4$ solutions of concentration up to 6 M; in 11 M $HClO_4$ it reacts very rapidly and reversibly forming $(H_2O)(py)_3Co(OH)_2Co(py)_3$-$(H_2O)^{4+}$ (188). The metacn complex does not undergo bridge cleavage in acidic (2 M $HClO_4$) or basic (2 M NaOH) solutions. This difference in behavior relative to the analogous tacn complex has been rationalized in terms of steric hindrance in both isomers of $(H_2O)(metacn)$-$Co(OH)_2Co(metacn)(H_2O)^{4+}$ (repulsion between two methyl groups or a methyl group and a water ligand). The Cr(II) reduction of the metacn complex was found to follow a rate law of the form $k_{obs} = a[Cr(II)]$ (97).

B. CHROMIUM(III)

Only a few trihydroxo-bridged chromium(III) species are known. The $(tacd)Cr(OH)_3Cr(tacd)^{3+}$ and the $(metacn)Cr(OH)_3Cr(metacn)^{3+}$ ions are unusually kinetically and thermodynamically stable in acidic solution, i.e., they do not undergo hydroxo-bridge cleavage in 1 M

$HClO_4$ (97, 98). In strong basic solution the metacn complex deprotonates, giving a di-μ-hydroxo–μ-oxo complex, which rather surprisingly has been reported not to undergo bridge cleavage even in 1 M NaOH. The tacd complex hydrolyzes in basic solution, giving a dihydroxo-bridged species by the mechanism shown in Eq. (63).

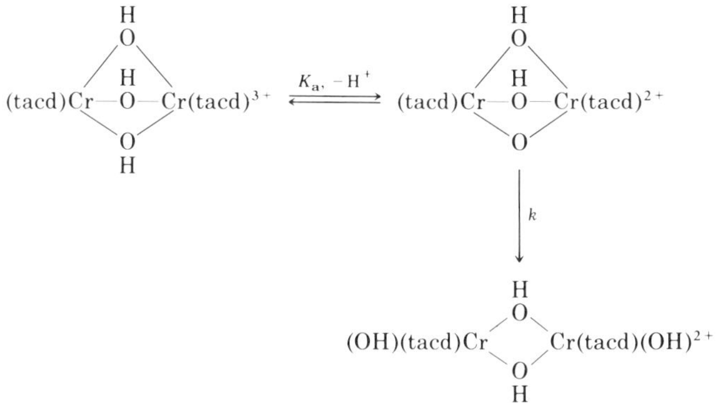

$$(tacd)Cr\underset{\underset{H}{O}}{\overset{\overset{H}{O}}{\diamondsuit}}Cr(tacd)^{3+} \underset{}{\overset{K_a, -H^+}{\rightleftharpoons}} (tacd)Cr\underset{O}{\overset{\overset{H}{O}}{\diamondsuit}}Cr(tacd)^{2+} \xrightarrow{k} (OH)(tacd)Cr\underset{\underset{H}{O}}{\overset{\overset{H}{O}}{\diamondsuit}}Cr(tacd)(OH)^{2+} \quad (63)$$

At 25°C and $I = 0.4$ M [(Li,H)ClO$_4$] the parameters in Eq. (63) are $K_a = 10^{-12.9}$ M and $k = 1.02$ sec^{-1} ($\Delta H^{\ddagger} = 71(4)$ kJ mol^{-1} and $\Delta S^{\ddagger} = -3(12)$ J mol^{-1} K^{-1}) (98). The difference in reactivity between the tacd and metacn complexes has been rationalized on the basis of steric hindrance, as discussed above for the metacn cobalt(III) complex.

C. RHODIUM(III)

Several trihydroxo-bridged rhodium(III) complexes are known, but their hydrolysis reactions have been studied only briefly. The tacn complex undergoes a rapid two-stage reaction in acidic solution to give trans-$(H_2O)(tacn)Rh(OH)_2Rh(tacn)(H_2O)^{4+}$ (76), whereas the metacn complex is stable in acid solution (97).

XII. Bridge Cleavage of Mixed Bridge Complexes

A. HYDROXO BRIDGE CLEAVAGE OF AMIDO-BRIDGED DICOBALT(III) COMPLEXES

1. The μ-Amido–μ-Hydroxo Complex

The hetero-bridged complex $(NH_3)_4Co(NH_2)(OH)Co(NH_3)_4^{4+}$ in acidic solution undergoes hydroxo bridge cleavage much faster than amido bridge cleavage. The bridge-cleavage reactions in acidic solution in the presence of neutral and anionic nucleophiles, X^{n-}, have been studied intensively by Sykes and co-workers (*372–375*). These reactions have led to the preparation of a series of new amido-bridged complexes of the types **IX**, **X**, and **XI**. The reactions with Cl^-, Br^-, SO_4^{2-}, SeO_4^{2-},

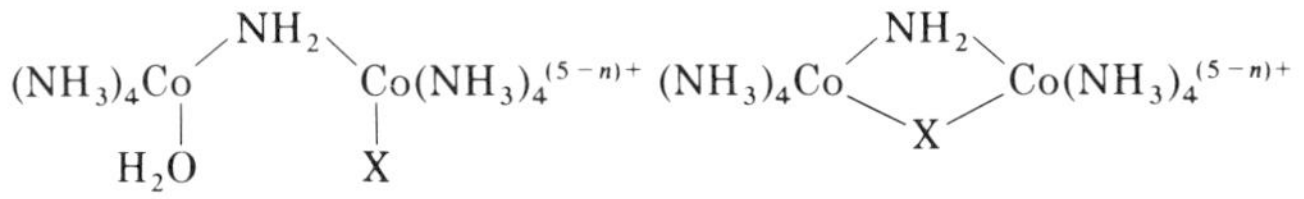

$$(NH_3)_4Co\underset{H_2O\quad X}{\overset{NH_2}{\diagdown\diagup}}Co(NH_3)_4^{(5-n)+} \quad (NH_3)_4Co\underset{X}{\overset{NH_2}{\diagdown\diagup}}Co(NH_3)_4^{(5-n)+}$$

$$\textbf{IX} \hspace{4cm} \textbf{X}$$

$$(NH_3)_4Co\underset{X\quad X}{\overset{NH_2}{\diagdown\diagup}}Co(NH_3)_4^{(5-2n)+}$$

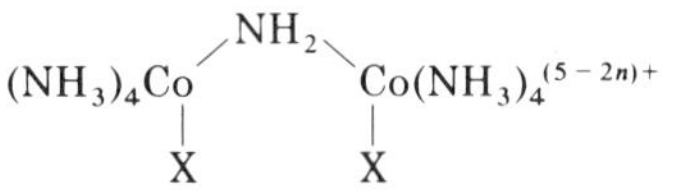

$$\textbf{XI}$$

$H_2PO_4^-$, and CH_3COOH in acidic solution give complexes of type **X**. As shown in Eq. (64), the conversions are reversible and involve intermediate formation of the corresponding singly bridged complexes **IX** (*376–386*).

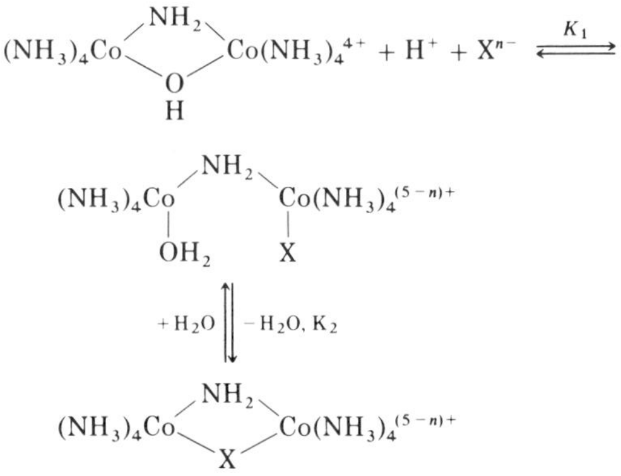

$$(NH_3)_4Co\underset{\underset{H}{O}}{\overset{NH_2}{\diagdown\diagup}}Co(NH_3)_4^{4+} + H^+ + X^{n-} \xrightleftharpoons{K_1}$$

$$(NH_3)_4Co\underset{OH_2\quad X}{\overset{NH_2}{\diagdown\diagup}}Co(NH_3)_4^{(5-n)+}$$

$$+H_2O \updownarrow -H_2O,\ K_2$$

$$(NH_3)_4Co\underset{X}{\overset{NH_2}{\diagdown\diagup}}Co(NH_3)_4^{(5-n)+} \tag{64}$$

The chloro- and bromo-bridged complexes were first prepared by Werner (*7*). Werner originally formulated these complexes as singly bridged complexes (structure **IX**), but the μ-amido–μ-chloro and μ-amido–μ-bromo structures have recently been demonstrated by crystallographic studies (*387, 388*). The equilibrium constant $K_{12} = K_1 \times K_2$ has been determined spectrophotometrically and the formation of structure **X** for $X^{n-} = Br^-$ is less favorable than its formation for the chloro analog (*376, 379*) (Table XXXVI). The equilibrium between

TABLE XXXVI

Hydroxo Bridge-Cleavage Reactions of $(NH_3)_4Co(NH_2)(OH)Co(NH_3)_4^{4+}$ at $25°C$[a]

X^{n-}	$10^4 \times k_0$ (M^{-1} sec^{-1})	$\Delta H^{\ddagger}$ (kJ mol^{-1})	$\Delta S^{\ddagger}$ (J mol^{-1} K^{-1})	$10^4 \times k_1$ (M^{-2} sec^{-1})	$\Delta H^{\ddagger}$ (kJ mol^{-1})	$\Delta S^{\ddagger}$ (J mol^{-1} K^{-1})	$K_{12} = K_1K_2$ (M^{-2})	Reference
SeO_4^{2-}	—	—	—	135	76(1)	−26(4)	265(5)	383
SO_4^{2-}	—	—	—	215	74(1)	−28(1)	2240(50)	382
$H_2PO_4^-$	—	—	—	11.8	82(3)	−27(10)	376(80)	384
$HSeO_4^-$	—	—	—	~2	—	—	—	383
HSO_4^-	—	—	—	1.45	75(1)	−67(3)	—	382
Br^-	1.4	108(8)	45(24)	4.33	72(3)	−68(11)	1.0(3)	379
Cl^-	1.2	105(15)	31(53)	9.6	68(2)	−76(8)	10.6(14)	376
NO_3^-	—	—	—	4.2[b]	57(3)	−119(8)	—	385
NCS^-	0.028	—	—	4.7	73(2)	−62(6)	—	377, 381
H_3PO_4	—	—	—	0.854	60(3)	−122(12)	—	384
CH_3COOH	0.014	111(10)	13(29)	0.065[c]	60(3)	−144(8)	—	386
H_2O	—	—	—	0.81[c,d]	52(5)	−148(15)	—	389

[a] Equation (64). Medium 2 M (Na,H)ClO$_4$, unless otherwise stated.
[b] 2 M NaNO$_3$.
[c] 2 M LiClO$_4$.
[d] Units of M^{-1} sec^{-1}.

structures **IX** and **X** is fast and lies considerably to the right, as shown by ^{1}H NMR measurements and potentiometric measurements ($K_2 > 20$ for both Cl$^-$ and Br$^-$) (378). At very high [Cl$^-$] or [Br$^-$], complexes of type **XI** was formed, although they have not been isolated as crystalline salts (376, 379).

Reaction of the μ-amido–μ-hydroxo complex with excess sulfate gives a quantitative yield of μ-amido–μ-sulfato complex, which can be isolated as salts with various anions (382). Similarly, salts of the μ-amido–μ-selenato species have been isolated (383). The reaction with phosphate yields a μ-amido–μ-phosphate complex, which has been isolated both in its monoprotonated and unprotonated forms (as the nitrate and bromide salts, respectively), and support for the phosphato-bridged structures was provided by spectral (IR/UV) studies (384). For $X^{n-} = SO_4^{2-}$, SeO_4^{2-}, or PO_4^{3-}, the intermediate structure **IX** has not been detected, and presumably never attains a high concentration. For $X^{n-} = SO_4^{2-}$ and SeO_4^{2-}, values for K_{12} have been determined spectrophotometrically (Table XXXVI), and kinetic studies (see below) support these values. For $X^{n-} = H_2PO_4^-$, the value for K_{12} given in Table XXXVI has been deduced from kinetic parameters.

Reaction with NO$_3^-$ yields an aqua nitrato complex (structure **IX**), and $K_1 = 0.022\ M^{-2}$ (25°C, 2 M NaNO$_3$) has been determined spectrophotometrically. At very high nitric acid concentration, evidence for another nitrato complex has been obtained and a μ-amido–μ-nitrato complex is considered more likely than, e.g., a dinitrato complex (385).

Reaction with carboxylic acids leads to formation of μ-amido–μ-carboxylato complexes, and complexes with structure **X** have been isolated for $X^{n-} = CH_3COO^-$ and HCOO$^-$. For equilibrium Eq. (65), a value of $K = 53\ M^{-1}$ (50°C, 2 M LiClO$_4$) has been determined from kinetic data (386).

$$(NH_3)_4Co\underset{OH}{\overset{NH_2}{\diagup\diagdown}}Co(NH_3)_4^{4+} + CH_3COOH \overset{K}{\rightleftharpoons}$$

$$(NH_3)_4Co\underset{O\diagdown_{\underset{CH_3}{C}}\diagup O}{\overset{NH_2}{\diagup\diagdown}}Co(NH_3)_4^{4+} + H_2O \qquad (65)$$

The kinetics of equilibrium Eq. (64) have been studied in detail for each of the nucleophiles mentioned above, i.e., for $X^{n-} = Cl^-$, Br$^-$, SO$_4^{2-}$, SeO$_4^{2-}$, H$_2$PO$_4^-$, NO$_3^-$, and CH$_3$COOH. For solutions containing initially $(NH_3)_4Co(NH_2)(OH)Co(NH_3)_4^{4+}$ and a large excess of the nucleophile X^{n-} a rate law of the form $k_{obs} = k_0[X^{n-}] + k_1[X^{n-}][H^+]$ [forward reaction in Eq. (64)] was observed. Only for $X^{n-} = Cl^-$, Br^-, SCN^-, or CH_3COOH does the uncatalyzed (k_0) path contribute significantly. Values for the rate constants for the uncatalyzed (k_0) and the acid-catalyzed (k_1) bridge-cleavage reactions are listed in Table XXXVI and are discussed later. Kinetic data for the reverse reaction, cleavage of $(NH_3)_4Co(OH)(X)Co(NH_3)_4^{(5-n)+}$, have also been reported, but will not be discussed further.

The reaction with thiocyanate yields a μ-amido–bis(tetraamminethiocyanatocobalt(III) complex (377). The kinetic data are consistent with a dominant reaction path such as in Eq. (66) (377, 381). There is

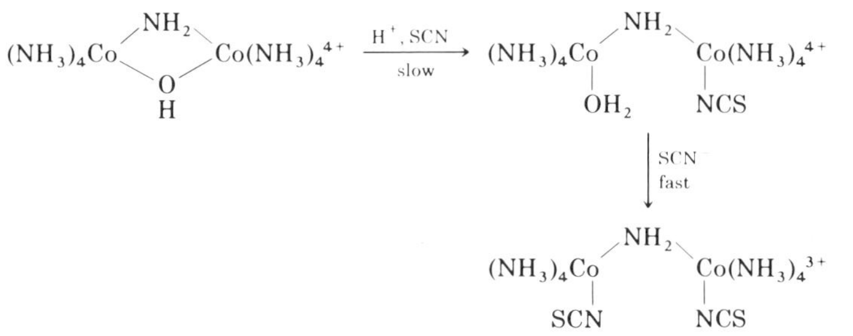

$$(NH_3)_4Co\diagup\mathop{\diagdown}\limits_{\substack{O \\ H}}^{NH_2}Co(NH_3)_4^{4+} \xrightarrow[\text{slow}]{H^+, SCN^-} (NH_3)_4Co\diagup\mathop{\diagdown}\limits^{NH_2}\mathop{\underset{OH_2}{|}}\mathop{\underset{NCS}{|}}Co(NH_3)_4^{4+}$$

$$\downarrow \text{SCN}^- \text{ fast}$$

$$(NH_3)_4Co\diagup\mathop{\diagdown}\limits^{NH_2}\mathop{\underset{SCN}{|}}\mathop{\underset{NCS}{|}}Co(NH_3)_4^{3+} \quad (66)$$

no evidence for buildup of the intermediate aqua thiocyanato complex or of a μ-thiocyanato complex, as would correspond to the results for the analogous reactions with Cl^- or Br^-. At $[H^+] \leq 0.4\ M$, the rate law is of the form $k_{obs} = k_0[NCS^-] + k_1[H^+][NCS^-]$. This corresponds to the dependence mentioned above, and k_0 and k_1 have been interpreted likewise (Table XXXVI). At $[H^+] > 0.4\ M$, k_{obs} becomes nonlinear in $[H^+]$.

The conversion of the μ-amido–μ-hydroxo complex to the diaqua μ-amido complex in Eq. (67) has been difficult to study because the spectra of the two species are similar and because the equilibrium constant, K_1, is too small to be measured in, e.g., $2\ M\ (Na,H)ClO_4$ (389–391).

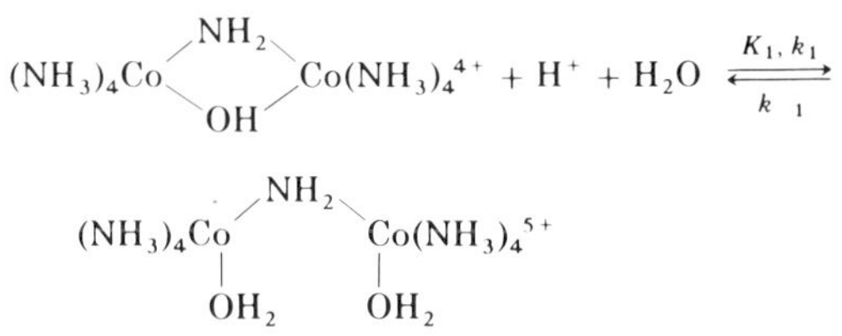

$$(NH_3)_4Co\diagup\mathop{\diagdown}\limits_{OH}^{NH_2}Co(NH_3)_4^{4+} + H^+ + H_2O \underset{k_{-1}}{\overset{K_1,\ k_1}{\rightleftarrows}}$$

$$(NH_3)_4Co\diagup\mathop{\diagdown}\limits^{NH_2}\mathop{\underset{OH_2}{|}}\mathop{\underset{OH_2}{|}}Co(NH_3)_4^{5+} \quad (67)$$

There is, however, spectroscopic evidence for the formation of the diaqua complex in approximately 6 M $HClO_4$, and a value of $K_1 = 0.05(2)$ M^{-1} (1.5°C) has been obtained from spectrophotometric measurements in 0–6 M $HClO_4$ (*391*). A further complication is the fact that ammonia hydrolysis [Eq. (68)] takes place at a rate which

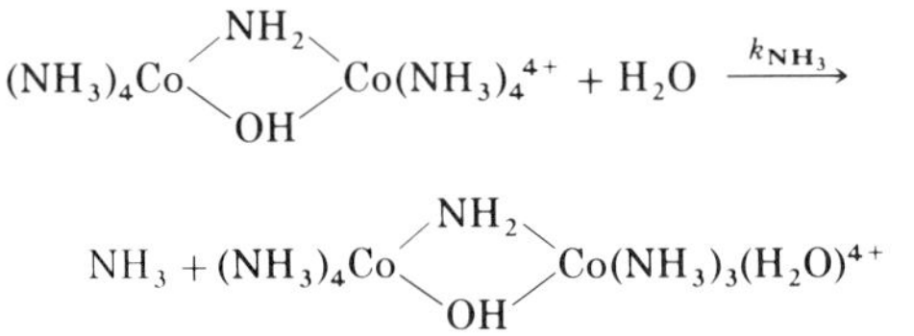

$$(NH_3)_4Co\underset{OH}{\overset{NH_2}{\diagdown\diagup}}Co(NH_3)_4{}^{4+} + H_2O \xrightarrow{k_{NH_3}}$$

$$NH_3 + (NH_3)_4Co\underset{OH}{\overset{NH_2}{\diagdown\diagup}}Co(NH_3)_3(H_2O)^{4+} \quad (68)$$

is comparable to that of acid-catalyzed hydroxo bridge cleavage. Both rate constants, k_1 and k_{NH_3}, have been determined from kinetic studies (*389*) of the vanadium(II) and chromium(II) reductions of the μ-amido–μ-hydroxo complexes in acidic solution, which gave $k_{NH_3} = 0.46 \times 10^{-4}$ sec^{-1} (25°C, 2 M $LiClO_4$) and k_1 as listed in Table XXXVI. Evidence for the formation of an aqua perchlorato as well as a μ-perchlorato complex at very high concentrations of perchloric acid (8–12 M) has been presented (*390*), and further characterization would be of interest.

The rate constants k_1 for acid-catalyzed cleavage of the μ-amido–μ-hydroxo complex given in Table XXXVI fall into three distinct groups, depending on the charge of the ligand X^{n-}. Rate constants (M^{-2} sec^{-1}) for doubly charged anions are about 10^{-2}, for singly charged anions from 10^{-4} to 10^{-3}, and for uncharged nucleophiles from 10^{-5} to 10^{-4}. The various nucleophiles are believed to react via a common mechanism, Eq. (69), involving protonation of the hydroxo bridge followed by rate-determining bridge cleavage and formation of structure **IX** (*372*). An S_N2 mechanism can be excluded since $k_{NCS}/k_{Cl} \sim 0.5$, which is much smaller than the value of about 50 found for S_N2 reactions of Cr^{3+}, V^{3+},

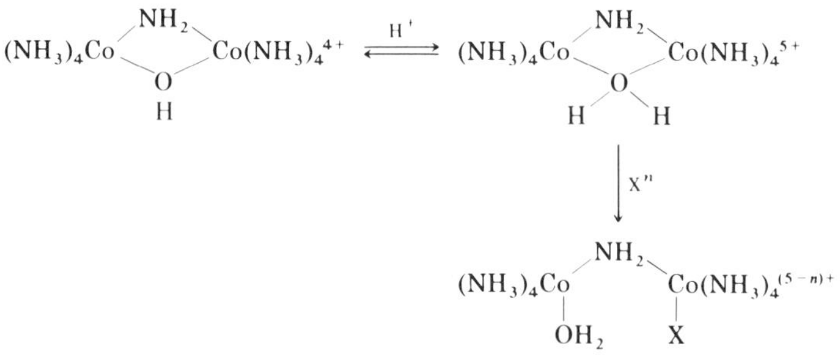

$$(NH_3)_4Co\underset{\underset{H}{O}}{\overset{NH_2}{\diagdown\diagup}}Co(NH_3)_4{}^{4+} \xrightleftharpoons{H^+} (NH_3)_4Co\underset{\underset{H\ \ H}{O}}{\overset{NH_2}{\diagdown\diagup}}Co(NH_3)_4{}^{5+}$$

$$\Big\downarrow X^{n-}$$

$$(NH_3)_4Co\underset{\underset{OH_2}{|}}{\overset{NH_2}{\diagdown\diagup}}\underset{X}{}Co(NH_3)_4{}^{(5-n)+} \quad (69)$$

and Mo^{3+}. The data are therefore more consistent with an S_N1 mechanism, and a mechanism of the "nonlimiting" type was preferred and it was discussed how such a mechanism could rationalize the kinetic data in terms of ion-pair and ion-triplet formation, in spite of the fact that the required less-than-first-order dependence on $[X^{n-}]$ is not observed. A "limiting" S_N1 mechanism was shown to require that a five-coordinated cobalt(III) center captures coordinated water by an intramolecular reaction [Eq. (70a)] much faster than it captures bulk water Eq. (70b)], and this was considered unlikely (*372*). In the light of

$$(NH_3)_4Co\underset{OH_2}{\overset{NH_2}{<}}Co(NH_3)_4{}^{5+} \rightleftarrows \begin{array}{c} (NH_3)_4Co\underset{H}{\overset{NH_2}{<}}\underset{H}{\overset{O}{>}}Co(NH_3)_4{}^{5+} \quad (70a) \\ \\ \underset{H_2O}{\overset{H_2O}{\longrightarrow}} \\ (NH_3)_4Co\underset{H_2O}{\overset{NH_2}{<}}Co(NH_3)_4{}^{5+} \quad (70b) \\ \underset{}{}\underset{OH_2}{} \end{array}$$

more recent studies, however, it could be argued that proximity effects would favor Eq. (70a) considerably. As discussed in Section IX,D, the $(H_2O)(NH_3)_4Rh(OH)Rh(NH_3)_4(H_2O)^{5+}$ ion forms the presumed μ-aqua–μ-hydroxo intermediate at a rate which is comparable to the rate of water exchange in cis-$Rh(NH_3)_4(H_2O)_2{}^{3+}$ (Table XXXIII). Therefore, a "limiting" S_N1 mechanism should not be entirely discounted on the basis of the present data.

2. *The μ-Amido–Di-μ-hydroxo Complex*

Acid cleavage of the hetero-bridged complex $(NH_3)_3Co(NH_2)$-$(OH)_2Co(NH_3)_3{}^{3+}$ yields $(H_2O)(NH_3)_3Co(NH_2)(OH)Co(NH_3)_3(H_2O)^{4+}$, which has been tentatively assigned a trans configuration on the basis of the expected similarity in behavior to the trihydroxo-bridged complexes discussed in Section XI (*259, 371*). The equilibrium constant $K = [(H_2O)Co(NH_2)(OH)Co(H_2O)^{4+}]/[Co(NH_2)(OH)_2\text{-}Co^{3+}][H^+]$ has been determined spectrophotometrically as $K = 62(3)\ M^{-1}$ (1 M LiClO$_4$, 25°C), which is close to the value 65.2 M^{-1} found for $(NH_3)_3Co(OH)_3Co(NH_3)_3{}^{3+}$. The kinetics of the equilibration between the amido-bridged complexes showed only one stage, the rate expression being $k_{obs} = k_a + k_b[H^+]$. In 1 M LiClO$_4$ and at 25°C, the kinetic parameters are $k_a = 1.96 \times 10^{-2}\ \text{sec}^{-1}$ and $k_b = 1.01\ M^{-1}\ \text{sec}^{-1}$ (*371*).

B. Other Complexes

In most other hetero-bridged systems studied the hydroxo bridge is kinetically more stable than the X^{n-} bridge. Base hydrolysis of $(en)_2Cr(OH)(X)Cr(en)_2^{(4-n)+}$ has been reported to give essentially X^{n-} bridge cleavage for $X^{n-} = SO_4^{2-}$, $RCH(NH_2)COO^-$, and CF_3COO^-. A similar result has been reported for the acid hydrolysis for $X^{n-} = CF_3COO^-$ (105, 107, 262–263). The complex $(tacn)Cr(OH)_2(CO_3)Cr(tacn)^{2+}$ and its rhodium(III) analog undergo carbonato bridge cleavage in acidic solution, and the kinetics of these reactions have been reported (100).

The kinetics of formation from and decomposition into Co(II) and O_2 of μ-hydroxo–μ-peroxo dicobalt(III) complexes have been studied in detail, and the mechanistic aspects have been discussed in a recent article by Fallab and Mitchell (119). Two different mechanisms have been invoked to explain the kinetics of the first bridge cleavage of $L_4Co(OH)(O_2)CoL_4^{3+}$ species. One mechanism involves hydroxo bridge cleavage and formation, as shown in Eq. (71). The parameter values

$$L_4Co(III)(OH)(O_2)Co(III)L_4^{3+} + H_2O \underset{k_{-1}}{\overset{k_1}{\rightleftharpoons}} L_4\underset{OH_2}{Co(III)}(O_2)\underset{OH}{Co(III)}L_4^{3+} \quad (71)$$

for $(tren)Co(OH)(O_2)Co(tren)^{3+}$ are $k_1 = 5 \times 10^{-5}$ sec^{-1} and $k_{-1} = 3.25$ sec^{-1} at 25°C (392) and similar values have been reported for other amine systems (119). However, doubt has been expressed concerning this mechanism, since it involves a comparatively fast substitution at a cobalt(III) center. Accordingly, an alternative reaction mechanism has been proposed, involving reductive peroxo bridge cleavage to give a hydroxo-bridged cobalt(II)–cobalt(III) intermediate containing terminally coordinated superoxide $Co(III)(OH)Co(II)O_2$ (119).

Preliminary studies seem to indicate that acid hydrolysis of corresponding rhodium(III) complexes leads to peroxo bridge cleavage rather than hydroxo bridge cleavage (124).

XIII. Concluding Remarks

Structural, thermodynamic, and kinetic studies have shown that hydroxo-bridged polynuclear complexes of chromium(III), cobalt(III), rhodium(III), and iridium(III) have many general features in common. Structurally, the four metal ions exhibit an almost identical pattern, and in particular the occurrence of many well-characterized oligomers of chromium(III) and cobalt(III) shows that the same structures are favored for these two ions. The nature of the nonbridging ligands is of great importance in influencing the formation of a specific oligomer (or isomer).

The thermodynamic and kinetic properties exhibit trends similar to those found for the "parent" mononuclear complexes, but properties specific to the polynuclear complexes have considerable influence. Important factors of the latter kind are high charge, proximity effects, and intramolecular hydrogen bond formation. Intramolecular hydrogen bonds can be formed in several ways and between a variety of acceptor and donor groups. A particularly strong effect has been observed in complexes in which the hydrogen bond is between terminally coordinated hydroxide and water ligands, and it is now well established that interaction of this kind may have a strong impact, e.g., on the aqua-ligand acidity, on the stability of different isomers, and on kinetic parameters.

Only a few thermodynamic studies have been made of intermolecular condensation reactions between mononuclear and relatively simple oligomeric species, and kinetic studies of such reactions have first been reported only quite recently. Charge effects seem to play a dominant role, although another important factor may be the formation of intermolecular hydrogen-bonded cation pairs.

A large body of thermodynamic as well as kinetic data for intramolecular bridge-cleavage and -formation reactions has been accumulated during the last two decades. These reactions are usually very fast and reversible, and in this sense are related to the preequilibria observed during the hydrolysis of bidentate ligands.

Bridge-cleavage and -formation reactions proceed via an uncatalyzed (k_0) (spontaneous) and an acid-catalyzed (k_H) path. It is generally accepted that the acid-catalyzed path involves the formation of a labile aqua-bridged intermediate. The aqua-bridged complex is believed to be a very strong acid (p$K \ll 0$), but it has not yet been identified. The uncatalyzed path is the dominant pathway in all the chromium(III) systems studied until now ($k_H/k_0 \lesssim 1\ M$ for the bridge-cleavage reactions) whereas the acid-catalyzed path dominates in the reactions of the three d^6 metals (k_H/k_0 being typically in the region of $10-10^3\ M$ for the bridge-cleavage reactions). The rate of the first bridge cleavage in mono-, di-, and trihydroxo-bridged complexes, respectively, increases in this order, which may well be a result of increasing strain in the bridge system.

Hydroxo bridge cleavage in the presence of anionic nucleophiles, which is a facile method for the synthesis of mixed bridge complexes, has been studied kinetically [particularly for cobalt(III)] and has been discussed in terms of different reactivities of protonated and unprotonated ion doublets and triplets. Bridge cleavage of complexes of the type $M(OH)(X)M^{(5-n)+}$ may occur by hydroxo bridge cleavage ($X^{n-} = NH_2^-$, O_2^{2-}) or by cleavage of the X^{n-} bridge ($X^{n-} = CO_3^{2-}$, SO_4^{2-}, $RCOO^-$, etc.).

Hydroxo-bridged complexes are of relevance to applied chemistry. One well-known example is the importance of hydroxo-bridged chromium(III) aqua oligomers in the chromium(III) tanning process. Mention should also be made of the application of hydroxo-bridged chromium(III) complexes in coating materials for magnetic toners used in electrostatic printing, in coatings for printing papers to improve their water resistance and printability, and in coatings for paper liners used in gypsum board manufacturing (393–395). Chromium(III) and cobalt(III) hydroxo-bridged oligomers have been used in silicon residue varnishes to improve the drying and hardening process (396, 397).

Finally, it should be noted that this chemistry may have biological relevance. Several metalloenzymes are believed to contain more than one metal ion bound at the active site. One relevant example is the glucose tolerance factor (GTF) which is important for the metabolic degradation of glucose (398–401). GTF is a low-molecular-weight protein which contains chromium(III). Its structure is not known, but it has been suggested that the active site contains a dinuclear chromium(III) complex (401). The fact that hydroxo-bridged dinuclear chromium(III) complexes exhibit reactions which are often very fast compared with those observed for the "parent" mononuclear species seems to support such a proposal.

XIV. Abbreviations for Ligands and Solvents

A. GENERAL

L_n, One or several ligands which coordinate at n sites

X or Y, Uni- or bidentate ligand

B. SPECIFIC

acac, Acetylacetonate
ala, Alaninate

bamp, 2,6-Bis(aminomethyl)pyridine
bipy, 2,2'-Bipyridine
bn, Butanediamine
bispicam, N,N'-Bis(2-pyridylmethyl)amine
bispicen, N,N'-Bis(2-pyridylmethyl)-1,2-ethanediamine
bispicpn, N,N'-Bis(2-pyridylmethyl)-1,2-propanediamine
bispictn, N,N'-Bis(2-pyridylmethyl)-1,3-propanediamine

chlorodipic, 4-Chloro-2,6-pyridinedicarboxylate
chxn, 1,2-Cyclohexanediamine

dien, Diethylenetriamine
dipic, 2,6-Pyridinedicarboxylate (dipicolinate)
DMF, Dimethyl formamide
DMSO, Dimethyl sulfoxide
dmpz, 3,5-Dimethylpyrazole
dpt, Di(3-aminopropyl)amine (dipropylenetriamine)

edda, N,N'-Ethylenediaminediacetate
en, Ethylenediamine (1,2-ethanediamine)

gly, Glycinate

hydroxodipic, 4-Hydroxo-2,6-pyridinedicarboxylate

ibn, Isobutylenediamine (2-methyl-1,2-propanediamine)

mal, Malonate
mepic, 1-(2-Pyridyl)ethylamine
metacn, 1,4,7-Trimethyl-1,4,7-triazacyclononane

nta, Nitrilotriacetate

ox, Oxalate

PBu_3, Tributylphosphine
phen, 1,10-Phenanthroline
pic, 2-Picolylamine [1-(2-pyridyl)methylamine]
pn, Propylenediamine (1,2-propanediamine)
pro, Prolinate
py, Pyridine

salen, N,N'-Ethylenebis(salicylideneiminate)

tacd, 1,5,9-Triazacyclododecane
tach, cis,cis-1,3,5-Cyclohexanetriamine
tacn, 1,4,7-Triazacyclononane
tame, 1,1,1-Tris(aminomethyl)ethane
tn, Trimethylenediamine (1,3-propanediamine)

tpyea, Tris(2-pyrazol-1-ylethyl)amine
trans-[14]diene, 5,7,7,12,14,14-Hexamethyl-1,4,8,11-tetraazacyclotetradeca-4,11-diene
tren, 2,2′,2″-Triaminotriethylamine
trien, Triethylenetetramine

ACKNOWLEDGMENTS

The author is grateful to Dr. Lone Melchior Larsen for her help with the literature search, to Dr. Martin Hancock for revising the English manuscript, and to Drs. Peter Andersen, Jørgen Glerup, and Erik Larsen for valuable comments. Rigmor Jensen and Per Jensen are also thanked for their assistance in the preparation of the manuscript.

REFERENCES

1. Jørgensen, S. M., *J. Prakt. Chem.* **25**, 321 (1882).
2. Jørgensen, S. M., *Z. Anorg. Allg. Chem.* **16**, 184 (1898).
3. Jørgensen, S. M., *J. Prakt. Chem.* **25**, 398 (1882).
4. Jørgensen, S. M., *J. Prakt. Chem.* **45**, 260 (1892).
5. Werner, A., *Ber. Dtsch. Chem. Ges.* **40**, 4834 (1907).
6. Werner, A., *Ber. Dtsch. Chem. Ges.* **40**, 2103 (1907).
7. Werner, A., *Liebigs Ann. Chem.* **375**, 1 (1910).
8. Werner, A., *Liebigs Ann. Chem.* **406**, 261 (1914).
9. Werner, A., *Ber.* **47**, 3087 (1914).
10. Werner, A., *Ber.* **47**, 1961 (1914).
11. Werner, A., *Ber.* **40**, 4817 (1907).
12. Werner, A., *Ber.* **40**, 4434 (1907).
13. Werner, A., *Ber.* **41**, 3879 (1908).
14. Bjerrum, N., "Studies on Basic Chromic Compounds. A Contribution to the Theory of Hydrolysis. Dissertation. Copenhagen, 1908.
15. Bjerrum, N., and Faurholt, C., *Z. Phys. Chem.* **130**, 584 (1927).
16. Pfeiffer, P., *Z. Anorg. Allg. Chem.* **58**, 272 (1908).
17. Pfeiffer, P., *Z. Anorg. Allg. Chem.* **56**, 261 (1907).
18. Dubsky, J. V., *J. Prakt. Chem.* **90**, 61 (1914).
19. Sillén, L. G., *Q. Rev. Chem. Soc.* **13**, 146 (1959).
20. Sillén, L. G., *Pure Applied Chem.* **17**, 55 (1968).
21. Sillén, L. G., *Coord. Chem.* **1**, 491 (1971).
22. Baes, C. F., and Mesmer, R. E., "The Hydrolysis of Cations." Wiley, New York, 1976.
23. Burgess, J., "Metal Ions in Solution." Horwood, Chichester, 1978.
24. Schwarzenbach, G., and Magyar, B., *Helv. Chim. Acta* **45**, 1425 (1962).
25. Ardon, M., and Linnenberg, A., *J. Phys. Chem.* **65**, 1443 (1961).
26. Thompson, M., and Connick, R. E., *Inorg. Chem.* **20**, 2279 (1981).
27. Finholt, J. E., Thompson, M. E., and Connick, R. E., *Inorg. Chem.* **20**, 4151 (1981).
28. Thompson, G., Ph.D. Thesis. Lawrence Radiation Laboratory Report, UCRL-11410, University of California, Berkely, 1964.
29. Laswick, J. A., and Plane, R. A., *J. Am. Chem. Soc.* **81**, 3564 (1959).
30. Ardon, M., and Plane, R. A., *J. Am. Chem. Soc.* **81**, 3197 (1959).

31. Stüntzi, H., and Marty, W., *Inorg. Chem.* **22,** 2145 (1983).
32. Springborg, J., and Toftlund, H., *J. Chem. Soc. Chem. Commun.* 422 (1975).
33. Springborg, J., and Toftlund, H., *Acta Chem. Scand.* **A30,** 171 (1976).
34. Toftlund, H., and Springborg, J., *J. Chem. Soc. Chem. Commun.* 1017 (1976).
35. Christensson, F., Springborg, J., and Toftlund, H., *Acta Chem. Scand.* **A34,** 317 (1980).
36. Christensson, F., and Springborg, J., *Acta Chem. Scand.* **A36,** 21 (1982).
37. Josephsen, J., and Schäffer, C. E., *Acta Chem. Scand.* **24,** 2929 (1970).
38. Veal, J. T., Hatfield, W. E., and Hodgson, D. J., *Acta Crystallogr. Sect. B* **29,** 12 (1973).
39. Scaringe, R. P., Singh, P., Eckberg, R. P., Hatfield, W. E., and Hodgson, D. J., *Inorg. Chem.* **14,** 1127 (1975).
40. Andersen, P., Damhus, T., Pedersen, F., and Petersen, A., *Acta Chem. Scand.* **A38,** 359 (1984).
41. Andersen, P., and Bang, E., *Acta Chem. Scand.* **A40,** 476 (1986).
42. Andersen, P., and Berg, T., *Acta Chem. Scand.* **A32,** 989 (1978).
43. Nakamoto, K., Fujita, J., and Murata, H., *J. Am. Chem. Soc.* **80,** 4817 (1958).
44. Chatt, J., Duncanson, L. A., Gatehouse, B. M., Lewis, J., Nyholm, R. S., Tobe, M. L., Todd, P. F., and Venanzi, L. M., *J. Chem. Soc.* 4073 (1959).
45. Ferraro, J. R., Driver, R., Walker, W. R., and Wozniak, W., *Inorg. Chem.* **6,** 1586 (1967).
46. Hewkin, D. J., and Griffith, W. P., *J. Chem. Soc. (A)* 472 (1966).
47. Hackbusch, W., Rupp, H. H., and Wieghardt, K., *J. Chem. Soc. Dalton Trans.* 2364 (1975).
48. Koine, N., Bianchini, R. J., and Legg, J. I., *Inorg. Chem.* **25,** 2835 (1986).
49. Wieghardt, K., Weiss, J., and Siebert, H., *Z. Anorg. Allg. Chem.* **383,** 151 (1971).
50. Wieghardt, K., Schmidt, W., Endres, H., and Wolfe, C. R., *Chem. Ber.* **112,** 2837 (1979).
51. Larsen, S., Michelsen, K., and Pedersen, E., *Acta Chem. Scand.* **A40,** 63 (1986).
52. Søtofte, I., and Bang, E., *Acta Chem. Scand.* **25,** 1164 (1971).
53. Thewalt, U., *Chem. Ber.* **104,** 2657 (1971).
54. Thewalt, U., and Ernst, J., *Z. Naturforsch.* **30B,** 818 (1975).
55. Yamamoto, Y., and Shimura, Y., *Bull. Chem. Soc. Jpn.* **53,** 2934 (1981).
56. Kudo, T., and Shimura, Y., *Bull. Chem. Soc. Jpn.* **52,** 1648 (1979).
57. Kudo, T., and Shimura, Y., *Bull. Chem. Soc. Jpn.* **52,** 3553 (1979).
58. Kudo, T., and Shimura, Y., *Bull. Chem. Soc. Jpn.* **53,** 1588 (1980).
59. Masuda, I., and Douglas, B. E., *J. Coord. Chem.* **1,** 189 (1971).
60. Andersen, P., and Berg, T., *J. Chem. Soc. Chem. Commun.* 600 (1974).
61. Bang, E., *Acta Chem. Scand.* **A38,** 419 (1984).
62. Flood, M. T., Marsh, R. E., and Gray, H. B., *J. Am. Chem. Soc.* **91,** 193 (1969).
63. Veal, J. T., Jeter, D. Y., Hempel, J. C., Eckberg, R. P., Hatfield, W. E., and Hodgson, D. J., *Inorg. Chem.* **12,** 2928 (1973).
64. Cline, S. J., Glerup, J., Hodgson, D. J., Jensen, G. S., and Pedersen, E., *Inorg. Chem.* **20,** 2229 (1981).
65. Hodgson, D. J., and Pedersen, E., *Inorg. Chem.* **19,** 3116 (1980).
66. Kaas, K., *Acta Crystallogr. Sect. B* **35,** 1603 (1979).
67. Galsbøl, F., Larsen, S., Rasmussen, B., and Springborg, J., *Inorg. Chem.* **25,** 290 (1986).
68. Engel, P., and Güdel, H. U., *Inorg. Chem.* **16,** 1589 (1977).
69. Cline, S. J., Hodgson, D. J., Kallesøe, S., Larsen, S., and Pedersen, E., *Inorg. Chem.* **22,** 637 (1983).
70. Kaas, K., *Acta Crystallogr. Sect. B* **32,** 2021 (1976).
71. Veal, J. T., Hatfield, W. E., Jeter, D. Y., Hempel, J. C., and Hodgson, D. J., *Inorg. Chem.* **12,** 342 (1973).

72. Hodgson, D. J., Veal, J. T., Hatfield, W. E., Jeter, D. Y., and Hempel, J. C., *J. Coord. Chem.* **2**, 1 (1972).
73. Vannerberg, N. G., *Acta Chem. Scand.* **17**, 85 (1963).
74. Prout, C. K., *J. Chem. Soc.* 4429 (1962).
75. Thewalt, U., and Zehnder, M., *Helv. Chim. Acta* **60**, 2000 (1977).
76. Wieghardt, K., Schmidt, W., Nuber, B., Prikner, B., and Weiss, J., *Chem. Ber.* **113**, 36 (1980).
77. Oki, H., and Yoneda, H., *Inorg. Chem.* **20**, 3875 (1981).
78. Ranger, G., and Beauchamp, A. L., *Acta Crystallogr. Sect. B* **37**, 1063 (1981).
79. Ou, C.-C., Borowski, W. J., Potenza, J. A., and Schugar, H. J., *Acta Crystallogr. Sect. B* **33**, 3246 (1977).
80. Cline, S. J., Kallesøe, S., Pedersen, E., and Hodgson, D. J., *Inorg. Chem.* **18**, 796 (1979).
81. Srdanov, G., Herak, R., Radanović, D. J., and Veselinović, D. S., *Inorg. Chim. Acta* **38**, 37 (1980).
82. Lethbridge, J. W., *J. Chem. Soc. Dalton Trans.* 2039 (1980).
83. Tanaka, I., Jiu-No, N., Kushida, T., Tsutsui, N., Ashida, T., Suzuki, H., Sakurai, H., Moro-Oka, Y., and Ikawa, T., *Bull. Chem. Soc. Jpn.* **56**, 657 (1983).
84. Larsen, S., and Sørensen, B., *Acta Chem. Scand.* **A35**, 105 (1981).
85. Heinrichs, M. A., Hodgson, D. J., Michelsen, K., and Pedersen, E., *Inorg. Chem.* **23**, 3174 (1984).
86. Fischer, H. R., Hodgson, D. J., Michelsen, K., and Pedersen, E., *Inorg. Chim. Acta* **88**, 143 (1984).
87. Andersen, P., Nielsen, K. M., and Petersen, A., *Acta Chem. Scand.* **A38**, 593 (1984).
88. Baur, W. H., and Wieghardt, K., *J. Chem. Soc. Dalton Trans.* 2669 (1973).
89. Oro, L. A., Carmona, D., Lamata, M. P., Apreda, M. C., Foces-Foces, C., Cano, F. H., and Maitlis, P. M., *J. Chem. Soc. Dalton Trans.* 1823 (1984).
90. Scaringe, R. P., Hatfield, W. E., and Hodgson, D. J., *Inorg. Chim. Acta* **22**, 175 (1977).
91. Hodgson, D. J., and Pedersen, E., *Inorg. Chem.* **23**, 2363 (1984).
92. Cline, S. J., Scaringe, R. P., Hatfield, W. E., and Hodgson, D. J., *J. Chem. Soc. Dalton Trans.* 1662 (1977).
93. Beutler, A., Güdel, H. U., Snellgrove, T. R., Chapuis, G., and Schenk, K. J., *J. Chem. Soc. Dalton Trans.* 983 (1979).
94. Thewalt, U., *Z. Anorg. Allg. Chem.* **412**, 29 (1975).
95. Searle, G. H., and Hambley, T. W., *Aust. J. Chem.* **35**, 1297 (1982).
96. Nutton, A., Bailey, P. M., and Maitlis, P. M., *J. Chem. Soc. Dalton Trans.* 1997 (1981).
97. Wieghardt, K., Chaudhuri, P., Nuber, B., and Weiss, J., *Inorg. Chem.* **21**, 3086 (1982).
98. Wieghardt, K., Guttmann, M., Ventur, D., and Gebert, W., *Z. Anorg. Allg. Chem.* **527**, 33 (1985).
99. Andersen, P., *Acta Chem. Scand.* **21**, 243 (1967).
100. Wieghardt, K., Schmidt, W., Eldik, R., Nuber, B., and Weiss, J., *Inorg. Chem.* **19**, 2922 (1980).
101. Churchill, M. R., Lashewycz, R. A., Koshy, K., and Dasgupta, T. P., *Inorg. Chem.* **20**, 376 (1981).
102. Mandel, G. S., Marsh, R. E., Schaefer, W. P., Mandel, N. S., and Wang, B.-C. *Acta Crystallogr. Sect. B* **33**, 3185 (1977).
103. Schaefer, W. P., *Acta Crystallogr. Sect. C* **39**, 1610 (1983).
104. Wieghardt, K., and Maas, G., *Z. Anorg. Allg. Chem.* **385**, 289 (1971).
105. Kaas, K., and Springborg, J., *Acta Chem. Scand.* Submitted.
106. Kaas, K., *Acta Crystallogr. Sect. B* **35**, 596 (1979).
107. Kaas, K., and Springborg, J., *Acta Chem. Scand.* **A40**, 515 (1986).

108. Fallab, S., Zehnder, M., and Thewalt, U., *Helv. Chim. Acta* **63**, 1491 (1980).
109. Thewalt, U., and Struckmeier, G., *Z. Anorg. Allg. Chem.* **419**, 163 (1976).
110. Churchill, M. R., Harris, G. M., Lashewycz, R. A., and Dasgupta, T. P., *Inorg. Chem.* **18**, 2290 (1979).
111. Michelsen, K., Pedersen, E., Wilson, S. R., and Hodgson, D. J., *Inorg. Chim. Acta* **63**, 141 (1982).
112. Yevitz, M., and Stanko, J. A., *J. Am. Chem. Soc.* **93**, 1512 (1971).
113. Urushiyama, A., Numora, T., and Nakahara, M., *Bull. Chem. Soc. Jpn.* **43**, 3971 (1970).
114. Vaira, M. D., and Mani, F., *Inorg. Chem.* **23**, 409 (1984).
115. Michelsen, K., *Acta Chem. Scand.* **A30**, 521 (1976).
116. Michelsen, K., and Pedersen, E., *Acta Chem. Scand.* **A32**, 847 (1978).
117. Thewalt, U., Jensen, K. A., and Schäffer, C. E., *Inorg. Chem.* **11**, 2129 (1972).
118. Mønsted, L., Mønsted, O., and Springborg, J., *Inorg. Chem.* **24**, 3496 (1985).
119. Fallab, S., and Mitchell, P. R., *Adv. Inorg. Bioinorg. Mech.* **3**, 311 (1984).
120. Lever, A. B. P., and Gray, H. B., *Acc. Chem. Res.* **11**, 348 (1978).
121. Springborg, J., and Zehnder, M., *Helv. Chim. Acta* **69**, 199 (1986).
122. Springborg, J., and Zehnder, M., in preparation.
123. Springborg, J., and Zehnder, M., *Acta Chem. Scand.* **A41**, 484 (1987).
124. Springborg, J., and Zehnder, M., *Helv. Chim. Acta* **67**, 2218 (1984).
125. Glerup, J., *Acta Chem. Scand.* **26**, 3775 (1972).
126. Glerup, J., Hodgson, D. J., and Pedersen, E., *Acta Chem. Scand.* **A37**, 161 (1983).
127. Güdel, H. U., *Comments Inorg. Chem. A* **3**, 189 (1984).
128. Hodgson, D. J., *NATO ASI Ser. C* **140**, 497 (1985).
129. Glerup, J., personal communication.
130. Siebert, H., and Feuerhake, H., *Chem. Ber.* **102**, 2951 (1969).
131. Springborg, J., and Schäffer, C. E., *Inorg. Synth.* **18**, 75 (1978).
132. Lindhard, M., and Siebert, H., *Z. Anorg. Allg. Chem.* **364**, 24 (1969).
133. Christensson, F., and Springborg, J., *Inorg. Chem.* **24**, 2129 (1985).
134. Pedersen, E., *Acta Chem. Scand.* **26**, 333 (1972).
135. Larsen, E., and La Mar, G. N., *J. Chem. Educ.* **51**, 633 (1974).
136. Damhus, T., *Mol. Phys.* **50**, 497 (1983).
137. Bolster, D. E., Gütlich, P., Hatfield, W. E., Kremer, S., Müller, E. W., and Wieghardt, K., *Inorg. Chem.* **22**, 1725 (1983).
138. Kremer, S., *Inorg. Chem.* **24**, 887 (1985).
139. Güdel, H. U., Furrer, A., Bührer, W., and Hälg, B., *Surf. Sci.* **106**, 432 (1981).
140. Furrer, A., and Güdel, H. U., *Phys. Rev. Lett.* **39**, 657 (1977).
141. Güdel, H. U., and Furrer, A. *Mol. Phys.* **33**, 1335 (1977).
142. Schläfer, H. L., Martin, M., Gausmann, H., and Schmidtke, H.-H., *Z. Phys. Chem. (Frankfurt am Main)* **76**, 61 (1971).
143. Flood, M. T., Barraclough, G. G., and Gray, H. B., *Inorg. Chem.* **8**, 1855 (1969).
144. Iwashita, T., Idogaki, T., and Uryû, N., *J. Phys. Soc. Jpn.* **30**, 1587 (1971).
145. Güdel, H. U., and Hauser, U., *J. Solid State Chem.* **35**, 230 (1980).
146. Güdel, H. U., and Hauser, U., *Inorg. Chem.* **19**, 1325 (1980).
147. Damhus, T., and Pedersen, E., *Inorg. Chem.* **23**, 695 (1984).
148. Güdel, H. U., Hauser, U., and Furrer, A., *Inorg. Chem.* **18**, 2730 (1979).
149. Morita, M., *J. Lumin.* **6**, 414 (1973).
150. Güdel, H. U., Furrer, A., and Murani, A., *J. Magn. Magn. Mater.* **15**, 383 (1980).
151. Sorai, M., and Seki, S., *J. Phys. Soc. Jpn.* **32**, 382 (1972).
152. Michelsen, K., and Pedersen, E., *Acta Chem. Scand.* **A37**, 141 (1983).

153. Kaizaki, S., Hidaki, J., and Shimura, Y., *Inorg. Chem.* **12**, 135 (1973).
154. Yamamoto, Y., and Shimura, Y., *Bull. Chem. Soc. Jpn.* **54**, 3351 (1981).
155. Kahi, S., Yamanari, K., and Shimura, Y., *Bull. Chem. Soc. Jpn.* **56**, 2276 (1983).
156. Mason, S. F., and Wood, J. W., *J. Chem. Soc. Chem. Commun.* 209 (1967).
157. Michelsen, K., *Acta Chem. Scand.* **A31**, 429 (1977).
158. Tauzher, G., and Costa, G., *J. Inorg. Nucl. Chem.* **34**, 2676 (1972).
159. Wong, C. F. C., and Kirk, A. D., *Can. J. Chem.* **53**, 3388 (1975).
160. Poulsen, K. G., and Garner, C. S., *J. Am. Chem. Soc.* **81**, 2615 (1959).
161. Hancock, M. P., Josephsen, J., and Schäffer, C. E., *Acta Chem. Scand.* **A30**, 79 (1976).
162. Michelsen, K., personal communication.
163. Fujii, Y., Kyuno, E., and Tsuchiya, R., *Bull. Chem. Soc. Jpn.* **42**, 1569 (1969).
164. Oki, H., and Otsuka, K., *Bull. Chem. Soc. Jpn.* **49**, 1841 (1976).
165. Oki, H., *Bull. Chem. Soc. Jpn.* **50**, 680 (1977).
166. Oki, H., *J. Sci. Hiroshima Univ., Ser. A* **45**, 447 (1982).
167. Eduok, E. E., Owens, J. W., and O'Connor, C. J., *Polyhedron* **3**, 17 (1984).
168. Hoggard, P. E., *Inorg. Chem.* **20**, 415 (1981).
169. Gillard, R. D., Laurie, S. H., Price, D. C., Phipps, D. A., and Weick, C. F., *J. Chem. Soc. Dalton Trans.* 1385 (1974).
170. Scaringe, R. P., Hatfield, W. E., and Hodgson, D. J., *Inorg. Chem.* **16**, 1600 (1977).
171. Ricard, A., and Souchay, P., *Rev. Chim. Min.* **8**, 859 (1971).
172. Chesick, J. P., and Doany, F., *Acta Crystallogr. Sect. B* **37**, 1076 (1981).
173. West, R., and Niu, H. Y., *J. Am. Chem. Soc.* **85**, 2589 (1963).
174. Schwarzenbach, G., Boesch, J., and Egli, H., *J. Inorg. Nucl. Chem.* **33**, 2141 (1971).
175. Rastogi, D. K., Sahni, S. K., Rana, V. B., and Dua, S. K., *J. Coord. Chem.* **8**, 97 (1978).
176. Palmer, W. G., "Experimental Inorganic Chemistry," p. 550. Cambridge Univ. Press, London. 1954.
177. Melcon, D. R., and Harris, G. M., *Inorg. Chem.* **16**, 434 (1977).
178. Thacker, M. A., and Higginson, W. C. E., *J. Chem. Soc. Dalton Trans.* 704 (1975).
179. Mori, M., Shibati, M., Kyuno, E., and Okubo, Y., *Bull. Chem. Soc. Jpn.* **31**, 940 (1958).
180. Uehara, A., Kyuno, E., and Tsuchiya, R., *Bull. Chem. Soc. Jpn.* **44**, 1552 (1971).
181. Inskeep, R. G., and Bjerrum, J., *Acta Chem. Scand.* **15**, 62 (1961).
182. House, D. A., and Garner, C. S., *J. Inorg. Nucl. Chem.* **28**, 904 (1966).
183. Mason, S. F., and Wood, J. W., *J. Chem. Soc. Chem. Commun.* 1512 (1968).
184. Mason, S. F., *Q. Rev. Chem. Soc.* **17**, 20 (1963).
185. Wieghardt, K., Schmidt, W., Nuber, B., and Weiss, J., *Chem. Ber.* **112**, 2220 (1979).
186. Kähler, H. C., Geier, G., and Schwarzenbach, G., *Helv. Chim. Acta* **57**, 802 (1974).
187. Jentsch, W., Schmidt, W., Sykes, A. G., and Wieghardt, K., *Inorg. Chem.* **16**, 1935 (1977).
188. Laier, T., and Springborg, J., *Acta Chem. Scand.* **A35**, 145 (1981).
189. Schäffer, C. E., and Andersen, P., *Proc. Symp. Theory Struct. Complex Compounds, Wrocław, Poland, 1962* p. 571.
190. White, C., Oliver, A. J., and Maitlis, P. M., *J. Chem. Soc. Dalton Trans.* 1901 (1973).
191. Maitlis, P. M., *Chem. Soc. Rev.* **10**, 1 (1981).
192. Nutton, A., and Maitlis, P. M., *J. Chem. Soc. Dalton Trans.* 2335 (1981).
193. Bertram, H., and Wieghardt, K., *Angew. Chem.* **90**, 218 (1978).
194. Neves. A., Herrman, W., and Wieghardt, K., *J. Am. Chem. Soc.* **106**, 5532 (1984).
195. Andersen, P., Berg, T., and Jacobsen, J., *Acta Chem. Scand.* **A29**, 381 (1975).
196. Andersen, P., Berg, T., and Jacobsen, J., *Acta Chem. Scand.* **A29**, 599 (1975).
197. Kolaczkowski, R. W., and Plane, R. A., *Inorg. Chem.* **3**, 322 (1964).
198. Ardon, M., and Stein, G., *J. Chem. Soc.* 2095 (1956).

199. Meyenburg, U., Siroký, O., and Schwarzenbach, G., *Helv. Chim. Acta.* **56,** 1099 (1973).
200. Stüntzi, H., Rotzinger, F. P., and Marty, W., *Inorg. Chem.* **23,** 2160 (1984).
201. Rotzinger, F. P., Stüntzi, H., and Marty, W., *Inorg. Chem.* **25,** 489 (1986).
202. Spiccia, L., and Marty, W., *Inorg. Chem.* **25,** 266 (1986).
203. Olivier, D., *C. R. Acad. Sci. Paris, Ser. C*, **264,** 1176 (1967).
204. Johnston, R. F., and Holwerda, R. A., *Inorg. Chem.* **22,** 2942 (1983).
205. Holwerda, R. A., and Petersen, J. S., *Inorg. Chem.* **19,** 1775 (1980).
206. Johnston, R. F., and Holwerda, R. A., *Inorg. Chem.* **24,** 153 (1985).
207. Johnston, R. F., and Holwerda, R. A., *Inorg. Chem.* **24,** 3176 (1985).
208. Johnston, R. F., and Holwerda, R. A., *Inorg. Chem.* **24,** 3181 (1985).
209. Ostrich, I., and Leffler, A. J., *Inorg. Chem.* **22,** 921 (1983).
210. Kauffman, G. B., and Pinnell, R. P., *Inorg. Synth.* **6,** 176 (1960).
211. Balt, S., and Kieviet, W., *Inorg. Chem.* **11,** 2251 (1972).
212. Fujii, S., Shibahara, T., and Mori, M., *Chem. Lett.* 1149 (1978).
213. Cannon, R. D., and Benjarvongkulchai, S., *J. Chem. Soc. Dalton Trans.* 1924 (1981).
214. Springborg, J., and Schäffer, C. E., *Inorg. Chem.* **15,** 1744 (1976).
215. Springborg, J., and Schäffer, C. E., *Acta Chem. Scand.* **A30,** 787 (1976).
216. Laier, T., Nielsen, B., and Springborg, J., *Acta Chem. Scand.* **A36,** 91 (1982).
217. Hancock, M., Nielsen, B., and Springborg, J., *Acta Chem. Scand.* **A36,** 313 (1982).
218. Hancock, M., Nielsen, B., and Springborg, J., *Inorg. Synth.* **24,** 220 (1986).
219. Hancock, M. P., *Acta Chem. Scand.* **A33,** 499 (1979).
220. Springborg, J., unpublished results.
221. Nambiar, P. R., and Wendtlandt, W. W., *Thermochim. Acta* **20,** 417 (1977).
222. Nambiar, P. R., and Wendtlandt, W. W., *Thermochim. Acta* **20,** 422 (1977).
223. Wendtlandt, W. W., and Fischer, J. K., *J. Inorg. Nucl. Chem.* **24,** 1685 (1962).
224. Tsuchiya, R., Uehara, A., and Kyuno, E., *Bull. Chem. Soc. Jpn.* **46,** 3737 (1973).
225. Wilmarth, W. K., Graff, H., and Gustin, S. T., *J. Am. Chem. Soc.* **78,** 2683 (1956).
226. Joyner, T. B., and Wilmarth, W. K., *J. Am. Chem. Soc.* **83,** 516 (1961).
227. Lindhard, M., and Weigel, M., *Z. Anorg. Allg. Chem.* **299,** 15 (1959).
228. Liston, D. J., Kennedy, B. J., Murray, K. S., and West, B. O., *Inorg. Chem.* **24,** 1561 (1985).
229. Liston, D. J., and West, B. O., *Inorg. Chem.* **24,** 1568, (1985).
230. Lancashire, R., and Smith, T. D., *J. Chem. Soc. Dalton Trans.* 693 (1982).
231. Yang, C.-H., and Grieb, M. W., *Inorg. Chem.* **12,** 663 (1973).
232. McLendon, G., MacMillan, D. T., Harihavan, M., and Martell, A. E., *Inorg. Chem.* **14,** 2322 (1975).
233. McLendon, G., Motekaitis, R. J., and Martell, A. E., *Inorg. Chem.* **14,** 1993 (1975).
234. Lawrance, G. A., and Lay, P. A., *J. Inorg. Nucl. Chem.* **41,** 301 (1979).
235. Zuberbühler, A., Kaden, Th., and Koechlin, F., *Helv. Chim. Acta* **54,** 1502 (1971).
236. Zehnder, M., and Fallab, S., *Helv. Chim. Acta* **55,** 1691 (1972).
237. Braun-Steinle, D., Mäcke, H., and Fallab, S., *Helv. Chim. Acta* **59,** 2032 (1976).
238. Zehnder, M., Mäcke, H., and Fallab, S., *Helv. Chim. Acta* **58,** 2306 (1975).
239. Zehnder, M., and Fallab, S., *Helv. Chim. Acta* **58,** 13 (1975).
240. Zehnder, M., and Fallab, S., *Helv. Chim. Acta* **58,** 2312 (1975).
241. Exnar, I., and Mäcke, H., *Helv. Chim. Acta* **60,** 2504 (1977).
242. Mäcke, H., Zehnder, M., Thewalt, U., and Fallab, S., *Helv. Chim. Acta* **62** 1804 (1979).
243. Zehnder, M., Thewalt, U., and Fallab, S., *Helv. Chim. Acta* **59,** 2290 (1976).
244. Davies, R., Mori, M., Sykes, A. G., and Weil, J. A., *Inorg. Synth.* **12,** 197 (1970).
245. Dixon, D. A., Marsh, R. E., and Schaefer, W. P., *Acta Crystallogr. Sect. B* **34,** 807 (1978).

246. Goodwin, H. A., Cyarfas, E. C., and Mellor, D. P., *Aust. J. Chem.* **11**, 426 (1958).
247. Kern, R. D., and Wentworth, R. A. D., *Inorg. Chem.* **6**, 1018 (1967).
248. Boucher, L. J., and Herrington, D. R., *J. Inorg. Nucl. Chem.* **33**, 4349 (1971).
249. Nishide, T., and Saito, K., *Bull. Chem. Soc. Jpn.* **50**, 2618 (1977).
250. Brigando, J., and Colaitis, D., *Bull. Soc. Chim. Fr.* 3440, 3445, 3449, 3453 (1969).
251. Castillo-Blum, S. E., Richens, D. T., and Sykes, A. G., *J. Chem. Soc. Chem. Commun.* 1120 (1986).
252. Hoppenjans, D. W., and Hunt, J. B., *Inorg. Chem.* **8**, 505 (1969).
253. Brauer, G., "Handbuch der Präparativen Anorganischen Chemie" Vol. 2, pp. 1191. Enke, Stuttgart 1962.
254. Po, H. N., Chung, Y.-H., and Davis, S. R., *J. Inorg. Nucl. Chem.* **35**, 2849 (1973).
255. Po, H. N., Chung, Y.-H., Davis, S. R., and Enomoto, H., *J. Inorg. Nucl. Chem.* **36**, 1349 (1974).
256. Schwarzenbach, G., and Magyar, B., *Helv. Chim. Acta* **45**, 1454 (1962).
257. Mori, M., Ueshiba, S., and Zamatera, H., *Bull. Chem. Soc. Jpn.* **32**, 88 (1959).
258. Matts, T. C., Moore, P., Ogilvie, D. M. W., and Winterton, N., *J. Chem. Soc. Dalton Trans.* 992 (1973).
259. Taylor, R. S., and Sykes, A. G., *J. Chem. Soc.* (A) 1426 (1971).
260. Springborg, J., *Acta Chem. Scand.* Submitted.
261. Kaas, K., and Springborg, J., in preparation.
262. Springborg, J., *Acta Chem. Scand.* **A32**, 231 (1978).
263. Springborg, J., and Toftlund, H., *Acta Chem. Scand.* **A33**, 31 (1979).
264. Galsbøl, F., Rasmussen, B., and Springborg, J., in preparation.
265. Siebert, H., und Wittke, G., *Z. Anorg. Allg. Chem.* **406**, 282 (1974).
266. Siebert, H., Tremmel, G., and Kramer, V., *Z. Anorg. Allg. Chem.* **383**, 158 (1971).
267. Bertram, H., and Wieghardt, K., *Chem. Ber.* **111**, 832 (1978).
268. Hery, M., and Wieghardt, K., *J. Chem. Soc. Dalton Trans.* 1536 (1976).
269. Kipling, B., Wieghardt, K., Hery, M., and Sykes, A. G., *J. Chem. Soc. Dalton Trans.* 2176 (1976).
270. Hyde, M. R., Scott, K. L., Wieghardt, K., and Sykes, A. G., *J. Chem. Soc. Dalton Trans.* 153 (1976).
271. Wharton, R. K., and Wieghardt, K., *Z. Anorg. Allg. Chem.* **425**, 145 (1976).
272. Hery, M., and Wieghardt, K., *Inorg. Chem.* **15**, 2315 (1976).
273. Spiecker, H., and Wieghardt, K., *Inorg. Chem.* **16**, 1290 (1977).
274. Bertram, H., and Wieghardt, K., *Inorg. Chem.* **18**, 1799 (1979).
275. Srinivasan, V. S., Singh, A. N., Wieghardt, K., Rajasekar, N., and Gould, E. S., *Inorg. Chem.* **21**, 2531 (1982).
276. Scott, K. L., Wieghardt, K., and Sykes, A. G., *Inorg. Chem.* **12**, 655 (1973).
277. Wieghardt, K., *J. Chem. Soc. Dalton Trans.* 2548 (1973).
278. Wieghardt, K., und Siebert, H., *Z. Anorg. Allg. Chem.* **374**, 186 (1970).
279. Wieghardt, K., *Z. Anorg. Allg. Chem.* **391**, 142 (1972).
280. Siebert, H., und Tremmel, G., *Z. Anorg. Allg. Chem.* **390**, 292 (1972).
281. Leupin, P., Sykes, A. G., and Wieghardt, K., *Inorg. Chem.* **22**, 1253 (1983).
282. Wieghardt, K., *Z. Naturforsch.* **29B**, 809 (1974).
283. Scott, K. L., and Sykes, A. G., *J. Chem. Soc. Dalton Trans.* 736 (1973).
284. Hyde, M. R., Wieghardt, K., and Sykes, A. G., *J. Chem. Soc. Dalton Trans.* 690 (1976).
285. Spiecker, H., and Wieghardt, K., *Inorg. Chem.* **15**, 909 (1976).
286. Hery, M., and Wieghardt, K., *Inorg. Chem.* **16**, 1287 (1977).
287. Wieghardt, K., Cohen, H., and Meyerstein, D., *Angew. Chem.* **90**, 632 (1978).
288. Bertram, H., Bölsing, E., Spiecker, H., and Wieghardt, K., *Inorg. Chem.* **17**, 221 (1978).

289. Hery, M., and Wieghardt, K., *Inorg. Chem.* **17**, 1130 (1978).
290. Huck, H.-M., and Wieghardt, K., *Inorg. Chem.* **19**, 3688 (1980).
291. Beitz, J. V., Miller, J. R., Cohen, H., Wieghardt, K., and Meyerstein, D., *Inorg. Chem.* **19**, 966 (1980).
292. Srinivasan, V. S., Rajasekar, N., Singh, A. N., Radlowski, C. A., and Heh, J. C.-K., *Inorg. Chem.* **21**, 2824 (1982).
293. Srinivasan, V. S., and Gould, E. S., *Inorg. Chem.* **21**, 3854 (1982).
294. Cohen, H., Efrima, S., Meyerstein, D., Nutkovich, M., and Wieghardt, K., *Inorg. Chem.* **22**, 688 (1983).
295. Hollaway, W. F., Srinivasan, V. S., and Gould, E. S., *Inorg. Chem.* **23**, 2181 (1984).
296. Wieghardt, K., Cohen, H., and Meyerstein, D., *Ber. Bunsenges. Phys. Chem.* **82**, 388 (1978).
297. Cohen, H., Nutkovich, M., Meyerstein, D., and Wieghardt, K., *J. Chem. Soc. Dalton Trans.* 943 (1982).
298. Baldea, I., Wieghardt, K., and Sykes, A. G., *J. Chem. Soc. Dalton Trans.* 78 (1977).
299. Thornton, A. T., Wieghardt, K., and Sykes, A. G., *J. Chem. Soc. Dalton Trans.* 147 (1976).
300. Gustavson, K. H., "The Chemistry of Tanning Processes." Academic Press, New York, 1956.
301. "Ullmanns Encyclopädie der Technischen Chemie," 4th Ed., Vol. 16. Verlag Chemie, Weinheim, 1978.
302. Stiasny, E. "Gerbereichemie." Steinkopff, Dresden und Leipzig, 1931.
303. Riess, W., *Leder* **16**, 102 (1965).
304. Erdmann, H., *Leder* **14**, 249 (1963).
305. Erdmann, H., *Leder* **15**, 181 (1964).
306. Spahrkäs, H., and Schmid, H., *Leder* **14**, 217 (1963).
307. Erdmann, H., *Leder* **16**, 262 (1965).
308. Erdmann, H., *Leder* **17**, 10 (1966).
309. Küntzel, A., and Mahdi, H., *Leder* **21**, 123 (1970).
310. Kawamura, A., and Wada, K., *J. Am. Chem. Leather Assoc.* **62**, 612 (1967).
311. Ellis, M. J., and Sykes, R. L., *J. Am. Chem. Leather Assoc.* **58**, 346 (1963).
312. Knorr, C. A., Münster, G., and Feigl, H., *Z. Elektrochem.* **63**, 59 (1959).
313. Bowes, J. H., Moss, J. A., and Young, F. S., *J. Am. Chem. Leather Assoc.* **59**, 136 (1964).
314. Ellis, M. J., Shuttleworth, S. G., and Sykes, R. L., *J. Am. Chem. Leather Assoc.* **58**, 358 (1963).
315. Finholt, J. E., Coulton, K., Kimball, K., and Uhlenhopp, E., *Inorg. Chem.* **7**, 610 (1968).
316. Indubala, S., and Ramaswamy, D., *J. Inorg. Nucl. Chem.* **35**, 2055 (1973).
317. Slabbert, N. P., *J. Inorg. Nucl. Chem.* **39**, 883 (1977).
318. Siebert, H., and Schiedermaier, R., *Z. Anorg. Allg. Chem.* **361**, 169 (1968).
319. Andersen, P., Døssing, A., and Nielsen, K. M., *Acta Chem. Scand.* **A40**, 142 (1986).
320. Riccieri, P., and Zinato, E., *Inorg. Chim. Acta* **7**, 117 (1973).
321. Ruminski, R. R., and Coleman, W. F., *Inorg. Chem.* **19**, 2185 (1980).
322. Stevenson, K. L., and Verdieck, J. F., *J. Am. Chem. Soc.* **90**, 2974 (1968).
323. Stevenson, M. B., and Sykes, A. G., *J. Chem. Soc. (A)* 2979 (1969).
324. Wolcott, D. C., Ph.D. dissertation, Catholic University of America, Washington, D.C., 1965.
325. Kaas, K., and Springborg, J., *Inorg. Chem.* **26**, 387 (1987).
326. Ardon, M., and Bino, A., *Struct. Bond.* **65**, 1–28 (1987).
327. Bino, A., and Gibson, D., *J. Am. Chem. Soc.* **103**, 6741 (1981).
328. Bino, A., and Gibson, D., *J. Am. Chem. Soc.* **104**, 4383 (1982).

329. Ardon, M., and Bino, A., *J. Am. Chem. Soc.* **105,** 7747 (1983).
330. Ardon, M., and Bino, A., *J. Am. Chem. Soc.* **106,** 3359 (1984).
331. Ardon, M., and Bino, A., *Inorg. Chem.* **24,** 1343 (1985).
332. Bjerrum, J., and Rasmussen, S. E., *Acta Chem. Scand.* **6,** 1265 (1952).
333. Larsen, S., Nielsen, K. B., and Trabjerg, I., *Acta Chem. Scand.* **A37,** 833 (1983).
334. Postmus, C., and King, E. L., *J. Phys. Chem.* **59,** 1208 (1955).
335. Josephsen, J., and Pedersen, E., *Inorg. Chem.* **16,** 2534 (1977).
336. Wolcott, D., and Hunt, J. B., *Inorg. Chem.* **7,** 755 (1968).
337. Mønsted, L., and Mønsted, O., *Acta Chem. Scand.* **A30,** 203 (1976).
338. Albers, M. O., Liles, D. C., Singleton, E., and Stead, J. E., *Acta Crystallogr. Sect. C* **42,** 46 (1986).
339. Turpeinen, U., *Finn. Chem. Lett.* 123 (1977).
340. Turpeinen, U., Ahlgrén, M., and Hämäläinen, R., *Finn. Chem. Lett.* 246 (1977).
341. Wei, C. H., and Jacobson, K. B., *Inorg. Chem.* **20,** 356 (1981).
342. Chauvel, C. C., Girard, J. J., Jeannin, Y., Kahn, O., and Lavigne, G., *Inorg. Chem.* **18,** 3015 (1979).
343. Turpeinen, U., Ahlgrén, M., and Hämäläinen, R., *Acta Crystallogr. Sect. B* **38,** 1580 (1982).
344. Ahlgrén, M., Turpeinen, U., and Hämäläinen, R., *Acta Chem. Scand.* **A32,** 189 (1978).
345. Bukowska-Strzyzewska, M., and Tosik, A., *Acta Crystallogr. Sec. C* **39,** 203 (1983).
346. Skibsted, L. H., and Ford, P. C., *Acta Chem. Scand.* **A34,** 109 (1980).
347. Andersen, P., private communication.
348. Galsbøl, F., and Rasmussen, B. S., *Acta Chem. Scand.* **A38,** 141 (1984).
349. Gamsjäger, H., and Beutler, P., *J. Chem. Soc. Dalton Trans.* 1415 (1979).
350. Finholt, J. E., Glasfeld, A., Bent, E., Weier, J., *Abstr. Int. Conf. Coord. Chem., Boulder* 13–9 (1984).
351. Xu. F.-C., Krouse, H. R., and Swaddle, T. W., *Inorg. Chem.* **24,** 267 (1985).
352. Liteplo, M. P., and Endicott, J. F., *Inorg. Chem.* **10,** 1420 (1971).
353. Grant, D. M., and Hamm, R. E., *J. Am. Chem. Soc.* **78,** 3006 (1956).
354. Po, H. N., and Enomoto, H., *J. Inorg. Nucl. Chem.* **35,** 2581 (1973).
355. Wharton, R. K., and Sykes, A. G., *J. Chem. Soc. Dalton Trans.* 439 (1973).
356. Buckingham, D. A., Marty, W., and Sargeson, A. M., *Inorg. Chem.* **13,** 2165 (1974).
357. Rotzinger, F. P., and Marty, W., *Inorg. Chem.* **24,** 1617 (1985).
358. Rotzinger, F. P., and Marty, W., *Helv. Chim. Acta* **68,** 1914 (1985).
359. El-Awady, A. A., and Hugus, Z. Z., *Inorg. Chem.* **10,** 1415 (1971).
360. Koshy, K., and Dasgupta, T. P., *J. Chem. Soc. Dalton Trans.* 2781 (1984).
361. Hoppenjans, D. W., Hunt, J. B., and Penzhorn, L., *Inorg. Chem.* **7,** 1467 (1968).
362. Mønsted, L., and Mønsted, O., *Acta Chem. Scand.* **A38,** 67 (1984).
363. Mønsted, L., and Mønsted, O., *Acta Chem. Scand.* **A34,** 259 (1980).
364. Monsted, L., *Acta Chem. Scand.* **A32,** 377 (1978).
365. Hoffman, A. B., and Taube, H., *Inorg. Chem.* **7,** 903 (1968).
366. DeMaine, M. M., and Hunt, J. B., *Inorg. Chem.* **10,** 2106 (1971).
367. Rasmussen, S. E., and Bjerrum, J., *Acta Chem. Scand.* **9,** 735 (1955).
368. Ellis, J. D., Scott, K. L., Wharton, R. K., and Sykes, A. G., *Inorg. Chem.* **11,** 2565 (1972).
369. Edwards, J. D., Ph.D. thesis, Leeds, 1974.
370. Hin-Fat, L., and Higginson, W. C. E., *J. Chem. Soc. (A)*, 2589 (1971).
371. Edwards, J. D., Wieghardt, K., and Sykes, A. G., *J. Chem. Soc. Dalton Trans.* 2198 (1974).
372. Taylor, R. S., and Sykes, A. G., *Inorg. Chem.* **13,** 2524 (1974).
373. Sykes, A. G., *Chem. Br.* 170 (1974).

374. Sykes, A. G., and Mast, R. D., *J. Chem. Soc.* (A) 784 (1967).
375. Sykes, A. G., and Weil, J. A., *Prog. Inorg. Chem.* **13,** 1 (1970).
376. Foong, S. W., Mast, R. D., Stevenson, M. B., and Sykes, A. G., *J. Chem. Soc.* (A) 1266 (1971).
377. Foong, S. W., Stevenson, M. B., and Sykes, A. G., *J. Chem. Soc.* (A) 1064 (1970).
378. Hyde, M. R., and Sykes, A. G., *J. Chem. Soc. Dalton Trans.* 1583 (1974).
379. Foong, S. W., and Sykes, A. G., *J. Chem. Soc. Dalton Trans.* 1453 (1974).
380. Stevenson, M. B., Mast, R. D., and Sykes, A. G., *J. Chem. Soc.* (A) 937 (1969).
381. Scott, K. L., Taylor, R. S., Wharton, R. K., and Sykes, A. G., *J. Chem. Soc. Dalton Trans.* 2119 (1975).
382. Stevenson, M. B., Sykes, A. G., and Taylor, R. S., *J. Chem. Soc.* (A) 3214 (1970).
383. Foong, S. W., and Sykes, A. G., *J. Chem. Soc. Dalton Trans.* 504 (1973).
384. Edwards, J. D., Foong, S. W., and Sykes, A. G., *J. Chem. Soc. Dalton Trans.* 829 (1973).
385. Stevenson, M. B., Taylor, R. S., and Sykes, A. G., *J. Chem. Soc.* (A) 1059 (1970).
386. Scott, K. L., and Sykes, A. G., *J. Chem. Soc. Dalton Trans.* 2364 (1972).
387. Barro, R., Marsh, R. E., and Schaefer, W. P., *Inorg. Chem.* **9,** 2131 (1970).
388. Kubicki, M. M., and Galdecki, Z., *Acta Crystallogr. Sect. B* **35,** 1898 (1979).
389. Taylor, R. S., and Sykes, A. G., *J. Chem. Soc.* (A) 1991 (1970).
390. Taylor, R. S., and Sykes, A. G., *J. Chem. Soc. Chem. Commun.* 1137 (1969).
391. Sykes, A. G., and Taylor, R. S., *J. Chem. Soc.* (A) 1424 (1970).
392. Mäcke, H., *Helv. Chim. Acta* **64,** 1579 (1981).
393. Cannon, K. K., *Jpn. Kokai Tokkyo Koho. Chem. Abstr.* **101,** 81604 d.
394. Megnin, P., *Chem. Abstr.* **82,** 87980 j.
395. Daniel, R., *Chem. Abstr.* **97,** 43224 w.
396. Tokyo Shibaura Electric Co., *Chem. Abstr.* P **66,** 19982 f.
397. Wada, T., and Ishizaka, M. (Tokyo Shibaura Electric Co.), *Chem. Abstr.* P **66,** 86726 p.
398. Andersen, R. A., *Sci. Total Environ.* **17,** 13 (1981).
399. Schwarz, K., and Mertz, W., *Arch. Biochem. Biophys.* **85,** 292 (1959).
400. Cooper, J. A., Blackwell, L. F., and Buckley, P. D., *Inorg. Chim. Acta* **92,** 23 (1984).
401. Gonzalez-Vergara, E., Hegenauer, J., Saltmann, P., Sabat, M., and Ibers, J. A., *Inorg. Chim. Acta* **66,** 115 (1982).

CATENATED NITROGEN LIGANDS PART II.[1] TRANSITION METAL DERIVATIVES OF TRIAZOLES, TETRAZOLES, PENTAZOLES, AND HEXAZINE

DAVID S. MOORE*[2] and STEPHEN D. ROBINSON**

*Department of Chemistry, Dover College, Dover, Kent CT17 9RL, England, and
**Department of Chemistry, King's College London, Strand, London WC2R 2LS, England

I. Introduction
II. Triazole and Triazolate Complexes
   A. Synthesis
   B. Structural Properties
   C. Spectroscopic Studies
   D. Group Survey
III. Tetrazole and Tetrazolate Complexes
   A. Synthesis
   B. Structural Properties
   C. Spectroscopic Studies
   D. Group Survey
IV. Pentazolate and Hexazine Complexes
   References

I. Introduction

This review, which complements an earlier one (Part I) dealing with transition metal complexes of triazenes, tetrazenes, tetraazadienes, and pentaazadienes, examines the coordination chemistry of related cyclic catenated nitrogen ligands. Six-membered rings containing three, four, or five adjacent nitrogen atoms—1,2,3-triazines, 1,2,3,4-tetrazines, and pentazines, respectively—are either unknown or are relatively unstable species whose coordination chemistry has yet to be explored.

---

[1] Part I. Transition metal derivatives of triazenes, tetrazenes, tetraazadienes, and pentazadienes. *Adv. Inorg. Chem. Radiochem.* **30**, 1 (1986).

[2] Present address: Department of Chemistry, St. Edward's School, Oxford OX2 7NN, England.

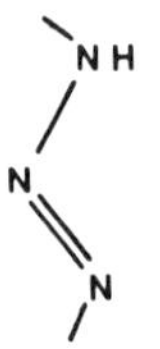

TRIAZENE 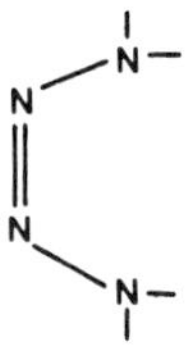　　TETRAZENE 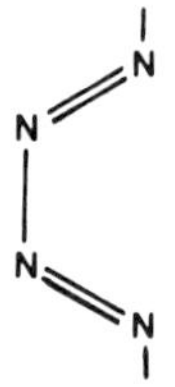　　TETRAAZADIENE 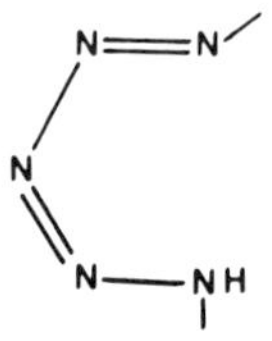　　PENTAAZADIENE

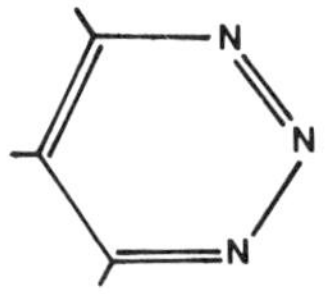

TRIAZINE 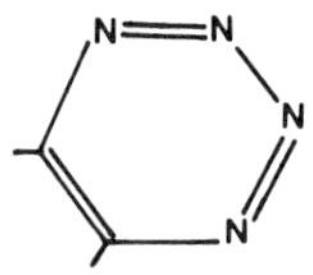　　TETRAZINE 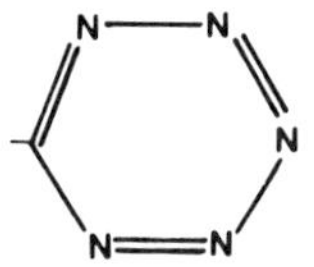　　PENTAZINE 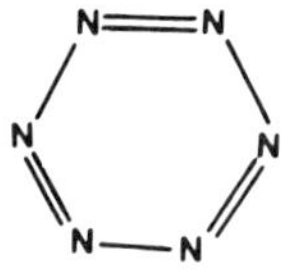　　HEXAZINE

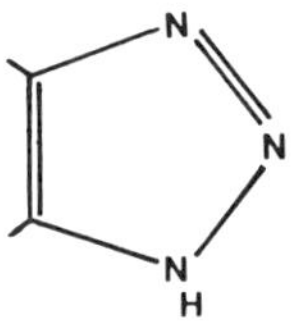

TRIAZOLE 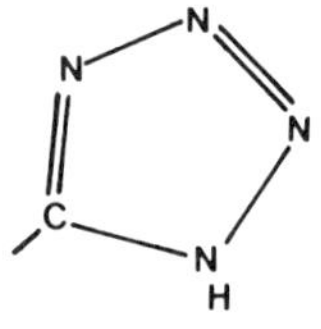　　TETRAZOLE 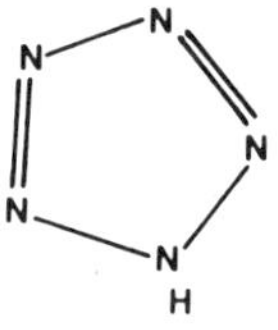　　PENTAZOLE

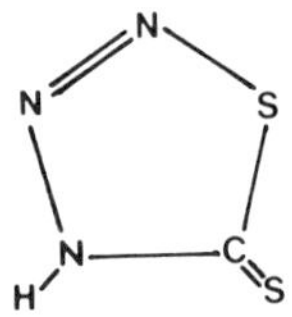

1,2,3,4-THIATRIAZOLINE-5-THIONE 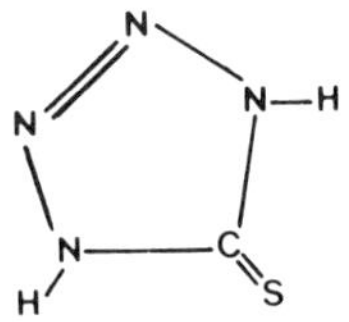　　1,2,3,4-TETRAZOLINE-5-THIONE

Consequently the area covered is essentially concerned with the chemistry of the more accessible five-membered 1,2,3-triazole and 1,2,3,4-tetrazole ring systems. Coverage of complexes containing related N,S heterocyclic ligands, notably 1,2,3,4-thiatriazoline-5-thione and 1,2,3,4-tetrazoline-5-thione, is selective since the coordination chemistry of these species has, in part, been reviewed elsewhere (168). Finally, prospects for the isolation of complexes containing higher nitrogen ligands, in particular the pentazolate anion ($cyclo\text{-}N_5^-$) and the hexazine molecule ($cyclo\text{-}N_6$), are briefly examined.

II. Triazole and Triazolate Complexes

This section covers ligands containing the 1,2,3-triazole ring system. These include, in addition to the parent triazole, various N- and/or C-substituted triazoles, benzotriazole, and a number of 8-azapurines. The coordination chemistry of 5-thio-1,2,3,4-thiatriazole is selectively reviewed. All of these molecules, with the exception of the N-substituted triazoles, are capable of coordinating in anionic as well as neutral form. 1,2,3-Triazole, first prepared by von Pechmann in 1888 (*215*), is a weak acid ($pK_a = 9.26$) (*88*) and exists as a mixture of the tautomeric forms (structures **1a** and **1b**). Benzotriazole (**2**), first correctly formulated

by Ladenburg in 1876 (*120*), is also a weak acid ($pK_a = 8.38$) (*88*) and displays similar tautomerism. The use of 1,2,3-triazoles—notably benzotriazole and naphthotriazole—as corrosion inhibitors for copper (*99*) and as quantitative precipitating agents for copper (*46*) and the platinum group metals (*13*) has contributed substantially to the interest in the coordination chemistry of these ligands. The 1,2,3,4-thiatriazoline-5-thionate anion, formed by 1,3-cycloaddition of carbon disulfide to the azide anion, was formulated as the azidodithiocarbonate anion $N_3CS_2^-$ (*35*) before being correctly identified by Lieber and co-workers in 1957 (*127*). The parent acid is thought to be a tautomeric mixture of thione (**3a** and **3b**) and thiol forms (**3c**), with the latter being dominant (*45*). The 8-azapurines (aza analogs of the purine bases), several of which have attracted attention by virtue of their antitumor properties (*108*), include 8-azaadenine (**4**), 8-azaguanine (**5**), and 8-azahypoxanthine (**6**).

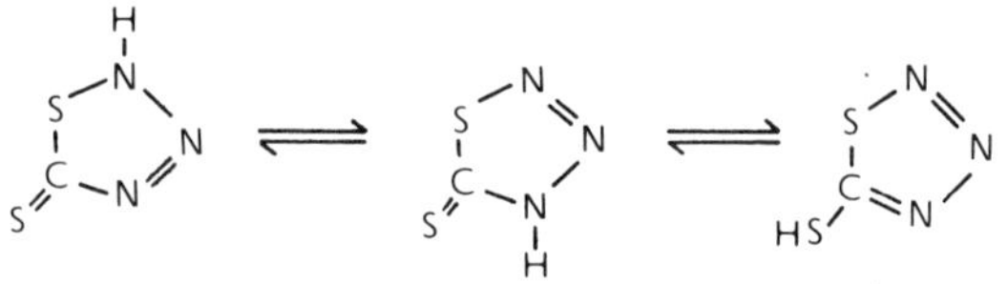

The generally accepted numbering schemes for 1,2,3-triazoles, thiatriazoline-5-thiones, and 8-azapurines, which are used throughout this section of the review, are given in (**7**), (**8**), and (**9**), respectively. Abbreviations used throughout this section to indicate neutral and anionic forms of the 1,2,3-triazole and benzotriazole ligands are taH/ta and btaH/bta, respectively.

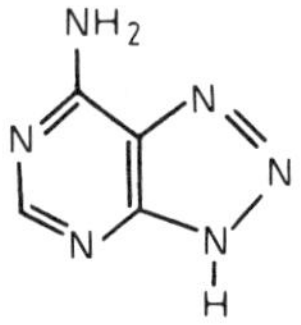

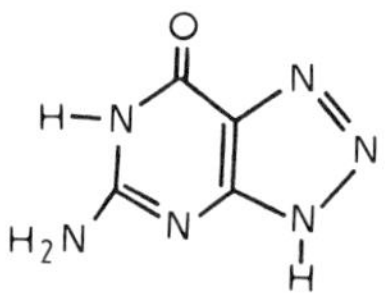

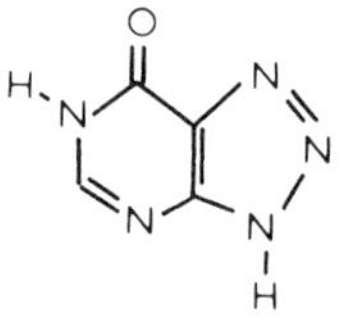

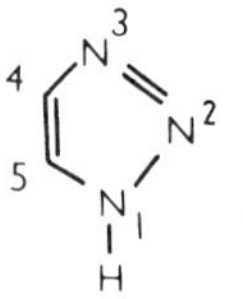

A. Synthesis

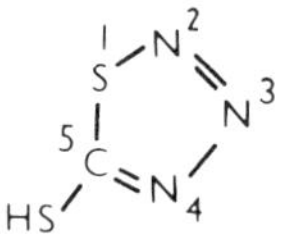

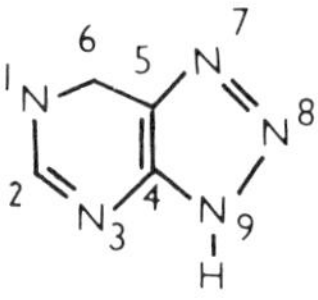

Adducts of triazoles with transition metal salts are usually prepared by direct reactions between the two components involved and frequently precipitate or crystallize spontaneously from the reaction mixture (*55, 172, 194, 202*). Complexes containing triazolate anions can usually be obtained from the corresponding transition metal halide, carboxylate, nitrate, or perchlorate complex and an alkali metal (*146, 147, 172*) or thallium(I) triazolate salt (*33*). Other routes to triazolate complexes include the direct reactions of metal halides with triazoles in the presence of a base (*201*) and the treatment of triazole/metal halide adducts with base (*156, 194*). Direct reactions of triazoles with metal carboxylates have also been employed (*172*). 1,3-Dipolar cycloaddition of electron-deficient nonterminal acetylenes to coordinated azide anions [Eq. (1)] is a well-established route to complexes of 4,5-disubstituted triazolate anions (*37, 109, 114, 181*). A similar 1,3-dipolar addition of carbon disulfide [Eq. (2)] provides a useful route to derivatives of 1,2,3,4-thiatriazoline-5-thione (*17, 64, 121, 234, 235*).

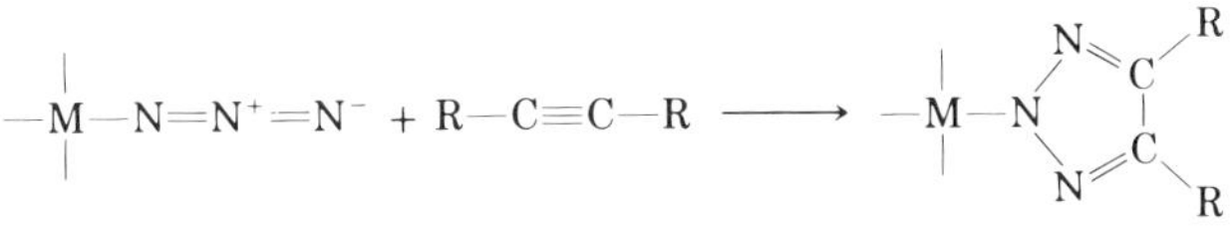

$$-M-N=N^+=N^- + R-C{\equiv}C-R \longrightarrow -M-N\underset{N}{\overset{N}{\underset{\|}{\underset{C}{\overset{C}{\diagdown}}}}}\begin{smallmatrix}R\\ \\R\end{smallmatrix} \qquad (1)$$

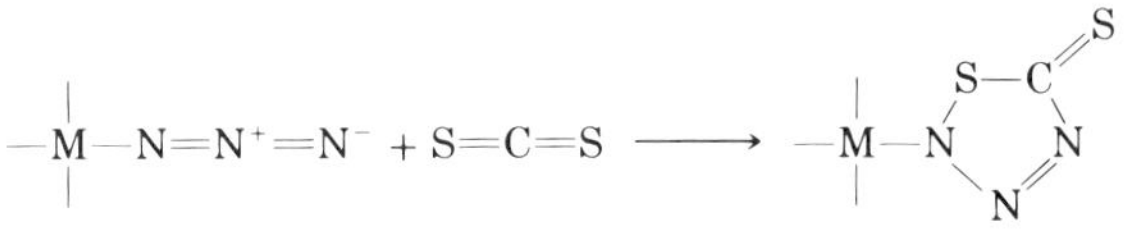

$$-M-N=N^+=N^- + S=C=S \longrightarrow -M-N\underset{N}{\overset{S-C}{\underset{\|}{\underset{N}{\diagdown}}}}\begin{smallmatrix}S\\ \\ \end{smallmatrix} \qquad (2)$$

B. STRUCTURAL PROPERTIES

1,2,3-Triazoles and their conjugate bases, the triazolate anions, are potentially versatile ligands offering three adjacent donor sites. Although the generally preferred coordination positions appear to be N-3 for the parent triazoles and N-1 for the triazolate anions, all three N donor sites display considerable basicity and participate in metal–ligand bonding.

The neutral triazoles coordinate in monodentate fashion through N-3 (*197, 198, 202, 209*) or occasionally N-2 (*209*) donor sites and have also been found to act as N-2/N-3-bridging ligands (*1*). Coordination via N-1 has been established for the square–planar iridium(I) benzotriazolate complex Ir(bta)(CO)(PPh$_3$)$_2$ (*33*), but in certain octahedral cobalt(III) derivatives of 4,5-disubstituted triazoles kinetically determined N-1 isomers readily convert to the thermodynamically preferred N-2 isomers in response to steric influences (*109*). Iron(II) triazolate complexes, Fe(triazolate)(C$_5$H$_5$)(CO)$_2$, undergo irreversible thermal N-1 → N-2 or N-2 → N-1 isomerization, depending upon the nature of the substituents occupying the 4 and 5 positions on the triazole ring (*146*). Triazolate anions usually bridge through N-1 and N-2 sites in bi- or polynuclear structures containing single (*107*), double (*201, 208*), or triple (*1, 57, 133, 199*) triazolate linkages. However, examples of binuclear (*91*) and polynuclear (*198*) complexes containing single

N-1/N-3 triazolate bridges are also known. There are in addition several X-shaped pentanuclear copper(I/II) or copper(II) benzotriazolate complexes in which the three nitrogen atoms of each triazolate anion are involved in bonding to three different metal atoms (*92, 113, 131*). A similar bonding arrangement has also been proposed for a tetranuclear rhodium benzotriazolate complex (*156*). By virtue of their cyclic nature, individual triazoles or triazolate anions are not capable of serving as chelate ligands. However H-bonded triazole–triazolate anion pairs have been shown to function as bidentate ligands, thereby generating six-membered chelate rings (*96*). Finally, since the triazolate anion is isoelectronic with the cyclopentadienide anion, there is the prospect, as yet unfulfilled, of η^3 or η^5 coordination of the triazolate ring to a metal center. Triazolate bonding modes are represented schematically in

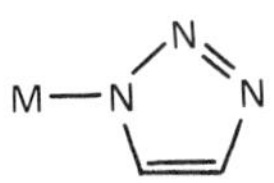

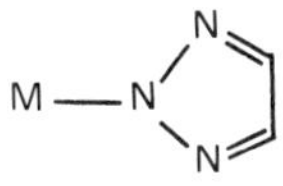

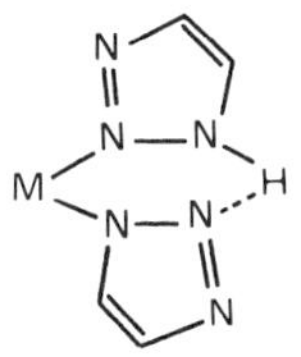

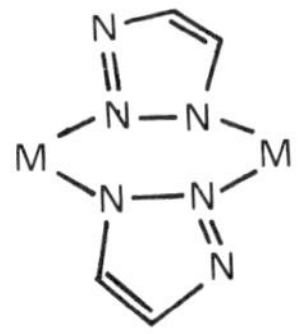

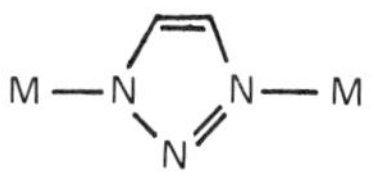

structures **10a–10f**. A list of crystallographically determined structures involving coordinated triazole moieties is given in Table I.

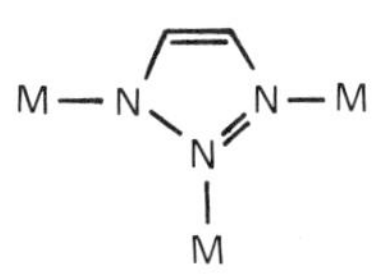

The tautomeric forms of the thiatriazoline-5-thionate anion offer three possible coordination sites, N-1, N-3, and S, and scope for N/S chelate or bridging structures. Spectroscopic evidence for N-1- and N-3-bonded isomers has been reported for the silver(I) complex

TABLE I

X-Ray Diffraction Studies on Triazole and Triazolate Complexes

Complex	Triazole- and Triazolate-bonding mode	Reference
Benzotriazole Derivatives		
$Mo_2(btaH)_3(CO)_6$	Bridging/N-2/N-3	1
$\{MnCl_2(btaH)_2\}_n$	Monodentate/N-3	202
$Fe(1\text{-}CH_2{=}CH{-}CH_2\text{-}bta)(CO)_3$	Chelate/N-2, $-CH{=}CH_2$	143
$OsH(bta)(btaH)(CO)(PPh_3)_2$	Chelate/bta-H···bta/N-1/N-2	96
$CoCl_2(btaH)_2$	Monodentate/N-3	191
$CoCl_2(1\text{-}CH_2{=}CH\text{-}bta)_2$	Monodentate/N-3	196
$[btaH_2]_2[CoCl_4]$	$btaH_2^+$ cations	198
$[Rh(bta)(C_8H_{12})]_2$	Bridging/N-1/N-2	208
$Ir(bta)(CO)(PPh_3)_2$	Monodentate/N-1	33
$NiCl_2(btaH)_2$	Monodentate/N-3	191
$Ni_3(bta)_6(NH_3)_6 \cdot 2Me_2CO \cdot 2H_2O$	Bridging/N-1/N-2	199
$Ni_3(bta)_6(NH_2 \cdot CH_2 \cdot CH{=}CH_2)_6(Ph_3PO)_2$	Bridging/N-1/N-2	133
$\{CuCl_2(btaH)_2\}_2 \cdot H_2O$	Monodentate/N-3	197
$CuSO_4(btaH)_3(H_2O) \cdot btaH$	Mondentate/N-3	97
$\{Cu(bta)_2(H_2O)\}_n$	Bridging/N-1/N-2	201
$Cu_5(bta)_6(Bu^tNC)_4$	Bridging/N-1/N-2/N-3	92
$Cu_2(bta)(tmbma)_2(NO_3)_3$ [a]	Bridging/N-1/N-3	91
$Ag(NO_3)(btaH)_2$	Monodentate/N-3; bridging/N-2/N-3	200
$ZnCl_2(btaH)_2$	Monodentate/N-3	198
$\{Zn_2(bta)_4\}_n$	Bridging/N-1/N-2	198
$\{CdCl_2(btaH)_2\}_n$	Monodentate/N-3	202
$\{HgMe(bta)\}_2$	Bridging/N-1/N-2	203
Miscellaneous Triazole Derivatives		
$[Fe(taH)(C_5H_5)(CO)_2][HSO_4]$	Monodentate/N-3	5
$Co(4,5\text{-}R_2\text{-}ta)(dmg)_2(PPh_3)$ [R = C(O)OMe]	Monodentate/N-2	142
$Rh_2(\mu\text{-}N_3)(\mu\text{-}4,5\text{-}R_2\text{-}ta)(4,5\text{-}R_2\text{-}ta)_2(C_5Me_5)_2$ (R = CF_3)	Monodentate/N-2; bridging/N-1/N-2	178
$Cu(4,5\text{-}R,R'\text{-}ta)_2(H_2O)_2 \cdot 2H_2O$ (R = CO_2, R' = CO_2H)	Chelate/N-3, $-CO_2^-$	155
$Cu(4,5\text{-}R,R'\text{-}taH)_2Cl_4$ [R = NH_2, R' = $C(NH_2){=}NH_2^+$]	Monodentate/N-2	167
8-Azapurine Derivatives [b]		
$Cu(3\text{-}methyl\text{-}8\text{-}azaxanthinato)_2(NH_3)_2$	Monodentate/N-8	168a
$ZnCl_3(8\text{-}azaadeninium)$	Monodentate/N-3	166
$Cd(8\text{-}azahypoxanthinato)_2(H_2O)_4$	Monodentate/N-7	165
$HgCl_2(8\text{-}azaadenine)_2$	Monodentate/N-3	85
$[HgMe(8\text{-}azaadenine)]NO_3$	Monodentate/N-9	192
$HgMe(8\text{-}azaadeninato) \cdot 4H_2O$	Monodentate/N-9	192
$[(HgMe)_2(8\text{-}azaadeninato)]NO_3 \cdot H_2O$	Bidentate/N-3/N-9	192
$[(HgMe)_3(8\text{-}azaadenine\text{-}H_2)]NO_3$	Tridentate/N-1/N-6/N-9	192
$Hg(8\text{-}azahypoxanthinato)_2 \cdot 4H_2O$	Monodentate/N-9	85

[a] tmbma, tris(*N*-Methylbenzimidazol-2-ylmethyl)amine.

[b] Note: Nitrogen numbering scheme different from that employed for simple triazoles (see p. 174).

Ag(CN$_3$S$_2$)(PPh$_3$) (235), and N/S-bridging thiatriazoline-5-thionate ligands have been proposed for some binuclear palladium(II) complexes (37).

C. Spectroscopic Studies

This section highlights some of the more important spectroscopic results obtained for triazole and triazolate complexes.

1. Vibrational Spectra

Infrared data have been tabulated for benzotriazole and a wide range of its transition metal complexes or adducts (172). Far infrared spectra have been recorded for copper(II) benzotriazole adducts and bands at ~ 270–320 cm^{-1} have been assigned to Cu–N vibrations (172). Infrared absorptions at approximately 825, 800, and 775 cm^{-1} in the spectra of cobalt(III)/4,5-disubstituted triazolate complexes have been attributed to triazolate ring vibrations (109). Infrared data have been reported and assignments made for palladium and platinum thiatriazoline-5-thionate complexes (37) and for the parent thione (127). Vibrational spectroscopy has been employed in an attempt to determine coordination sites for a range of 8-azapurine complexes (108).

2. Electronic Spectra

Relatively little has been reported on the electronic spectra of triazole and triazolate complexes. Copper(II) benzotriazolate adducts and benzotriazolate complexes show ligand-field maxima in the range 12.5–15.9 kK, in agreement with the proposed octahedral coordination geometry (172). Electronic spectra have also been reported for rhodium(I) and iridium(I) benzotriazolate complexes (33).

3. Nuclear Magnetic Resonance Spectra

Proton and ^{13}C NMR spectra have been used to distinguish between symmetrically (N-2) and asymmetrically (N-1) coordinated benzotriazole (209) or 4,5-disubstituted triazolate (109, 146) ligands, and to measure equilibrium constants for the N-1/N-2-coordinated benzotriazole system [Fe(CN)$_5$(btaH)]$^{3-}$ (209). Hydrogen bond formation and proton transfer reactions in [Fe(taH)(C$_5$H$_5$)(CO)$_2$][HSO$_4$] have been followed by variable-temperature ^{1}H and ^{13}C NMR (5, 146).

4. Electron Paramagnetic Resonance Spectra

EPR data have been reported for iron triazole complexes (101) and a particularly wide range of copper(II) benzotriazole derivatives (172). The EPR spectra of the copper(II) complexes all show either isotropic g values or indications of axial or rhombic splittings; in no case were copper hyperfine splittings observed (172). A more detailed examination of $[CuCl_2(btaH)_2]_2 \cdot H_2O$ gave a g value of 2.045 ± 0.003 and produced evidence of a weak paramagnetic interaction indicative of a triplet ground state lying 1.8 ± 0.2 cm^{-1} below the singlet excited state (93). The EPR spectrum (9, 24, and 54 GHz) of the copper(I/II) complex $Cu_5(bta)_6(Bu^tNC)_4$ has been recorded at temperatures ranging from that of liquid nitrogen to room temperature. The appearance of Cu(II) hyperfine structure in almost all the spectra confirmed the expectation that the large Cu(II)–Cu(II) separation (12.858 Å) would be sufficient to ensure that the effects of exchange interactions are virtually non-existent and that electronic dipolar broadening is small. At low temperatures the EPR spectra are quite normal and consistent with a $d_{x^2-y^2}$ type of ground state. Above 128 K a "reversed spectrum" is observed and is interpreted in terms of a two-dimensional dynamical Jahn–Teller effect (113). A later study of the Jahn–Teller effect in $Cu_5(bta)_6(Bu^tNC)_4$ places emphasis on the temperature-dependent quantitative dynamical changes which occur as the critical temperature is approached from above and suggests a possible transformation of localized molecular vibration into lattice phonons (10). For the closely related copper(II) complex $Cu_5(bta)_6(acac)_4$ the much closer approach of adjacent copper(II) centers (~ 3.5–4.0 Å) leads to dipolar and electron-exchange interactions which are substantially larger than the Cu(II) hyperfine interactions, hence a broad structureless resonance is observed at $g \approx 2$. An additional weak resonance at $g = 4$ confirms the presence of spin-coupled pairs (131). Q-Band and X-band EPR spectra show typical $S = 1$ signals for $[Cu_2(tmbma)_2(bta)][NO_3]_3 \cdot 4H_2O$ [tmbma = tris(N-methylbenzimidazol-2-ylmethyl)amine] and variable-temperature magnetic susceptibility measurements indicate antiferromagnetic interactions with J values of -10 to -12 cm^{-1} (91). A more recent single-crystal EPR study has established that even with a Cu(II)–Cu(II) distance of 5.536 Å an exchange contribution to the zero-field splitting is operative and that ferromagnetic exchange interactions can be propagated through triazolate bridges connecting ions separated by that distance (21).

D. GROUP SURVEY

To date no triazole complexes appear to have been reported for the scandium, yttrium, and lanthanum group of metals.

1. *Titanium, Zirconium, and Hafnium*

A trans structure with N-3-coordinated benzotriazole ligands has been proposed for the air- and moisture-sensitive orange adduct $TiCl_4(btaH)_2$ obtained by addition of the free ligand to a solution of $TiCl_4$ in dry dichloromethane (*172*).

2. *Vanadium, Niobium, and Tantalum*

The green-brown paramagnetic complex $VO(bta)_2 \cdot H_2O$ obtained when vanadyl sulfate reacts with benzotriazole in aqueous solution at pH 7–8 (*58*) probably has a polymeric structure with N-1/N-2-bridging bta ligands. The reaction of TaF_5 with benzotriazole in CH_2Cl_2/CH_3CN solution affords the salt *trans*-$[TaF_4(btaH)_2][TaF_6]$, which has been characterized by ^{19}F NMR (*98*).

3. *Chromium, Molybdenum, and Tungsten*

The red complex, $Mo_2(btaH)_3(CO)_6 2 \cdot 5THF$, obtained by treatment of $Mo(CH_3CN)_3(CO)_3$ with benzotriazole in tetrahydrofuran (THF) has been shown by X-ray diffraction methods to contain binuclear btaH-bridged units (Fig. 1) linked by H-bonded THF molecules. N-1-Methylbenzotriazole affords an analogous product. Deprotonation of $Mo_2(btaH)_3(CO)_6$ with NaH generates the yellow-green salt $Na_3[Mo_2(bta)_3(CO)_6]2 \cdot 5THF$ (*1*). A purple substitution product, $MoCl(C_5H_5)(btaH)(CO)_2$, has been obtained from $MoCl(C_5H_5)(CO)_3$ and benzotriazole in hexane (*111*).

4. *Manganese, Technetium, and Rhenium*

A mononuclear structure with N-1 benzotriazolate ligands is proposed for the yellow complex, $Mn(bta)(CO)_5$, obtained from Li(bta) and $MnBr(CO)_5$ in dry THF solution (*172*). High-spin ($\mu_{eff} = 5.92$–6.05 BM) manganese(II) complexes $MnCl_2(btaH)$, $MnX_2(btaH)_2 \cdot nH_2O$ (X = Cl, Br, I, NCS, or NO_3; $n = 0$–2.5), and $Mn(O_2CCH_3)(bta)$ deposit as creamy white precipitates when the appropriate manganese salts are treated with benzotriazole in hot alcohol solution. Neutralization of the acetate reaction mixture to pH 7–8 with alcoholic ammonia af-

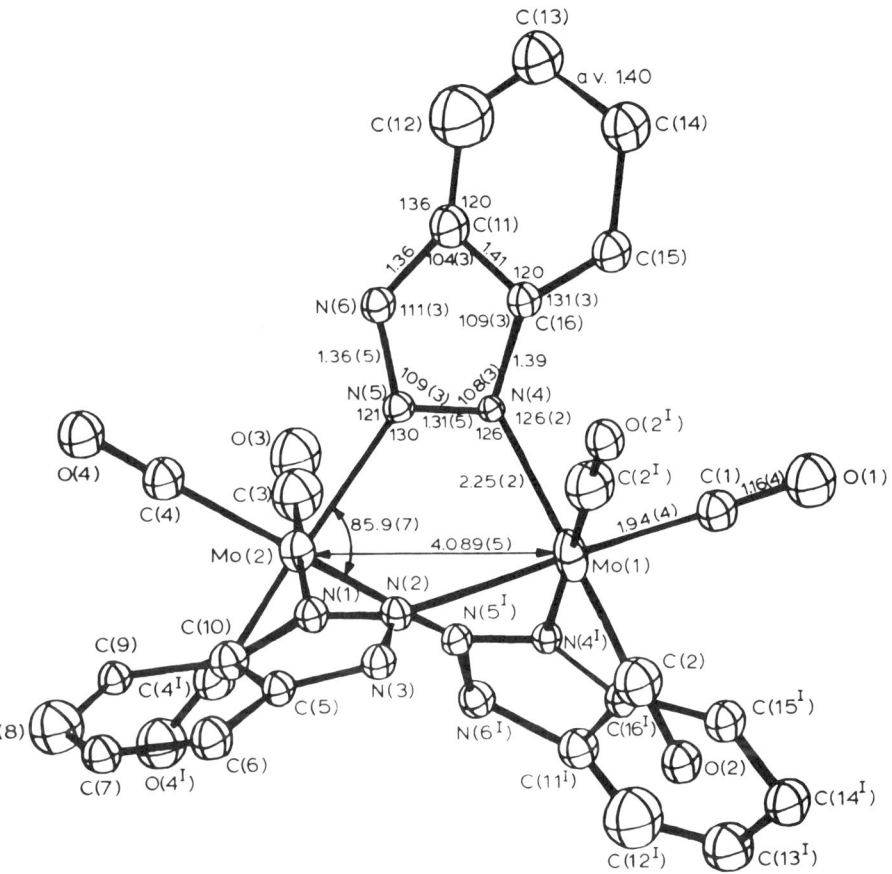

FIG. 1. Molecular structure of $Mo_2(btaH)_3(CO)_6$.

fords $Mn(bta)_2 \cdot 2.5H_2O$. All the hydrates lose water at $\sim 120°C$ (183). More recently, crystals of $MnCl_2(btaH)_2$ (191), obtained by mixing $MnCl_2 \cdot 4H_2O$ and benzotriazole in methanol, have been described as rose colored and have been shown to possess a chloride-bridged ribbon structure with octahedral coordination about each manganese center completed by a trans pair of N-3-bound btaH ligands (Fig. 2) (202).

5. Iron, Ruthenium, and Osmium

Thermogravimetric analysis of the yellow salt $(btaH)_2H_2[Fe(CN)_6]$, deposited from $K_4[Fe(CN)_6]$/btaH solutions on acidification with HCl, reveals o-aminobenzonitrile to be the major organic product (226).

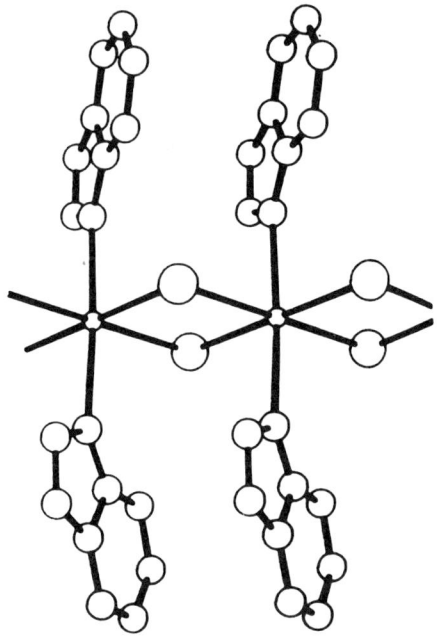

FIG. 2. Coordination about the metal atoms in [MCl$_2$(btaH)$_2$]$_2$ (M = Mn or Cd).

Benzotriazole reacts with the complex anion [Fe(CN)$_5$(H$_2$O)]$^{3-}$ to afford a mixture of N-2- and N-3-coordinated isomers [Fe(CN)$_5$(btaH)]$^{3-}$; characterization by ^{1}H and ^{13}C NMR spectroscopy yielded an equilibrium constant $K = 1.9 \pm 0.3$ favoring the less symmetric N-3 isomer (209). Relaxation times obtained from cyclic voltammetry data indicate an intramolecular mechanism for the (N-2)$\rightleftharpoons$(N-3) isomerization process (210). The N-2-coordinated triazolate complex Fe(ta)-(C$_5$H$_5$)(CO)$_2$, obtained from FeX(C$_5$H$_5$)(CO)$_2$ (X = Cl, Br, or I) and Na(ta) in THF at 25°C, undergoes irreversible thermal isomerization in neutral or acidic media to afford the N-1 isomer (144, 146). Treatment of the N-1-bonded isomer with acids affords H-bonded products (11). Dibasic acids (H$_2$SO$_4$ and H$_2$C$_2$O$_4$) form insoluble adducts (145, 146), one of which [Fe(taH)(C$_5$H$_5$)(CO)$_2$][HSO$_4$] has been

$$(C_5H_5)(CO)_2Fe^+ - N \overset{N}{\underset{}{\diagup\kern-0.4em\diagdown}} N-H \cdots X^-$$

11

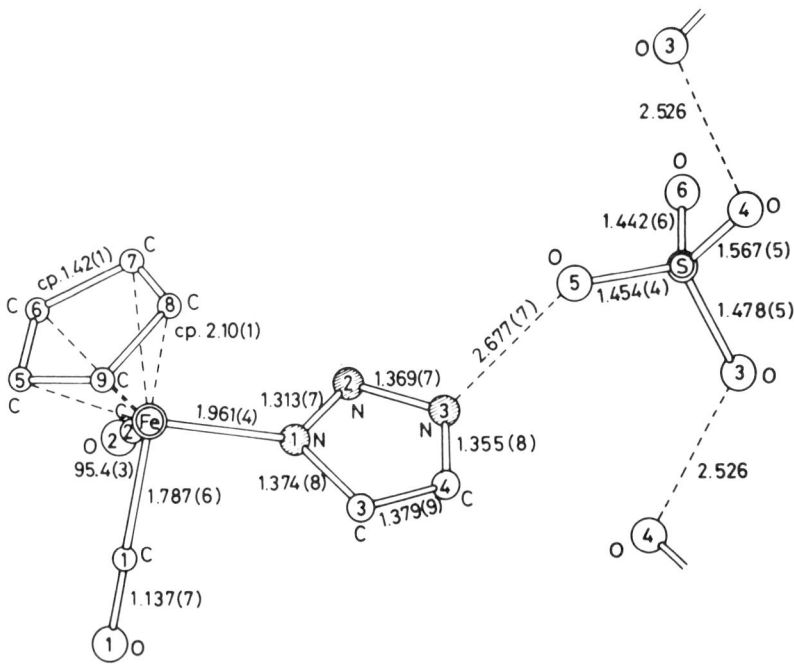

FIG. 3. The structure of 1-*N*-(1,2,3-triazole)(cyclopentadienyl)dicarbonyliron bisulfate.

characterized by diffraction methods (Fig. 3) (5). With aqueous NaBPh$_4$ the solvate initially formed decays to liberate triphenylboron, which in turn acts as a Lewis acid to generate the adduct (**12**) (*145*). Benzotriazole displays a similar behavior pattern but 4-phenyl-1,2,3-triazole (4-Ph-taH) initially forms an N-1-bonded complex (**13a**) which subsequently transforms to the N-2 isomer (**13b**). This difference in

$(C_5H_5)(CO)_2Fe-N\underset{\diagdown}{\overset{N=N}{\diagup}}N\to BPh_3$

12

$(C_5H_5)(CO)_2Fe-N\underset{\diagdown}{\overset{N=N}{\diagup}}{}_{Ph}$ $(C_5H_5)(CO)_2Fe-N\underset{\diagdown}{\overset{N}{\diagup}}\overset{N}{\underset{Ph}{\diagdown}}$

13a   13b

behavior is attributed to a kinetically controlled initial step involving primary attack at the N atom with highest electron density (N-2 for btaH and taH; N-1 for 4-Ph-taH) followed by thermal isomerization to the thermodynamically preferred isomers (N-1 for btaH and taH; N-2 for 4-Ph-taH) (*146*). The N-2 to N-1 isomerization observed for Fe(4-Ph-ta)(C_5H_5)(CO)$_2$ under acid conditions (H_2SO_4/THF) is attributed to stabilization of the N-1 isomer by H bonding to the sulfuric acid (*146*). Hydrogen bond formation and proton transfer reactions have been investigated by variable-temperature ^{1}H NMR and mass spectra have been recorded (*146*). On the basis of the X-ray crystal structure determined for [Fe(taH)(C_5H_5)(CO)$_2$][HSO$_4$] (Fig. 3) and variable-temperature ^{13}C NMR data for this and related complexes, a bifunctional concerted mechanism is proposed for proton exchange in these systems (*5*). α-Ferrocenyl alkylation of the benzotriazolate anion has been achieved (*112*) by the reaction of the salts [Fe(C_5H_5)-(C_5H_4CHR)]$^+$BF$_4^-$ with Na(bta). Diiron nonacarbonyl reacts with 1- and 2-allylbenzotriazoles to give isomeric species (**14**) and (**15**),

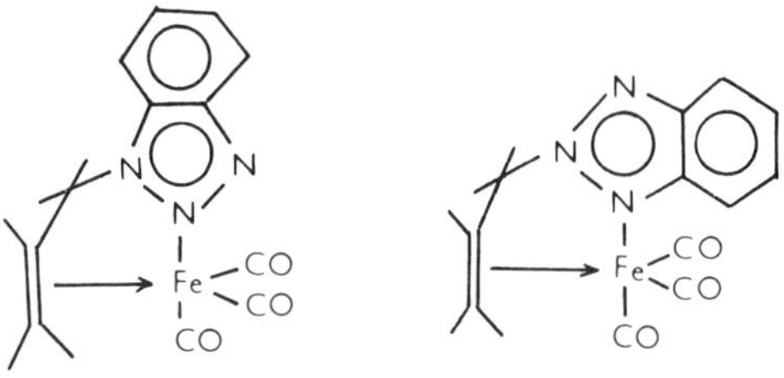

14   **15**

respectively (*20, 143*). The former product is also obtained when FeI(C_3H_5)(CO)$_3$ is treated with Na(bta) (*20*), and has been characterized by X-ray diffraction methods (*143*). In marked contrast, Fe$_2$(CO)$_9$ reacts with 4-benzoyl- and 4-butyryl-1,2,3-triazole in THF to afford air- and moisture-sensitive rose-violet crystals of stoichiometry Fe(4-RCO-ta)$_2$ (R = Ph or C_3H_7). Structures involving delocalized π or localized Fe–N σ bonds have been considered; the most probable structure appears to be one analogous to that reported for bis(1,4-diazobutadiene)iron. These triazolate complexes are attacked by air or moisture to give brown products similar to those previously obtained from ferrous chloride and the sodium salts of the corresponding triazoles. The latter products, which analyze for Fe(4-RCO-ta)$_2 \cdot n$H$_2$O (n = 1 or 2, R = Ph or C_3H_7), are insoluble in noncoordinating organic solvents and presumably possess polymeric triazole bridged structures (147). Addition of dimethyl acetylenedicarboxylate to the azide Fe(N$_3$)(C$_5$H$_5$)(CO)$_2$ yields the corresponding N-1-coordinated triazole dicarboxylate derivative Fe(4,5-R$_2$-ta)(C$_5$H$_5$)-(CO)$_2$ (R = COOMe) (181). EPR data have been recorded for an iron nitrosyl/1,2,3-triazole complex of indeterminate stoichiometry (101).

In a series of articles published between 1956 and 1967, R. F. Wilson and co-workers described numerous benzotriazole derivatives of the platinum group metals. Spectroscopic studies on ruthenium/benzotriazole solutions established the formation of reddish-orange (1:4) and green (1:2) complexes in solution (221). Treatment of RuCl$_3$·3H$_2$O with benzotriazole in aqueous alcohol solution followed by addition of alkali or alkaline earth metal salts afforded insoluble purple compounds which were formulated as oxyruthenium(III) species M^I[Ru(O)$_2$(btaH)$_2$] (M^I = Li, Na, or K) or M^{II}[Ru(O)$_2$(btaH)$_2$]$_2$ (M^{II} = Mg, Sr, or Pb). When warmed in acetic acid in a steam bath the purple sodium salt converted to an isomeric green product. Treatment with acids HX in aqueous solution afforded purple (X = Cl or Br) or dark green crystalline products of stoichiometry RuX(O)(btaH)$_2$ (227). Since oxy complexes of ruthenium are usually confined to higher oxidation states it seems probable that the oxy–ruthenium(III) formulations originally proposed are incorrect, and that the complexes are really polymeric species containing octahedral ruthenium(III) centers linked by bridging bta, oxide, and/or halide ligands. The feasibility of such structures has been demonstrated by the results of X-ray diffraction studies on several other transition metal/benzotriazole derivatives (see p. 196). More recently Kukushkin et al. have demonstrated that solid-phase dehydrohalogenation (Anderson rearrangement) of the complexes [btaH$_2$]$_2$[RuCl$_6$] and [btaH$_2$]$_2$-[Ru(NO)Cl$_5$] leads to formation of the benzotriazole complexes RuCl$_4$-(btaH)$_2$, RuCl$_3$(NO)(btaH)$_2$, and [btaH$_2$][RuCl$_4$(NO)(btaH)] (117). Complex formation between benzotriazole and ruthenium(IV) perchlorate in solution has been reported (38).

A spectrometric study of the OsO$_4$/btaH/NaOH system in aqueous solution has established the formation of red and green osmium benzotriazole complexes with metal/ligand ratios of 1:4 and 1:2, respectively (224). Addition of benzotriazole to OsO$_4$ in ethanol affords a red stock solution which on acidification precipitates a yellow product of stoichiometry "Os(OH)$_3$(btaH)$_3$." This in turn converts to the green (X = Cl) or purple (X = Br) halides "OsX$_3$(btaH)$_3$" on heating in NaX/HX solution at 100°C. Treatment of the same stock solutions with metal salts leads to precipitation of bright red or yellow complex salts "$M_3[Os(O)_3(btaH)_4(H_2O)_2]$" (M = Na, $\frac{1}{2}$Ba, or $\frac{1}{2}$Ca), "$M_3[Os(O)_3(btaH)_4]$" (M = K or $\frac{1}{2}$Pb), or "$M_2[Os(O)_2(OH)(btaH)_3]$" (M = Ag or $\frac{1}{2}$Zn) (222). In the absence of magnetic and spectroscopic data it is not possible to establish the true nature of these products, although probably polymeric benzotriazolate-, oxide-, or halide-bridged structures are involved, similar to those postulated above for the related ruthenium complexes. Thermogravimetric studies reveal that the hydrated complexes lose water between 130 and 150°C and decompose completely in the range 380–560°C (223). The use of $Os(OH)_3(btaH)_3$ as a precipitate for the gravimetric determination of osmium has been described (225). More recently the complex $OsH(bta)(btaH)(CO)(PPh_3)_2$, isolated from the reaction of $OsH_2(CO)(PPh_3)_3$ with benzotriazole, has been shown to contain an H-bonded chelate bta-H···bta ligand (Fig. 4) (96).

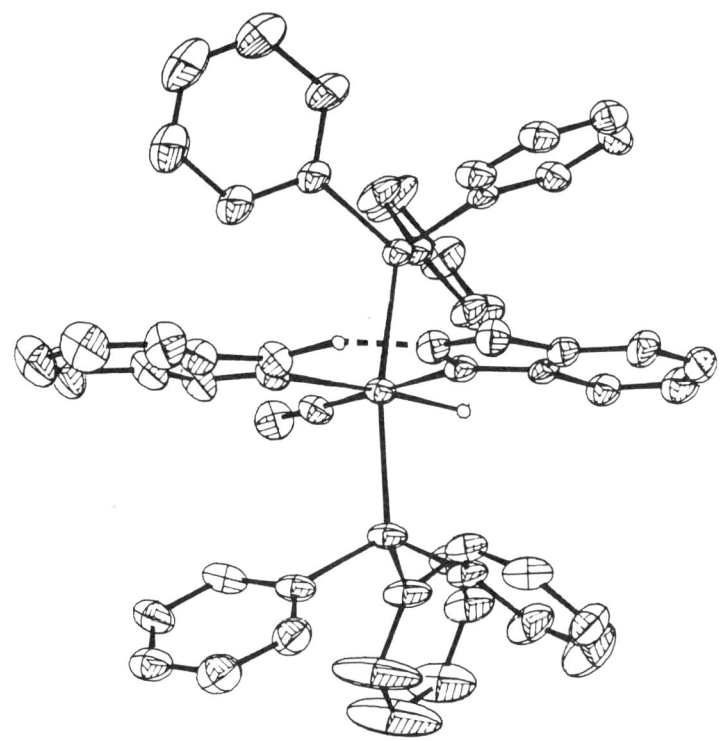

FIG. 4. Molecular structure of $OsH)(bta-H···bta)(CO)(PPh_3)_2$.

6. Cobalt, Rhodium, and Iridium

The blue crystalline salt [btaH$_2$]$_2$[CoCl$_4$] obtained by addition of benzotriazole to CoCl$_2$·6H$_2$O in 2 M HCl solution has been shown by X-ray diffraction methods to contain planar btaH$_2^+$ cations (protons at N-1 and N-3) and almost regular tetrahedral CoCl$_4^{2-}$ anions (198). The synthesis and X-ray diffraction study of the adduct CoCl$_2$-(btaH)$_2$ have been reported (191). Reaction of cobalt(III) salts with benzotriazole are reported by Cambi et al. to involve reduction to cobalt(II) (41); complexes isolated include Co(bta)$_2$·H$_2$O (40, 42) and Co$_3$(bta)$_8$ (40). Cobalt(III) ammine and ethylenediamine complexes afford [Co(bta)(NO$_2$)(NH$_3$)$_4$]Cl, Co(NO$_2$)$_3$(btaH)(NH$_3$)$_2$ (40), Co(bta)(NO$_2$)$_2$(NH$_3$)$_3$, and [Co(bta)Cl(en)$_2$]Cl (137). Magnetic data have been reported for some of these products (39). Other cobalt(III) benzotriazole derivatives include the species Co(nqo)$_2$(NO$_2$)(btaH) (nqoH = 2-nitro-1-naphthol or 1-nitro-2-naphthol) (43). Sodium salts of 4-butyroyl- and 4-benzoyl-1,2,3-triazole react with cobalt(II) halides to afford Co(4-RCO-ta)$_2$ (R = C$_3$H$_7$ or Ph) (147). Absorption of oxygen by solutions of CoCl$_2$(PPh$_3$)$_2$ in allylamine in the presence of benzotriazole leads to formation of the novel trinuclear complex Co$_3$(bta)$_6$-(CH$_2$=CH—CH$_2$NH$_2$)$_6$(Ph$_3$PO)$_2$, which has a linear chain structure analogous to that determined for the corresponding nickel(II) complex (see page 191) (57). 1-Vinylbenzotriazole ligands are coordinated through N-3 to a slightly distorted tetrahedral cobalt(II) center in CoCl$_2$(1-CH$_2$=CH-bta)$_2$ (196). An extensive range of cobalt(III) triazolate complexes have been prepared by 1,3-dipolar cycloaddition of the electron-poor acetylene CH$_3$O$_2$CC≡CCO$_2$CH$_3$ to cobalt(III) azides CoN$_3$(chel)$_2$L (chel = dmg or acac; chel$_2$ = salphen or salen; L = P or N donor ligand). Initial products are thought to be kinetically controlled N-1 isomers which rapidly rearrange to thermodynamically preferred N-2 isomers (109). The crystal structure of one such product, Co(4,5-R$_2$-ta)(dmg)$_2$(PPh$_3$) (R = MeOCO), confirms the trans stereochemistry of the complex and the N-2 coordination of the triazolate ligand (142). Reactions of alkenes with cobalt(III) azides afford Δ^2-triazoline derivatives which are too unstable to isolate (109). Addition of carbon disulfide to [AsPh$_4$]$_2$[Co(N$_3$)$_4$] affords a transient thiatriazoline-5-thionato complex [AsPh$_4$]$_2$[Co(N$_3$CS$_2$)$_4$] which rapidly decomposes to the isothiocyanate [AsPh$_4$]$_2$[Co(NCS)$_4$] (64). The orange complex CoCl$_2$(8-azahypoxanthine)$_2$, obtained by treatment of CoCl$_2$·6H$_2$O with the free ligand in aqueous solution, is thought to possess a polymeric structure with bridging 8-azahypoxanthine ligands (108). Rose-colored precipitates of stoichiometry Rh(bta)$_3$(H$_2$O)$_3$ have been obtained from reactions between $RhCl_3 \cdot 3H_2O$ (229), Na_3RhCl_6 (129), or $Rh(NO_3)_3$ (232) and benzotriazole. Under similar conditions rhodium sulfate and rhodium halides afford $Rh(OH)(bta)_2(H_2O)_3$ and $RhX_2(bta)(btaH)_n$ (X = Br, n = 2; X = I, n = 1), respectively (232). The low solubility of these products suggests the presence of polymeric structures with bridging halide and bta⁻ or btaH ligands. Osmometric molecular weight data imply a hexameric formulation for the yellow complex $[Rh(bta)(CO)(PPh_3)_2]_n$ obtained from $RhCl(CO)(PPh_3)_2$ and Tl(bta) (83). The benzotriazole adduct $RhCl(btaH)(C_8H_{12})$ carbonylates to give $RhCl(btaH)(CO)_2$ and reacts with base (NEt_3) to form $[Rh(\mu\text{-}bta)(C_8H_{12})]_2$ (156). Carbonylation of the latter product affords $[Rh(\mu\text{-}bta)(CO)_2]_2$, which in turn reacts with triphenylphosphine to generate $[Rh(\mu\text{-}bta)(CO)(PPh_3)]_2$ (156). The N-1/N-2-bridging benzotriazolate ligands are proposed for the binuclear complexes (156) and this arrangement has been confirmed for $[Rh(\mu\text{-}bta)(C_8H_{12})]_2$ by an X-ray diffraction study (208). $[Rh(bta)(n\text{-}C_3H_5)_2]_3$ has been assigned a trinuclear benzotriazolato-bridged structure on the basis of X-ray data for related trinuclear azolate complexes (156a). A novel tetranuclear structure (**16**) involving benzotriazolate ligands coordinated

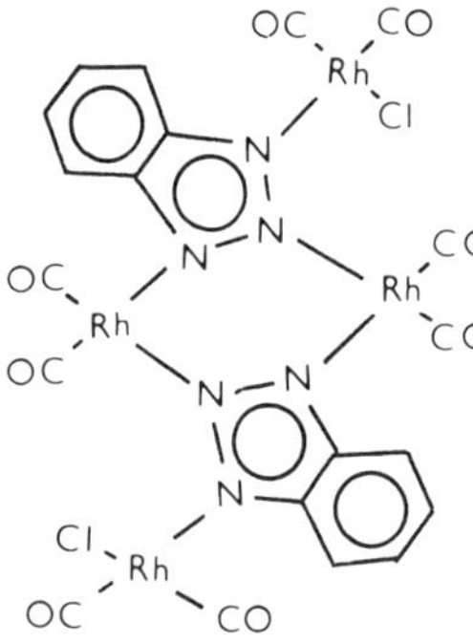

16

through all three nitrogen atoms has been proposed for the complex $Rh_4Cl_2(bta)_2(CO)_8$ obtained by mixing solutions of $RhCl(btaH)(CO)_2$ and $Rh(acac)(CO)_2$ (156). Related tetranuclear complexes form when $[RhCl(C_8H_{12})]_2$ is added to $[Rh(\mu\text{-}bta)(C_8H_{12})]_2$ and when $Rh(acac)(C_8H_{12})$ is added to $RhCl(btaH)(C_8H_{12})$ (156). The complex $RhCl(btaH)(CO)_2$ has also been obtained from $RhCl(benzothiadiazole)(CO)_2$ and benzotriazole (116). 2-Aryl-4,5-dimethyl-1,2,3-triazoles react with $RhCl_3 \cdot 3H_2O$ in boiling 2-methoxyethanol to afford the cyclometallated products $[RhCl(C-N)_2]_2$ (C—N = **17**), which are cleaved by

                    17

pyridine and tributylphosphine to form RhCl(C—N)$_2$py and RhCl-(C—N)$_2$(PBu$_3$), respectively (*152*). Rhodium(III) azides react with hexafluorobut-2-yne to yield products containing N-2-coordinated and/or N-1/N-2-bridging bis(trifluoromethyl)triazolate ligands (Scheme 1) (*178*). An X-ray diffraction study on the complex Rh$_2$(μ-N$_3$){μ-4,5-(CF$_3$)$_2$-ta}{4,5-(CF$_3$)$_2$-ta}$_2$(C$_5$Me$_5$)$_2$ has confirmed the proposed structure (*178*). An unstable 1,2,3,4-thiatriazole-5-thiolate complex of rhodium formed from Rh(C$_5$Me$_5$)(N$_3$)$_2$(PPh$_3$) and CS$_2$ easily breaks down to form Rh$_2$(C$_5$Me$_5$)$_2$(SCN)$_4$ (*178*).

Reactions of Na$_3$IrCl$_6$ with benzotriazole afford Ir(bta)$_3$·3H$_2$O and Ir(bta)$_3$(btaH); "bromobenzotriazole" gives analogous products (*129*). Thallium benzotriazolate reacts with [Ir(CO)(Me$_2$CO)(PPh$_3$)$_2$][PF$_6$] and IrCl(CO)(PPh$_3$)$_2$ to yield tetrameric [Ir(bta)(CO)(PPh$_3$)]$_4$ and bimetallic Ir(bta)$_2$(CO)(PPh$_3$)Tl·C$_6$H$_6$, respectively. The latter product breaks down in methanol to form *trans*-Ir(bta)(CO)(PPh$_3$)$_2$, which has been shown by diffraction methods to contain N-1-coordinated benzotriazolate ligands bound to square–planar iridium(I) (*33*). *mer*-IrH$_3$(PPh$_3$)$_3$ reacts with benzotriazole to form IrH$_2$(bta–H···bta)-(PPh$_3$)$_2$, an octahedral iridium(III) complex with an H-bonded bta–H···bta chelate ligand (*96*). The cyclometallated iridium(III) complex [IrCl(C—N)$_2$]$_2$ (C—N = **17**) has been obtained from iridium(IV) chloride and 2-*p*-tolyl-4,5-dimethyl-1,2,3-triazole (*152*).

7. Nickel, Palladium, and Platinum

Benzotriazole, 1-ethylbenzotriazole, and 1-vinylbenzotriazole form 1:1 or 2:1 adducts with nickel(II) chloride (*55*). X-Ray crystallographic data have been recorded for NiCl$_2$(btaH)$_2$ (*191*). Benzotriazole reacts with nickel acetate to form Ni(bta)$_2$·H$_2$O; 1,2,3-triazole gives a product of uncertain nature (*42*). 4-Butyroyl- and 4-benzoyl-1,2,3-triazolate complexes of nickel (II), Ni(4-RCO-ta)$_2$·nH$_2$O (R = C$_3$H$_7$, n = 1; R = Ph, n = 2), have been obtained as lilac precipitates by treatment of NiCl$_2$·6H$_2$O with the sodium salts of the triazoles (*147*). A light brown product, [Ni(bta)(btaH)][BF$_4$], of unknown structure forms

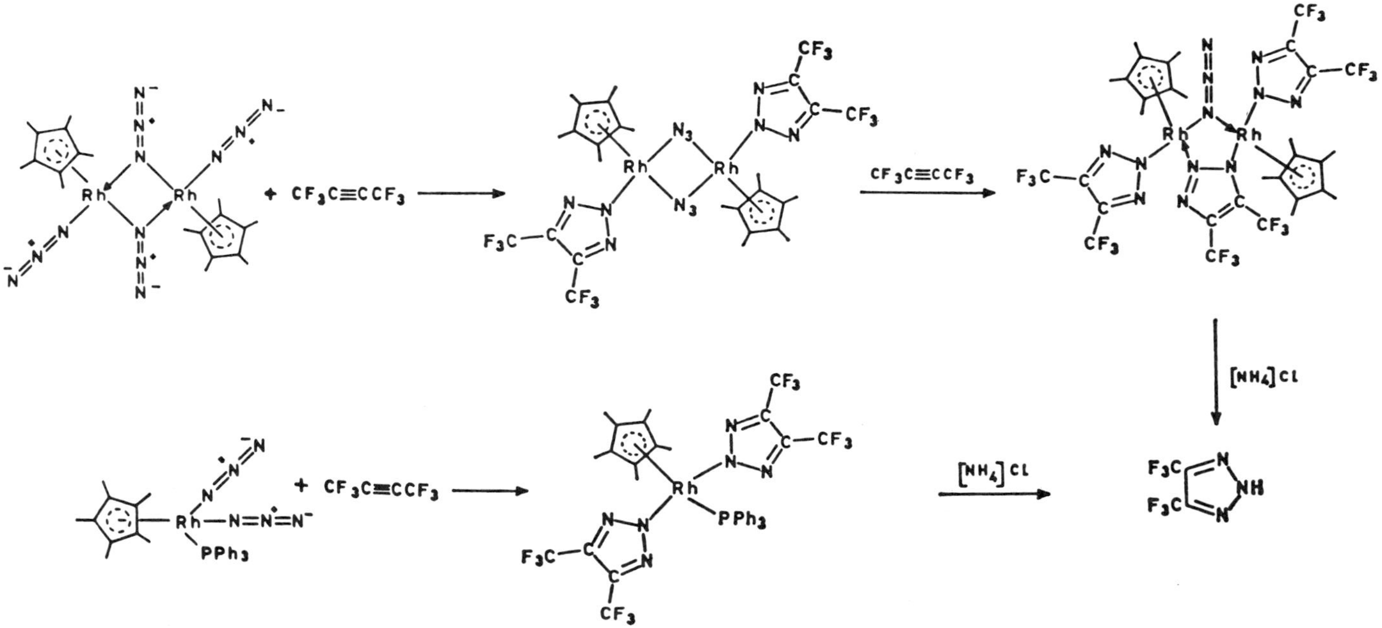

SCHEME 1. Reactions between $CF_3C\equiv CCF_3$ and some rhodium azides.

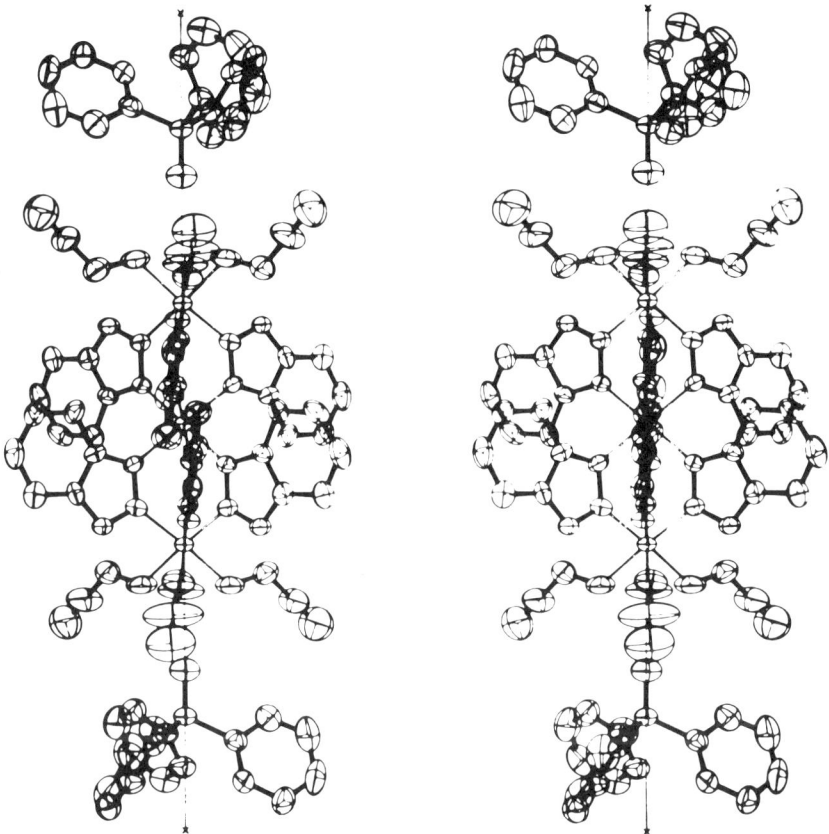

Fig. 5. Stereoscopic view of the structure of $Ni_3(bta)_6(CH_2=CH-CH_2-NH_2)_6\cdot(OPPh_3)_2$.

when benzotriazole and triethylamine is added to a suspension of $[Ni(C_5H_5)(C_7H_8)][BF_4]$ in THF (172). Mixtures of $NiI_2(PPh_3)_2$ and benzotriazole in neat allylamine react under nitrogen to form blue crystals of the trinuclear complex $Ni_3(bta)_6(CH_2=CH-CH_2NH_2)_6\cdot(PPh_3)_2$, which are oxidized by air to yield the closely related phosphine oxide adduct $Ni_3(bta)_6(CH_2=CH-CH_2NH_2)_6(OPPh_3)_2$ (57). The latter product has been shown by X-ray diffraction methods to possess a novel linear trinuclear structure (Fig. 5) and to be isostructural with the corresponding cobalt(II) derivative (57, 133). A rather similar linear trinuclear structure (Fig. 6) has been reported for the pale purple complex $Ni_3(bta)_6(NH_3)_6\cdot 2Me_2CO\cdot 2H_2O$ obtained by mixing solutions of $NiCl_2\cdot 6H_2O$ in 7 M ammonia and benzotriazole in

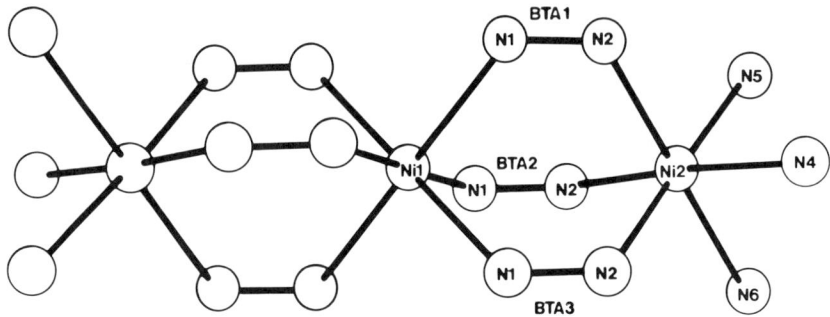

Fig. 6. Coordination about the nickel atoms in $Ni_3(bta)_6(NH_3)_6 \cdot 2Me_2CO \cdot 2H_2O$.

acetone (*199*). An N-bonded thiatriazoline-5-thionato complex of nickel(II), $Ni(CS_2N_3)(C_5H_5)(PBu^n_3)$, obtained by addition of carbon disulfide to $[Ni(C_5H_5)(PBu^n_3)_2][N_3]$ pyrolyzes to form the corresponding isothiocyanate $Ni(C_5H_5)(NCS)(PBu^n_3)$ (*184*).

Benzotriazole adducts with palladium(II) salts include $PdCl_2(btaH)$ (*230*), $PdCl_2(btaH)_2$ (*95, 172, 219, 230*), $PdI_2(btaH)_2$ (*95*), $Pd(NO_3)_2$-$(btaH)_2$ (*95*), $PdCl_2(1\text{-}Bz\text{—}btaH)_2$, and $PdCl_2(1\text{-}CH_2\text{=}CH\text{—}btaH)_2$ (*216, 219*). Far infrared data [$\nu(Pd-Cl)$ 370 and 320 cm^{-1}] indicate a cis configuration for $PdCl_2(btaH)_2$; the solvate $PdCl_2(btaH)_2(DMSO)_2$ is thought to contain dimethyl sulfoxide (DMSO) hydrogen bonded to the btaH ligands (*172*). Quantitative estimation of palladium by precipitation as $PdCl_2(btaH)_2$ (*229*) or $PdCl_2(5\text{-}Br\text{—}btaH)_2$ (*130*) and by amperometric titration of $PdCl_2$ with btaH (*229*) has been described.

Buffered solutions of $Pd(NO_3)_2$ and $Pd(SO_4)$ react with benzotriazole to form gray, beige, or yellow insoluble complexes $Pd(NO_3)(bta)$ (*231*), $Pd(NO_3)(bta)(btaH)$ (*95, 231*), $Pd_2(SO_4)(bta)_2$, and $[Pd(bta)(btaH)]_2SO_4$ (*231*). Possible structures for these complexes and their platinum(II) analogs are described below. Cleavage of $[PdCl(\eta^3\text{-}C_3H_5)]_2$ with benzotriazole affords $PdCl(\eta^3\text{-}C_3H_5)(btaH)$, which in dimethyl sulfoxide solution converts to the fluxional η^1-allyl $PdCl(\eta^1\text{-}C_3H_5)(btaH)(DMSO)$, similar behavior is reported for 1-benzyl- and 1-vinylbenzotriazoles (*219*). 2-Aryl-1,2,3-triazoles undergo cyclometallation reactions with palladium and platinum chlorides to form binuclear complexes $[MCl(C\text{—}N)]_2$ (C—N = **17**) (*3, 152*); chlorination of the palladium products yields 2-(o-chloroaryl)-1,2,3-triazoles (*3*). 1,3-Dipolar cycloaddition of $RC\equiv CR$ (R = CO_2Me) (*114*) and CS_2 (*17, 64*) to $Pd(N_3)_2(PPh_3)_2$ generates products with structures **18** and **19**, respectively. The latter complex has also been obtained from $Pd(NO_3)_2$, $Na(CN_3S_2)$, and triphenylphosphine (*17, 64*). Crystallization of **18** in the presence of free triphenylphosphine affords monomeric $Pd(R_2C_2N_3)_2(PPh_3)_2$ (*114*). Azido-bridged complexes of palladium(II) and platinum(II), $[M(N_3)(diene\text{-}OMe)]_2$ and $[Pd(N_3)(\eta^3\text{-allyl})]_2$, undergo 1,3-cycloaddition reactions with CS_2 to form thiatriazolate-bridged species, $[M(CN_3S_2)(diene\text{-}OMe)]_2$ and $[Pd(CN_3S_2)(\eta^3\text{-allyl})]_2$, respectively (*37*). Reactions between platinum(II) or platinum(IV) halides and benzotriazole under pH-controlled conditions have afforded a wide variety of insoluble precipitates (see Table II). The rather intractable nature of these products has prevented collection of evidence concerning their structures (*228*). However, given the strong preference of platinum(II) and platinum(IV) for four and six coordination, respectively, it is possible to speculate, on the basis of stoichiometry alone, that these complexes contain bridging bta and/or chelating bta–H···bta ligands. Thus products 1 and 8 in Table II can be tentatively assigned structures **20** and **21**, respectively. Similar structures can be advanced for the related palladium complexes mentioned above. The 8-azapurine adducts $PdCl_2$(8-azaguanine) and $PtCl_2$(8-azaadenine) have been synthesized and their infrared spectra reported (*108*).

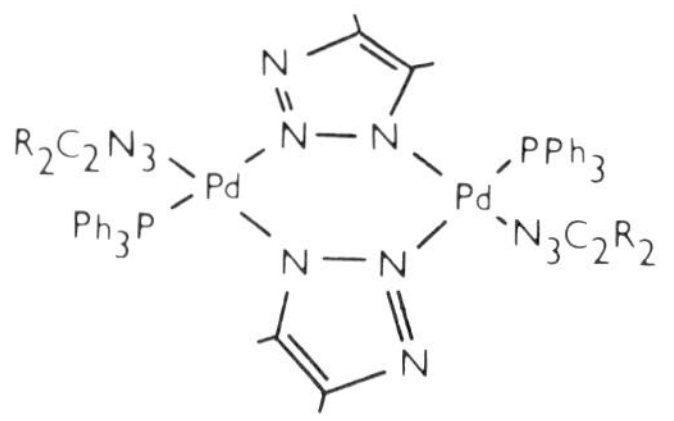

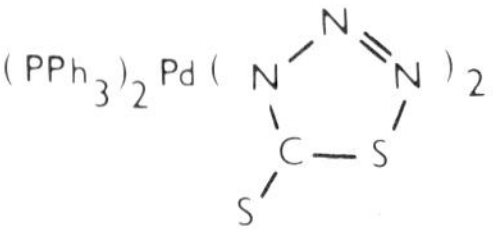

TABLE II

PLATINUM–BENZOTRIAZOLE COMPLEXES

	Reagents	pH	Product	Color
1	$[PtCl_4]^{2-}$/btaH	6.5	$Pt(bta)_2(btaH)_2$	White
2	$[PtCl_4]^{2-}$/btaH	1	$PtCl(bta)(btaH)_3$	Greenish white
3	$[PtI_6]^{2-}$/btaH	1	$PtI(bta)(btaH)H_2O$	Beige
4	$[PtI_6]^{2-}$/btaH	6	$Pt(bta)_2(H_2O)_2$	Gray
5	$[PtBr_6]^{2-}$/btaH	6	$Pt(bta)_4Pt(bta)_2(btaH)_2$	Beige
6	$[PtCl_6]^{2-}$/btaH	—	$PtCl(bta)_3(btaH)_2$	Yellow
7	$[PtCl_6]^{2-}$/btaH	6.7	$Pt(bta)_4(btaH)_2H_2O$	Gray-white
8	$[PtCl_6]^{2-}$/btaH	1	$PtCl_2(bta)_2(btaH)_2$	Yellow
9	$[PtBr_6]^{2-}$/btaH	1	$PtBr(bta)_3$	Orange

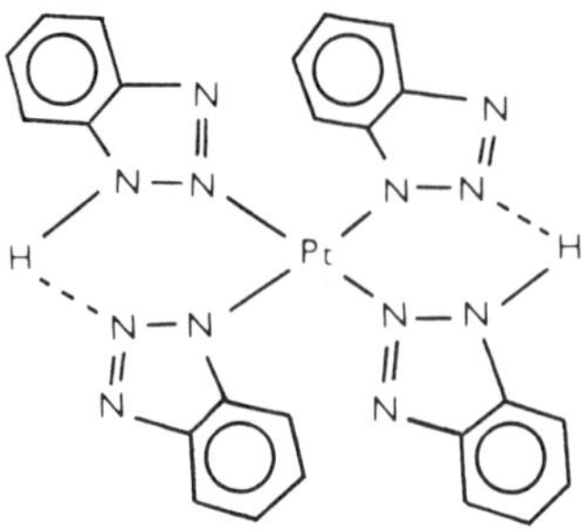

20 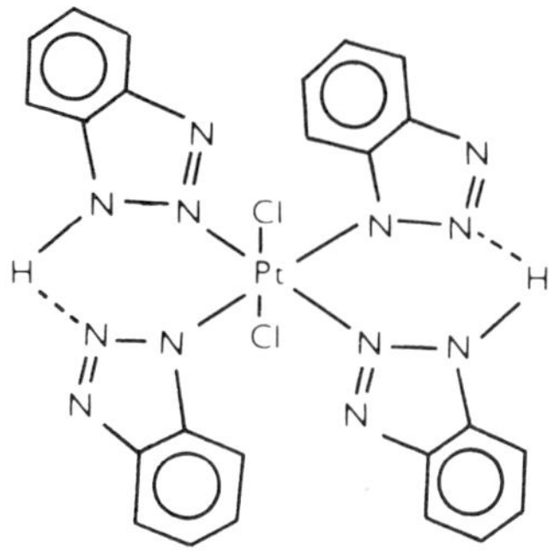   21

8. Copper, Silver, and Gold

The technological importance of triazoles in general and benzotriazole in particular as anticorrosion agents for copper and its alloys (*99*) has generated much interest in copper triazole complexes. The protective action of benzotriazole is thought to involve attack on surface oxide or hydroxide, leading to formation of an impenetrable layer of copper(I) [or possibly copper(II)] benzotriazolate salts, which serve as a barrier to further reaction (*2, 182, 233*). For a full discussion of the technological aspects of this field interested readers are referred to recent papers (*2, 171, 172, 182, 217*) and references therein. Many copper(I) and/or copper(II) benzotriazole complexes have been reported and several of these have been found to display novel structural features. The yellow tetrameric copper(I) adduct [CuCl(btaH)]$_4$, obtained from CuCl and benzotriazole in dichloromethane solution, is converted by triethylamine to the insoluble polymeric copper(I) complex $\{Cu(bta)\}_n$. The copper(I) chloride adduct displays intense yellow luminescence ($\lambda_{max}^{excit} = 381$ nm, $\lambda_{max}^{emiss} = 563$ nm); for a series of copper(I) halide adducts λ_{max}^{emiss} displays a bathochromic shift I < Br < Cl < CN, indicating emission from an excited state produced by a metal → ligand charge transfer process (*194*). Benzotriazole attacks copper surfaces to give a yellow surface layer with the characteristics of $\{Cu(bta)\}_n$ (*193, 194*). Acid solutions of CuX/KX (X = Cl, Br, or I) and Cu$_2$SO$_4$ react with benzotriazole to deposit insoluble, diamagnetic pale yellow to orange precipitates of the adducts Cu$_2$X$_2$(btaH) and Cu$_2$(SO$_4$)(btaH), respectively (*106*).

Benzotriazole adducts of copper(II) halides include CuCl$_2$(btaH) (*172*), CuCl$_2$(btaH)$_2$ (*191*), CuCl$_2$(btaH)$_2$·0.5H$_2$O, CuCl$_2$(btaH)$_3$, CuCl$_2$(btaH)$_3$·H$_2$O, CuBr$_2$(btaH), and CuBr$_2$(btaH)·0.25H$_2$O (*172*). Similar

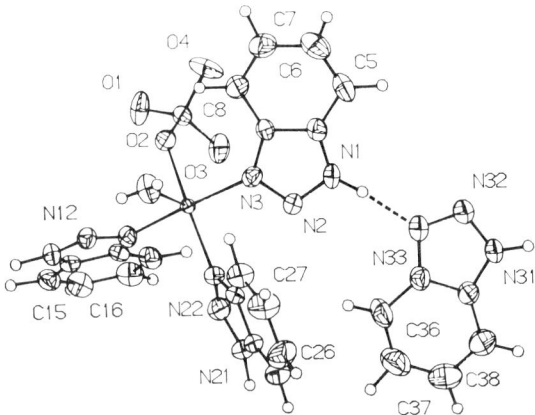

FIG. 7. Structure of $Cu(SO_4)(btaH)_3(H_2O) \cdot btaH$.

adducts have been reported for 1-methyl- and 5-nitrobenzotriazole (172). Green crystals of $CuCl_2(btaH)_2 \cdot 0.5H_2O$ have been shown to contain binuclear chloride-bridged units. The geometry around each copper atom is approximately trigonal–bipyramidal with equatorial chlorine atoms and axial N-3-coordinated benzotriazole ligands (197). The complex exhibits a weakly ferromagnetic interaction indicative of a triplet ground state lying 1.8 ± 0.2 cm^{-1} below the singlet excited state (93). A copper sulfate adduct, "$CuSO_4(btaH)_4 \cdot 2H_2O$," prepared from the components in aqueous sulfuric acid at pH 1 (105), has been reformulated as a monohydrate and shown to adopt a structure (Fig. 7) comprising tetragonal–pyramidal $Cu(SO_4)(btaH)_3 H_2O$ units and noncoordinated btaH groups linked in a three-dimensional array by H bonds (97). On attempted dissolution in water the complex converts to an insoluble green product of stoichiometry "$[Cu(btaH)_2(OH)(H_2O)]OH$" (105).

Green polymeric $Cu(bta)_2$, obtained from copper(II) carboxylates and benzotriazole in anhydrous toluene, is insoluble in dimethyl formamide and dimethyl sulfoxide but forms a yellow solution in pyridine (172). It has been reported that copper can be quantitatively separated and gravimetrically estimated as $Cu(bta)_2$ (46). The monohydrate $Cu(bta)_2 H_2O$, obtained as blue crystals from $CuCl_2 \cdot 2H_2O$ and benzotriazole in 7 M ammonia solution, has been shown to possess a polymeric structure (Fig. 8) with tetragonal–bipyramidal copper(II) centers linked by bridging benzotriazole and water ligands (201). A dihydrate $Cu(bta)_2 \cdot 2H_2O$ has also been described (42). The adduct

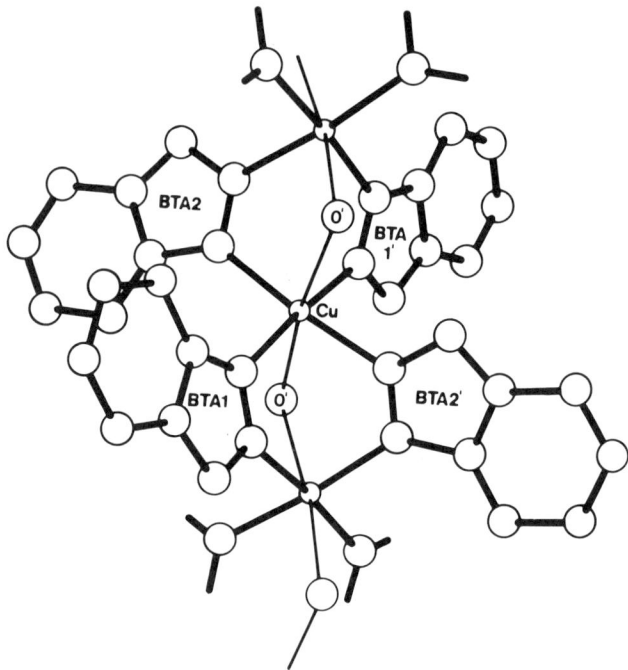

FIG. 8. Coordination about copper atoms in $[Cu(bta)_2(H_2O)]_n$.

$Cu(bta)_2(o\text{-phen})_2$ has been obtained as green nodules by adding triethylamine to a blue solution of $[Cu(o\text{-phen})_2][BF_4]_2$ and benzotriazole in methyl cyanide (172). Mixtures of $CuX_2 \cdot 2H_2O$ (X = Cl or Br) and benzotriazole in ethanol deposit green precipitates of stoichiometry CuX(bta) (100, 172, 194); an acetate, $Cu(O_2CMe)(bta)$, has also been reported (42). Under similar conditions $Cu(NO_3)_2 \cdot 3H_2O$, $Cu(BF_4)_2$, and $Cu(SO_4) \cdot 5H_2O$ afforded green $Cu_2(NO_3)_2(bta)_2(btaH)$, turquoise $Cu_2(bta)_3(BF_4)$, and dull green $Cu_2(SO_4)(bta)_2(btaH)$, respectively (172). A remarkable pentanuclear structure (22) has been proposed for the dark green copper(II) complex $Cu_5(bta)_6(acac)_4$ obtained from equimolar amounts of $Cu(NO_3)_2 \cdot 3H_2O$, acetylacetone, and benzotriazole in ammoniacal methanol or from $Cu(acac)_2$ and benzotriazole in dichloromethane solution (131). A closely related pentanuclear structure (Fig. 9) has been established by X-ray diffraction methods for the copper(I/II) complex $Cu_5(bta)_6(CNBu^t)_4$; the central copper(II) is octahedrally coordinated whereas the peripheral copper(I) ions are in an approximately tetrahedral environment (92, 113). The form of the

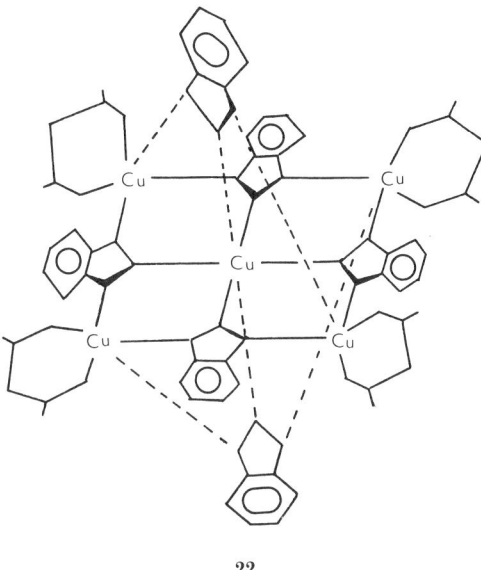

22

copper(II) hyperfine structure in the EPR spectrum is consistent with the conclusion that the large Cu(II)–Cu(II) distance (12.858 Å) is sufficient to ensure that the effects of exchange interactions are virtually nonexistent and that electronic dipolar broadening is small

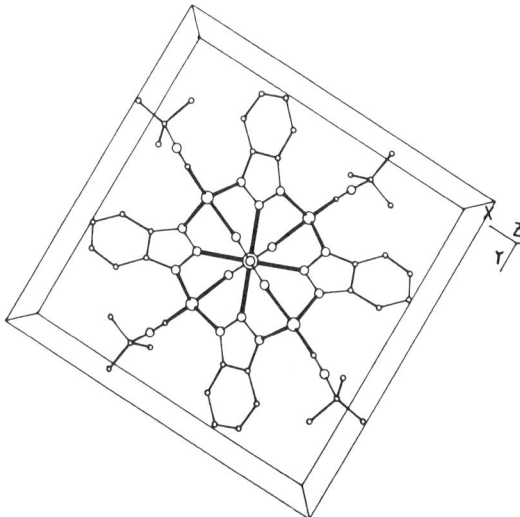

FIG. 9. Structure of $Cu_5(bta)_6(CNBu^t)_4$ viewed down a crystallographic $\bar{4}$ axis.

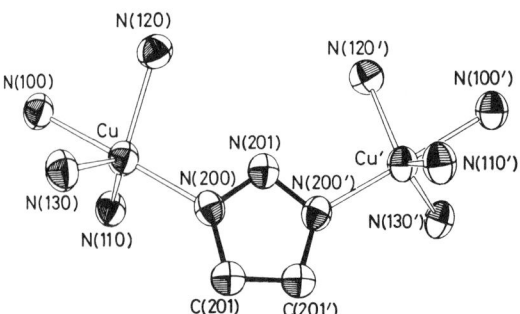

FIG. 10. Coordination geometry of the copper atoms in [Cu$_2$(tmbma)$_2$(bta)][NO$_3$]$_3$.

(*113*). Related complexes Cu$_5$(bta)$_6$(RNC)$_4$ (R = c-C$_6$H$_{11}$, Pr^i, and Ph) have also been studied by EPR methods (*10, 113*). Studies on models for superoxide dismutase have afforded a green binuclear copper complex [Cu$_2$(μ-bta)(tmbma)$_2$][NO$_3$]$_3$ [tmbma = tris(N^1-methylbenzimidazol-2-ylmethyl)amine] in which the copper centers [Cu$\cdots$Cu = 5.536(2) Å] are linked by an N-1/N-3-bridging benzotriazolate ligand (Fig. 10) (*91*). Variable-temperature magnetic susceptibility measurements indicate antiferromagnetic interactions with J values of approximately -12 cm^{-1}. Q- and X-band EPR spectra show typical $S = 1$ signals (*91*). A more detailed single-crystal EPR study has established that even with a Cu$\cdots$Cu distance of 5.536 Å an exchange contribution to the zero-field splitting is operative (*21*). An N,O-chelate structure (Fig. 11) has been

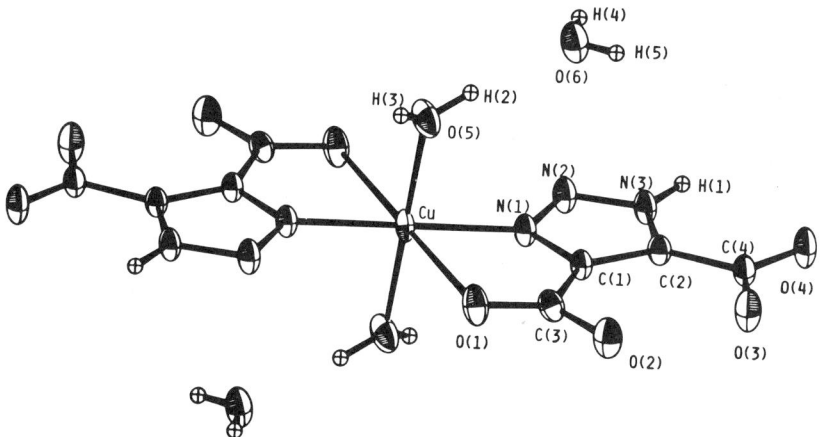

FIG. 11. ORTEP view of diaquabis(1,2,3-triazole-4,5-dicarboxylato)copper(II)dihydrate.

found for the copper(II) derivative of 1,2,3-triazole-4,5-dicarboxylic acid; related structures **23** and **24** are proposed for complexes of 1,2,3-triazole-4-carboxylic acid and 1-methyl-1,2,3-triazole-4-carboxylic acid (*155*). A polymeric structure with the 8-azaguanine bridging by coordination through any two of the nitrogen atoms, N-3, N-7, or N-8, is proposed for Cu(8-azaguanine)(OH)$_2 \cdot$5H$_2$O (*108*). The reaction of 8-azaadenine with CuCl$_2$ in 0.36 M HCl is accompanied by ring opening at C-2, leading to formation of tetrachlorobis-2-[(5-amino-4-carboxamidinium)-1,2,3-triazole]copper(II) monohydrate, which has been shown to possess a tetragonally distorted structure (Fig. 12) with axial Cu–Cl (2.967 Å) and equatorial Cu–Cl and Cu–N distances of 2.258 and 2.049 Å, respectively (*167*). Copper(I) azides Cu(N$_3$)(PPh$_3$)$_2$ (*234*), Cu$_2$(N$_3$)$_2$(dppe)$_3$ (*234, 235*), and Cu(N$_3$)(o-phen)(PPh$_3$) (*121*) react with carbon disulfide to form N- or N,S-coordinated 5-thio-1,2,3,4-thiatriazolato complexes Cu(CN$_3$S$_2$)(PPh$_3$), Cu$_2$(CN$_3$S$_2$)$_2$(dppe)$_3$, and Cu(CN$_3$S$_2$)(o-phen)(PPh$_3$), respectively, which undergo thermal or photolytic decomposition to the corresponding N thiocyanates.

Silver(I) salts of 1,2,3-triazole (*9*) and various substituted triazoles

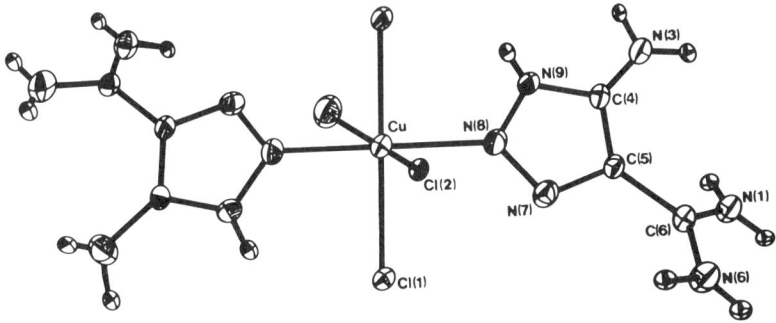

FIG. 12. View of tetrachlorobis-2-[(5-amino-4-carboxamidinium)[1,2,3]triazole]copper-(II) molecule.

(*8, 26, 82, 86, 207*) were reported by von Pechmann and others soon after the discovery of the parent compounds. Silver(I) benzotriazole derivatives include the salt Ag(bta) (*59, 206*) and the complex nitrate [Ag(btaH)$_2$][NO$_3$] (*200*). A structure composed of alternate layers of two-coordinate [Ag(btaH)$_2$]$^+$ cations and NO$_3$$^-$ anions has been established for the latter product (*200*). Benzotriazole (*44, 173, 206*) and 5-bromobenzotriazole (*206*) have been proposed as precipitants for silver(I). Silver(I) 5-thio-1,2,3,4-thiatriazolate has been isolated as a white, slightly photosensitive solid (*126*). 1,3-Dipolar addition of carbon disulfide to silver(I) azides Ag(N$_3$)(PAr$_3$)$_2$ affords 5-thio-1,2,3,4-thiatriazolate complexes Ag(CN$_3$S$_2$)(PAr$_3$)$_2$, which are thought to exist in two isomeric forms, structures **25** and **26** (*235*). A related gold(I)

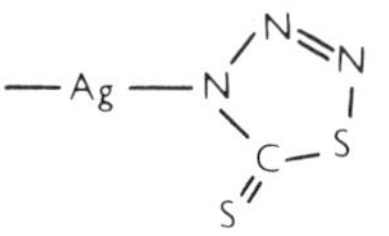

— Ag — N—N=N—C—S (S)   — Ag — N—N=N—S—C—S

25   **26**

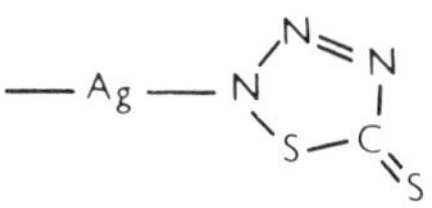

monophosphine derivative Au(CN$_3$S$_2$)(PPh$_3$) has been similarly prepared (*235*). By comparison with the simple salts M(CN$_3$S$_2$) (M = Cu, Ag, or Au) (*195*), these phosphine-containing complexes are relatively stable. The gold(I) complexes AuCl(PPh$_3$) and AuCl(SMe$_2$) react with benzotriazole in alcoholic KOH solution to yield Au(bta)(PPh$_3$) and {Au(bta)}$_3$, respectively (*28, 135*). ^{197}Gold Mössbauer spectra for the latter product consist of a single doublet, indicating that the three gold atoms in each molecule are equivalent and thus favoring a symmetrical nine-membered ring structure with N-1/N-2-bridging benzotriazolate ligands (*107*). The azaguanine derivative Au(azaguaninate)(PPh$_3$) has been similarly prepared (*29*). 8-Azahypoxanthine and 8-azaguanine adducts of gold(I) and gold(II) chlorides AuCl(8-azahypoxanthine)·H$_2$O, AuCl$_3$(8-azahypoxanthine), and AuCl$_3$(8-azaguanine) have been synthesized and their infrared spectra reported (*108*).

9. Zinc, Cadmium, and Mercury

The white crystalline adduct ZnCl$_2$(btaH)$_2$ (*191*) obtained by mixing the components in ethanol (*172*) or in aqueous HCl/NH$_3$ solution (*198*) has been shown to possess the expected tetrahedral structure with N-3-coordinated benzotriazole ligands (*198*). The same components react together in 7 *M* ammonia solution to afford colorless, polymeric

[Zn$_2$(bta)$_4$]$_n$ in which tetrahedral Zn^{II} centers are linked by N-1/N-3 benzotriazolate bridges (*198*). The salt [btaH$_2$]$_2$[ZnCl$_4$], obtained from ZnCl$_2$ and benzotriazole in 2 M HCl solution, is isostructural with the corresponding cobalt salt (see p. 187) (*198*). Complex formation between 5-thio-1,2,3,4-thiatriazolate anions and Zn(II) or Cd(II) has been studied by potentiometric (*149*) and polarographic (*148*) methods.

Cadmium dichloride is reported to form a 1:1 adduct with benzotriazole in aqueous ethanol (*172*). However, a colorless 1:2 adduct [CdCl$_2$(btaH)$_2$]$_n$ has recently been obtained by a similar procedure and shown to possess an infinite $>$CdCl$_2$CdCl$_2$Cd$<$ ribbon structure, with octahedral coordination about each cadmium completed by N-3-coordinated benzotriazole ligands (see Mn^{II} analog, p. 181) (*198*). Under their original formulation as salts of azidodithiocarbonic acid (HS$_2$CN$_3$), 5-thio-1,2,3,4-thiatriazolate derivatives of cadmium Cd(CN$_3$S$_2$)$_2$·2H$_2$O, mercury(I) Hg(CN$_3$S$_2$), and mercury(II) Hg(CN$_3$S$_2$)$_2$ have been synthesized and found to be shock sensitive (*195*).

A chloride-bridged structure has been proposed for the 1:1 adduct, [HgCl$_2$(btaH)]$_2$ (*172*). Mercury(II) benzotriazolate, Hg(bta)$_2$, has been obtained from mercury(II) perchlorate, benzotriazole, and triethylamine in ethanol (*172*), and from mercuric oxide and benzotriazole (*158*). Mercuric acetate and chloride react with benzotriazole to form mixed mercury(II) salts, HgX(bta) (X = MeCO$_2$ or Cl) (*139*). A dimeric structure (Fig. 13) with T-shaped coordination at mercury has been determined by X-ray diffraction methods (*203*) for the methyl derivative Hg(Me)(bta) (*172*). Cyclooctane-1,2-dione dihydrazide reacts with HgO

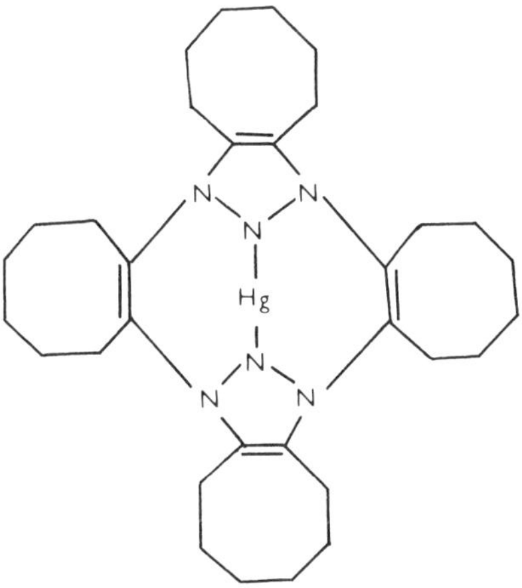

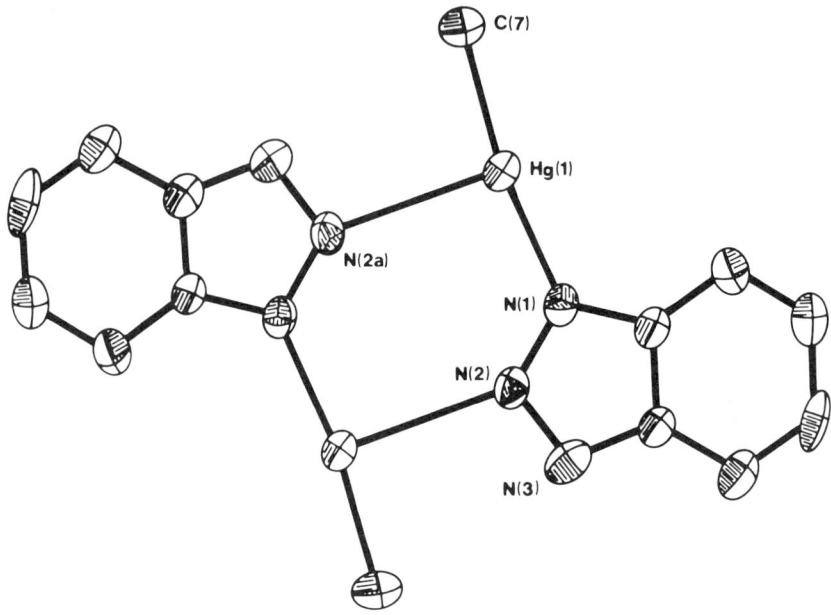

FIG. 13. ORTEP view of centrosymmetric [HgMe(bta)]$_2$ dimer.

to form the product, **27**, which is converted by aqueous HCl into the N-2-coordinated cyclooctane triazolate complexes, **28** and **29** (*139*).

28   **29**

A series of papers by Hodgson and co-workers describe the synthesis and structural characterization of zinc, cadmium, and mercury derivatives of the 8-azapurines—8-azaadenine and 8-azahypoxanthine. Coordination through N-3 occurs in the 8-azaadenine complexes ZnCl$_3$(8-azaadeninium) (Fig. 14) (*166*) and HgCl$_2$(8-azaadenine)$_2$ (Fig. 15) (*85*), but through N-7 and N-9 in Cd(8-azahypoxanthinato)$_2$(H$_2$O)$_4$ (Fig. 16) (*165*) and Hg(8-azahypoxanthinato)$_2$(H$_2$O)$_4$ (Fig. 17) (*85*), respectively. An 8-azaadenine adduct of zinc chloride, ZnCl$_2$(8-azaadenine)·2H$_2$O, has also been reported (*108*). X-Ray diffraction

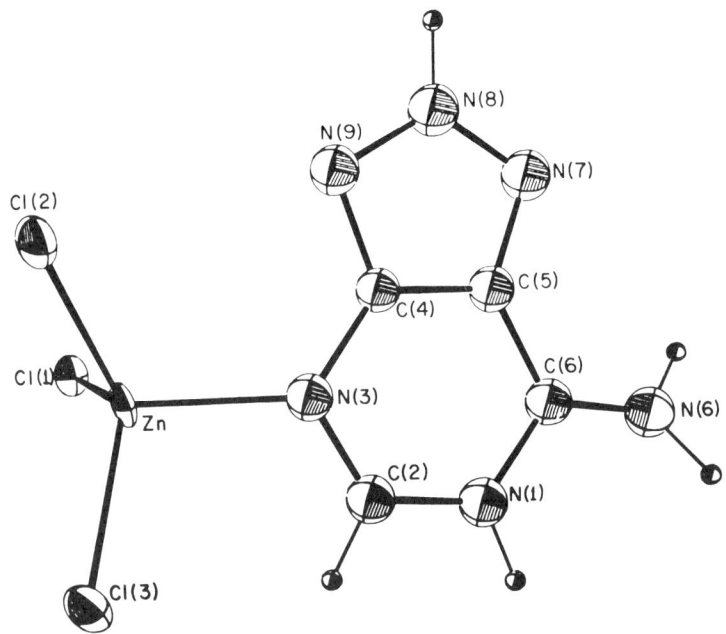

FIG. 14. View of a single molecule of trichloro(8-azaadeninium)zinc(II).

studies on four complexes isolated from the reaction of methylmercury(II) hydroxide with 8-azaadenine (Scheme 2) reveal MeHg groups bound at N-1, N-3, and N-9 as well as at the primary (N-6) metal coordination site (192).

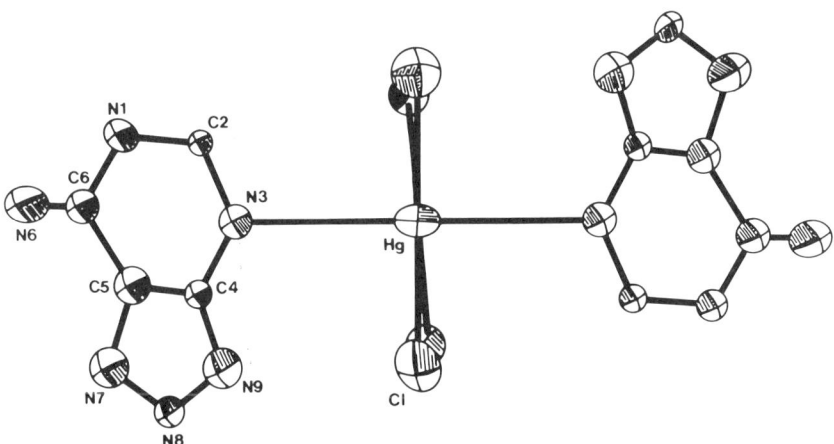

FIG. 15. View of coordination about mercury in dichlorobis(8-azaadenine)mercury(II).

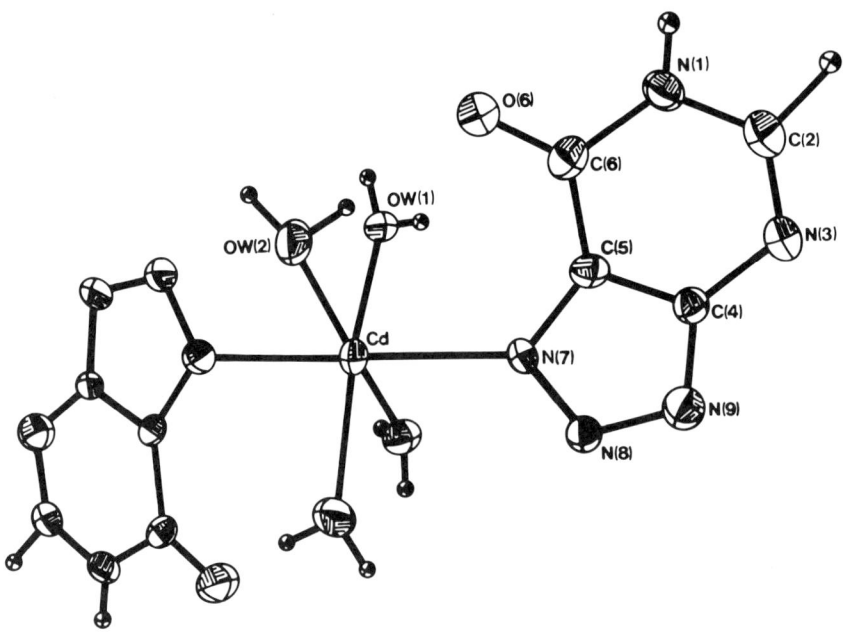

FIG. 16. View of the coordination about cadmium in tetraaquabis(8-azahypoxanthinato)cadmium(II).

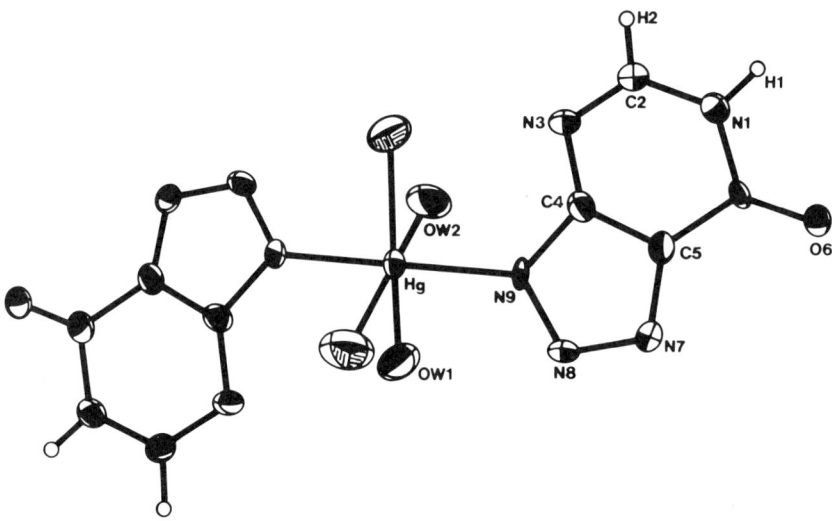

FIG. 17. View of the coordination about mercury in tetraaquabis(8-azahypoxanthinato)mercury(II).

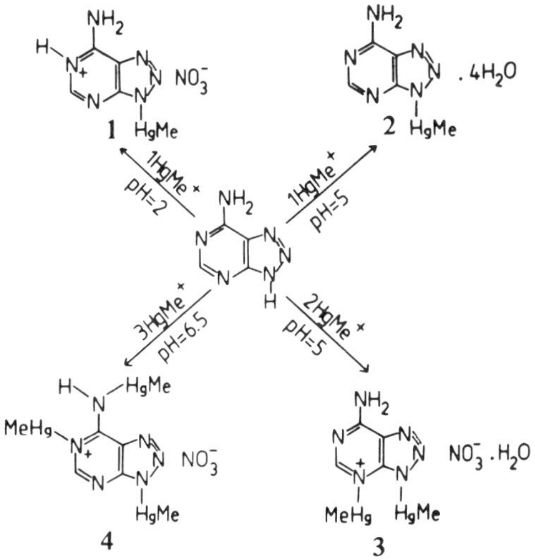

SCHEME 2. Reactions between methylmercury(II) hydroxide and 8-azaadenine.

III. Tetrazole and Tetrazolate Complexes

In this section ligands containing the tetrazole ring system are considered. These include the parent tetrazole, various nitrogen- and/or carbon-substituted tetrazoles, 1,5-pentamethylenetetrazole, and the cyclized tautomer of 2-azidopyridine. Although a few examples are mentioned in passing, thiones such as 1,2,3,4-tetrazoline-5-thione and dehydrodithione are generally excluded since they most commonly function as sulfur donors and because they have recently been extensively covered in a review by Raper (*168*) dealing with heterocyclic thione donors. Substituted tetrazoles were first prepared in 1885 by Bladin (*24*), who went on to isolate the parent tetrazole and a series of silver salts in 1892 (*25*). Tetrazoles are quite strong acids; the parent tetrazole has a pK_a value of ~4.8 (*134*), similar to that of acetic acid, and readily forms salts with transition metal ions. Tetrazolate salts and complexes are frequently heat and shock sensitive, and several have found commercial use as explosives or detonators. The parent tetrazole exists in two tautomeric forms (structures **30a** and **30b**), of which the 1H form (structure **30a**) is dominant. Monosubstituted tetrazoles exist in three isomeric forms (structures **31**, **32**, and **33**) with substituents on carbon, nitrogen-1, and nitrogen-2, respectively. The carbon-substituted forms, like the parent, exist as tautomeric mixtures of 1H

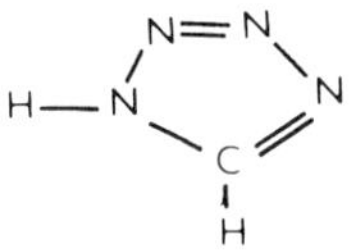

30a 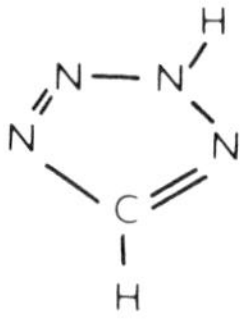 **30b**

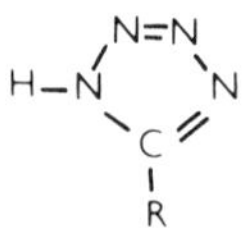

31 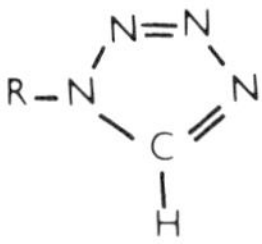 **32** 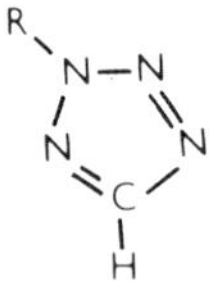 **33**

and 2H forms. In addition to simple alkyl and aryl groups, substituents include a variety of functional groups, notably CF_3, halide, NH_2, CN, $CONH_2$, and COOH. Disubstituted tetrazoles, which lack an acidic proton and can only coordinate as neutral molecules, are characterized by rather weak base strength (*162*). Examples include the 1,5-polymethylenetetrazoles, of which 1,5-pentamethylenetetrazole (structure **34**) is the most important example, and 2-azidopyridine

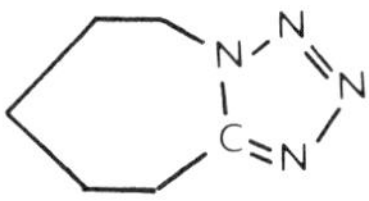

34

(structure **35a**), which coordinates as the cyclized tautomer, structure **35b**. 1,5-Pentamethylenetetrazole (metrazole or cardiazole), first reported by Schmidt in 1925 (*189*), has important medicinal uses as a

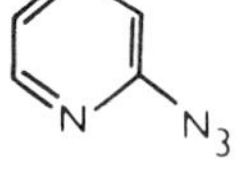

35a 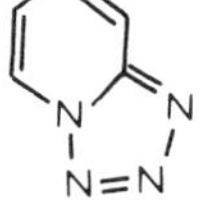 **35b**

stimulant for the central nervous system; its coordination chemistry was reviewed in 1969 (*162*). Early work on the synthesis and properties of tetrazoles has been summarized in a review by F. R. Benson (*22*). The generally accepted numbering scheme for tetrazoles which is used throughout this section is given in structure **36**. Abbreviations employed for the various ligands are tetrazole (ttaH), tetrazolate anion (tta), 1,5-pentamethylenetetrazole (1,5-pmtta), and the cyclic form of 2-azidopyridine (pytta).

36

A. SYNTHESIS

Tetrazoles combine readily with transition metal halides to form adducts, many of which are polymeric and deposit from solution as insoluble precipitates (*32, 50, 73, 83*). Reactions between hydrated metal perchlorates or tetrafluoroborates and tetrazoles—1,5-pentamethylenetetrazole and 1- or 2-substituted tetrazoles—usually in the presence of a dehydrating agent, lead to formation of the complex salts $[M(1,5\text{-pmtta})_6][Y]_2$, $[M(1\text{-R-ttaH})_6][Y]_2$, and $[M(2\text{-R-ttaH})_6][Y]_2$ (M = Mn, Fe, Co, Ni, Cu, or Zn; Y = ClO_4 or BF_4) (*51, 70, 213*). Tetrazole complexes of chromium, molybdenum, and tungsten have been obtained by thermal or photolytic displacement of CO or weakly coordinated MeCN and THF ligands from the hexacarbonyls and their substitution products (*79, 80, 160, 220*). Complex formation between transition metal carbonyls or halides and 2-azidopyridine is accompanied by cyclization of the ligand to the tetrazole form (*160*). Tetrazole and 5-substituted tetrazoles are acidic and readily form alkali metal salts which react with transition metal halides to form tetrazolate complexes (*18, 204*). The latter products can frequently be obtained directly by treatment of the metal salts with the free tetrazole in the presence of base, usually hydrazine (*141*), triethylamine (*172*), or potassium hydroxide (*136*). In some instances the use of added base is unnecessary (*69, 77, 119*). Salts of 1- or 2-substituted tetrazoles, prepared from the free tetrazoles and lithium alkyls, also react with transition metal halides to afford tetrazolate complexes (*75*).

Reactions between 5-cyanotetrazole and transition metals, when performed in boiling acetone, lead to hydrolysis of the cyano group and formation of 5-carbamyl tetrazolate complexes (68). Complexes containing 1- or 5-substituted tetrazolate anions can also be obtained by 1,3-dipolar cycloaddition of organic isonitriles (RNC) (15) or nitriles (RCN) (61), respectively, to coordinated azide ligands [Eqs. (3) and (4)].

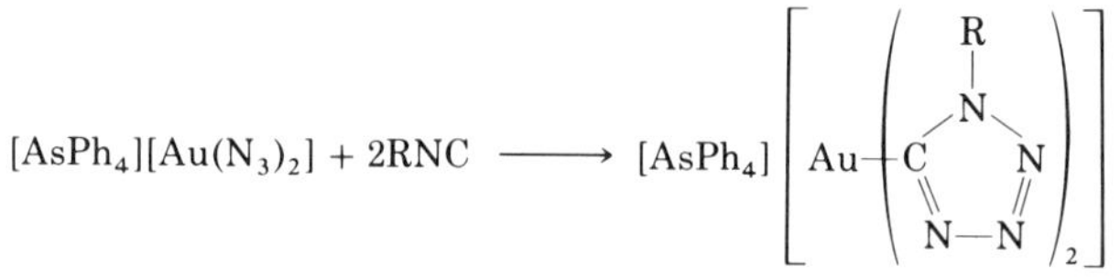

$$[AsPh_4][Au(N_3)_2] + 2RNC \longrightarrow [AsPh_4]\left[Au\left(\begin{array}{c}R\\|\\N\\C\diagdown N\\\|\quad\|\\N-N\end{array}\right)_2\right] \quad (3)$$

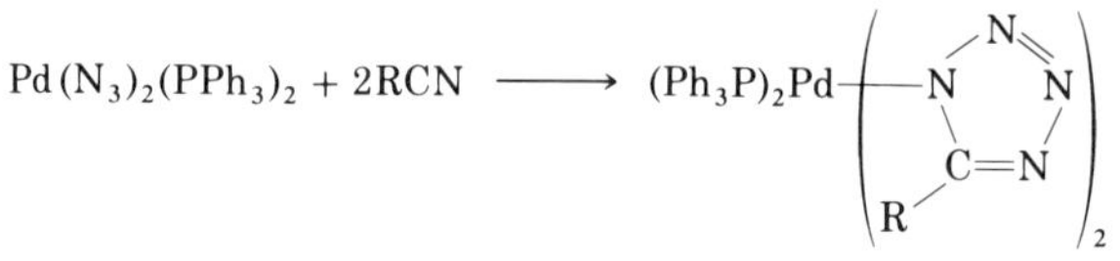

$$Pd(N_3)_2(PPh_3)_2 + 2RCN \longrightarrow (Ph_3P)_2Pd\left(\begin{array}{c}N\diagdown\\N\quad N\\C=N\\R\end{array}\right)_2 \quad (4)$$

Alternatively, the azide anion can be added to coordinated organic isonitrile (211) or nitrile (60) ligands to give the same products. Coordination of RCN ligands greatly increases their susceptibility to attack by azide anions; reactions typically go to completion in 15 minutes at 25°C (60).

B. STRUCTURAL PROPERTIES

Early workers, noting the isoelectronic nature of tetrazolate and cyclopentadienide anions, looked for evidence of structures containing η^5-tetrazolate ligands. Initial claims of success (89) were later withdrawn (90) and to date all established structures involve localized M–N bonds between metal centers and tetrazoles or tetrazolate anions. However, η^5-coordinated tetrazolate ligands have been proposed as intermediates in the N-1 to N-2 isomerization of some cobalt(III) tetrazolate complexes (164). Choice of coordination positions on the tetrazole rings is governed by electronic and steric considerations, which in turn are dictated by the nature and position of the substituent groups on the rings. Although complexes containing the parent tetrazole (ttaH) and tetrazolate anion (tta$^-$) are known, there is very little firm evidence concerning the point of coordination of these ligands. 1-Substituted tetrazoles are thought to coordinate through the N-4 position (70) and this arrangement has been confirmed by X-ray diffraction methods for $ZnCl_2(1\text{-Me-ttaH})_2$ (6). The triple tetrazole bridges in the complexes $(OC)_3Mo(\mu\text{-}1\text{-}R\text{-}ttaH)_3Mo(CO)_3$ (R = Me, cyclohexyl) are thought to involve N-3/N-4 coordination (74). 1-Substituted tetrazolate anions, which are usually formed *in situ* by the interaction of azide and isocyanide groups, must coordinate through the ring carbon (15, 65) and this mode of bonding has been confirmed for $[AsPh_4][Au(1\text{-}Pr^i\text{-}tta)_4]$ by diffraction methods (66). Relatively little is known about the coordination preferences of 2-substituted tetrazoles and tetrazolate anions. However, an X-ray diffraction study on the complex salt $[Ni(2\text{-}Me\text{-}ttaH)_6][BF_4]_2$ has established N-4 coordination for the 2-methyltetrazole ligands (213).

There also appears to be very little hard evidence concerning the coordination sites on 5-substituted tetrazoles. However, much is known about structures of complexes containing 5-substituted tetrazolate anions. Molecular orbital calculations indicate that N-1 and N-2 coordination arrangements are virtually equivalent electronically and energetically (109, 141), with perhaps a slight preference for N-1 coordination (204). However, steric factors favor N-2 coordination particularly if the ring substituent is large or the coordination site is crowded. Thus X-ray diffraction studies reveal N-1 coordination for square-planar *cis*-$Pd(5\text{-}Me\text{-}tta)_2(PMe_2Ph)_2$ (4) but N-2 coordination for *trans*-$Pd(5\text{-}Ph\text{-}tta)_2(PPh_3)_2$ (115) and several octahedral cobalt(III) complexes, including $[Co(5\text{-}Me\text{-}tta)(NH_3)_5][ClO_4]_2$ (157) and $[Co(5\text{-}CN\text{-}tta)(NH_3)_5][ClO_4]_2$ (84), where the coordination sphere is more congested. In several instances syntheses afford mixtures of N-1- and N-2-coordinated linkage isomers (141) and in others N-1 → N-2 isomerizations are observed (60, 164). Structures involving bridging through the least sterically hindered N-2 and N-3 positions of the 5-Ar-tta$^-$ ligands have been proposed for a series of polymeric 5-aryl tetrazolate complexes of iron (176), cobalt (177), and nickel (175), and a structure of this type has been confirmed by X-ray diffraction methods for the binuclear silver(I) complex $[Ag(5\text{-}CF_3\text{-}tta)(PPh_3)_2]_2$ (159). The N-3 site is the least sterically hindered coordination point on the 1,5-pentamethylenetetrazole ligand (51) and is likely to be the one used in the crowded octahedral complex cations $[M(1,5\text{-}pmtta)_6]^{2+}$. However, an X-ray diffraction study has established N-4 coordination for the monodentate 1,5-pmtta ligands in the binuclear complex $[Ag(NO_3)(1,5\text{-}pmtta)_2]_2$ (27). The bridging 1,5-pmtta ligands in the same complex are coordinated through N-3 and N-4 positions (27). For complexes of 2-azidopyridine (160), coordination at the N-3 position on the cyclized ligand seems probable on steric grounds.

Generally monodentate N-1-, N-2-, or C-5-coordinated tetrazolate anions appear to adopt planar, regular pentagonal structures with

TABLE III

X-Ray Diffraction Studies on Tetrazole and Tetrazolate Complexes

Complex	Tetrazole- and tetrazolate-bonding mode	Reference
Tetrazole Complexes		
[Ni(2-Me-ttaH)$_6$][BF$_4$]$_2$	Monodentate/N-4	213
[Cu(1-Me-ttaH)$_6$][BF$_4$]$_2$	Monodentate/N-4	220a
{Ag(NO$_3$)(1,5-pmtta)$_2$}$_2$	Monodentate/N-4; bridging/N-3/N-4	27
ZnCl$_2$(1-Me-ttaH)$_2$	Monodentate/N-4	6
Tetrazolate Complexes		
Ni(5-O$_2$N-tta)$_2$(H$_2$O)$_4$	Monodentate/N-2	43a
[Co(5-Me-tta)(NH$_3$)$_5$][ClO$_4$]$_2$	Monodentate/N-2	67
		157
[Co(5-CN-tta)(NH$_3$)$_5$][ClO$_4$]$_2$	Monodentate/N-2	84
[Co(5-R-tta)(NH$_3$)$_5$][Br]$_2$ (R = NH$_2$—C=NH)	Chelate/N-1, =NH	84
Co(5-CF$_3$-tta)(dmg)$_2$(PBu$_3^n$)	Monodentate/N-2	204
cis-Pd(5-Me-tta)$_2$(PPhMe$_2$)$_2$	Monodentate/N-1	4
trans-Pd(5-Ph-tta)$_2$(PPh$_3$)$_2$	Monodentate/N-2	115
Cu$_2$(5-CF$_3$-tta)$_2$(dppe)$_3$	Monodentate/N-2	81
{Ag(5-CF$_3$-tta)(PPh$_3$)$_2$}$_2$	Bridging/N-3/N-4	159
[AsPh$_4$][Au(1-Pr^i-tta)$_4$]	Monodentate/C-5	66

mean bond lengths of ~1.32 Å, indicating extensive electron delocalization within the ring (*4, 66, 204*). However, the complexes Cu$_2$(5-CF$_3$-tta)$_2$(dppe)$_3$ (*81*) and {Ag(5-CF$_3$-tta)(PPh$_3$)$_2$}$_2$ (*159*) contain monodentate and bridging tetrazolate ligands, respectively, in which there are significant variations in interatomic distances within the rings. Metal–carbon and metal–nitrogen bond lengths are consistent with the presence of single metal–ligand σ bonds (*66*).

A list of crystallographically determined structures containing coordinated tetrazole moieties is given in Table III.

C. Spectroscopic Studies

1. Vibrational Spectra

Infrared spectra have been recorded for the parent tetrazole (ttaH), its sodium salt, and a range of transition metal derivatives (*94*). In subsequent work the infrared spectra of sodium tetrazolate and of the copper complex Cu(tta)$_2$·H$_2$O were assigned and a vibrational analysis was reported for the tetrazolate anion (*77*). Infrared data have been assigned for 1-methyltetrazole, its lithium salt, and the nickel(II) complex, Ni(1-Me-tta)$_2$; v(Ni–C) and v(Ni–N) occur at 456 and 298 cm^{-1}, respectively (76). Characteristic infrared frequencies have been recorded for 1-substituted tetrazolate ligands in a range of palladium, platinum, and gold complexes (15). Vibrational spectra have been recorded and assigned for a series of alkyltetrazole complex salts [M(1-R-ttaH)$_6$][BF$_4$]$_2$ (M = Mn, Co, Ni, Cu, and Zn) (70, 138) and adducts CuX$_2$(1 or 2-R-ttaH)$_2$ (X = Cl, NCS) (50a). Bands assigned to v(M–N) follow the expected Irving–Williams sequence (70). Infrared spectra (4000–200 cm^{-1}) have been recorded and assigned for 1-alkyl-tetrazoles and the complex iron(II) cations, [Fe(1-R-ttaH)$_6$]$^{2+}$, including high- and low-spin forms of [Fe(1-Pr-ttaH)$_6$][BF$_4$]$_2$ (71). Similar data have been reported for 5-phenyltetrazole, its sodium salt, and complexes with chromium(III), cobalt(II), nickel(II), copper(II), and zinc(II); assignments are given for the free tetrazole and the sodium salt (119). The presence of a band at $\sim$1250 cm^{-1}, arising from bending of an unsubstituted =N—N=N— group, is reported to be a diagnostic test for N-1 coordination of 5-R-tta$^-$ ligands (87). The infrared spectra of 5-aminotetrazole, its sodium salt, and a copper(II) derivative have been recorded and assigned (36, 103). Bi- and tridentate bonding modes have been proposed on the basis of infrared data for 5-cyanotetrazolate ligands in the complexes M(5-CN-tta)$_2$·3H$_2$O and M(5-CN-tta)$_2$(5-CN-ttaH)·2H$_2$O (M = Fe, Co, Ni, and Cu) (69). Infrared data have been tabulated for sodium salts and iron(II) complexes of the 5-substituted tetrazolate anions, 5-R-tta$^-$ (R = Cl, NO$_2$, or CF$_3$) (89). Vibrational spectra (5000–180 cm^{-1}) have been reported for a range of 1,5-pentamethylenetetrazole complex salts [M(1,5-pmtta)$_6$][ClO$_4$]$_2$ (M = Mn, Fe, Co, Ni, Cu, and Zn), [Cu(1,5-pmtta)$_4$][ClO$_4$]$_2$, and [Cu(1,5-pmtta)$_2$][ClO$_4$] (51, 53). Similar data have been recorded for the free 1,5-pmtta ligand and its silver(I) complex [Ag(1,5-pmtta)$_2$][NO$_3$] (163).

Far infrared data have been reported and v(M–N) frequencies assigned for a variety of cobalt, nickel, copper, zinc, cadmium, and mercury tetrazolate complexes (153). An inverse relationship between the basicity of the tetrazolate anions and the values of v(M–N) has been interpreted as evidence of significant M → L π bonding (153).

The complexes cis-Pd(5-CF$_3$-tta)$_2$(PR$_3$)$_2$ show v(Pd–N) and v(Pd–P) absorptions in the ranges 300–380 and 370–460 cm^{-1}, respectively (170).

2. Electronic Spectra

High D_q values, comparable with those of 2,2'-dipyridyl, found for the complex cations [M(1-R-ttaH)$_6$]$^{2+}$ indicate very high ligand-field strengths for monodentate 1-alkyltetrazoles (70). The iron(II) spin crossover compound [Fe(1-Pr-ttaH)$_6$][BF$_4$]$_2$ has been particularly thoroughly investigated (49, 71). The thermally induced high-spin ($^5T_{2g}$) to low-spin ($^1A_{1g}$) transition and the phenomenon of light-induced excited state spin state trapping have been followed and analyzed using data from single-crystal spectra measured between room temperature and 8 K (49, 90a). Similar behavior has been noted for the mixed crystals [Zn$_{1-x}$Fe$_x$(1-Pr-ttaH)$_6$][BF$_4$]$_2$ (90a). The D_q values of 806 and 833 cm^{-1} reported for 1-methyl and 1-cyclohexyl tetrazolate anions in the nickel(II) complexes Ni(1-R-tta)$_2$ are less than the values for H$_2$O and NH$_3$ in [Ni(H$_2$O)$_6$]$^{2+}$ and [Ni(NH$_3$)$_6$]$^{2+}$, respectively (76). However, a D_q value of 1140 cm^{-1} found for Ni(5-CF$_3$-tta)$_2\cdot$4H$_2$O places the 5-trifluoromethyl tetrazolate anion above NH$_3$ in the spectrochemical series (90). 5-Phenyl and 5-benzyl tetrazolate anions are similar in ligand-field strength (119). Electronic spectra have been recorded and assigned for copper(II) complexes with a variety of 5-substituted tetrazolate anions (119). The D_q values of 2203 and 2257 cm^{-1} calculated for N-1- and N-2-coordinated isomers of [Co(5-Me-tta)(NH$_3$)$_5$][ClO$_4$]$_2$ appear to confirm that, although the nitrogen N-1 of tetrazolate anions is more nucleophilic than nitrogen N-2 (204), steric hindrance at the N-1 position arising from the presence of the 5-methyl substituent favors N-2 coordination (87).

Electronic spectra have been recorded for the 1,5-pentamethylenetetrazole adducts MX$_2$(1,5-pmtta) (M = Mn, Fe, Co, Ni, and Cu; X = Cl and Br) (32) and for the complex salts [M(1,5-pmtta)$_6$][ClO$_4$]$_2$ (M = Fe, Co, Ni, and Cu) (51, 52). Data for the complex salts indicate that 1,5-pentamethylenetetrazole is a stronger ligand than ethylenediamine in the nickel(II) system but only slightly stronger than water in cobalt(II) complexes (51).

3. Nuclear Magnetic Resonance Spectra

The isomeric purity of the N-2-coordinated cobalt(III) tetrazolate complexes Co(5-R-tta)(dmg)$_2$(PR$_3'$) has been confirmed by ^{1}H, ^{13}C, ^{19}F, and ^{31}P{^{1}H} NMR. Evidence of interannular conjugation, implying coplanarity of aryl and tetrazolate rings, is provided by the proton NMR spectra of the aryl tetrazolate derivatives Co(5-Ar-tta)(dmg)$_2$-(PR$_3'$) (Ar = Ph, 3-FC$_6$H$_4$, and 4-FC$_6$H$_4$) (204). The N-2-coordinated complexes [Co(5-R-tta)(NH$_3$)$_5$][ClO$_4$]$_2$ have been characterized by ^{1}H, ^{13}C, and ^{15}N NMR using samples enriched selectively at each ring position (7). Formation and interconversion of N-1- and N-2-coordinated isomers of the palladium and platinum complexes M(5-R-tta)$_2$(PR$_3'$)$_2$

and PtH(5-R-tta)(PR$'_3$)$_2$ (R = Me or CF$_3$) have been investigated by ^{1}H and ^{19}F NMR (141, 170).

4. Electron Paramagnetic Resonance Spectra

EPR data have been reported for 5-aryl tetrazolate salts of copper(II); $\langle g \rangle$ values are in the range 2.12–2.15 and the $g_\perp$ and $g_\parallel$ values are similar to those recorded for copper(II) complexes of other N donors. Related cobalt and nickel complexes gave complex or unresolved spectra (119). EPR data have been analyzed for the complex salts [Mn(1,5-pmtta)$_6$][ClO$_4$]$_2$ and [Cu(1,5-pmtta)$_n$][ClO$_4$]$_2$ (n = 4 and 6). For the copper complexes, copper and nitrogen hyperfine splittings were observed (118).

An electron spin resonance study (4–300 K) of [Cu(1-Me-ttaH)$_6$]-[BF$_4$]$_2$ reveals that the two crystallographically inequivalent Cu(II) sites are subject to different Jahn–Teller effects arising from differences in the symmetries of the two sites (220a).

5. Mössbauer Spectra

The ^{57}Fe Mössbauer spectra of the iron(II) spin-crossover complex [Fe(1-Pr-ttaH)$_6$][BF$_4$]$_2$ have been recorded; the high-spin form shows a quadrupole doublet at room temperature, the low-spin form generates a single-line spectrum. Similar data have been obtained using other 1-alkyl tetrazoles (138). The absence of quadrupole splitting and the small isomer shift observed in the Mössbauer spectra of the iron(II) complexes Fe(5-R-tta)$_2$2H$_2$O (R = CF$_3$, Cl, or NO$_2$) were originally interpreted as evidence of a sandwich structure with metal–ligand π bonding (89). ^{197}Gold Mössbauer isomer shifts for [AsPh$_4$][Au(1-Cy-tta)$_4$] and [AsPh$_4$][Au(1-Cy-tta)$_2$] occur at the very low end of the scale, near those found for gold halogen complexes and far below those recorded for gold cyanide and methyl complexes. These results suggest that, unlike cyanide and methyl groups, C-bonded tetrazolate anions are poor σ donors and do not participate to any great extent in metal to ligand π bonding (11).

D. Group Survey

To date no tetrazole complexes appear to have been reported for the following triads: scandium, yttrium, and lanthanum; titanium, zirconium, and hafnium; and vanadium, niobium, and tantalum.

1. Chromium, Molybdenum, and Tungsten

The air- and moisture-sensitive light blue-green paramagnetic ($\mu_{\text{eff}} = 4.55$ BM) 1,5-pentamethylenetetrazole adduct, $CrCl_2(1,5\text{-pmtta})_2$, obtained from the free ligand and anhydrous $CrCl_2$, is oxidized to chromium(III) chloride even in dry air (32). Reactions of chromium(III) perchlorate with sodium salts of 5-aryltetrazoles afford chromium(III) products $Cr(5\text{-Ar-tta})_2(OH) \cdot 4H_2O$. Magnetic moments (3.06–5.33 BM), electronic spectra, and EPR data are consistent with the presence of octahedral chromium(III) ions (119). Ligand substitution reactions of $Cr(CO)_6$ and $Cr(CO)_5(THF)$ afford $Cr(5\text{-Ph-ttaH})(CO)_5$ (220), $Cr(1\text{-Ph-ttaH})(CO)_5$ and $Cr(1\text{-Ph-ttaH})_2(CO)_4$ (79), and $Cr(\text{pytta})(CO)_5$ (160). Chromium(III) tetrazolate complexes, $[Cr(5\text{-R-tta})(H_2O)_5][ClO_4]_2$, are formed when the cobalt(III) complexes $[Co(5\text{-R-tta})(NH_3)_5][ClO_4]_2$ are reduced by $[Cr(H_2O)_6]^{2+}$ (Scheme 3) (7). An attempt to prepare η^5-tetrazole complexes by treatment of $fac\text{-Mo}(CO)_3(MeCN)_3$ with 1-aryl- or 1-alkyltetrazoles, 1-R-ttaH (R = Me, Cy, or Ph), gave instead the

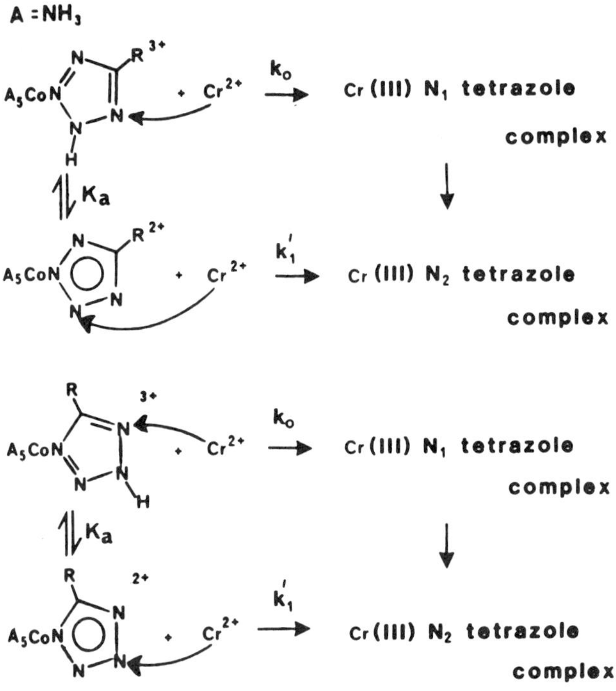

SCHEME 3. Mechanisms for reduction of N-1- and N-2-coordinated tetrazolate complexes $[Co(5\text{-R-tta})(NH_3)_5][ClO_4]_2$ by chromium(II).

tetrazole-bridged binuclear complexes $(OC)_3Mo(1\text{-}R\text{-}ttaH)_3Mo(CO)_3$; a mononuclear product, $Mo(CO)_3(1\text{-}Ph\text{-}ttaH)_3$, was also described (74). The photolytic reaction of $W(CO)_6$ with 1-phenyltetrazole yields $W(1\text{-}Ph\text{-}ttaH)(CO)_5$ and $W(1\text{-}Ph\text{-}ttaH)_2(CO)_4$ (79, 80). The carbonyls $M(CO)_6$ (M = Cr, Mo, or W) react with sodium tetrazolates $Na[5\text{-}R\text{-}tta]$ to afford the salts $Na[M(5\text{-}R\text{-}tta)(CO)_5]$ (R = CF_3, MeS, or Ph) (220). The reaction of 2-azidopyridine with $W(CO)_5THF$ is accompanied by cyclization of the ligand to yield the pyridinotetrazole complex $W(CO)_5(pytta)$ (160).

2. Manganese, Technetium, and Rhenium

Salts $[Mn(1\text{-}R\text{-}ttaH)_6][BF_4]_2$ have been isolated by treatment of $Mn(BF_4)_2$ with the free tetrazoles, 1-R-ttaH (R = Me, Et, or Pr^i) (70); the related pentamethylenetetrazole derivative, $[Mn(pmtta)_6][ClO_4]_2$, is similarly obtained from the free ligand and anhydrous manganese perchlorate (51). The latter product has a magnetic moment μ_{eff} = 5.90 BM (51) and displays an EPR spectrum indicative of almost regular octahedral symmetry and highly ionic (91%) metal–ligand bonds (118). Far infrared bands at approximately 302–285 and 238–198 cm^{-1} have been assigned to M–L stretch and L–M–L deformation modes, respectively (53). The stable, isolable adducts $MnX_2(pmtta)$ display electronic spectra and magnetic data (μ_{eff} = ~5.5 BM) consistent with octahedral coordination and presumably possess polymeric structures with bridging halide (X = Cl or Br) and/or pentamethylenetetrazole ligands (32). The 1:2 adduct obtained from $MnCl_2$ and 2-azidopyridine contains the cyclized (tetrazole) form of the ligand (160). The 5-cyanotetrazole complexes $Mn(5\text{-}CN\text{-}tta)_2 \cdot 2H_2O$ and $Mn(5\text{-}CN\text{-}tta)_2(5\text{-}CN\text{-}ttaH) \cdot 2H_2O$ are both thought to contain octahedral MnN_6 chromophores with the tetrazole functioning as a bridging bi- or tridentate ligand (69). Reactions of manganese(II) salts with 5-cyanotetrazole in acetone gave the 5-carbamyltetrazole derivatives $Mn\{5\text{-}NH_2 \cdot C(O)\text{-}tta\}_2$ and $Mn\{5\text{-}NH_2 \cdot C(O)\text{-}tta\}_2 \cdot 2H_2O$ (68).

To date no tetrazole complexes of technetium and rhenium have been reported.

3. Iron, Ruthenium, and Osmium

Ferrous tetrafluoroborate reacts with 1-alkyltetrazoles, 1-R-ttaH (R = Me, Et, Pr^n, or Pr^i), to yield octahedral iron(II) salts $[Fe(1\text{-}R\text{-}ttaH)_6][BF_4]_2$, which undergo reversible transitions from white, high-spin ($^5T_{2g}$) to purple, low-spin ($^1A_{1g}$) form at temperatures ranging from

140 to 80 K (71). Variable-temperature Mössbauer, magnetic susceptibility, and far infrared studies have been reported for these salts and have shown that the spin transition in the n-propyltetrazole derivative induces a first-order phase transition with structural reordering in the vicinity of the iron centers (138). The phenomena of thermally induced spin transition and light-induced excited state spin trapping in $[Fe(1-Pr^n-ttaH)_6][BF_4]_2$ have been investigated by single-crystal optical absorption spectroscopy. The magnetic susceptibility of the light-induced high-spin state has been measured (49). Dehydration of ferrous perchlorate by 2,2-dimethoxypropane in the presence of pentamethylenetetrazole affords the related salt $[Fe(pmtta)_6][ClO_4]_2$ (51), for which the magnetic moment ($\mu_{eff} = 5.20$ BM), electronic spectra (51), and far infrared data (53) have been reported. Magnetic moment and electronic (reflectance) spectra data for the insoluble yellow pentamethylenetetrazole adduct $FeCl_2(pmtta)_2$ indicate the presence of a polymeric structure with octahedral iron centers linked by bridging tetrazole ligands (32). Anhydrous ferrous chloride complexes 2-azidopyridine to form the pyridinotetrazole adduct $FeCl_2(pytta)_3$ (160).

Nonstoichiometric iron(III) tetrazolate precipitates, prepared by mixing ferrous salts and tetrazole in water, methanol, or ethyl acetate, react very vigorously with water after dehydration over P_2O_5 (94). In an attempt to prepare a tetrazole analog of ferrocene, hydrated ferrous chloride was allowed to react with sodium tetrazolate salts Na[5-R-tta] containing electron-withdrawing groups (R = NO_2, CF_3, or Cl). The products $Fe(5-R-tta)_2 \cdot 2H_2O$ isolated from these reactions were first assigned sandwich structures involving tetrahedral coordination of iron(II) to π electron density associated with four N–N double bonds (89). However, subsequent consideration of reflectance spectra led to their reformation as polymeric species with octahedral iron centers coordinated to terminal aquo and bridging tetrazolate ligands (90). Rather similar structures involving bi- or tridentate 5-cyanotetrazolate ligands have been proposed for the iron(III) complexes $Fe(5-CN-tta)_2 \cdot 2H_2O$ and $Fe(5-CN-tta)_2(5-CN-ttaH) \cdot 2H_2O$ (69). Viscous red solutions that form when dilute anaerobic solutions of "$Fe(NH_4)_2SO_4 \cdot 6H_2O$" are treated with a large excess of Na(5-Ar-tta) (Ar = Ph or p-tolyl) have been shown to contain complex anions of stoichiometry $[Fe(5-Ar-tta)_3]^-$, which are believed to be polymeric species with triple N-2/N-3 tetrazolate bridges (176). 2-Azidopyridine reacts with $RuCl_2(SbPh_3)_3$ and $RuX_2(PPh_3)_3$ (X = Cl or Br) to yield pyridinotetrazole derivatives $RuCl_2(pytta)_2(SbPh_3)_2$ and $RuX_2(pytta)_2(PPh_3)_2$; the latter products afford $RuX_2(pytta)(CO)(PPh_3)_2$ on carbonylation (160). The ruthenium tetrazolate complex Ru(5-CF$_3$-tta)$_2$(Ph$_2$PCH$_2$PPh$_2$)$_2$ has been obtained from the corresponding dichloride and Na(5-CF$_3$-tta) (14, 18), and by the 1,3-cycloaddition of CF$_3$CN to the azide Ru(N$_3$)$_2$-(Ph$_2$PCH$_2$PPh$_2$)$_2$ (18). Labeling experiments indicate that a transient cyclic $\overline{\text{Ru—N=N—N=N}}$—O intermediate is involved in the reaction between [RuCl(NO)(py)$_4$]$^{2+}$ and N$_3^-$ (30). Reactions between [RuCl(NO)(Me$_2$AsC$_6$H$_4$AsMe$_2$)$_2$]Cl and azide anions or between Ru(N$_3$)Cl(Me$_2$AsC$_6$H$_4$AsMe$_2$)$_2$ and NO$^+$ proceed via similar cyclic N$_4$O intermediates (56).

To date there appear to be no examples of osmium tetrazolate complexes.

4. Cobalt, Rhodium, and Iridium

Stability constants of cobalt(II) complexes with 5-methyltetrazole have been determined by potentiometry (125). A series of papers report solution studies on cobalt(II) complexes of aminoalkyltetrazoles (62, 63, 122). The complex salts [Co(1-R-ttaH)$_6$][BF$_4$]$_2$ (R = Me, Et, or Pr) (70) and [Co(pmtta)$_6$][ClO$_4$]$_2$ (51) have been prepared, and characterized by magnetic and spectroscopic methods. Magnetic moment and electronic (reflectance) spectra recorded for cobalt(II) pentamethylenetetrazole adducts confirm octahedral and tetrahedral structures for pink [CoCl$_2$(pmtta)]$_n$ and blue CoBr$_2$(pmtta)$_2$, respectively (32). Magnetic data and vibrational and electronic spectra have been reported for the cobalt(II) adducts CoCl$_2$(ttaH)$_{1.5}$, CoCl$_2$(5-NH$_2$-ttaH)$_2$, and Co(NO$_3$)$_2$ (5-NH$_2$-ttaH)$_2$ (124, 153); for the first of these a polymeric structure has been proposed (124). The vibrational spectrum of an impure sample of Co(tta)$_2$ has been reported and assigned (94). The yellow amorphous precipitate, Co(5-CF$_3$-tta)$_2$·6H$_2$O, obtained from Co(NO$_3$)$_2$·6H$_2$O and sodium 5-trifluoromethyl tetrazolate (104) is believed to contain at least one coordinated 5-CF$_3$-tta anion, since the D_q value recorded is considerably larger than that of the [Co(H$_2$O)$_6$]$^{2+}$ cation (90). Formation constants for the adducts CoCl$_2$(1-R-ttaH) and CoCl$_2$(1-R-ttaH)$_2$ in THF are reported to be 135 ± 5 and 3.3 ± 0.1 × 10^3, respectively, for R = Me, and 145 ± 25 and 3.3 ± 0.5 × 10^3, respectively, for R = Cy (83). Cobalt(II) tetrazolate complexes synthesized, and characterized by vibrational and electronic spectroscopy include the monohydrate Co(5-R-tta)$_2$·H$_2$O (R=p-ClC$_6$H$_4$, p-MeOC$_6$H$_4$, o-ClC$_6$H$_4$, or p-Cl-C$_6$H$_4$CH$_2$) (119), the trihydrate Co(5-CN-tta)$_2$·3H$_2$O (69), the adduct Co(5-CN-tta)$_2$(5-CN-ttaH)·2H$_2$O (69), and the 5-carbamyl tetrazolate complexes Co(5'-R-tta)$_2$ and Co(5'-R-tta)$_2$·2H$_2$O [R' = C(O)NH$_2$] (68). Viscous yellow solutions that form when dilute solutions of

$Co(NO_3)_2 \cdot 6H_2O$ are treated with a large excess of sodium 5-phenyl tetrazolate are thought to contain a polymeric complex $[Co(5\text{-Ph-tta})_2(H_2O)_2]_n$ with a molecular weight in the region 10^4 to 10^5. A rigid linear chain structure with double tetrazolate bridges is proposed (*177*).

Reactions between cobalt(III) complexes $CoCl(dmg)_2L$ and sodium salts of tetrazoles $Na(5\text{-R-tta})$ do not always give isolable products. However, yellow or brown crystalline complexes $Co(5\text{-R-tta})(dmg)_2L$ (L = PBu_3^n, R = Me, CF_3, Bz, Ph, *m*- or *p*-$F\text{-}C_6H_4$, or NMe_2; L = $PMePh_2$, R = Me, Bz, or *p*-$F\text{-}C_6H_4$; L = py, R = Ph) were obtained by this route and were shown by NMR to involve exclusively N-2-coordinated tetrazolate anions. This mode of coordination was confirmed for $Co(5\text{-}CF_3\text{-tta})(dmg)_2(PBu_3^n)$ by an X-ray diffraction study (*204*). Since quantum mechanical (MINDO/3) calculations reveal that N-1 is more nucleophilic than N-2, stereochemical factors must be decisive in determining the tetrazolate coordination mode (*204*). In marked contrast to all previous alkylations of tetrazolate complexes, alkylation of the above-mentioned cobalt(III) derivatives gave exclusively 1,5-disubstituted tetrazoles with no evidence of the 2,5 isomers (*204, 205*). Kinetic and mechanistic studies on the alkylation process indicate nucleophilic attack by the alkyl halide on the coordinated tetrazolate to form an intimately associated charged intermediate which subsequently undergoes a dissociative interchange of halide and 1,5-disubstituted tetrazole producing $CoX(dmg)_2PBu_3^n$ and the free tetrazole (*205*). Octahedral cobalt(III) azides $CoN_3(L_4)PPh_3$ (L_4 = tetradentate Schiff base, L_2 = dmg) undergo 1,3-dipolar cycloadditions with organic isonitriles (*65*) and electron-deficient nitriles (*109*) to form C-coordinated 1-alkyl tetrazolate (**37**) and N-2-coordinated 5-alkyl tetrazolate

37   **38**

(**38**) complexes, respectively. Anation reactions between $[Co(NH_3)_5(H_2O)][ClO_4]_3$ and the anions 5-R-tta$^-$ in aqueous solution afford the explosive perchlorate salts [Co(5-R-tta)(NH$_3$)$_5$][ClO$_4$]$_2$ (R = H, Me, CN, or CONH$_2$), which can be converted by ion exchange into the rather safer tetrafluoroborates [Co(5-R-tta)(NH$_3$)$_5$][BF$_4$]$_2$ (7). NMR (^{1}H, ^{13}C, and ^{15}N) and electronic spectra establish that tetrazolate coordination occurs through N-2 (7) and this coordination mode has been confirmed for [Co(5-Me-tta)(NH$_3$)$_5$][X]$_2$ (67, 157) and [Co(5-CN-tta)-(NH$_3$)$_5$][ClO$_4$]$_2$ (84). The N-1-bonded isomers [Co(5-R-tta)(NH$_3$)$_5$]-[ClO$_4$]$_2$ (R = Me or Ph) have been prepared by controlled reactions between the nitrile complexes [Co(RCN)(NH$_3$)$_5$][ClO$_4$]$_3$ and azide anions and were precipitated from solution as iodides [Co(5-R-tta)-(NH$_3$)$_5$]I$_2$·H$_2$O (60). Prolonged stirring of the reaction mixture prior to precipitation of the iodides leads to formation of the N-2-bonded isomers (60). N-1- or N-2-bonded tetrazolate complexes [Co(5-Ar-tta)-(NH$_3$)$_5$]I$_2$ have been obtained by treating the corresponding organonitrile complexes [Co(ArCN)(NH$_3$)$_5$][ClO$_4$]$_3$ successively with azide and iodide anions. Azide attack on the coordinated nitrile has been shown to be first order in complex and first order in azide anion (87). Mechanisms involving concerted 1,3-dipolar cycloaddition of azide anion across the nitrile bond or azide attack on the nitrile carbon atom to form an imidoylazide intermediate which subsequently cyclizes have been advanced (87). The kinetics of reduction of N-1- and N-2-coordinated tetrazolate complexes [Co(5-R-tta)(NH$_3$)$_5$][ClO$_4$]$_2$ by Cr^{II} have been investigated. The reactions are thought to proceed by inner-sphere mechanisms leading to formation of N-1- and N-2-coordinated chromium(III) tetrazolate complexes, [Cr(5-R-tta)(H$_2$O)$_5$][ClO$_4$]$_2$ (Scheme 3) (7). Kinetics of the linkage isomerization of N-1-coordinated [Co(5-Me-tta)-(NH$_3$)$_5$]$^{2+}$ to the N-2-bonded form have been studied as a function of temperature and pH. Rate constants for isomerization of the protonated (+3) and nonprotonated (+2) complex were found to differ appreciably. A base catalyzed pathway was established for the +2 complex (164). The use of [Co(5-CN-tta)(NH$_3$)$_5$][ClO$_4$]$_2$ (N-2 isomer) as an explosive has led to the development of a purification technique based upon capillary tube isotachophoresis (123). Chemical technology associated with the use of [Co(5-CN-tta)(NH$_3$)$_5$][ClO$_4$]$_2$ as an explosive has been reviewed (128, 132). A chelate structure (N-1 and exocyclic =NH coordination) has been found by diffraction methods for the 5-amidinotetrazolate complex [Co(5-R-tta)(NH$_3$)$_4$]Br$_2$ [R = —C(NH$_2$)=NH] (84).

The first rhodium tetrazolate complex Rh(5-CF$_3$-tta)(PPh$_3$)$_3$ was obtained in 1969 by treatment of RhCl(PPh$_3$)$_3$ with Na(5-CF$_3$-tta) (18). The complex Rh$_2$(μ-5-CF$_3$-tta)(μ-N$_3$)(5-CF$_3$tta)$_2$(C$_5$Me$_5$)$_2$, obtained by addition of CF$_3$CN to the rhodium azide Rh$_2$(μ-N$_3$)$_2$(N$_3$)$_2$(C$_5$Me$_5$)$_2$, contains terminal (N-2) and bridging (N-2/N-3) tetrazolate ligands (*178*). The rhodium azide $Rh(N_3)(CO)(PPh_3)_2$ undergoes 1,3-cycloaddition reactions with amino-, phosphino-, and thionitriles to form the corresponding tetrazolates $Rh(5\text{-R-tta})(CO)(PPh_3)_2$ and a range of chelated and/or bridged products (structures **39**, **40**, and **41**) (*61*). Similar

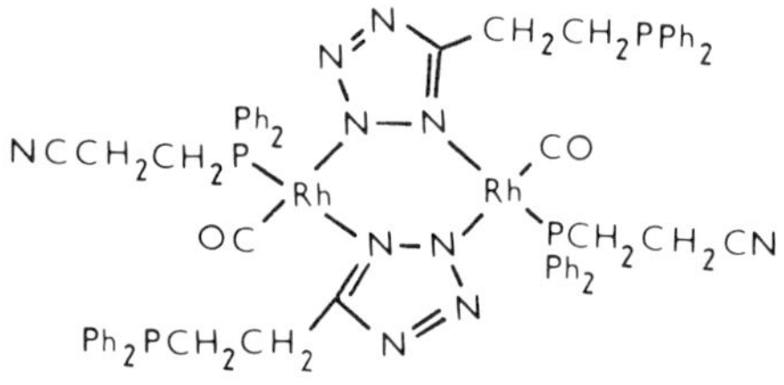

39

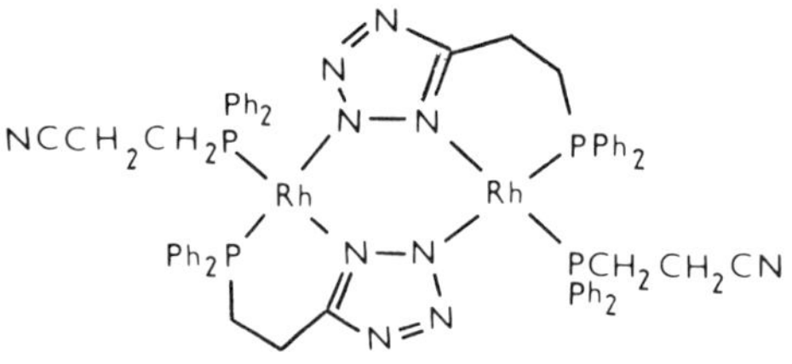

40

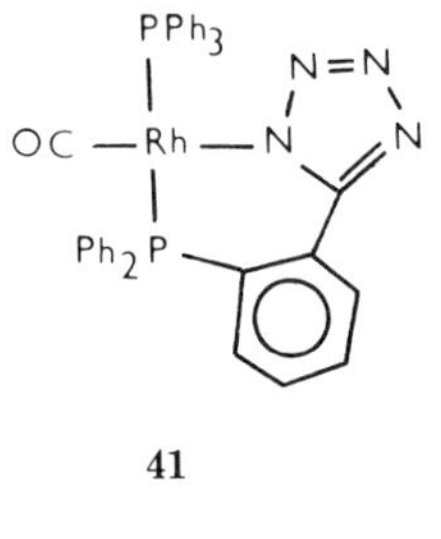

41

reactions with $Ir(N_3)(CO)(PPh_3)_2$ have furnished the iridium analogs $Ir(5\text{-R-tta})(CO)(PPh_3)_2$ (*61, 114*). Other rhodium tetrazolate complexes include $Rh_3(\mu\text{-tta})_3(\eta^3\text{-}C_3H_5)_6$ and $Rh_3(\mu\text{-tta})(\mu\text{-Cl})Cl(\eta^3\text{-}C_3H_5)_2(CO)_4$ (*156a*).

5. Nickel, Palladium, and Platinum

A series of papers describe solution studies on nickel(II) complexes of aminoalkyltetrazoles (*62, 63, 122*). Stepwise stability constants for the Ni(ClO$_4$)$_2$/5-Me-ttaH system in aqueous solution range from 5.02 (log β_1) to 19.1 (log β_6) (125). Disodium ethyl bis(5-tetrazolylazo)acetate trihydrate has been used as a colorimetric agent for nickel (102). Vibrational and electronic spectra have been reported for the complex salts [Ni(5-R-ttaH)$_6$][BF$_4$]$_2$ (R = Me, Et, or Pr^n) (70) and the related 1,5-pentamethylenetetrazole derivative [Ni(1,5-pmtta)$_6$][ClO$_4$]$_2$ (51, 53). The crystal structure of the first 2-substituted tetrazole complex [Ni(2-Me-ttaH)$_6$][BF$_4$]$_2$ reveals coordination at the N-4 position (213). Complex formation between nickel(II) salts and various tetrazoles 5-R-ttaH (R = Me, NH$_2$, Ph, p-MeOC$_6$H$_4$, and p-ClC$_6$H$_4$) has been studied by spectroscopic methods, but only impure products could be isolated from the blue solutions formed (48). Spectroscopic evidence for the formation of a 2:1 tetrazole:nickel perchlorate adduct in dimethyl formamide solution has been reported; stability constants are $\beta_1 = 2.3 \pm 1.5$ and $\beta_2 = 13.1 \pm 0.9$ (94). Infrared and stability constant data have been reported for the nickel(II) chloride adducts NiCl$_2$(5-R-ttaH)$_2$ (R = Me or Cy) (83). Vibrational and electronic spectra have been recorded for the adducts NiCl$_2$(5-R-ttaH)·0.5H$_2$O (R = H and Ph·CH=N—NH—) and NiCl$_2$(5-NH$_2$-ttaH)$_2$ (124, 153); a polynuclear structure is proposed for NiCl$_2$(ttaH)·0.5H$_2$O (124). Polymeric structures with bridging halide ligands and octahedral coordination about the nickel centers have also been proposed for the 1:1 nickel(II) halide:1,5-pentamethylenetetrazole adducts NiX$_2$(pmtta) (X = Cl or Br) (32). 2-Azidopyridine adopts the tautomeric tetrazole structure (pytta) in the polymeric adducts NiX$_2$(pytta)$_2$ (X = Cl, or Br) and NiCl$_2$(pytta), which form when the free ligand reacts with nickel(II) phosphine complexes NiX$_2$(PPh$_3$)$_2$. Magnetic and spectroscopic data indicate the presence of octahedral nickel(II) (160). A range of blue-violet nickel(II) tetrazolates, Ni(5-R-tta)$_2$·H$_2$O (R = H, NH$_2$, or aryl), prepared from nickel(II) perchlorate and the appropriate sodium tetrazolates, give magnetic moments and electronic and vibrational spectra which do not distinguish unambiguously between octahedral and tetrahedral coordination (94, 119). A similar reaction involving the sodium salt of 5-trifluoromethyltetrazole affords a pink product Ni(5-CF$_3$-tta)$_2$·4H$_2$O, which, after drying over P$_2$O$_5$, becomes violet in color (104). Electronic spectra are consistent with octahedral nickel(II) σ bonded to H$_2$O and tetrazolate ligands (90). N-2-Coordinated tetrazolate ligands have been confirmed for $trans$-Ni(5-O$_2$N-tta)$_2$(H$_2$O)$_4$ by X-ray diffraction methods (43a). The blue precipitate initially formed on addition of Na(5-Ph-tta) to aqueous solutions of Ni(NO$_3$)$_2$·6H$_2$O dissolves on standing to give clear, viscous, pink-purple solutions from which a pink-purple solid of stoichiometry Na[Ni(5-Ph-tta)$_3$]7H$_2$O was recovered on slow evaporation to dryness. Studies on the viscous solutions indicate the presence of a polymeric anionic species of stoichiometry $[Ni(5\text{-Ph-tta})_3]^-$ with a molecular weight in the range $0.75-1.5 \times 10^5$ (175). A rodlike structure with triple tetrazolate bridges is proposed (175). The reactions of nickel(II) salts with 5-cyanotetrazole in aqueous or aqueous ethanol solutions afford the expected 5-cyanotetrazolate derivatives $Ni(5\text{-CN-tta})_2 \cdot 3H_2O$ and $Ni(5\text{-CN-tta})_2(5\text{-CN-ttaH}) \cdot 2H_2O$ (69), whereas similar reactions in boiling acetone yield 5-carbamyl tetrazolate complexes $Ni(5\text{-NH}_2 \cdot CO\text{-tta})_2$ and $Ni(5\text{-NH}_2 \cdot CO\text{-tta})_2 \cdot 3H_2O$ (68). In both types of complex (68, 69) the tetrazolate anions are thought to function as bridging tridentate ligands, binding through two tetrazole nitrogens and the cyanide or carbamyl nitrogen. Green nickel(II) complexes containing C-coordinated 1-alkyl tetrazolate anions $Ni(1\text{-R-tta})_2$ (R = Me or Cy) have been obtained from $NiCl_2(PEt_3)_2$ and $Li(1\text{-R-tta})$ under strictly anhydrous and anaerobic conditions (75, 76). Magnetic and spectroscopic data are consistent with polymeric structures containing octahedral or tetrahedral nickel(II) centers with briding tetrazolate anions coordinated through C-1, N-4, and N-2 or N-3 (76). An N-bonded 1-phenyltetrazoline-5-thionato complex $Ni(PhN_4CS)(C_5H_5)(PBu_3)$ has been obtained by addition of PhNCS to the azide $Ni(N_3)(C_5H_5)(PBu_3)_2$ (185).

The palladium(II) adducts, $PdCl_2(1,5\text{-pmtta})_2$, react with nucleosides to yield mixed ligand complexes $[Pd(1,5\text{-pmtta})_2(\text{nucl})_2]Cl_2$ (nucl = inosine, guanosine, or cytidine) (161). 2-Azidopyridine is bound as the tetrazole tautomer (pytta) in the adduct cis-$PdCl_2(\text{pytta})_2$ (160). The square-planar palladium(II) tetrazolate complexes $Pd(5\text{-R-tta})_2L_2$ (R = Me, CF_3, cyclo-Pr, or Ph; L = tertiary phosphine) have been prepared by four different routes—metathesis of the corresponding chlorides with sodium tetrazolate salts, reactions of PdL_4 with free tetrazoles, 1,3-dipolar addition of nitriles RCN to the corresponding azides, and "reduction" of $PdCl_2L_2$ with hydrazine in the presence of free tetrazole (14, 17, 18, 141, 170). Spectroscopic measurements, in particular ^1H and ^{19}F NMR, confirm the ambidentate nature of the tetrazolate anion; N-1/N-1, N-2/N-2, and mixed N-1/N-2 isomers can be detected in solution. The mechanisms of linkage and geometric (cis/trans) isomerization in these complexes has been studied by spectroscopic methods (170). X-Ray diffraction studies reveal cis-phosphine ligands and N-1-coordinated tetrazolate anions for $Pd(5\text{-Me-tta})_2(PPhMe_2)_2$ (4) (Fig. 18), but trans-phosphine ligands and N-2-coordinated tetrazolate anions for $Pd(5\text{-Ph-tta})_2(PPh_3)_2$ (115) (Fig. 19). The complexes $Pd(5\text{-R-tta})_2(PPh_3)_2$ are cleaved by HCl and R'COCl under mild conditions to yield monosubstituted (5-R-ttaH) and disubstituted

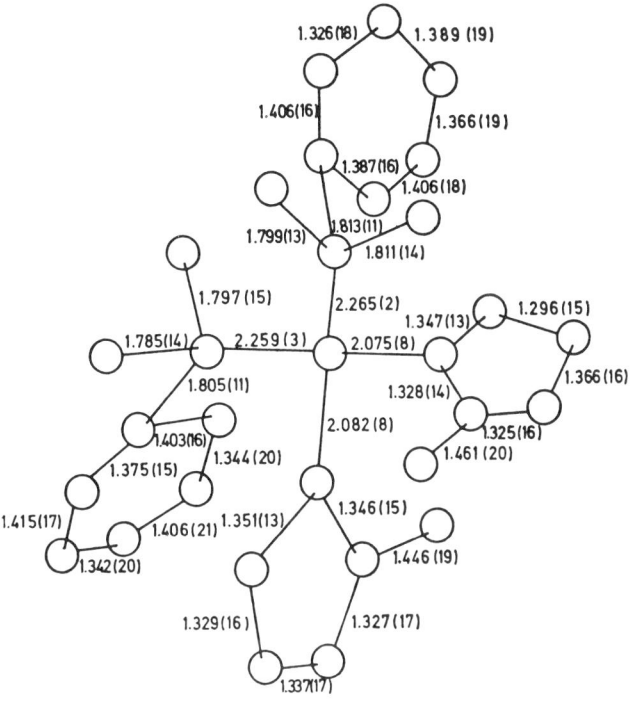

FIG. 18. Molecular structure of *cis*-bis(dimethylphenylphosphine)bis(5-methyltetrazolato)palladium(II).

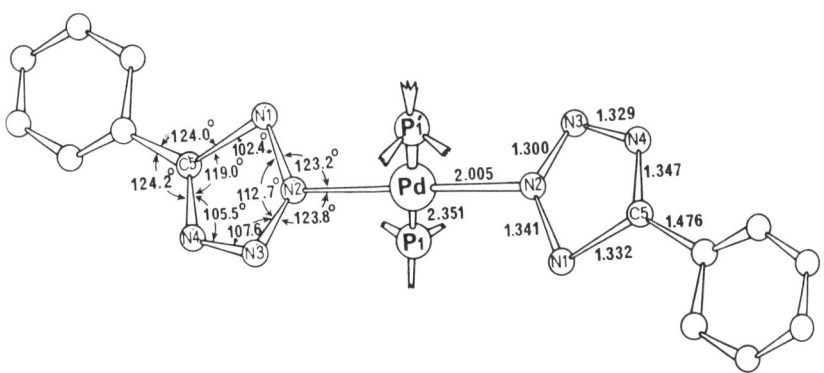

FIG. 19. Molecular structure of *trans*-bis(5-phenyltetrazolato)bis(triphenylphosphine)palladium(II).

(3-R'CO-5-R-ttaH) tetrazoles, respectively (*114*). The azide Pd(N$_3$)$_2$-(PPh$_3$)$_2$ also undergoes 1,3-cycloaddition reactions with organic isonitriles (RNC) (*15*) and isothiocyanates (RNCS) (*114*) to afford the complexes **42** and **43** (M = Pd), respectively. N-1-Coordinated tetrazolate chelate complexes **44** and **45** are formed by treatment of

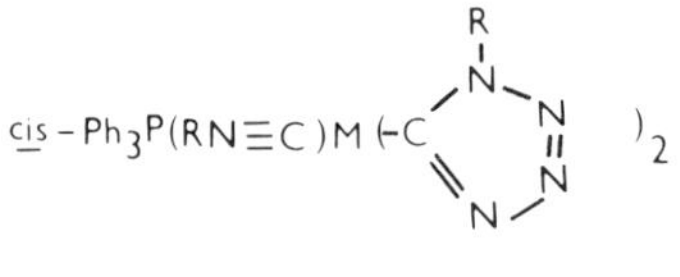

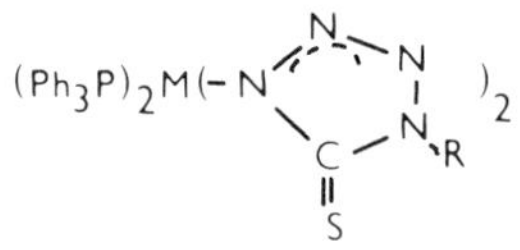

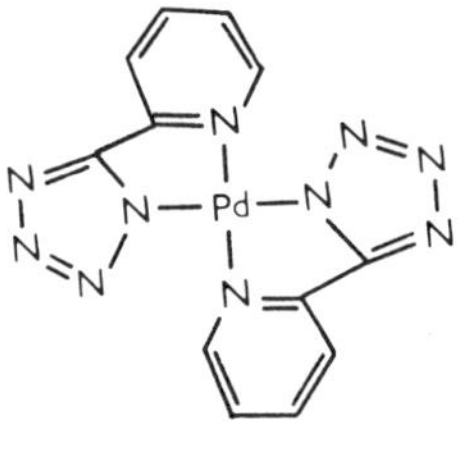

the azide Pd(N$_3$)$_2$(PPh$_3$)$_2$ with *o*-cyanopyridine and 2-cyanoethyldiphenylphosphine, respectively (*61*). Chelate ring opening to accommodate bis(diphenylphosphino)ethane ligands converts structure **44** to structure **46** (*61*). A variety of functional organonitriles react with Pd(N$_3$)$_2$(PPh$_3$)$_2$ and [Pd(N$_3$)(μ − N$_3$)(PPh$_3$)]$_2$ to form terminal

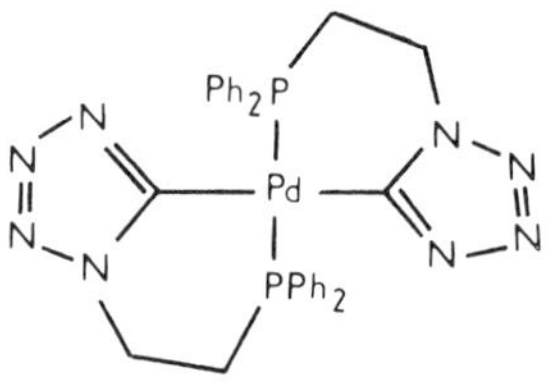

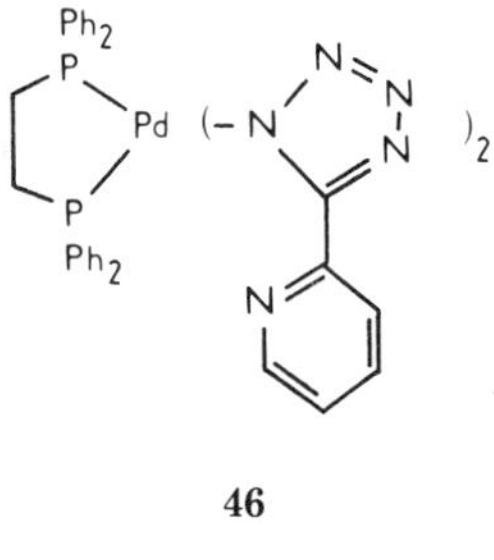

and bridging tetrazolate derivatives (structures **47** and **48**), respectively (*61*). Internal 1,3-cycloaddition converts the azide Pd(N$_3$)-

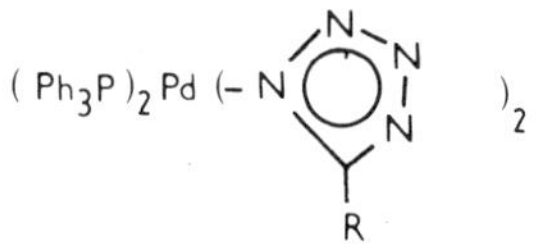

48   49

(CH$_2$C$_6$H$_4$CN-o)(dppe) to the chelate tetrazolate complex (structure **49**) (*179*). The azide-bridged complex [Pd(N$_3$)(η^3-C$_3$H$_5$)]$_2$ undergoes 1,3-cycloaddition reactions with CF$_3$CN and PhNCS to afford the corresponding 5-CF$_3$tta and PhN$_4$CS bridged complexes (*37*). Anions of 1-substituted 1,2,3,4-tetrazoline-5-one and -5-thione bond through nitrogen (N-2 or N-4) and sulfur, respectively, in palladium(II) complexes (*16*).

The platinum(II) adduct PtCl$_4$(Et-ttaH)$_2$, reported in 1910, provided an early example of a transition metal tetrazole complex (*154*). Other platinum(II) tetrazole adducts include PtCl$_2$(1-R-ttaH)$_2$ (R = Me, Cy, Ph, p-MeC$_6$H$_4$, and p-ClC$_6$H$_4$) (*23, 83*) or PtX$_2$(1,5-pmtta)$_2$ (X = Cl, Br, or I) (*161*). The latter products (X = Cl) add nucleosides (L = cytidine, adenosine, or guanine) to form salts [Pt(1,5-pmtta)$_2$L$_2$]Cl$_2$ (*161*). The platinum(II) complex Pt(5-CF$_3$-tta)$_2$, obtained from K$_2$PtCl$_4$ and Na(5-CF$_3$-tta), adds triphenylphosphine to form Pt(5-CF$_3$-tta)$_2$(PPh$_3$)$_2$; the same product forms directly from PtCl$_2$(PPh$_3$)$_2$ and Na(5-CF$_3$-tta) (*18*). Reactions of PtCl$_2$(PPh$_3$)$_2$ with tetrazoles, 5-R-ttaH, in the presence of N$_2$H$_4$ afford Pt(5-Cy-tta)$_2$(PPh$_3$)$_2$ and the hydrides *trans*-PtH(5-R-tta)(PPh$_3$)$_2$ (R = Ph, Cl, or Br) (*141*). Oxidative addition of 5-Ph-ttaH to Pt(PPh$_3$)$_4$ generates Pt(5-Ph-tta)$_2$(PPh$_3$)$_2$ (*141*). Treatment of PtCl$_2$(dppe) with tetrazole in methanolic KOH affords Pt(tta)$_2$(dppe) (*136*). NMR studies on several of these complexes indicate the presence of N-1- and N-2-coordinated linkage isomers (*141*). The N-1- and N-2-bonded linkage isomers Pt(5-Me-tta)(CN)(PPh$_3$)$_2$, obtained by addition of MeCN to Pt(N$_3$)(CN)(PPh$_3$)$_2$, display fluxional N-1 $\rightleftharpoons$ N-2 $\rightleftharpoons$ N-3 $\rightleftharpoons$ N-4 exchange above 150°C (*19*). Carbon-bonded tetrazolato complexes of platinum(II) have been obtained by azide ion attack on coordinated isocyanide ligands (*211*) and by interaction of azide ligands with free isocyanides (*15*). Confirmation of structure is provided by ^{1}H NMR and infrared spectra, and by hydrolysis reactions which liberate 1-substituted tetrazoles (*15*). Reactions of functional cyanides, 2-CN-py and $Ph_2PCH_2CH_2CN$, with platinum azide complexes give chelate tetrazolate derivatives analogous to those reported for palladium (vide supra) (61). A chelate tetrazolate product is also formed by the thermally induced intramolecular cycloaddition reaction of cis-$Pt(N_3)(CH_2C_6H_4CN$-$o)(PPh_3)_2$ (179, 180). Addition of methyl thiocyanate and methyl isothiocyanate to $Pt(N_3)_2(PPh_3)_2$ affords the isomeric complexes 50 and 43 (M = Pt), respectively (114).

$$(Ph_3P)_2Pt(N\underset{C}{\overset{N-N}{\underset{|}{\bigcirc}}}{\overset{|}{\underset{N}{N}}})_2$$
$$\overset{|}{SMe}$$

50

6. Copper, Silver, and Gold

Solution studies on copper(II) complexes with aminoalkyltetrazoles have been described (62, 63, 122). Disodium ethyl bis(5-tetrazolylazo)-acetate trihydrate has been employed as a colorimetric agent for copper (102). Copper(II) adducts of 1-alkyl- and 2-alkyltetrazoles CuX_2-$(1$-R-$ttaH)_2$ and $CuX_2(2$-R-$ttaH)_2$ (X = Cl or SCN; R = Me, Et, vinyl, allyl, hexyl, or octyl) contain N-4- and N-1-coordinated tetrazoles, respectively (50, 73). Formation of the copper(II) chloride adducts is an effective method for the isolation and purification of tetrazoles, and for the separation of 1-alkyl and 2-alkyl isomers (72). The adducts have been characterized by infrared, EPR, and thermal analysis techniques (50, 73). 2-Azidopyridine is coordinated as the tetrazole tautomer in its 1:1 copper chloride adduct (160). Other copper(II) adducts include $CuCl_2(ttaH)_{1.5}$, $CuCl_2(5$-$PhCH$=N—NH—$ttaH)$ (124, 153), and the 1,5-pentamethylenetetrazole derivatives $CuX_2(1,5$-$pmtta)$ (X = Cl or Br) (32, 174). The reactions of $Cu(ClO_4)_2 \cdot 6H_2O$ with 1,5-pentamethylenetetrazole in acetic acid or 2,2-dimethoxypropane solution afford the copper(I) complex $[Cu(1,5$-$pmtta)_2][ClO_4]$ and the copper(II) complexes $[Cu(1,5$-$pmtta)_n][ClO_4]_2$ (n = 4 or 6). The ease with which the hexacoordinated copper(II) salt forms was unexpected; however, spectroscopic evidence is in favor of this structure rather than the alternative $[Cu(1,5$-$pmtta)_4][ClO_4]_2 \cdot 2(1,5$-$pmtta)$, and the complex is isomorphous with other octahedral $[M(1,5$-$pmtta)_6][ClO_4]_2$ salts (52). Far infrared spectra have been reported for all three salts and tentative assignments have been given for metal–ligand stretching and deformation modes (53). Electron spin resonance spectra including copper hyperfine splittings have been recorded for $[Cu(1,5$-$pmtta)_4]$-

[ClO$_4$]$_2$ and [Cu(1,5-pmtta)$_6$][ClO$_4$]$_2$; for the latter complex nitrogen hyperfine splittings are also seen. The data indicate that the Jahn–Teller distortion resonates along all three axes for diluted samples, but gives rise to a permanent tetragonal distortion in the undiluted powder (*118*). Octahedral copper(II) salts [Cu(1-R-ttaH)$_6$][BF$_4$]$_2$ (R = Me, Et, or Pr^n) have been synthesized and electronic and vibrational spectra have been reported (*70*). The crystal structure and EPR spectrum (see Section III,C,4) of the 1-methyltetrazole complex have been reported (*220a*). Copper(II) salts react with tetrazole and 5-substituted tetrazoles to yield a wide variety of copper(II) tetrazolate complexes including Cu(tta)$_2$·H$_2$O (*77*), Cu(5-Ar-tta)$_2$ (*119*), Cu(5-NH$_2$-tta)$_2$, Cu(5-NH$_2$-tta)$_2$·0.5H$_2$O (*36*), Cu(5-Ph-tta)$_2$·H$_2$O, Cu(5-Ph-tta)OH (*47, 119*), Cu(5-Ph-tta)Cl, and the sulfates Cu$_2$(5-Ar-tta)$_2$(SO$_4$)$_2$·2H$_2$O (*47*). These products are insoluble and tend to precipitate in an impure state, frequently contaminated by the parent copper salt (*36, 47*). Vibrational, electronic (*36, 47, 77, 119*), and EPR (*119*), spectra have been reported for many of these products. Copper(II) salts react with 5-cyanotetrazole in aqueous or alcoholic solution to afford the complexes Cu(5-CN-tta)$_2$·3H$_2$O and Cu(5-CN-tta)$_2$(5-CN-ttaH)·2H$_2$O (*69*). However, in boiling acetone solution the same reagents yield the 5-carbamyltetrazole derivatives Cu(5-NH$_2$-CO-tta)$_2$ and Cu(5-NH$_2$-CO-tta)$_2$·2H$_2$O (*68*). Electronic spectra and physical properties of the complex Cu(5-CF$_3$-tta)$_2$·H$_2$O are consistent with a polymeric structure in which each octahedral copper center is bound to at least five nitrogens (*90, 104*). A binuclear N-2/N-3 tetrazole-bridged structure, similar to that found for the corresponding silver(I) complex (vide infra), has been proposed for the copper complex Cu(5-CF$_3$-tta)(PPh$_3$)$_2$, prepared by 1,3-dipolar addition of CF$_3$CN to the azide Cu(N$_3$)(PPh$_3$)$_2$ (*235*). The same product has been obtained from CuCl(PPh$_3$)$_2$ and Na(5-CF$_3$-tta) (*18*). An X-ray diffraction study (*81*) on the complex Cu$_2$(5-CF$_3$-tta)$_2$(dppe)$_3$, prepared from the corresponding azide and CF$_3$CN (*235*), revealed a pair of tetrahedral copper(I) centers linked by a bridging dppe ligand, with each coordinated to a bidentate dppe ligand and a monodentate (N-2) tetrazolate anion (Fig. 20). The binuclear copper(I) derivatives of 1-substituted 1,2,3,4-tetrazoline-5-one and -5-thione [Cu(YCN$_4$R)(PPh$_3$)]$_2$ (Y = O or S; R = Me or Ph) contain N-2/N-4-bridging and S- or S/N-4-bridging ligands, respectively (*16*).

The silver(I) salt of tetrazole, Ag(tta), was mentioned by Bladin in 1892 (*25*); more recently, silver(I) tetrazolates Ag(5-R-tta) have been obtained from silver nitrate and the free tetrazole or 5-substituted tetrazole (*78, 94, 134*). The silver salt of 5-nitrotetrazole, Ag(5-O$_2$N-tta),

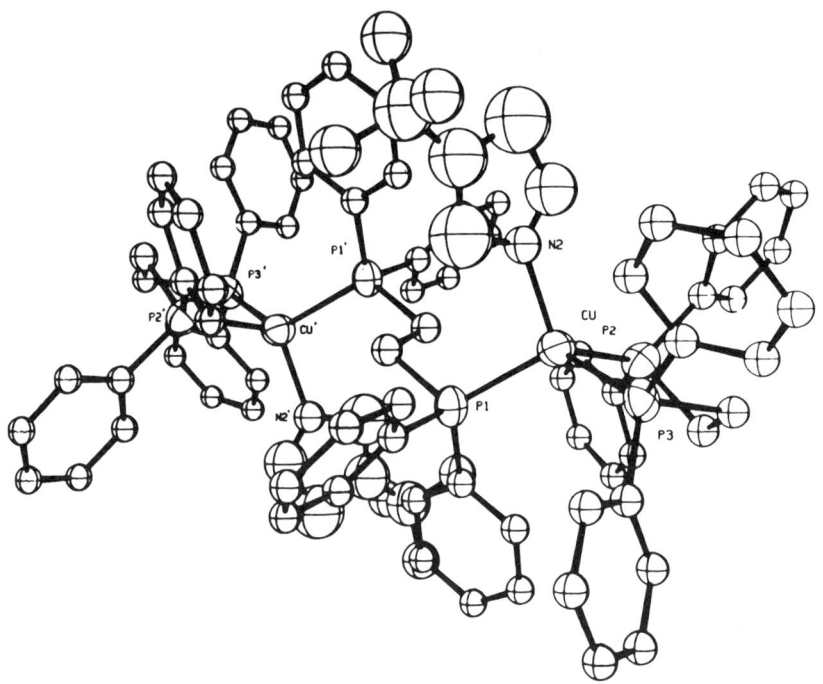

FIG. 20. Perspective view of the molecular structure of $Cu_2(CF_3\text{-tta})_2$-$(Ph_2PCH_2CH_2PPh_2)_3$.

has been proposed as an alternative to lead azide for use in detonators (12) and its thermal decomposition characteristics have been examined (34). The silver complex $[Ag(5\text{-}CF_3\text{-tta})(PPh_3)_2]_2$, obtained by 1,3-dipolar addition of CF_3CN to $[Ag(N_3)(PPh_3)_2]_2$ (235), is binuclear (structure **51**) with N-2/N-3-bridging tetrazolate anions (159). Stability

51

constants have been measured for a wide range of silver(I) complex cations [Ag(tetrazole)$_2$]$^+$ (tetrazoles include 8-alkyl-1,5-pmtta, 7,9-dialkyl-1,5-pmtta, 1-cy-5-Me-tta, and various 1,5-polymethylenetetrazoles) (54, 163). The 1,5-pentamethylenetetrazole derivative has been isolated as the nitrate salt [Ag(1,5-pmtta)$_2$][NO$_3$] (163), which has been shown (27) to possess a binuclear structure with monodentate (N-4) and bridging (N-3/N-4) tetrazole ligands, and monodentate nitrate anions surrounding distorted tetrahedral silver(I) centers. A 1:1 adduct Ag(NO$_3$)(1,5-pmtta) has also been reported (174). Silver(I) complexes of 1,5-dimethyltetrazole and 1,4-bis(1-methyl-5-tetrazolyl)butane have been studied in nitromethane and methyl cyanide solution, and a salt [Ag(1,5-Me$_2$-tta)$_2$][ClO$_4$] has been isolated (31). Silver(I) salts of 1-substituted tetrazoline-5-ones and -5-thiones are analogous to the corresponding copper(I) derivatives (vide supra) (16).

The gold(I) and gold(III) azido complexes Au(N$_3$)(PPh$_3$), [AsPh$_4$][Au(N$_3$)$_2$], and [AsPh$_4$][Au(N$_3$)$_4$] undergo cycloaddition with isonitriles (RNC) (R = Me, Pr^i, Cy, Ph, Bz, or p-MeOC$_6$H$_4$) under mild conditions to afford C-bonded tetrazolate complexes Au(1-R-tta)(PPh$_3$), [AsPh$_4$][Au(1-R-tta)$_2$], and [AsPh$_4$][Au(1-R-tta)$_4$], respectively (15, 17). Proton NMR and infrared spectra as well as acid hydrolysis to liberate 1-substituted tetrazoles are consistent with the presence of C-5-bound tetrazolate ligands (15). Confirmatory evidence has been supplied by an X-ray diffraction study on the square–planar gold(III) complex [AsPh$_4$][Au(1-Pr^i-tta)$_4$] (66). Cycloaddition of CF$_3$CN to Au(N$_3$)-(PPh$_3$)$_2$ affords Au(5-CF$_3$-tta)(PPh$_3$) (235). Tetrazoles react with Au(O$_2$CMe)(PPh$_3$) to form the complexes Au(5-R-tta)(PPh$_3$) (R = H, CF$_3$, NH$_2$, NMe$_2$, or Ph). With 5,5'-bitetrazole and bis(5-tetrazolyl)methane, Ph$_3$P·Au·CN$_4$·CN$_4$·Au·PPh$_3$ and Ph$_3$P·Au·CN$_4$·CH$_2$·CN$_4$·Au·PPh$_3$, respectively, were obtained (110). Anions of 1-substituted tetrazoline-5-ones and -5-thiones coordinate through nitrogen (N-2 or N-4) and sulfur, respectively, in a range of gold(I) and gold(III) complexes (16).

7. Zinc, Cadmium, and Mercury

Solution studies on zinc(II) complexes with aminoalkyltetrazoles have been reported (62, 63, 122). Zinc halides form 1:2 adducts with 1-alkyl- or 1-aryltetrazoles, ZnCl$_2$(1-R-ttaH)$_2$ (R = Me, Cy, Ph, or p-tolyl) (23, 83), and 1,5-pentamethylenetetrazole, ZnX$_2$(1,5-pmtta)$_2$ (X = Cl, Br, or I) (32, 174). An X-ray diffraction study on the 1-methyltetrazole adduct ZnCl$_2$(1-Me-ttaH)$_2$ reveals distorted tetrahedral geometry with N-4-coordinated tetrazole ligands (6). 1-Alkyltetrazoles, 1-R-ttaH (R = Me, Et, or Pr), and 1,5-pentamethylenetetrazole react with zinc salts to form the cationic complexes [Zn(1-R-ttaH)$_6$][BF$_4$]$_2$ (70) and [Zn(1,5-pmtta)$_6$][ClO$_4$]$_2$ (51), respectively. Far infrared data have been reported and assignments made (53 70); ν(Zn–N) occurs at $\sim$185 cm^{-1} for the 1-alkyltetrazole complexes (70). Stability constants for zinc(II) complexes with the 5-methyl tetrazolate anion have been determined by the potentiometric method (log β_1 = 1.71, log β_6 = 11.9) (125). Zinc complexes, Zn(5-Ar-tta)$_2$, deposit as white precipitates on addition of sodium tetrazolates to aqueous solutions of zinc perchlorate (119). 5-Cyanotetrazole reacts with zinc salts in cold aqueous or alcoholic solution to afford the complexes Zn(5-CN-tta)$_2$·3H$_2$O and Zn(5-CN-tta)$_2$(5-CN-ttaH)·2H$_2$O (69), whereas in boiling acetone solution hydrolysis of the cyanide group occurs and 5-carbamyltetrazole derivatives Zn(5-NH$_2$·CO-tta)$_2$ and Zn(5-NH$_2$·CO-tta)$_2$·2H$_2$O are formed (68).

Cadmium tetrazole complexes examined by far infrared spectroscopy include CdCl$_2$(ttaH)$_2$, CdCl$_2$(5-NH$_2$-ttaH)$_2$, and Cd(5-NH$_2$-tta)$_2$·0.5H$_2$O (124, 153). The synthesis and infrared spectrum of Cd(tta)$_2$·1.5H$_2$O have also been reported (94). The cadminum(II) adducts CdX$_2$-(1,5-pmtta)$_n$ (X = Cl, n = 2; X = Br or I, n = 1), Cd(SCN)$_2$(1,5-pmtta), and Cd(ClO$_4$)$_2$(1,5-pmtta) have been described (174, 236).

Interest in mercury(II) tetrazolates, notably Hg(5-O$_2$N-tta)$_2$, as detonators (12, 190) has prompted thermal decomposition studies on mercury salts of 5-nitrotetrazole (34) and 5,5'-azotetrazole (169). Cycloaddition of methyl isocyanide to Hg(N$_3$)$_2$(PPh$_3$)$_2$ under mild conditions yields the C-bonded tetrazolate complex Hg(1-Me-tta)$_2$ (15). Mercury(II) adducts HgX$_2$(1,5-pmtta) (X = Cl, Br, or CN) and Hg(NO$_3$)$_2$(1,5-pmtta)$_2$ have been described (174, 236).

IV. Pentazolate and Hexazine Complexes

Although no complexes with cyclic ligands containing five or more adjacent nitrogen atoms are currently known, it seems probable that such species will be synthesized in the future. With a few notable exceptions, coordination to a transition metal imparts additional stability to catenated nitrogen molecules and it therefore seems not unlikely that entities such as the pentazolate anion, *cyclo*-N$_5^-$, and the hexazine molecule, *cyclo*-N$_6$—which are isoelectronic with the well-known cyclopentadienyl and benzene ligands, respectively—might be stabilized in this manner. Indeed, recent *ab initio* calculations on the pentazolate anion N$_5^-$ (151) and on the hypothetical complexes [M(η^5-N$_5$)(CO)$_3$]^n (M = Cr^0, n = –1; M = Mn^I, n = 0; M = Fe^{II}, n = +1) (150) suggest combination of the [M(CO)$_3$]$^{(n+1)+}$ and *cyclo*-N$_5^-$ moieties

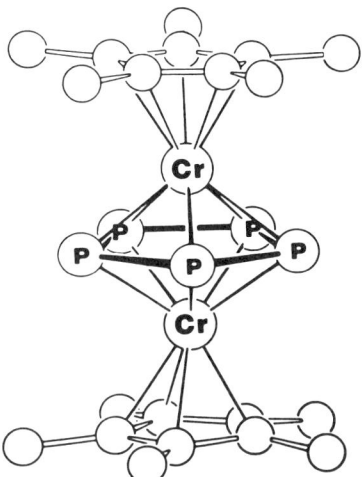

FIG. 21. Molecular structure of $(C_5Me_5)Cr(\eta$-$P_5)Cr(C_5Me_5)$.

should be thermodynamically favorable and that the N_5^- ligand should not be susceptible to 1,3-dipolar cycloreversion to form azide anion and liberate dinitrogen. Moreover, the success of organic chemists in synthesizing para-substituted phenylpentazoles, p-XC_6H_4-N_5 (140, 212), one of which, X = NMe_2, has been characterized by X-ray diffraction methods (218), suggests that suitable sources of the N_5^- ligand may be at hand. Further encouragement for those prepared to accept the challenge of the η^5-N_5 ligand has recently been provided by the successful synthesis of the binuclear complex $(\eta^5$-$C_5Me_5)Cr(\eta^5$-$P_5)Cr(\eta^5$-$C_5Me_5)$, which has been shown by X-ray diffraction method to contain a bridging η^5-P_5 ligand (Fig. 21) (187).

Stabilization of the hexazine, cyclo-N_6, molecule by coordination to transition metals also appears to be a realistic medium-term prospect for synthetic chemists. Self-consistent field (SCF) theory calculations suggest that the hypothetical D_{6h} "hexaazabenzene" molecule is likely to be a classic aromatic species, and give a value of 1.288 Å for the equilibrium N—N distance (186). Meanwhile, studies on the photolytic decomposition of the platinum(II) azide, cis-$Pt(N_3)_2(PPh_3)_2$, have provided the first experimental evidence for the transient existence of hexaazabenzene (214). Finally, the recent synthesis and characterization of a binuclear complex $(C_5Me_5)Mo(\eta^6$-$P_6)Mo(C_5Me_5)$, in which an η^6-cyclo-P_6 ligand bridges the two metal atoms (Fig. 22) (188), must raise hopes for the eventual isolation of complexes containing the analogous η^6-N_6 ligand.

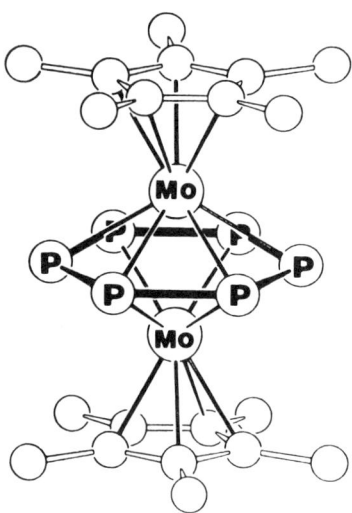

FIG. 22. Molecular structure of $(C_5Me_5)Mo(\mu\text{-}P_6)Mo(C_5Me_5)$.

REFERENCES

1. Aleksandrov, G. G., Babin, V. N., Kharchevnikov, A. P., Struchkov, Y. T., and Kochetkova, N. S., *J. Organomet. Chem.* **266,** 109 (1984).
2. Al-Kharafi, F. M. Al-Hajjar, F. H., and Katrib, A., *Corros. Sci.* **26,** 257 (1986).
3. Allison, J. A. C., El. Khadem, H. S., and Wilson, C. A., *J. Heterocyclic Chem.* **12,** 1275 (1975).
4. Ansell, G. B., *J. Chem. Soc. Dalton Trans.* 371 (1973).
5. Antipin, M. Yu., Aleksandrov, G. G., Struchkov, Yu. T., Belousov, Yu. A., Babin, V. N., and Kochetkova, N. S., *Inorg. Chim. Acta* **68,** 229 (1983).
6. Baenziger, N. C., and Schultz, R. J., *Inorg. Chem.* **10,** 661 (1971).
7. Balahura, R. J., Purcell, W. L., Victoriano, M. E., Lieberman, M. L., Loyola, V. M., Fleming, W., and Fronabarger, J. W., *Inorg. Chem.* **22,** 3602 (1983).
8. Baltzer, O., and von Pechmann, H., *Liebigs Ann. Chem.* **262,** 302 (1891).
9. Baltzer, O., and von Pechmann, H., *Liebigs Ann. Chem.* **262,** 314 (1891).
10. Baranowski, J., Padula, F., Goldstein, C., Kokoszka, G., and Siedle, A. R., *J. Phys. Chem.* **89,** 1976 (1985).
11. Bartunik, H. D., Potzel, W., Mossbauer, R. L., and Kaindl, G., *Z. Phys.* **240,** 1 (1970).
12. Bates, L. R., and Jenkins, J. M., *Proc. Int. Conf. Res. Primary Explosives E.R.D.E.* Vol. 2, Pap. 14, (1975).
13. Beamish, F. E., *Talanta* **1,** 1 (1958); **8,** 85 (1961); **13,** 773 (1966).
14. Beck, W., Bauder, M., Fehlhammer, W. P., Pöllmann, P., and Schachl, H., *Inorg. Nucl. Chem. Lett* **4,** 143 (1968).
15. Beck, W., Burger, K., and Fehlhammer, W. P., *Chem. Ber.* **104,** 1816 (1971).
16. Beck, W., Burger, K., and Keubler, M., *Z. Anorg. Allgem. Chem.* **428,** 173 (1977).
17. Beck, W., and Fehlhammer, W. P., *Angew. Chem. Int. Ed. Engl.* **6,** 169 (1967).

18. Beck, W., Fehlhammer, W. P., Bock, H., and Bauder, M., *Chem. Ber.* **102**, 3637 (1969).
19. Beck, W., and Schorpp, K., *Chem. Ber.* **108**, 3317 (1975).
20. Belousov, Yu. A., Gumenyuk, V. V., Babin, V. N., Kochetkova, N. S., and Solodovnikov, S. P., *Koord. Khim.* **9**, 819 (1983); *Sov. J. Coord. Chem.* **9**, 467 (1983).
21. Bencini, A., Gatteschi, D., Reedijk, J., and Zanchini, C., *Inorg. Chem.* **24**, 207 (1985).
22. Benson, F. R., *Chem. Rev.* **41**, 1 (1947).
23. Biefield, R. M., and Gilbert, G. L., *J. Inorg. Nucl. Chem.* **33**, 3947 (1971).
24. Bladin, J. A., *Ber.* **18**, 1544, 2907 (1885).
25. Bladin, J. A., *Ber.* **25**, 1411 (1892).
26. Bladin, J. A., *Ber.* **26**, 545 (1893).
27. Bodner, R. L., and Popov, A. I., *Inorg. Chem.* **11**, 1410 (1972).
28. Bonati, F., Burini, A., Felici, M., and Pietroni, B. R., *Gazz. Chim. Ital.* **113**, 105 (1983).
29. Bonati, F., Burini, A., Pietroni, B. R., *Z. Naturforsch.* **40B**, 1749 (1985).
30. Bottomley, F., and Mukaida, M., *J. Chem. Soc. Dalton Trans.* 1933 (1982).
31. Bowers, D. M., Erlich, R. H., Policec, S., and Popov, A. I., *J. Inorg. Nucl. Chem.* **33**, 81 (1971).
32. Bowers, D. M., and Popov, A. I., *Inorg. Chem.* **7**, 1594 (1968).
33. Brown, L. D., Ibers, J. A., and Siedle, A. R., *Inorg. Chem.* **17**, 3026 (1978).
34. Brown, M. E., and Swallowe, G. M., *Thermochim. Acta* **49**, 333 (1981).
35. Browne, A. W., Smith, G. B. L., and Wilcoxon, F., *J. Am. Chem. Soc.* **45**, 2604 (1923).
36. Brubaker, C. H., *J. Am. Chem. Soc.* **82**, 82 (1960).
37. Busetto, L., Palazzi, A., and Ros, R., *Inorg. Chim. Acta* **13**, 233 (1975).
38. Busev, A. I., Lomakina, L. N., Mikheeva, M. N., and Ignat'eva, T. I., *Vestn. Mosk. Univ. Khim.* **16**, 581 (1975); *Chem. Abstr.* **85**, 71470 (1976).
39. Cambi, L., *Atti Accad. Nazl. Lincei, Rend. Cl. Sci. Fis. Mat. Nat.* **18**, 581 (1955); *Chem. Abstr.* **50**, 5347e (1956).
40. Cambi, L., Canonica, L., and Sironi, C., *Atti Accad. Nazl. Lincei, Rend. Cl. Sci. Fis. Mat. Nat.* **18**, 583 (1955); *Chem. Abstr.* **50**, 3940 (1956).
41. Cambi, L., and Paglia, E., *Atti Accad. Nazl. Lincei, Rend. Cl. Sci. Fis. Mat. Nat.* **20**, 735 (1956); *Chem. Abstr.* **51**, 7216 (1957).
42. Canonica, L., DeLeone, R., and Bersano, C., *Rend. Ist. Lomb. Sci. Cl. Sci. Mat. Nat.* **87**, 276 (1954); *Chem. Abstr.* **49**, 15592 (1955).
43. Charalambous, J., Soobramanien, G., Betts, A., and Bailey, J., *Inorg. Chim. Acta* **60**, 151 (1982).
43a. Charalambous, J., Georgiou, G. C., Henrick, K., Bates, L. R., and Healey, M., *Acta Cryst.* **C43**, 659 (1987).
44. Cheng, K. L., *Anal. Chem.* **26**, 1038 (1954).
45. Conti, M., Franco, D. W., and Trsic, M., *Inorg. Chim. Acta* **113**, 71 (1986).
46. Curtis, J. A., *Ind. Eng. Chem. Anal. Ed.* **13**, 349 (1941).
47. Daugherty, N. A., and Brubaker, C. H., *J. Am. Chem. Soc.* **83**, 3779 (1961).
48. Daugherty, N. A., and Brubaker, C. H., *J. Inorg. Nucl. Chem.* **22**, 193 (1961).
49. Decurtins, S., Gutlich, P., Hasselbach, K. M., Hauser, A., and Spiering, H., *Inorg. Chem.* **24**, 2174 (1985).
50. Degtyarik, M. M., Gaponik, P. N., Lesnikovich, A. I., and Vrublevskii, A. I., *Zh. Obshch. Khim.* **55**, 516 (1985); *Chem. Abstr.* **103**, 47180u (1985).
50a. Degtyarik, M. M., Gaponik, P. N., Naumenko, V. N., Lesnikovich, A. I., and Nikanovich, M. V., *Spectrochem. Acta* **43A**, 349 (1987).
51. D'Itri, F. M., and Popov, A. I., *Inorg. Chem.* **5**, 1670 (1966).
52. D'Itri, F. M., and Popov, A. I., *Inorg. Chem.* **6**, 597 (1967).
53. D'Itri, F. M., and Popov, A. I., *Inorg. Chem.* **6**, 1591 (1967).

54. D'Itri, F. M., and Popov, A. I., *J. Inorg. Nucl. Chem.* **31**, 1069 (1969).
55. Domnina, E. S., Shergina, N. I., Chipanina, N. N., Belousova, L. V., Frolov, Y. L., and Skvortsova, G. G., *Zhur. Obshch. Khim.* **41**, 1102 (1971); *J. Gen. Chem. USSR* **41**, 1106, (1971).
56. Douglas, P. G., and Feltham, R. D., *J. Am. Chem. Soc.* **94**, 5254 (1972).
57. Drapier, J., and Hubert, A. J., *J. Organomet. Chem.* **64**, 385 (1974).
58. Dutta, R. L., and Lahiry, S., *Sci. Cult. (Calcutta)* **26**, 139 (1960); *Chem. Abstr.* **55**, 9141h (1961).
59. Elbs, K., Hirschel, O., Wagner, F., Himmler, K., Turk, W., Henrich, A., and Lehmann, E., *J. Prakt. Chem.* **108**, 209 (1924).
60. Ellis, W. R., and Purcell, W. L., *Inorg. Chem.* **21**, 834 (1982).
61. Erbe, J., and Beck, W., *Chem. Ber.* **116**, 3867 (1983).
62. Ermakova, M. I., Shikhova, I. A., Ignatenko, N. K., Latosh, N. I., *Zh. Obsch. Khim.* **53**, 1364 (1983); *Chem. Abstr.* **99**, 77724 (1983).
63. Ermakova, M. I., Shikhova, I. A., Sinitsyna, T. A., and Latosh, N. I., *Zh. Obshch. Khim.* **49**, 1387 (1979); *Chem. Abstr.* **91**, 97541 (1979).
64. Fehlhammer, W. P., and Beck, W., *Z. Naturforsch.* **38B**, 546 (1983).
65. Fehlhammer, W. P., Beck, W., and Kemmerich, T., *Chem. Ber.* **112**, 468 (1979).
66. Fehlhammer, W. P., and Dahl, L. F., *J. Am. Chem. Soc.* **94**, 3370 (1972).
67. Fleming, W., Fronabarger, J. W., Leiberman, M. L., and Loyola, V. M., *Abstr. Chem. Conf. N. Am. Continent, 2nd, Las Vegas 1980; Amer. Chem. Soc. Washington D.C. Abstr. Inorg.* **13**.
68. Franke, P. L., and Groeneveld, W. L., *Inorg. Chim. Acta* **40**, 111 (1980).
69. Franke, P. L., and Groeneveld, W. L., *Transition Metal Chem.* **5**, 240 (1980).
70. Franke, P. L., and Groeneveld, W. L., *Transition Metal Chem.* **6**, 54 (1981).
71. Franke, P. L., Haasnoot, J. G., and Zuur, A. P., *Inorg. Chim. Acta* **59**, 5 (1982).
72. Gaponik, P. N., Degtyarik, M. M., Karavai, V. P., Komarov, V. F., Lesnikovich, A. I., and Sviridov, V. V., *Dokl. Akad. Nauk. BSSR* **28**, 543 (1984); *Chem. Abstr.* **101**, 110833k (1984).
73. Gaponik, P. N., Degtyarik, M. M., and Sviridov, V. V., *Dokl. Akad. Nauk. BSSR* **26**, 716, (1982); *Chem. Abstr.* **97**, 119541v (1982).
74. Garber, L. L., *Inorg. Chem.* **21**, 3244 (1982).
75. Garber, L. L., and Brubaker, C. H., *J. Am. Chem. Soc.* **88**, 4266 (1966).
76. Garber, L. L., and Brubaker, C. H., *J. Am. Chem. Soc.* **90**, 309 (1968).
77. Garber, L. L., Sims, L. B., and Brubaker, C. H., *J. Am. Chem. Soc.* **90**, 2518 (1968).
78. Garbrecht, W. L., and Herbst, R. M., *J. Org. Chem.* **18**, 1003 (1953).
79. Garnovskii, A. D., Kolobova, N. E., Osipov, O. A., Anisimov, K. N., Zlotina, I. B., Mitina, G. K., and Kolodyazhnyi, Yu. V., *Zh. Obshch. Khim.* **42**, 929 (1972); *Chem. Abstr.* **77**, 96372 (1972).
80. Garnovskii, A. D., Kolobova, N. E., Zlotina, I. B., Mitina, G. K., Anisimov, K. N., and Osipov, O. A., *Izv. Akad. Nauk. SSSR Ser. Khim.* 629 (1972).
81. Gaughan, A. P., Bowman, K. S., and Dori, Z., *Inorg. Chem.* **11**, 601 (1972).
82. Gel'fman, M. I., and Kustova, N. A., *Zhur. Neorg. Khim.* **16**, 2335 (1971); *Russ. J. Inorg. Chem.* **16**, 1245 (1971).
83. Gilbert, G. L., and Brubaker, C. H., *Inorg. Chem.* **2**, 1216 (1963).
84. Graeber, E. J., Morosin, B., *Acta Crystallogr. C* **39**, 567 (1983).
85. Graves, B. J., and Hodgson, D. J., *Inorg. Chem.* **20**, 2223 (1981).
86. Grishkevich-Trokhimovskii, E., *Zh. Russ. Fiz. Khim. Obshch.* **55**, 548 (1924); *Chem. Abstr.* **19**, 2810 (1925).
87. Hall, J. H., Lopez de la Vega, R., and Purcell, W. L., *Inorg. Chim. Acta* **102**, 157 (1985).

88. Hansen, L. D. West, B. D., Baca, E. J., and Blank, C. L., *J. Am. Chem. Soc.* **90,** 6588 (1968).
89. Harris, A. D., Herber, R. H., Jonassen, H. B., and Wertheim, G. K., *J. Am. Chem. Soc.* **85,** 2927 (1963).
90. Harris, A. D., Jonassen, H. B., and Archer, R. D., *Inorg. Chem.* **4,** 147 (1965).
90a. Hauser, A., Gütlich, P., Spiering, H., *Inorg. Chem.* **25,** 4245 (1986).
91. Hendriks, H. M. J., Birker, P. J. M. W. L., Verschoor, G. C., and Reedijk, J., *J. Chem. Soc. Dalton Trans.* 623 (1982).
92. Himes, V. L., Mighell, A. D., and Siedle, A. R., *J. Am. Chem. Soc.* **103,** 211 (1981).
93. Hodgson, D. J., and Pedersen, E., *Acta Chem. Scand.* **36A,** 281 (1982).
94. Holm, R. D., and Donnelly, P. L., *J. Inorg. Nucl. Chem.* **28,** 1887 (1966).
95. House, J. E., and Lau, P. S., *J. Inorg. Nucl. Chem.* **36,** 223 (1974).
96. Hursthouse, M. B., Olby, B. G., Short, R. L., and Robinson, S. D., unpublished results.
97. Hursthouse, M. B., Short, R. L., and Robinson, S. D., *Polyhedron* **5,** 1573 (1986).
98. Ignatov, M. E., Il'in, E. G., Garnovskii, A. D., and Buslaev, Yu. A., *Koord. Khim.* **8,** 1368 (1982); *Sov. J. Coord. Chem.* **8,** (1982); *Chem. Abstr.* **98,** 45919w (1983).
99. Imperial Chemical Industries, Brit. Pat. 1,031,503.
100. Inoue, M., Kishita, M., and Kubo, M., *Inorg. Chem.* **4,** 626 (1965).
101. Jezowska-Trzebiatowska, B., Jezierski, A., and Kozlowski, H., *Bull. Acad. Pol. Sci. Ser. Sci. Chim.* **22,** 11 (1974); *Chem. Abstr.* **80,** 65284s (1974).
102. Jonassen, H. B., Chamblin, V. C., Wagner, V. L., and Henry, R. A., *Anal. Chem.* **30,** 1660 (1958).
103. Jonassen, H. B., Paukert, T., and Henry, R. A., *Appl. Spectrosc.* **21,** 89 (1967).
104. Jonassen, H. B., Terry, J. O., and Harris, A. D., *J. Inorg. Nucl. Chem.* **25,** 1239 (1963).
105. Karwecka, Z., and Pajdowski, L., *Rocz. Chem.* **50,** 1251 (1976).
106. Karwecka, Z., and Pajdowki, L., *Rocz. Chem.* **51,** 2017 (1977).
107. Katada, M., Sato, K., Uchida, Y., Iijima, S., Sano, H., Wei, H. H., Sakai, H., and Maeda, Y., *Bull. Chem. Soc. Jpn.* **56,** 945 (1983).
108. Katsaros, N., and Grigoratos, A., *Inorg. Chim. Acta* **108,** 173 (1985).
109. Kemmerich, T., Nelson, J. H., Takach, N. E., Boehme, H., Jablonski, B., and Beck, W., *Inorg. Chem.* **21,** 1226 (1982).
110. Kieft. R. L., Peterson, W. M., Blundell, G. L., Horton, S., Henry, R. A., and Jonassen, H. B., *Inorg. Chem.* **15,** 1721 (1976).
111. King, R. B., and Chen, K. N., *Inorg. Chem.* **16,** 3372 (1977).
112. Kochetkova, N. S., Boev, V. I., Popova, L. V., and Babin, V. N., *Izvest. Akad. Nauk. SSSR Ser. Khim.* **34,** 1397 (1985); *Bull. Akad. Sci. USSR. Ser. Chem.* **34,** 1278 (1985).
113. Kokoszka, G. F., Baranowski, J., Goldstein, C., Orsini, C., Mighell, A. D., Himes, V. L., and Siedle, A. R., *J. Am. Chem. Soc.* **105,** 5627 (1983).
114. Kreutzer, P. H., Weis, J. C., Bock, H., Erbe, J., and Beck, W., *Chem. Ber.* **116,** 2691 (1983).
115. Kreutzer, P., Weis, C., Boehme, H., Kemmerich, T., Beck, W., Spencer, C., and Mason, R., *Z. Naturforsch.* **27B,** 745 (1972).
116. Kukushkin, Y. N., Krylov, V. K., and Romanova, M. Y., *Zhur. Obshch. Khim.* **51,** 178 (1981).
117. Kukushkin, Y. N., Maslov, E. I., and Ryabkova, T. P., *Zhur. Neorg. Khim.* **28,** 2858 (1983); *Russ. J. Inorg. Chem.* **28,** 1622 (1983).
118. Kuska, H. A., D'Itri, F. M., and Popov, A. I., *Inorg. Chem.* **5,** 1272 (1966).
119. Labine, P., and Brubaker, C. H., *J. Inorg. Nucl. Chem.* **33,** 3383 (1971).
120. Ladenburg, A., *Chem. Ber.* **9,** 219 (1876).
121. La Monica, G., Ardizzoia, G., Cenini, S., and Porta, F., *J. Organomet. Chem.* **273,** 263 (1984).

122. Latosh, N. I., Ermakova, M. I., and Shikhova, I. A., *Zh. Obshch. Khim.* **48,** 2287 (1978); *Chem. Abstr.* **90,** 62034 (1979).
123. Lavoie, J. M., and Back, P. S., *J. Chromatogr.* **264,** 329 (1983).
124. Lavrenova, L. G., Larionov, S. V., Ikorskii, V. N., and Grankina, Z. A., *Zh. Neorg. Khim.* **30,** 964 (1985); *Russ. J. Inorg. Chem.* **30,** 542 (1985).
125. Lenarcik, B., Badyoczek-Grzonka, M., and Grzonka, Z., *Rocz. Chem.* **45,** 2023 (1971).
126. Lieber, E., Oftedahl, E., and Rao, C. N. R., *J. Org. Chem.* **28,** 194 (1963).
127. Lieber, E., Pillai, C. N., Ramachandran, J., and Hites, R. D., *J. Org. Chem.* **22,** 1750 (1957).
128. Lieberman, M. L., *Ind. Eng. Chem. Prod. Res. Dev.* **24,** 436 (1985); *Chem. Abstr.* **103,** 107115p (1985).
129. Lomakina, L. N., and Alimarin, I. P., *Zhur. Neorg. Khim.* **12,** 409 (1967); *Russ. J. Inorg. Chem.* **12,** 210 (1967).
130. Lomakina, L. N., and Tarasevich, N. I., *Vest. Moskov Univ. Ser. Mat. Mekh. Astron. Fiz. Khim.* 217 (1957), *Chem. Abstr.* **52,** 4405e (1958).
131. Marshall, J. H., *Inorg. Chem.* **17,** 3711 (1978).
132. Massis, T. M., Morenus, P. K., Hukisson, D. H., and Merrill, R. M., *J. Hazardous Mater.* **5,** 309 (1982).
133. Meunier-Piret, J., Piret, P., Putzeys, J-P., and Van Meerssche, M., *Acta Crystallogr.* B **32,** 714 (1976).
134. Mihina, J. S., and Herbst, R. M., *J. Org. Chem.* **15,** 1082 (1950).
135. Minghetti, G., Banditelli, G., and Bonati, F., *Inorg. Chem.* **18,** 658 (1979).
136. Minghetti, G., Banditelli, G., and Bonati, F., *J. Chem. Soc. Dalton Trans.* 1851 (1979).
137. Mishra, L. K., and Bhushan, H., *J. Indian Chem. Soc.* **59,** 1092 (1982).
138. Müller, E. W., Ensling, J., Spiering, H., and Gutlich, P., *Inorg. Chem.* **22,** 2074 (1983).
139. Müller, E., and Meier, H., *Liebigs Ann. Chem.* **716,** 11 (1968).
140. Muller, R., Wallis, J. D., and von Philipsborn, W., *Angew. Chem. Int. Ed. Engl.* **24,** 513 (1985).
141. Nelson, J. H., Schmitt, D. L., Henry, R. A., Moore, D. W., and Jonassen, H. B., *Inorg. Chem.* **9,** 2678 (1970).
142. Nelson, J. H., Takach, N. E., Bresciani-Pahor, N., Randaccio, L., and Zangrando, E., *Acta Crystallogr.* C **40,** 742 (1984).
143. Nesmeyanov, A. N., Aleksandrov, G. G., Antipin, M. Y., Struchkov, Y. T., Belousov, Y. A., Babin, V. N., and Kochetkova, N. S., *J. Organomet. Chem.* **137,** 207 (1977).
144. Nesmeyanov, A. N., Babin, V. N., Kochetkova, N. S., Mysov, E. I., Belousov, Y. A., and Fedorov, L. A., *Dokl. Akad. Nauk. SSSR* **200,** 1112 (1971); *Proc. Acad. Sci. USSR* **200,** 838 (1971).
145. Nesmeyanov, A. N., Belousov, Y. A., Babin, V. N., Aleksandrov G. G., Struchkov, Y. T., and Kochetkova, N. S., *Inorg. Chim. Acta* **23,** 155 (1977).
146. Nesmeyanov, A. N., Belousov, Y. A., Babin, V. N., Kochetkova, N. S., Sil'vestrova, S. Y., and Mysov, E. I., *Inorg. Chim. Acta* **23,** 173 (1977).
147. Nesmeyanov, A. N., Sheinker, Y. N., Kochetkova, N. S., Rybinskaya, M. I., Senyavina, L. B., and Materikova, R. B., *Zhur. Organ. Khim.* **3,** 403 (1967); *J. Org. Chem. USSR* **3,** 386 (1967).
148. Neves, E. A., and Franco, D. W., *J. Inorg. Nucl. Chem.* **37,** 277 (1975).
149. Neves, E. A., Milcken, N., and Franco, D. W., *J. Inorg. Nucl. Chem.* **43,** 2081 (1981).
150. Nguyen, M. T., McGinn, M. A., Hegarty, A. F., and Elguéro, J., *Polyhedron* **4,** 1721 (1985).
151. Nguyen, M. T., Sana, M., Leroy, G., and Elguéro, J., *Can. J. Chem.* **61,** 1435 (1983).
152. Nonoyama, M., and Hayata, C., *Transition Metal Chem.* **3,** 366 (1978).

153. Oglezneva, I. M., and Lavrenova, L. G., *Zh. Neorg. Khim.* **30,** 1473 (1985); *Russ. J. Inorg. Chem.* **30,** 840 (1985).
154. Oliveri-Mandala, E., and Algna, B., *Gazz. Chim. Ital.* **40 (2),** 411, 441 (1910).
155. Olson, J. R., Yamauchi, M., and Butler, W. M., *Inorg. Chim. Acta* **99,** 121 (1985).
156. Oro, L. A., Pinillos, M. T., and Tejel, C., *J. Organomet. Chem.* **280,** 261 (1985).
156a. Oro, L. A., Pinillos, M. T., Tejel, C., Foces-Foces, C., and Cano, F. H., *J. Chem. Soc. Dalton Trans.* 2193 (1986).
157. Ortega, R., Campana, C. F., and Morosin, B., *Proc. Am. Crystallogr. Assoc. Meet. Calgary* 1980, **8,** 24.
158. Peringer, P., *Monatsh. Chem.* **110,** 1123 (1979).
159. Pierpont, C. G., Eisenberg, R., Ziolo, R. F., Gaughan, A. P., and Dori, Z., unpublished results cited in ref. 235.
160. Pizzotti, M., Cenini, S., Porta, F., Beck, W., and Erbe, J., *J. Chem. Soc. Dalton Trans.* 1155 (1978).
161. Pneumatikakis, G., *Inorg. Chim. Acta* **46,** 243 (1980).
162. Popov, A. I., *Coord. Chem. Rev.* **4,** 463 (1969).
163. Popov, A. I., and Holm, R. D., *J. Am. Chem. Soc.* **81,** 3250 (1959).
164. Purcell, W. L., *Inorg. Chem.* **22,** 1205 (1983).
165. Purnell, L. G., Estes, E. D., and Hodgson, D. J., *J. Am. Chem. Soc.* **98,** 740 (1976).
166. Purnell, L. G., and Hodgson, D. J., *J. Am. Chem. Soc.* **99,** 3651 (1977); *Biochim. Biophys. Acta* **447,** 117 (1976).
167. Purnell, L. G., Shepherd, J. C., and Hodgson, D. J., *J. Am. Chem. Soc.* **97,** 2376 (1975).
168. Raper, E. S., *Coord. Chem. Rev.* **61,** 115 (1985).
168a. Ravichandran, V., Ruban, G. A., Chacko, K. K., Molina, M. A. R., Rodriguez, E. C., Salas-Peregrin, J. M., Aoki, K., and Yamazaki, H., *J. Chem. Soc. Chem. Commun.* 1780 (1986).
169. Reddy, G. O., and Chatterjee, A. K., *Thermochim. Acta* **66,** 231 (1983).
170. Redfield, D. A., Nelson, J. H., Henry, R. A., Moore, D. W., and Jonassen, H. B., *J. Am. Chem. Soc.* **96,** 6298 (1974).
171. Reedijk, J., Roelofsen, G., Siedle, A. R., and Spek, A. L., *Inorg. Chem.* **18,** 1947 (1979).
172. Reedijk, J., Siedle, A. R., Velapoldi, R. A., and Van Hest, J. A. M., *Inorg. Chim. Acta* **74,** 109 (1983).
173. Remington, W. J., and Moyer, H. V., *Diss. Abstr. Ohio State Univ. Press, Columbus, Ohio* p. 24 (1937).
174. Rheinboldt, M., and Stettiner, H., *Bol. Fac. Filosof. Cien. Let. Univ. Sao Paulo, Quim.* **14,** 27 (1942); *Chem. Abstr.* **40,** 1502 (1946).
175. Richards, L., Bow, S. N., Richards, J. L., and Halton, K., *Inorg. Chim. Acta* **25,** L113 (1977).
176. Richards, L., Koufis, I., Chan, C. S., Richards, J. L., and Cotter, C., *Inorg. Chim. Acta* **105,** L21 (1985).
177. Richards, L., LaPorte, M., Maguire, R., Richards, J. L., and Diaz, L., *Inorg. Chim. Acta* **28,** 119 (1978).
178. Rigby, W., Bailey, P. M., McCleverty, J. A., and Maitlis, P. M., *J. Chem. Soc. Dalton Trans.* 371 (1979).
179. Ros, R., Michelin, R. A., Boschi, T., and Roulet, R., *Inorg. Chim. Acta* **35,** 43 (1979).
180. Ros, R., Renaud, J., and Roulet, R., *J. Organomet. Chem.* **104,** 393 (1976).
181. Rosan, A., and Rosenblum, M., *J. Organomet. Chem.* **80,** 103 (1974).
182. Rubim, J., Gutz, I. G. R., Sala, O., and Orville-Thomas, W. J., *J. Mol. Struct.* **100,** 571 (1983).

183. Sahai, R. B. N., Chaturbedi, A. P., and Mishra, L. K., *Ind. J. Chem. Sect. A* **14**, 360 (1976).
184. Sato, F., Etoh, M., and Sato, M., *J. Organomet. Chem.* **37**, C51 (1972).
185. Sato, F., Etoh, M., and Sato, M., *J. Organomet. Chem.* **70**, 101 (1974).
186. Saxe, P., and Schaefer, H. F., *J. Am. Chem. Soc.* **105**, 1760 (1983).
187. Scherer, O. J., Schwalb, J., Wolmershauser, G., Kaim, W., and Gross, R., *Angew. Chem. Int. Ed. Engl.* **25**, 363 (1986).
188. Scherer, O. J., Sitzmann, H., and Wolmershäuser, G., *Angew. Chem. Int. Ed. Engl.* **24**, 351 (1985).
189. Schmidt, K. F., *Klin. Wochenschr.* **4**, 1678 (1925).
190. Scott, C. L., *Proc. Int. Conf. Res. Primary Explosives, E.R.D.E.* **2**, Pap. 15, (1975).
191. Semenishin, D. I., Yurchak, A. V., and Slobodyan, Z. V., *Visn. L'viv. Politekh. Inst.* **149**, 7 (1981); *Chem. Abstr.* **96**, 114882b (1982).
192. Sheldrick, W. S., and Bell, P., *Inorg. Chim. Acta* **123**, 181 (1986).
193. Siedle, A. R., Velapoldi, R. A., and Erickson, N., *Appl. Surf. Sci.* **3**, 229 (1979).
194. Siedle, A. R., Velapoldi, R. A., and Erickson, N., *Inorg. Nucl. Chem. Lett.* **15**, 33, (1979).
195. Smith, G. B. L., Warttman, P., and Browne, A. W., *J. Am. Chem. Soc.* **52**, 2806 (1930).
196. Sokol, V. I., Zefirov, Y. V., and Porai-Koshits, *Koord. Khim.* **5**, 1249 (1979); *Sov. J. Coord. Chem.* **5**, 985 (1979).
197. Sotofte, I., and Nielsen, K., *Acta Chem. Scand.* **35A**, 733 (1981).
198. Sotofte, I., and Nielsen, K., *Acta Chem. Scand.* **35A**, 739 (1981).
199. Sotofte, I., and Nielsen, K., *Acta Chem. Scand.* **35A**, 747 (1981).
200. Sotofte, I., and Nielsen, K., *Acta Chem. Scand.* **37A**, 891 (1983).
201. Sotofte, I., and Nielsen, K., *Acta Chem. Scand.* **38A**, 253 (1984).
202. Sotofte, I., and Nielsen, K., *Acta Chem. Scand.* **38A**, 257 (1984).
203. Spek, A. L., Siedle, A. R., and Reedijk, J., *Inorg. Chim. Acta* **100**, L15 (1985).
204. Takach, N. E., Holt, E. M., Alcock, N. W., Henry, R. A., and Nelson, J. H., *J. Am. Chem. Soc.* **102**, 2968 (1980).
205. Takach, N. E., and Nelson, J. H., *Inorg. Chem.* **20**, 1258 (1981).
206. Tarasevich, N. I., *Vestn. Moskov Univ.* **3**, 161 (1948); *Chem. Abstr.* **44**, 5759 (1950).
207. Thiele, I., and Schleussner, K., *Liebigs Ann. Chem.* **295**, 129 (1897).
208. Tiripicchio, A., private communication cited in ref. 156.
209. Toma, H. E., Giesbrecht, E., and Espinoza Rojas, R. L., *Can. J. Chem.* **61**, 2520 (1983).
210. Toma, H. E., Giesbrecht, E., and Espinoza Rojas, R. L., *Quim. Nova* **6**, 72 (1983).
211. Treichel, P. M., Knebel, W. J., and Hess, R. W., *J. Am. Chem. Soc.* **93**, 5424 (1971).
212. Ugi, I., Perlinger, H., and Behringer, L., *Chem. Ber.* **91**, 2324 (1958) and references therein.
213. Van Den Heuvel, E. J., Franke, P. L., Verschoor, G. C., and Zuur, A. P., *Acta Crystallogr. C* **39**, 337 (1983).
214. Vogler, A., Wright, R. E., and Kunkely, H., *Angew. Chem. Int. Ed. Engl.* **19**, 717 (1980).
215. Von Pechmann, H., *Chem. Ber.* **21**, 2751 (1888).
216. Voropaev, V. N., Domnina, E. S., Skvortsova, G. G., Teterin, Yu. A., Baev, A. S., and Voropaeva, I. K., *Koord. Khim.* **10**, 1543 (1984).
217. Walker, R., *Corrosion (Houston)* **32**, 414 (1976).
218. Wallis, J. D., and Dunitz, J. D., *J. Chem. Soc. Chem. Commun.* 910 (1983).
219. Watanabe, Y., Mitsudo, T.-A., Tanaka, M., Yamamoto, K., and Takegami, Y., *Bull. Chem. Soc. Jpn.* **45**, 925 (1972).
220. Weiss, J. C., and Beck, W., *Chem. Ber.* **105**, 3203 (1972).
220a. Wijnands, P. E. M., Maaskant, W. J. A., and Reedijk, J., *Chem. Phys. Lett.* **130**, 536 (1986).

221. Wilson, R. F., *Z. Anorg. Allgem. Chem.* **318,** 233 (1962).
222. Wilson, R. F., and Baye, L. J., *J. Am. Chem. Soc.* **80,** 2652 (1958).
223. Wilson, R. F., and Baye, L. J., *J. Inorg. Nucl. Chem.* **9,** 140 (1959).
224. Wilson, R. F., and Baye, L. J., *J. Inorg. Nucl. Chem.* **13,** 91 (1960).
225. Wilson, R. F., and Baye, L. J., *Talanta* **1,** 351 (1958).
226. Wilson, R. F., and James, J., *Z. Anorg. Allgem. Chem.* **315,** 235 (1962).
227. Wilson, R. F., and Joe, F. L., *J. Inorg. Nucl. Chem.* **15,** 255 (1960).
228. Wilson, R. F., and Larry, D., *J. Inorg. Nucl. Chem.* **17,** 244 (1961).
229. Wilson, R. F., and Wilson, L. E., *Anal. Chem.* **28,** 93 (1956).
230. Wilson, R. F., and Wilson, L. E., *J. Am. Chem. Soc.* **77,** 6204 (1955).
231. Wilson, R. F., Wilson, L. E., and Baye, L. J., *J. Am. Chem. Soc.* **78,** 2370 (1956).
232. Wilson, R. F., and Womack, C. M., *J. Am. Chem. Soc.* **80,** 2065 (1958).
233. Yamashita, M., Mizuta, H., Okamoto, K., and Takemura, H., *Fushoku Boshoku Bumon Iinkai Shiryo (Nippon Zairyo Gakkai)* **23,** 1 (1984); *Chem. Abstr.* **101,** 200144r (1984).
234. Ziolo, R. F., and Dori, Z., *J. Am. Chem. Soc.* **90,** 6560 (1968).
235. Ziolo, R. F., Thich, J. A., and Dori, Z., *Inorg. Chem.* **11,** 626 (1972).
236. Zwikker, J. J. L., *Pharm. Weekbl.* **71,** 1170 (1934).

THE REDOX CHEMISTRY OF NICKEL

A. GRAHAM LAPPIN* and ALEXANDER McAULEY**

*Department of Chemistry, University of Notre Dame,
Notre Dame, Indiana 46556, and
**Department of Chemistry, University of Victoria,
Victoria, British Columbia, Canada V8W 2Y2

I. Introduction
II. Steric and Electronic Requirements
III. Probes of Structure
IV. Oxidation of Nickel(II)
   A. Coordination Environments and Structural Chemistry
   B. Kinetic Studies
V. Reduction of Nickel(II)
   A. Coordination Environments and Structural Chemistry
   B. Kinetic Studies
VI. List of Abbreviations
    References

I. Introduction

The discovery of nickel(III) (1–7) and nickel(I) (8) in methanogenic bacteria and in other biological systems has focused attention on the redox chemistry of nickel. For many years, this aspect of the coordination chemistry of nickel was largely overlooked but facile oxidation and reduction extends from formal nickel(IV) species to formal nickel(I) species, encompassing structural requirements of the electronic configurations d^6–d^9. There exists a wealth of information on the requirements for stabilization of these electronic configurations for nickel and the area has been reviewed relatively recently (9, 10). In this article, the most recent advances are emphasized, particularly those where structural characterization is relatively unambiguous and where the work has been pursued to some conclusion. The area covered is fairly selective, thus ligands with phosphorus, arsenic, and dithiolate coordination have been excluded, as they are reviewed in detail elsewhere (9).

The article is organized around a discussion of the structural and electronic requirements of the four oxidation states, I, II, III, and IV, and of the primary physical techniques and probes used. A description of the most commonly used ligand types follows, together with an indication of recent trends in the area. Each section concludes with a review of the recent mechanistic chemistry of the oxidation states in turn. A glossary of ligand abbreviations is included at the end of the review.

II. Steric and Electronic Requirements

The starting point for most of the redox chemistry considered in this review is the nickel(II) ion. The nickel(II) ion has a d^8 electronic configuration and, with weak-field ligands such as H_2O, it forms a six-coordinate ion with approximately octahedral symmetry and a paramagnetic (two unpaired electrons) $^3A_2$ ground state. The characteristic solution chemistry of six-coordinate nickel(II) is well documented and, in particular, the substitution behavior has been extensively studied and is the subject of recent reviews (*11, 12*). It is a labile ion with solvent exchange rates around 10^4 sec^{-1} at 25°C and activation parameters are consistent with dissociatively activated interchange behavior (*13*).

Thermodynamic stability of the six-coordinate complexes generally increases as the ligand-field strength increases. For stronger field chelating ligands such as phen or bpy, the tris complexes can be resolved into optical isomers (*14*). Rates of racemization correspond to rates of ligand exchange, 1.6×10^{-4} sec^{-1} for [Ni(phen)$_3$]$^{2+}$ at 45°C, indicating that complete dissociation of a ligand is required before rearrangement can occur (*15*).

Addition of ligand systems capable of producing a strong square–planar ligand field results in formation of the diamagnetic $^1A_{1g}$ ground state which shows extensive ligand-field stabilization and consequent sluggish substitution behavior by mechanisms which tend to be associative in nature (*11*).

The importance of these two dominant complex geometries for nickel(II) becomes apparent on examination of the stereoelectronic requirements for the oxidized and reduced metal species. Both nickel(I), d^9, and nickel(III), d^7, which is generally found as a low-spin species because of the increased charge on the metal ion, are subject to Jahn–Teller distortion and hence have a strong preference for tetragonal geometry, intermediate between octahedral and square–planar geometry and readily accessible from both. In contrast, nickel(IV), d^6, again generally low spin, has a strong preference for octahedral geometry, analogous to cobalt(III), and is difficult to form from a square–planar nickel(II) complex. Thus consideration of the geometry of the nickel(II) complex can lead to control of the ultimate oxidation state of the species on oxidation or reduction.

Clearly different ligand types will favor different oxidation states. Higher oxidation states prefer hard acid donor atoms, generally first-row p-block elements, rich in electron density and capable of strong σ donation. A further provision is that they should resist oxidation. Common donor chromophores which have been used are amines $\geqslant$N, imides (including oximes and imines)$>$N$^-$, oxides —O$^-$ and fluorides F$^-$. Second- and third-row p-block donors have also been used, forming bonds which are more covalent in character and creating special problems, as discussed below.

The nickel(I) state is favored by soft donors capable of π back bonding. Although there is a rich chemistry associated with second- and third-row p-block donors to nickel(I), examples in this review will be restricted again to, primarily, nitrogen and oxygen as donor atoms.

On addition or removal of an electron from a nickel(II) complex, there is frequently a problem in deciding whether redox has occurred at the metal center in which case the ligand is an "innocent" bystander to the process or whether ligand centered redox has occurred, so-called "noninnocent" behavior. This problem has plagued the literature, particularly dealing with the higher oxidation states of nickel and there are proposals for deciding between the two possibilities, some of which are outlined in the next section. Clearly, however, this is a question of semantics. Every "metal-centered" oxidation or reduction involves accommodation on the part of the ligand. Some charge redistribution undoubtedly occurs. A good rule of thumb is that the products of ligand-centered oxidation or reduction will show characteristics of radicals while the products of metal-centered oxidation will show characteristics expected of isoelectronic metal ion complexes. Thus nickel(IV) chemistry should be similar to the chemistry of cobalt(III), nickel(III) chemistry should resemble that of low-spin cobalt(II), and nickel(I) that of copper(II). For the types of donors considered in this review, these similarities are indeed borne out.

III. Probes of Structure

The X-ray crystal structures of a variety of high- and low-oxidation-state nickel complexes are now known. These are introduced throughout the text in Section IV,A and the important points are summarized in

Section IV,A,5. In a number of instances, structures of the corresponding nickel(II) complexes are also known, allowing direct comparisons of metal–ligand bond lengths, and bond lengths and angles within the ligand, important in deciding whether redox is metal centered or ligand centered. If, on oxidation, metal–ligand bond lengths contract but those within the ligand are largely unaffected, then oxidation is thought to be primarily metal centered.

The technique most widely applied as a structural probe in the chemistry of both nickel(III) and nickel(I) is electron paramagnetic resonance. The g value is a function of the spin-orbit coupling constant and the ligand-field splitting, and is anisotropic. From frozen glasses of the complexes, the order of the g values gives detailed structural information (see Fig. 1), allowing distinction between tetragonal and square–planar geometry (16). In a square–planar geometry $g_{zz} > g_{xx}, g_{yy}$, similar to that found in a tetragonally compressed geometry but quite different from the situation for a tetragonally elongated geometry where $g_{xx}, g_{yy} > g_{zz} \approx 2$. For nickel(I) the situation is effectively reversed (17) since the unpaired electron is in a $d_{x^2-y^2}$ orbital in both tetragonally elongated and square–planar geometries, and in a d_{z^2} orbital for the tetragonally compressed case. Hyperfine interactions can also be used in assigning ligand donor atoms.

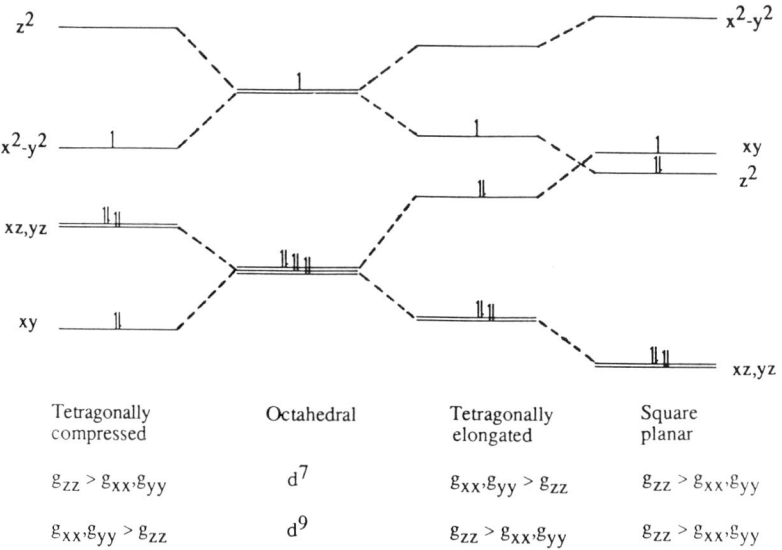

FIG. 1. The splitting of the d orbitals in tetragonally compressed, tetragonally elongated, and square–planar geometric for a d^7 ion and the resulting relative g values for d^7 and d^9 configurations.

With first-row p-block donor atoms, ligand radical species produced by ligand-centered redox processes tend to have isotropic g values close to the g value for a free electron, 2.0023, because the atoms in which the electron resides have no orbital contribution to paramagnetism (spin-orbit coupling is small compared with differences in the electronic energy levels). Thus, both the anisotropy and the magnitude of the g values of nickel(III) and nickel(I) complexes can be taken to indicate the degree of metal-centered oxidation or reduction. For nickel(III) complexes, Drago (18) has suggested (somewhat arbitrarily) a cutoff of 2.06 for the average g values, $\langle g \rangle$, to distinguish between metal-centered and ligand-centered oxidation. Nickel(IV) complexes are not amenable to study by EPR methods since, in general, they are diamagnetic. Nuclear magnetic resonance studies are possible, but thus far, results are few.

There is no generalized treatment of the UV/visible spectra of these complexes since they tend to be dominated by low-energy charge transfer bands which obscure details of the $d-d$ structure. In a few instances, circular dichroism data have been reported, but detailed interpretation has not been attempted (19).

IV. Oxidation of Nickel(II)

A. Coordination Environments and Structural Chemistry

1. Amines, Imines, and Oximes

Saturated amines provide a ligand field significantly greater than that of water and allow oxidation of nickel(II) under a variety of conditions. Early studies in this area involved pulse radiolysis (20, 21) and electrochemical (22) oxidation of nickel(II) in solutions of ammonia and 1,2-diaminoethane (en). The bis complex [Ni^{II}(en)$_2$Cl$_2$] can be oxidized by chlorine in methanolic solution to give [trans-Ni^{III}(en)$_2$Cl$_2$]Cl, which has a magnetic moment of 1.90 BM and shows (23) an axial EPR Spectrum (Table I), consistent with tetragonally elongated nickel(III). Hyperfine coupling from two axially bound chlorine nuclei is resolved in HCl solutions (24) and, interestingly, sulfate ion has a marked stabilizing effect on the complex. The chemistry of [Ni^{III}(en)$_2$Cl$_2$]Cl resembles the chemistry of the nickel(III) complexes of tetraazamacrocycles discussed in Section IV,A,2.

Under milder oxidation conditions a different product is formed (25, 26), best formulated as [Ni^{II}(en)$_2$Ni^{IV}(en)$_2$Cl$_2$]Cl$_4$. The magnetic moment is reduced to 0.76 BM and there is an intense intervalence band

TABLE I

Selected EPR Data for Nickel(III) Complexes

Complex	Medium	g_{xx}	g_{yy}	g_{zz}	A_{xx} (G)	A_{yy} (G)	A_{zz} (G)	Reference
[Ni^{III}(en)$_2$Cl$_2$]$^+$	Aq. HCl, 77 K	2.165	2.165	2.017	—	—	30	24
[Ni^{III}(bpy)$_3$]$^{3+}$	CH$_2$CN, 77 K	2.137	2.137	2.027	15	15	22.2	31
[Ni^{III}(phen)$_3$]$^{3+}$	CH$_3$CN, 77 K	2.136	2.137	2.027	16	16	23.2	31
[Ni^{III}(bpo)$_3$]	Solid	2.14	2.10	2.08	—	—	—	18
[Ni^{III}(dmg)$_3$]$^{3-}$	Base, 77 K	2.17	2.17	2.03	—	—	23.6	36
[Ni^{III}Me$_2$L]$^+$	Aq., 77 K	2.155	2.155	2.033	—	—	—	56
[Ni^{III}Me$_2$LH]$^{2+}$	Single crystal	2.1521	2.1289	2.0464	—	—	—	59
[Ni^{III}Me$_2$L']	Aq., 77 K	2.134	2.134	2.030	—	—	—	60
[Ni^{III}DOHDOpnCl$_2$]	Doped solid	2.155	2.143	2.026	11	11	65	61
[Ni^{III}[12]aneN$_4$]$^{3+}$	CH$_3$CN, 77 K	2.06	2.06	2.17	—	—	—	73
[Ni^{III}[14]aneN$_4$]$^{3+ a}$	CH$_3$CN, 77 K	2.20	2.20	2.03	—	—	41	73
[Ni^{III}[14]aneN$_4$]$^{3+ a}$	CH$_3$CN, 77 K	2.2148	2.2148	2.0250	—	—	—	75
[Ni^{III}[14]aneN$_4$]$^{3+ a}$	H$_2$O, 77 K	2.2193	2.2193	2.0332	—	—	—	75
[Ni^{III}[14]aneN$_4$]$^{3+ a}$	DMSO, 77 K	2.242	2.242	2.026	—	—	—	74
[Ni^{III}[14]aneN$_4$(Cl)$_2$]$^+$	DMSO, 77 K	2.180	2.180	2.022	<5	<5	28	74
[Ni^{III}[15]aneN$_5$]$^{3+ b}$	CH$_3$CN, 77 K	2.17	2.17	2.03	—	—	34	73
[Ni^{III}[18]aneN$_6$]$^{3+}$	CH$_3$CN, 77 K	2.16	2.16	2.06	—	—	—	73
[Ni^{III}([9]aneN$_3$)$_2$]$^{3+}$	Single crystal	2.12	2.12	2.03	—	—	—	33
[Ni^{III}H$_{-2}$[14]dioxoaneN$_4$]$^+$	H$_2$O, 77 K	2.23	2.23	2.02	—	—	—	95
[Ni^{III}H$_{-2}$[16]dioxoaneN$_5$]$^+$	H$_2$O, 77 K	2.154	2.154	2.016	—	—	21.7	102
[Ni^{III}H$_{-2}$G$_3$]	H$_2$O, 100 K	2.242	2.295	2.015	—	—	—	118
[Ni^{III}H$_{-3}$G$_4$]$^-$	H$_2$O, 100 K	2.297	2.278	2.010	—	—	—	118
[Ni^{III}H$_{-3}$G$_3$a]	H$_2$O, 100 K	2.310	2.281	2.006	—	—	—	118
[Ni^{III}H$_{-3}$G$_3$a(NH$_3$)]	H$_2$O, 100 K	2.217	2.217	2.011	—	—	23.4	118
[Ni^{III}H$_{-3}$G$_3$a(NH$_3$)$_2$]	H$_2$O, 100 K	2.178	2.178	2.019	—	—	19	118
[Ni^{III}(H$_{-1}$G$_2$)$_2$]$^-$	H$_2$O, 100 K	2.07	2.07	2.22	—	—	—	124
[Ni^{III}(H$_{-2}$G$_3$)$_2$]$^{3-}$	Base, 100 K	2.151	2.151	2.021	—	—	19.5	125
[Ni^{III}(edta)]$^-$	H$_2$O, 77 K	2.14	2.14	2.34	—	—	—	129
[Ni^{III}(bpyO$_2$)$_3$]$^{3+}$	CH$_3$CN, 77 K	2.220	2.155	2.060	—	—	—	135

^{a} Other N$_4$ macrocycles are similar. ^{b} Other N$_5$ macrocycles are similar.

around 15,000 cm^{-1}. Resonance Raman studies (27) support a structure with an alternating Ni–Cl chain, consistent with the high electrical conductivity of the material. Interconversion of [Ni^{III}(en)$_2$Cl$_2$]Cl and the mixed-valence form takes place in the presence of moisture (28).

The higher valent nickel complexes with saturated amine donors have little more than a transient existence as independent species in solution and consequently have a limited chemistry. Complexes with stronger donor ligands have a more extensive chemistry.

Oxidation of [Ni^{II}(bpy)$_3$]$^{2+}$ and other tris(polypyridyl) complexes is readily achieved electrolytically in 2 M HClO$_4$ (29) or in anhydrous acetonitrile (30, 31). The product, [Ni^{III}(bpy)$_3$]$^{3+}$, has a reduction potential of 1.72 V (versus normal hydrogen electrode, nhe) (Table II) showing an increasing trend with increasingly electronegative substituents on the polypyridine ligands. A weak absorption at 620 nm ($\varepsilon = 275$ M^{-1} cm^{-1}) and a more intense shoulder at 400 nm ($\varepsilon_{350} = 5300$ M^{-1} cm^{-1}) give the complex a lime-green color. The EPR data (31) with $g_{xx} = g_{yy} > g_{zz}$ (Table I) are interpreted in terms of a tetragonal elongation from octahedral symmetry with the unpaired electron in a d_{z^2} orbital and it is a little disconcerting that the X-ray structure of the complex (32) has two axial bonds 0.09 Å shorter than the average of the other four. Wieghardt and co-workers (33) have commented on this result, ruling out a dynamic in-plane distortion in favor of crystal packing forces. Whatever the explanation, it would appear that the energy barrier between elongated and compressed tetragonal distortion is not large. There is a significant (0.09–0.17 Å) shortening (32, 34) in the Ni–N bond lengths on going from [Ni^{II}(bpy)$_3$]$^{2+}$ to [Ni^{III}(byp)$_3$]$^{3+}$, while the ligand bond lengths and angles are little changed, indicating that oxidation is predominantly metal centered. These changes in Ni–N bond lengths contribute to the barrier to electron transfer between nickel(II) and nickel(III).

There is no evidence that oxidation of [Ni^{III}(bpy)$_3$]$^{3+}$ to give a nickel(IV) species can be effected. Likewise, when one of the pyridine rings is replaced by a benzoyl oxime as in syn-2-benzoylpyridine oxime (bpoH), oxidation, using persulfate in basic solution, is limited to

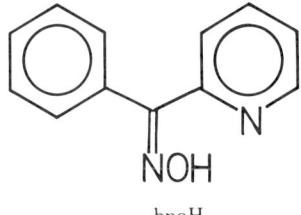

bpoH

TABLE II

REDUCTION POTENTIALS OF SELECTED NICKEL(III) COMPLEXES

Complex	$E^{0\ a}$			Reference
	CH_3CN^b	H_2O^c	$0.5\ M\ Na_2SO_4$	
$[Ni^{III}(bpy)_3]^{3+}$	1.71	—	—	31
$[Ni^{III}(phen)_3]^{3+}$	1.73	—	—	31
$[Ni^{III}Me_2L]^{2+}$	—	0.65	—	45, 56
$[Ni^{III}Me_2L]^+$	—	0.42	—	45, 56
$[Ni^{III}Me_2LH]^{2+}$	—	0.64	—	56
$[Ni^{III}[12]aneN_4]^{3+}$	1.66	—	—	73
$[Ni^{III}[13]aneN_4]^{3+}$	1.28–1.48	—	—	64, 73
$[Ni^{III}[14]aneN_4]^{3+}$	1.28, 1.25	1.05, 0.99, 0.74^d	0.744	64, 73, 96, 146
$[Ni^{III}iso[14]aneN_4]^{3+}$	1.42	—	—	73
$[Ni^{III}[15]aneN_4]^{3+}$	1.59, 1.48	—	1.014	64, 73, 96
$[Ni^{III}[16]aneN_4]^{3+}$	1.69	—	0.904	73, 102
$[Ni^{III}[15]aneN_5]^{3+}$	1.315	—	—	73
$[Ni^{III}[16]aneN_5]^{3+}$	1.35	—	—	73
$[Ni^{III}[17]aneN_5]^{3+}$	1.395	1.039	—	73, 108
$[Ni^{III}[18]aneN_6]^{3+}$	1.483	—	—	73
$[Ni^{III}([9]aneN_3)_2]^{3+}$	1.23	0.947, 1.015	—	108, 146
$[Ni^{III}([10]aneN_3)_2]^{3+}$	1.33	0.997, 1.014	—	108, 146

Complex			Ref.
[Ni^{III}([11]aneN$_3$)$_2$]$^{3+}$	1.54	—	108
[Ni^{III}H$_{-2}$[12]dioxoaneN$_4$]$^+$	—	0.86	96
[Ni^{III}H$_{-2}$[13]dioxoaneN$_4$]$^+$	—	1.14	96
[Ni^{III}H$_{-2}$[14]dioxoaneN$_4$]$^+$	1.09, 0.89	1.05	96
[Ni^{III}H$_{-2}$[15]dioxoaneN$_4$]$^+$	1.16	0.86	96
[Ni^{III}H$_{-1}$[16]oxoaneN$_5$]$^{2+}$	—	0.70	102
[Ni^{III}H$_{-2}$[16]dioxoaneN$_5$]$^+$	—	0.48	102
[Ni^{III}[14]aneN$_4$CH$_2$CH$_2$py]$^{3+}$	—	0.74	96
[Ni^{III}H$_{-2}$[13]dioxoaneN$_4$CH$_2$CH$_2$py]$^+$	—	1.14	96
[Ni^{III}H$_{-2}$[13]dioxoaneN$_4$CH$_2$CH$_2$pyO]$^+$	—	0.87	96
[Ni^{III}H$_{-2}$[14]dioxoaneN$_4$CH$_2$CH$_2$py]$^+$	—	1.10	96
[Ni^{III}[16]dioxoaneN$_5$CH$_2$CH$_2$py]$^{3+}$	—	0.48	102
[Ni^{III}H$_{-2}$G$_3$]	0.85	—	113
[Ni^{III}H$_{-3}$G$_4$]$^-$	0.79	—	113
[Ni^{III}H$_{-3}$G$_3$a]	0.83	—	113
[Ni^{III}(H$_{-2}$G$_3$)$_2$]$^{3-}$	<0.24	—	123
[Ni^{III}(bpyO$_2$)$_3$]$^{3+}$	1.81	—	135

[a] Versus a normal hydrogen electrode. Various ionic media are used; see references for details.
[b] Corrected for normal hydrogen electrode from Ag/Ag$^+$ by the addition of 0.578 V (Mann, C., and Barnes, K., "Electrochemical Reactions in Nonaqueous Systems." Dekker, New York, 1970.
[c] Generally with weakly coordinating media.
[d] [Ni^{III}[14]aneN$_4$(Cl)$_2$]$^+$.

formation of a neutral nickel(III) complex [Ni^{III}(bpo)$_3$] in which all three oxime protons are dissociated (*18*). In this case, a well-defined rhombic EPR signal is recorded with $g_{zz} > g_{xx} \approx g_{yy}$ consistent with tetragonally compressed geometry. When the second pyridine is replaced by an oxime group, as in dimethylglyoxime (dmgH$_2$) or diphenylglyoxime (dpgH$_2$), the situation changes and both nickel(III) and nickel(IV) complexes are readily accessible, depending on the mode of preparation.

The reasons for this change in behavior on going from bpy to bpo$^-$ and dmg^{2-} are not readily apparent since reduction potential data which would allow comparisons of the complexes are not available for all members of the series. The most obvious change is that the ligands differ in charge and that deprotonation parallels the ability of the complex to accept a loss of electrons.

The best characterized (*35–38*) of the oxidized nickel dimethylglyoximate species is [Ni^{IV}(dmg)$_3$]$^{2-}$, which can be obtained as the diamagnetic potassium, sodium, or barium salt. Preparation of

HON   NOH

dmgH$_2$

2

[Ni^{IV}(dmg)$_3$]$^{2-}$ is best carried out by addition of dmg^{2-} to a basic suspension of the higher valent oxides or hydroxides in the presence of an oxidant, thereby avoiding formation of the insoluble, square–planar [Ni(dmgH)$_2$]. With weaker oxidants, varying amounts of less well-characterized nickel(III) species are produced (*39*).

Alkaline solutions of [Ni(dmg)$_3$]$^{2-}$ are relatively long lived and exhibit absorption maxima at 460, 365, and 285 nm ($\varepsilon = 9600$, 7200, and 4.8×10^4 M^{-1} cm^{-1}, respectively) (*33, 34*). The complex is inert to substitution as expected for a low-spin d^6 ion with a decomposition rate of 2.6×10^{-7} sec^{-1} at 35°C, but is sensitive to acid and protonates with a pK_a of 10.8 (*37*). A two-electron potential of 0.54 V (versus nhe) is reported (*38*) in 1.5 M NaOH but the data are questionable. One-electron reduction at pH 12.4 gives a paramagnetic intermediate (*36, 40*), presumably [Ni^{III}(dmg)$_3$]$^{3-}$, with absorption maxima (*41*) at 520 and 280 nm ($\varepsilon = 3700$ and 2.95×10^4 M^{-1} cm^{-1}), a different species from that formed (*42*) in the one-electron oxidation of [Ni^{II}(dmgH)$_2$], which is likely to have a lower coordination number. The EPR spectrum of the proposed [Ni^{III}(dmg)$_3$]$^{3-}$ is axial (*36*) with $g_{xx} = g_{yy} > g_{zz}$, consistent with an elongated tetragonal distortion. A similar signal has been identified in the iodine oxidation of the nickel(II) complex in pyridine solution (*39*). However, in such oxidation experiments, a large number of different EPR spectra are claimed, depending on experimental conditions, and details of the structures have not been elucidated. It is of interest to note that for the series [Ni^{III}(bpy)$_3$]$^{3+}$, [Ni^{III}(bpo)$_3$], and [Ni(dmg)$_3$]$^{3-}$, the $\langle g \rangle$ value shows a modest increase from 2.10 to 2.12, perhaps signifying a slight increase in the degree of metal-centered oxidation. However, it is unlikely that this is important in increasing the accessibility of nickel(IV).

dapdH$_2$

3

In combination with other nitrogenous donor atoms, oximes have provided a particularly fruitful area for higher oxidation-state nickel chemistry. Tridentate coordination (*43*) in 2,6-diacetylpyridine dioxime (dapdH$_2$), which in its deprotonated form, dapd^{2-}, is a very strong field ligand, is used to suppress the formation of square–planar nickel(II). Persulfate oxidation of [Ni^{II}(dapd)$_2$]$^{2-}$ in basic media yields violet needles of the diamagnetic [Ni^{IV}(dapd)$_2$]. There is no evidence for the corresponding nickel(III) complex. The absorption spectrum of [Ni^{IV}(dapd)$_2$] shows peaks at 630 and 450 nm. The complex is inert to substitution as evidenced by the observation that it can be dissolved in concentrated HNO$_3$. In the X-ray structure (*44*), the ligand distorts to accommodate the small nickel(IV) ion and the Ni–N bond distances are significantly shorter than in comparable complexes of nickel(II). A simple molecular orbital treatment (*43*) suggests that the concentration of negative charge in the donor nitrogen and oxime oxygens of dapd^{2-} contributes to its ability to stabilize the higher oxidation states.

The most widely studied of the higher oxidation-state complexes incorporating oxime groups are those involving derivatives of the sexidentate ligand (*45*, *46*) 3,14-dimethyl-4,7,10,13-tetraazahexadeca-

Me₂LH₂

4

3,13-diene-2,15-dione dioxime, Me$_2$LH$_2$. A variety of structural variations have been prepared (47–50), but they share the donor system comprising two oximes, two imines, and two amine nitrogen donors, and only the parent complex will be considered.

The nickel(II) complex [Ni^{II}Me$_2$LH$_2$]$^{2+}$ is a typical six-coordinate, high-spin species with a magnetic moment of 3.12 BM and undergoes successive deprotonation of the oxime oxygens with pK_a values of 5.90 and 7.80 (45, 46). Chemical or electrochemical oxidation results in the highly colored diamagnetic nickel(IV) species [Ni^{IV}Me$_2$L]$^{2+}$ in which both oxime groups are deprotonated. The X-ray structures of both [Ni^{IV}Me$_2$L](ClO$_4$)$_2$ and [Ni^{II}Me$_2$LH$_2$](ClO$_4$)$_2$ have been determined (51–53) and show significant changes in the Ni–N bond lengths. For the former complex, average Ni–N bond lengths are 1.955 (amine), 1.871 (imine), and 1.953 (oxime) Å, while for the latter they are longer, 2.098, 2.055, and 2.119 Å, respectively. Small changes in the ligand bond lengths and angles are also noted.

Wrapping the sexidentate ligand around the metal (Fig. 2) creates a pair of optical isomers and the nickel(IV) complexes are sufficiently inert to allow optical resolution (54, 55). UV visible and circular dichroism spectra are shown in Fig. 3. Further evidence of the inert-

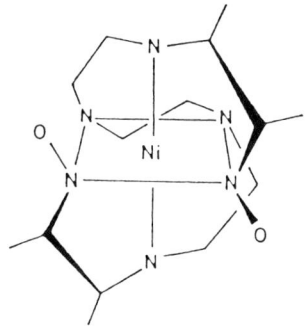

FIG. 2. [Ni^{IV}Me$_2$L]$^{2+}$.

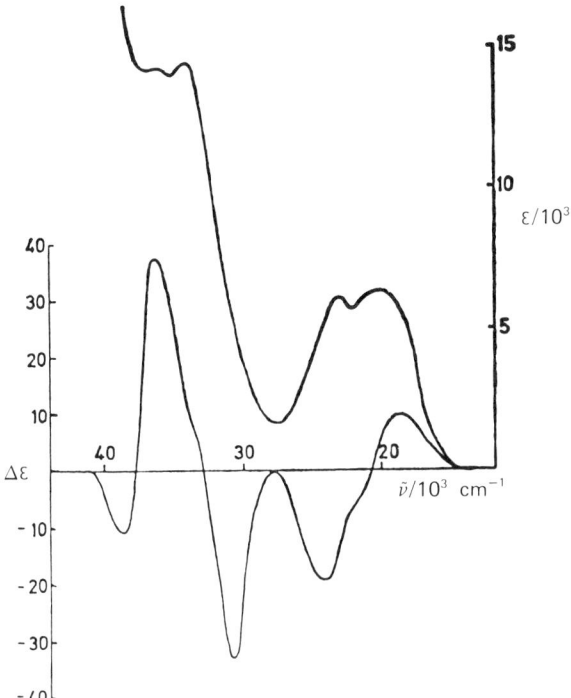

FIG. 3. Absorption and circular dichroism spectra of an aqueous solution of [Ni^{IV}Me$_2$L]$^{2+}$; from Ref. 54 by permission of the authors and the Royal Society of Chemistry.

ness to substitution of the nickel(IV) complex is provided by the fact that the ligand will not unwrap in concentrated HNO$_3$.

Above pH 5, the corresponding nickel(III) complex, [Ni^{III}Me$_2$L]$^+$, is thermodynamically stable (56, 58). It has an absorption spectrum similar to but less intense than that of [Ni^{IV}Me$_2$L]$^{2+}$, with maxima at 505 and 398 nm (ε = 2890 and 3000 M^{-1} cm^{-1}) (56). The EPR spectrum is typical of a tetragonally elongated complex with $g_{xx} = g_{yy} = 2.16$ and $g_{zz} = 2.04$. At lower pH a protonated form (pK_a = 4.05) [Ni^{III}Me$_2$LH]$^{2+}$ has been detected as a thermodynamically unstable kinetic transient with a broad absorption maximum at 490 nm (ε = 2980 M^{-1} cm^{-1}). Single-crystal EPR data (59) for this species are available and again a tetragonally elongated geometry is consistent with the data. The g values show a slight rhombic distortion with g_{xx} = 2.1521, g_{yy} = 2.1289, and g_{zz} = 2.0464, in good agreement with other determinations for this complex.

When one of the oxime–imime groups is replaced by two amine donors in the ligand Me$_2$L'H, oxidation is restricted to the trivalent state (*60*), leading to the conjecture that a minimum of two oxime–imine chromophores are required to stabilize nickel(IV). The complex [Ni^{III}Me$_2$L′]$^+$ is paramagnetic, showing an axial EPR spectrum with $g_{xx} = g_{yy} = 2.134$ and $g_{zz} = 2.030$. The reduction potential is slightly higher than that of the bis(oxime–imine) system. Incorporation of the bis(oxime–imine) chromophore in a ligand (DOH)$_2$pn, where the

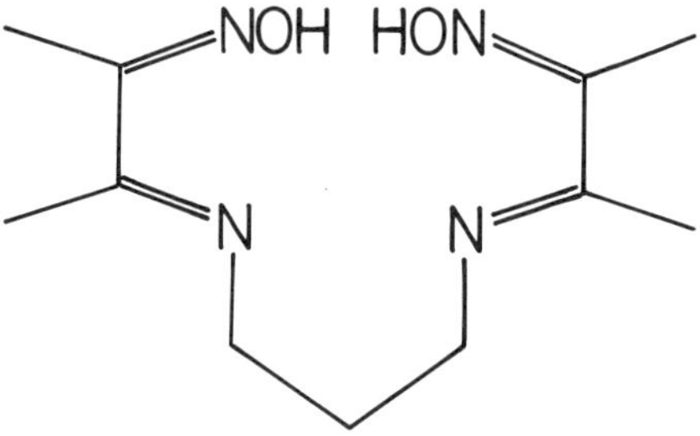

(DOH)$_2$pn

5

strong donors are restricted to a planar geometry, is also insufficient to stabilize nickel(IV) (*61*). The resulting nickel(III) complexes, [Ni^{III}(DOH)(DO)pnCl$_2$] and [Ni^{III}(DOH)(DO)pnBr$_2$], are tetragonally elongated and show axial hyperfine coupling from the halide ions.

2. Macrocycles

Complexes with macrocyclic ligands have a rich chemistry for both higher and lower oxidation states of nickel (*62–66*). The smallest members of the series, the triaza macrocycles, show no redox chemistry for 1:1 complexes with nickel(II), but there are recent reports of stable bis complexes which will be discussed later in this section.

The restricted geometry of the tetraaza macrocyclic ligands favors the formation of low-spin, square–planar nickel(II). An equilibrium exists between the blue, high-spin, six-coordinate form and the yellow, substitution-inert, square–planar form [Eq. (1)]. Intercoversion

$$[\text{Ni}^{II}[14]\text{aneN}_4(\text{H}_2\text{O})_2]^{2+} \rightleftharpoons [\text{Ni}^{II}[14]\text{aneN}_4]^{2+} + 2\text{H}_2\text{O} \quad (1)$$

blue, high spin   yellow, low spin

between the forms is dependent on structure (*67–70*), as shown in Table III, and appears to reflect a balance between exothermic solvent release and endothermic Ni–N bond shortening. Typical Ni–N

TABLE III

Equilibrium Constants and Thermodynamic Parameters for the Blue-Yellow Interconversion of Nickel(II) Complexes

Ligand	K_{eq}	ΔG^0 (kcal mol^{-1})	ΔH^0 (kcal mol^{-1})	ΔS^0 (cal K^{-1} mol^{-1})	Reference
2,3,2-tet	0.3	0.7	3.4	9	68
2,2,2-tet	0.01	2.6	3.4	3	67
[14]aneN$_4$	1.6	−0.26	5.3	18.7	68
iso[14]aneN$_4$	2.3	−0.5	5.4	20	70
[13]aneN$_4$	68	−1.14	7.5	30	69
[12]aneN$_4$	0.8	0.1	1.7	5.5	69
Me$_2$[14]aneN$_4$	0.4	0.5	—	—	140
[12]dioxoaneN$_4$ [a]	6.7	−1.13	8.7	33	94
[13]dioxoaneN$_4$	0.5	0.4	4.2	12.8	94

[a] 1,2-Dioxo derivative.

bond lengths for high-spin nickel(II) are 2.1–2.2 Å and for low-spin nickel(II) are 1.9–2.0 Å. Rates of interconversion between the two forms, estimated from studies with linear tetraamines, are rapid (71), $>10^5$ sec^{-1} at 5°C.

Early studies of the oxidation of nickel(II) in saturated tetraaza macrocycles were carried out using electrochemical methods in acetonitrile solutions (62, 64). The nickel(III) state is readily accessible but removal of a second electron is not possible except in a few instances where ligand oxidation is involved. The reduction potentials of a variety of nickel(III) macrocyclic complexes are shown in Table II. One important feature in stabilizing nickel(III) is the hole size supported by the macrocycle (64, 72, 73). The complex with the symmetric 14-membered ring ligand, [14]aneN$_4$, is most thermodynamically stable, with an ideal Ni–N bond length of 2.07 Å. However, the asymmetric isomer, iso[14]aneN$_4$, is significantly less stable (70), illustrating the importance of ring configuration. It is noteworthy that the square–planar component of the ligand-field splitting parameter is smaller for [Ni^{II}iso[14]aneN$_4$]$^{2+}$ than for [Ni[14]aneN$_4$]$^{2+}$.

Most structural characterization of these complexes has been carried out by EPR methods (64, 73–76) and, with a few notable exceptions, the nickel(III) species show the expected elongated tetragonal geometry with axial coordination of solvent or counterions confirmed by hyperfine interactions (Table I). The smallest member of the tetraaza macrocycle series, [12]aneN$_4$, is unusual in that the ligand folds to give cis coordination because the metal ion is too large to allow a planar

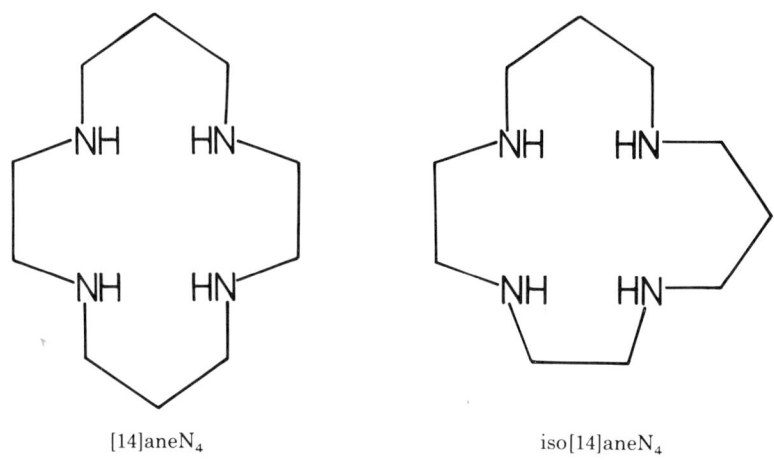

[14]aneN$_4$ iso[14]aneN$_4$

6

arrangement, explaining the EPR parameters for this species (73). Stability constants for addition of anions to nickel(III) planar macrocyclic complexes have been determined (75, 77–80), as shown in Table IV, and the kinetics and mechanisms of substitution reported (75).

In confirmation of the EPR studies, the X-ray structures of the complexes [Ni^{III}[14]aneN$_4$(Cl)$_2$]Cl (81) (Fig. 4) and [Ni^{III}[14]aneN$_4$(NCS)$_2$] (82) show a tetragonally elongated geometry. For the former complex, equatorial Ni–N bond lengths are 1.970 Å, shorter than the 2.058 Å found for the corresponding nickel(II) species (83). The axial Ni–Cl bonds are also shorter but by a smaller amount. However, bond lengths

TABLE IV

STABILITY CONSTANTS FOR FORMATION OF NICKEL(III) ADDUCTS

Reaction	Medium (moles/liter)	K	Reference
[Ni^{III}[14]aneN$_4$]$^{3+}$ + Cl$^-$	1.0, 25°C	210 M^{-1}	75
[Ni^{III}[14]aneN$_4$]$^{3+}$ + Br$^-$	1.0, 25°C	34 M^{-1}	75
[Ni^{III}[14]aneN$_4$]$^{3+}$ + SO$_4^{2-}$	0.3, 22°C	5×10^4 M^{-1}	77
[Ni^{III}[14]aneN$_4$]$^{3+}$ + SO$_4^{2-}$	0.3, 22°C	3×10^3 M^{-1}	77
[Ni^{III}Me$_6$[14]aneN$_4$]$^{3+}$ + 2SO$_4^{2-}$	0.5, 22°C	5×10^6 M^{-2}	80
[Ni^{III}Me$_6$[14]aneN$_4$]$^{3+}$ + 2Cl$^-$	0.3, 22°C	250 M^{-2}	80
[Ni^{III}Me$_6$[14]aneN$_4$]$^{3+}$ + 2(C$_8$H$_4$O$_4^{2-}$)	0.3, 22°C	$\geq 10^7$ M^{-2}	80
[Ni^{III}H$_{-3}$G$_3$a] + NH$_3$	1.0, 25°C	270 M^{-1}	122
[Ni^{III}H$_{-3}$G$_4$]$^-$ + NH$_3$	1.0, 25°C	160 M^{-1}	122
[Ni^{III}H$_{-3}$G$_3$a] + imidazole	1.0, 25°C	650 M^{-1}	122
[Ni^{III}H$_{-3}$G$_3$a] + pyridine	1.0, 25°C	55 M^{-1}	122

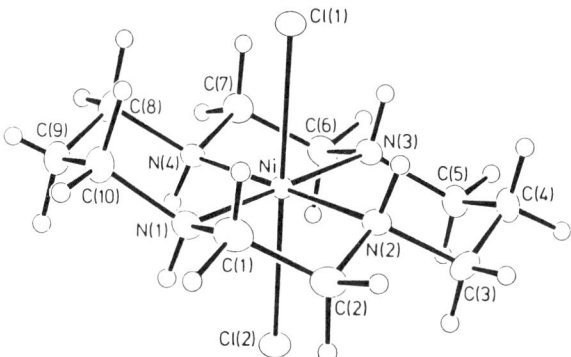

FIG. 4. X-Ray structure of $[Ni^{III}[14]aneN_4(Cl)_2]^+$; from Ref. 81 by permission of the authors and the Japanese Chemical Society.

and angles within the macrocyclic ligand are normal, supporting the concept of metal-centered oxidation.

Substituents on the macrocyclic backbone affect the redox characteristics by restricting axial ligand binding which would stabilize the higher oxidation state (77–80). This stabilization is not only thermodynamic but also kinetic in nature. It has been shown (80) that decomposition of $[Ni^{III}Me_6[14]aneN_4X_2]^{(3-2n)+}$ in aqueous media containing coordinating anions, X^{n-}, proceeds primarily through the unsubstituted species $[Ni^{III}Me_6[14]aneN_4(H_2O)_2]^{3+}$. The mechanism involves ligand oxidation by proton abstraction and the formation of imine groups. A similar mechanism is proposed for decomposition in acetonitrile (84), though some details of the intermediates involved are in dispute. Extensive use of the pulse radiolysis technique has been made to elucidate these species (85–88).

In general, unsaturation in a neutral macrocyclic ligand makes oxidation less favorable. Imine donors are less capable of stabilizing the higher oxidation states, particularly when they are in conjugation. However, dianionic deprotonated diimine donors are readily oxidized at low potentials (64) to give the corresponding nickel(III) complexes, which are determined from EPR parameters to be square–planar. This is most likely due to a strong π bonding interaction in which the metal ion d_{z^2} orbital is stabilized such that the unpaired electron is removed from a d_{xy} orbital. Deprotonation phenomena have been studied (89, 90) with related complexes and, surprisingly, the pK_a of the nickel(III) complex exceeds that of nickel(II), possibly as a result of coordination number changes. The deprotonated form decomposes through a ligand radical which readily dimerizes (91).

The ease of formation of higher oxidation-state nickel complexes with oligopeptide ligands has prompted incorporation of amide donors into the macrocyclic framework (92–98). Dioxo derivatives of the tetraaza macrocycles in which there are two amine and two deprotonated amide (imide) donors stabilize tetragonally elongated nickel(III) complexes in aqueous media. In the presence of coordinating anions such as sulfate, reduction potentials are somewhat higher than those of the corresponding saturated tetraaza macrocycles (92, 96), a consequence of weaker axial ligand binding which stabilizes the higher oxidation states. However, in the presence of poorly coordinating anions, the order of the potentials is reversed (95). Somewhat surprisingly, the small, 12-membered macrocycle (97, 98) has the lowest reduction potential of those represented.

The thermodynamic and kinetic stabilization of nickel(III) macrocyclic complexes by axial coordination has prompted a number of new approaches. Studies with tetraaza macrocyclic ligands with pendant donors acting as potential fifth ligands have had some success (96, 99, 100). Oxidation of $[Ni^{II}[14]aneN_4CH_2CH_2py]^{2+}$ in aqueous solution

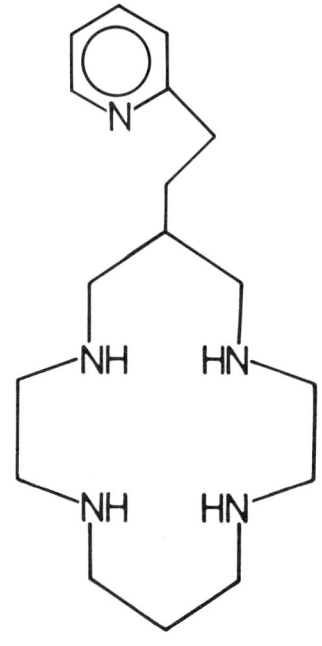

$[14]aneN_4CH_2CH_2py$

7

yields EPR evidence for axial coordination of the pyridine nitrogen, which appears to have little effect on the reduction potential. This may be the result of comparable axial coordination of both oxidation states of the complex; however, reduction potentials were measured in sulfate media, known to stabilize the higher oxidation state, so that this apparent lack of stabilization is misleading. In the case of the corresponding bis oxo complex [Ni^{III}H$_{-2}$[14]dioxoaneN$_4$CH$_2$CH$_2$py]$^+$ in which H$_{-2}$ indicates deprotonation at both amide nitrogens, there is

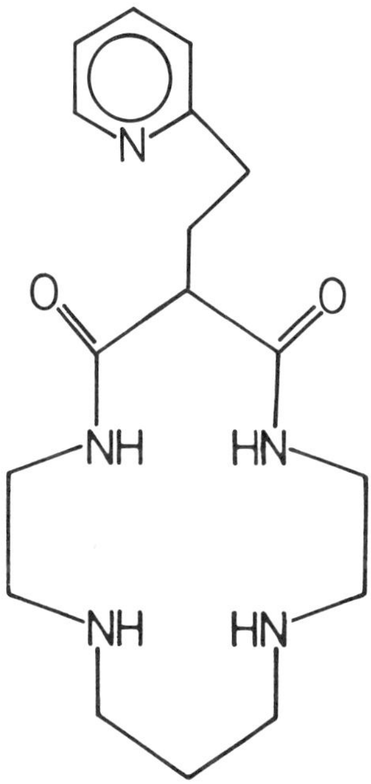

[14]dioxoaneN$_4$CH$_2$CH$_2$py

8

no evidence for axial coordination. Only for the 13-membered species with a pendant pyridine N oxide is there a significant decrease in reduction potentials (*96*).

The situation with pendant phenol groups is much less ambiguous. The crystal structure of the nickel(II) complex [Ni^{II}H$_{-1}$[14]ane-N$_4$C$_6$H$_5$OH]$^+$ (Fig. 5) shows it to be a square–pyramidal, high-spin species and the reduction potential of the nickel(III) complex is 0.15 V lower than that of the parent saturated 14-membered macrocyclic complex (*100*). Strong coordination by phenolate oxygen is even more

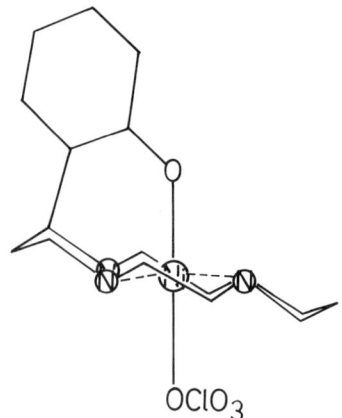

FIG. 5. X-Ray structure of [Ni^{III}H$_{-1}$[14]aneN$_4$C$_6$H$_5$OH(ClO$_4$)]; from Ref. *100* by permission of the authors and the American Chemical Society.

dramatically shown in the complex [Ni^{II}H$_{-1}$[13]aneN$_4$C$_6$H$_5$OH]$^+$, where deprotonation of the phenol group combined with the smaller ring size changes the ligand geometry from a square–planar low-spin species with planar tetraaza coordination to a high-spin six-coordinate species in which the macrocycle is bound in folded form (*99*).

A second approach to provide axial coordination with macrocyclic ligands has been the use of pentaaza and hexaaza macrocyclic ligands wherein the fifth and sixth ring nitrogens may be available for axial binding (*101*). Use of the saturated pentaaza macrocycles [15]aneN$_5$, [16]aneN$_5$, and [17]aneN$_5$ gives no dramatic stabilization of the higher oxidation state (*72, 73*). Reduction potentials of the nickel(III) complexes are more positive than for [Ni^{III}[14]aneN$_4$]$^{3+}$, although the values have been measured in a medium (CH$_3$CN) where axial stabilization of tetragonal nickel(III) has been established, masking the desired effect. In aqueous media, the oxo macrocycles show lower potentials (*102*). The species [Ni^{II}H$_{-2}$[16]dioxoaneN$_5$] is high spin, five coordinate with two imine and two amine nitrogen donors in the equatorial plane and the fifth ring nitrogen coordinated axially (*103*). The angle between the perpendicular to the N$_4$ plane and the axially coordinated nitrogen decreases from 18.4 to 7.7° on oxidation from nickel(II) to nickel(III) as a result of a contraction in the planar Ni–N bond lengths (*104*). These 16-membered dioxopentaaza macrocyclic complexes have been shown to activate molecular O$_2$ in the oxidation of benzene to phenol (*102, 105*). It is proposed that this takes place by formation of an adduct with O$_2$, for which there is some evidence, the most convincing being the appearance of additional hyperfine coupling in the nickel(III) EPR spectrum when $^{17}O_2$ is present compared with $^{16}O_2$ (*102*).

The six-membered macrocycle [18]aneN$_6$ forms a high-spin six-coordinate complex with nickel(II) which can be oxidized in acetonitrile solution to give a paramagnetic nickel(III) complex with an axial EPR spectrum indicating a tetragonally compressed octahedral geometry (*73*). Six-coordinate geometry is also possible with formation of bis complexes with the triaza macrocycle [9]aneN$_3$, and these remarkable species have attracted much interest (*106, 107*). The small nine-membered macrocycle forms a nickel(III) complex more thermodynamically stable than either [10]aneN$_3$ or [11]aneN$_3$ (*108*). The crystal structure reveals (*33*) a distorted octahedron in which two triaza macrocyclic ligands are coordinated facially and the Ni–N bond lengths are in two groups, four with the short distance of 1.971 Å and two with 2.110 Å, comparable with the Ni–N distance in the corresponding octahedral nickel(II) complex (*109*). Although the X-band EPR of frozen glasses of the complex yield spectra which are interpreted as rhombic, detailed Q-band studies reveal a spectrum consistent with the X-ray structure (*33*).

3. Amino Acids and Peptides

There are reports of transient nickel(III) complexes formed in pulse radiolysis experiments with high-spin nickel(II) in glycine solutions (*21*), but oligopeptide complexes containing deprotonated peptide linkages provide a much more extensive chemistry and have been examined in greater detail. At pH > 8, oligopeptide ligands, such as tetraglycine G$_4$, form low-spin diamagnetic, square–planar complexes with nickel(II) in which peptide hydrogens are ionized from the coordinated groups (*110*). The resulting complex (*111*)[Ni^{II}H$_{-3}$G$_4$]$^{2-}$ is inert to substitution, with a rate constant for cleavage of the terminal Ni–N bond of 1.6×10^{-5} sec^{-1} at 25°C.

The observation (*112–115*) that neutral aqueous solutions of [Ni^{II}H$_{-3}$G$_4$]$^{2-}$ consume molecular oxygen with the appearance of a strongly absorbing transient at 350 nm lead to detailed investigations and discovery of nickel(III)–peptide complexes (*113*). The oxidized nickel complexes have absorption maxima around 325 and 240 nm ($\varepsilon = 5240$ and $\sim 11{,}000\ M^{-1}\ cm^{-1}$, respectively for [Ni^{III}H$_{-3}$G$_4$]$^{-}$). Reduction potentials (*116*) (Table II), measured by cyclic voltammetry, show a small dependence on ligand structure which can be correlated

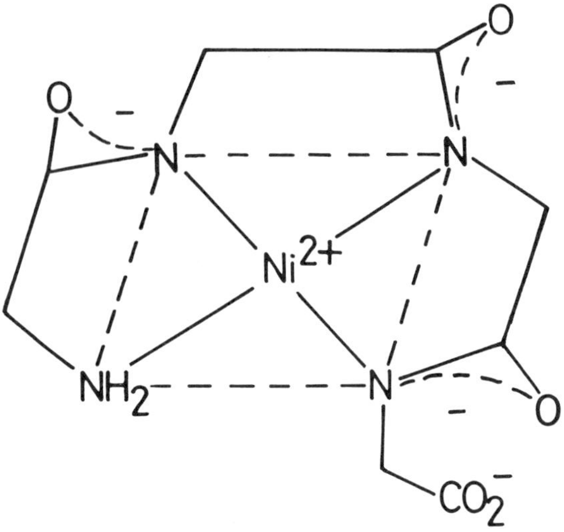

[Ni^II H_{-3}G_4]^{2-}

9

with the equatorial ligand-field stabilization energy. However, the dependence on the ligand field is smaller than expected on the basis of square-planar geometry for the nickel(III) complex and suggests axial coordination of the solvent. Temperature-dependence studies (*117*) reveal a positive entropy change for the reduction of 15 cal K^{-1} mol^{-1} corresponding to the release of two water molecules.

$$[Ni^{III}H_{-3}G_4(H_2O)_2]^- + e \rightleftharpoons [Ni^{II}H_{-3}G_4]^{2-} + 2H_2O \qquad (2)$$

Tetragonally elongated geometry is confirmed from EPR measurements (*118–121*) and there is a trend in the g values (*118*) with an increasing orbital contribution in the order $N^- > -NH_2 > -Im \sim CO_2^-$, reflecting the donor strength of the groups. Addition of ligands which bind axially to the complexes can be monitored by the hyperfine interaction (Fig. 6), or by the effect on the reduction potentials of the complexes (*122*) (Fig. 7), allowing evaluation of stability constants (Table IV). The rates of axial substitution are fast, $>4 \times 10^6$ sec^{-1} for the reaction of imidazole with [Ni^{III}H$_{-2}$Aib$_3$], faster than comparable reactions with macrocyclic complexes due to the stronger in-plane ligand field.

Axial substitution has both a thermodynamic and a kinetic effect in stabilizing the trivalent complex. In the presence of an excess of the oligopeptide ligands, long-lived bis complexes form (*118*). Reaction of

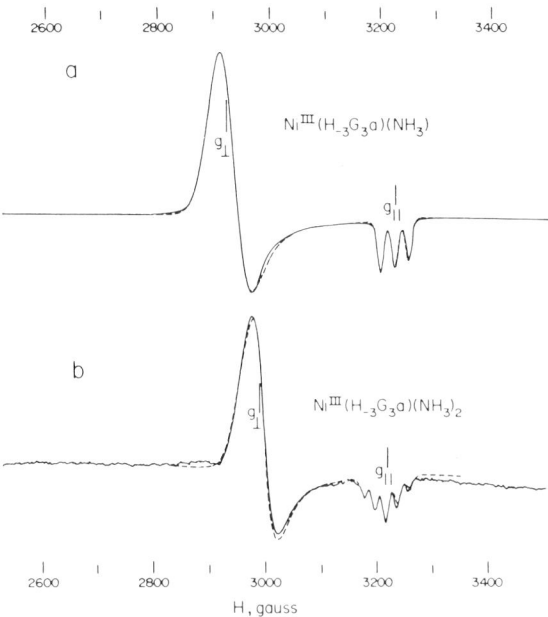

FIG. 6. EPR spectra of (a) [Ni^{III}H$_{-3}$G$_3$a(NH$_3$)] and (b) [Ni^{III}H$_{-3}$G$_3$a(NH$_3$)$_2$] in aqueous glasses at 100 K; from Ref. *118* by permission of the authors and the American Chemical Society.

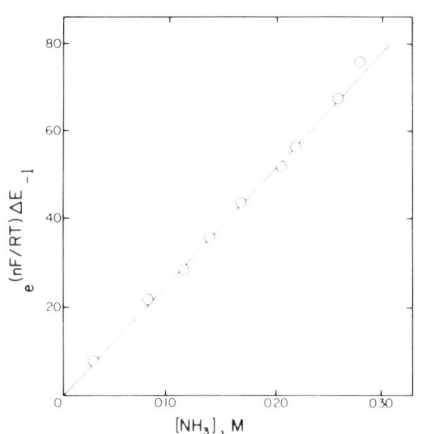

FIG. 7. Plot of exp[$\Delta E(nF/RT)$] − 1 versus NH$_3$ for the interaction of [Ni^{III}H$_{-3}$G$_3$a] with NH$_3$; from Ref. *117* by permission of the authors and the American Chemical Society.

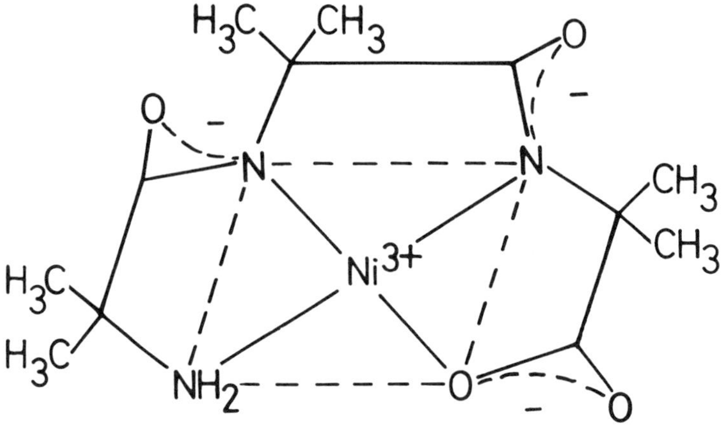

[Ni^{III}H$_{-2}$Aib$_3$]

10

[Ni^{III}H$_{-2}$G$_3$] with excess ligand G$_3^-$ leads (123) rapidly to the formation of [Ni^{III}(H$_{-2}$G$_3$)(H$_{-1}$G$_3$)]$^{2-}$, a tetragonal species with $g_{xx} = g_{yy} > g_{zz}$ and five nitrogens coordinated. Above pH 11 the six-coordinate complex [Ni^{III}(H$_{-2}$G$_3$)$_2$]$^{3-}$ is detected. Thermodynamic arguments suggest that this complex has a reduction potential less than 0.24 V. The tridentate ligand diglycine forms a high-spin six-coordinate nickel(II) complex with two trans deprotonated peptide groups (111) that can be oxidized to the corresponding nickel(III) species in which the unique axis contains both deprotonated peptide nitrogens in a compressed tetragonal geometry (124). The complex is violet-black in color with an absorption maximum in the visible region at 560 nm. On acidification, the carboxylate groups are protonated and the complex rearranges to give a yellow tetragonal species with two amine and two deprotonated peptide ligands in a plane.

Acid treatment (125) of [Ni^{III}H$_{-3}$G$_4$(H$_2$O)$_2$]$^-$ results in cleavage of the terminal Ni–N deprotonated peptide bond with a rate constant of 0.2 sec^{-1} at 25°C, faster than the corresponding rate for [Ni^{II}H$_{-3}$G$_4$]$^{2-}$. Further dissociation of the tridentate tetraglycine ligand is much slower and the intermediate can be trapped by the addition of terpy to give a stable, six-coordinate nickel(III) mixed-ligand complex (126). It is notable that the calculated reduction potential for the mixed complex is lower than for either [Ni^{III}H$_{-3}$G$_4$]$^-$ or for [Ni^{III}(terpy)$_2$]$^{3+}$.

The sexidentate amino acid edta^{4-} forms a high-spin nickel(II) complex which can be oxidized to a tetragonally compressed nickel(III) complex (127–130) with $g_{zz} = 2.337$ and $g_{xx} = g_{yy} = 2.139$. Though sensitive to O_2, it is relatively long lived. A related amino acid based on the triaza macrocycle 1,4,7-triazacyclononane-N,N', N''-triacetate (TACNTA) forms a very stable complex with nickel(II) which aerobically oxidized to the pink nickel(III) species in dilute nitric acid (131).

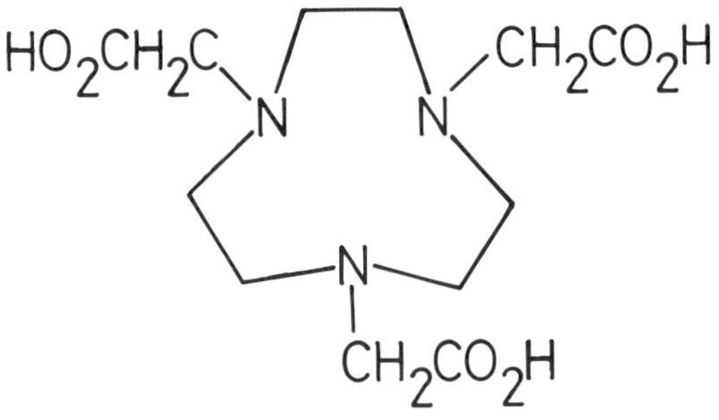

TACNTAH$_3$

11

Metal–nitrogen and metal–oxygen bond lengths in [Ni^{III}TACNTA] are 1.93 and 1.91 Å, respectively, both somewhat shorter than in the corresponding nickel(II) ion.

4. Other Ligands

The earliest high-oxidation-state complex of nickel reported was the heteropoly(molybdate) (132, 133) complex [Ni^{IV}Mo$_9$O$_{32}$]$^{6-}$, which contains nickel(IV) in an octahedral NiO$_6$ coordination environment. There is no evidence for the corresponding nickel(III) species but further work on nickel(IV) complexes of this type has been reported recently (134). Nickel(III) can be prepared in a six-coordinate oxygen donor environment (135) as a tris chelate with 2,2'-bipyridine-1,1'-dioxide (bpyO$_2$). The complex has a rhombic EPR spectrum and a reduction potential of 1.7 V, from which an estimate of the reduction potential of the ion [Ni^{III}(H$_2$O)$_6$]$^{3+}$ of 2.5 V (versus nhe) has been calculated.

One recent, important development has been (136) the electrolytic preparation from [Ni^{II}(CN)$_4$]$^{2-}$ of [Ni^{III}(CN)$_4$(H$_2$O)$_2$]$^-$, which has a half-life of 11 minutes in acidic solution at 25°C and serves as a precursor for the incorporation of nickel(III) in other ligand system. The complex is tetragonally elongated from its EPR spectrum and can add other anions axially. The bis aquo species has an absorption maximum at 225 nm ($\varepsilon = 1.16 \times 10^4 \, M^{-1} \, cm^{-1}$) and a reduction potential of 1.19 V (versus nhe).

5. Remarks on Structural Probes of Higher Oxidation-State Nickel

It is worthwhile making some comment on the primary data used for characterization of nickel(III) and nickel(IV). Reference to Table I reveals that EPR is an effective probe of nickel(III) structure, but without detailed analysis it gives little indication of the electronic distribution. What is becoming clearer as data on a variety of structures are reported is the finding of both tetragonally elongated and tetragonally compressed geometries for nickel(III), perhaps suggesting that there can be rapid interchange between the two Jahn–Teller distorted forms.

The extensive X-ray data on higher order oxidation state complexes is also worthy of comment. It has already been mentioned that Ni–N(amine) bond lengths for high-spin nickel(II) are in the range 2.1–2.2 Å and for low-spin nickel(II) are 1.9–2.0 Å. For nickel(III) they vary (33, 80–82, 131) from 1.93 to 2.11 Å, the large range a function of the Jahn–Teller distortion and the finding that tetragonally elongated structures predominate. On a small sample of nickel(IV) complexes (51, 52) the Ni–N(amine) lengths are 1.955 and 2.006 Å, a little larger than might be expected. For the Ni–N(imine) systems examined the situation is more in line with that expected with 2.01–2.09 Å for nickel(II) (34, 51, 53), 1.92–2.02 Å for nickel(III) (32), and 1.84–1.87 Å for nickel(IV) (44, 51, 52). Imide bond lengths are also shorter for nickel(III) (103) than for nickel(II) (103, 111).

In Table II the reduction potentials of a variety of nickel(III) complexes are reported. The values have been corrected as well as is possible for standard conditions and it is revealing that the potentials, particularly of complexes which have a strong square–planar ligand, show marked variations depending on the donor ability of the solvent and coordinating anions. This has itself evolved into a structural probe (108, 117) but clearly it makes comparisons difficult and overinterpretation somewhat dangerous.

B. KINETIC STUDIES

1. Complex Formation Reactions of Nickel(III)

Electrochemical studies (62, 64, 106, 137) have shown that nickel(III) complexes with macrocycles, peptides, and diimine ligands are relatively strong oxidants ($E^0 = 0.9-1.3$ V versus nhe), and in acidic solutions many of these species are sufficiently long lived to enable studies of both complex formation and redox reactivity. It has been suggested (86) that the role of the ligands is such that not only is the energy of the antibonding orbital [containing the electron to be removed on oxidation from nickel(II)] raised through strong M–N in-plane interactions, but also encapsulation and modification of the ligand pK values reduces the reactions with solvent.

Relatively few studies are available for complex formation reactions of nickel(III) macrocycles. For the tetraaza macrocycles, the octahedral structure of the ion requires that the two axial sites are coordinated by solvent. In aqueous media, reaction of [Ni^{III}[14]aneN$_4$(OH$_2$)$_2$]$^{3+}$ with halide ions may be monitored at $\lambda \sim 320-350$ nm. Owing to the relatively facile hydrolysis and redox decomposition of the nickel(III)–diaquo macrocycles—[Ni^{III}[14]aneN$_4$(OH$_2$)$_2$]$^{3+}$ decomposes within 1 hour at pH 2–3 (and more rapidly with increasing pH)—the early kinetic studies involving nickel(III) were made using pulse radiolysis (86) or flash photolysis techniques (138). However, the presence of axially coordinated chloride (75, 80), sulfate (79, 80), or phthalate (80) renders the nickel(III) center much more stable kinetically. Pulse radiolysis studies (86) of the diaquo-*trans*-(Me$_6$[14]dieneN$_4$)nickel(III), [Ni^{III}Me$_6$[14]dieneN$_4$(H$_2$O)$_2$]$^{3+}$, with Br$^-$

$$[\text{Ni}^{III}\text{Me}_6[14]\text{dieneN}_4(\text{H}_2\text{O})_2]^{3+} + \text{Br}^- \underset{k_b}{\overset{k_f}{\rightleftharpoons}} [\text{Ni}^{III}\text{Me}_6[14]\text{dieneN}_4(\text{H}_2\text{O})\text{Br}]^{2+} \quad (3)$$

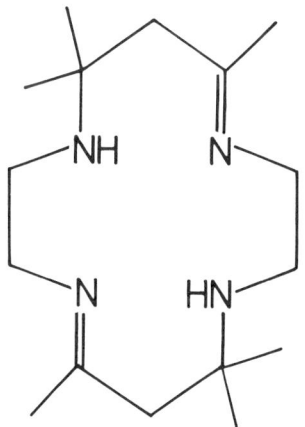

Me$_6$[14]dieneN$_4$

12

showed that at pH ~ 3, $k_f \sim 1300$ M^{-1} sec^{-1} ($I = 0.01$ M) with $k_b = 120$ sec^{-1}. The strong oxidizing power of the unsaturated ligand complex (~ 1.30 V) results in oxidation of the bromide to bromine. In the case of [14]aneN$_4$ (75) or its substituted derivatives (139) the lower E^0 values (0.96–1.1 V) are such that while formation reactions of stable chloride and bromide species may be monitored, the iodo complexes rapidly undergo redox reactions (140).

Rate constants derived from studies at a variety of hydrogen ion concentrations reveal no discernible kinetic effect from this source. Data for formation and dissociation of monohalo complexes are presented in Table V. It may be seen that for forward and reverse substitution processes [Eq. (3)] the rates increase in the order [14]aneN$_4$ < meso-Me$_2$[14]aneN$_4$ < meso-Et$_2$[14]aneN$_4$ ≪ rac-Me$_2$[14]aneN$_4$. All four of these complexes have identical chair conformations in the macrocyclic ring, differing only in the presence at the asymmetric centers of two equatorial methyl (or ethyl) groups in the meso-Me$_2$[14]aneN$_4$ and meso-Et$_2$[14]aneN$_4$ ligands and one equatorial and one axial methyl group in the rac-Me$_2$[14]aneN$_4$ complex. With the exception of the rac-Me$_2$[14]aneN$_4$ complex, the formation rates fall within an order of magnitude, suggesting a dissociative character to the reaction. There are no data available on solvent exchange rates of nickel(III). It is considered that the increase in rate for the racemic complex derives from steric crowding on one face of the molecule modifying the reactivity and favoring dissociation in the axial direction. Of interest is the observation that for the monochloro and monobromo complexes, there is a slower, second halide-independent rearrangement to a five-coordinate nickel(III) species ($k \sim 0.4$ sec^{-1}).

TABLE V

RATES OF FORMATION AND DISSOCIATION OF MONOHALO COMPLEXES OF NICKEL(III) MACROCYCLES[a]

Ligand	Cl$^-$		Br$^-$		I$^-$	
	$10^{-3} k_f$ (M^{-1} sec^{-1})	k_b (sec^{-1})	$10^{-3} k_f$ (M^{-1} sec^{-1})	k_b (sec^{-1})	$10^{-3} k_f$ (M^{-1} sec^{-1})	k_b (sec^{-1})
meso-Me$_2$[14]aneN$_4$	2.2	9.4	0.9	21.8	4.0[a]	—
meso-Et$_2$[14]aneN$_4$	3.1	22	2.8	25.6	9.8[b]	—
rac-Me$_2$[14]aneN$_4$	42	66	>10	≥100	29	0.8
[14]aneN$_4$	—	0.9	4.3	0.21	6.1[b]	3.1
Me$_6$[14]dieneN$_4$	—	—	1.3	120	—	—

[a] From Refs. 86, 139, and 140.
[b] Inner-sphere redox reaction; see text.

Although direct complex formation is observed kinetically (stopped flow) and spectrophotometrically, where X = Br or Cl, the reaction with I⁻ results in an oxidation of the halide. The reactions are rapid and there is the question of inner- or outer-sphere electron transfer, for the [14]aneN$_4$ complex. However, further studies (*140*) using ligand substituted (dimethyl) complexes reveal that for the *rac*-Me$_2$[14]aneN$_4$ isomer, two processes are observed, $k_f = 2.9 \times 10^4$ M^{-1} sec^{-1} and a subsequent redox step, $k_{red} = 5.5 \times 10^3$ M^{-1} sec^{-1}, both of which are iodide dependent. The mechanism proposed involves the formation of an octahedral complex which further reacts with a second mole of I⁻ in the redox step:

$$[\text{Ni}^{III}rac\text{-Me}_2[14]\text{aneN}_4(\text{OH}_2)_2]^{3+} + \text{I}^- \underset{k_f, k_b}{\rightleftharpoons} [\text{Ni}^{III}rac\text{-Me}_2[14]\text{aneN}_4(\text{I})(\text{OH}_2)]^{2+} \quad (4)$$

$$[\text{Ni}^{III}rac\text{-Me}_2[14]\text{aneN}_4(\text{I})(\text{OH}_2)]^{2+} + \text{I}^- \xrightarrow{k_{red}} [\text{Ni}^{II}rac\text{-Me}_2[14]\text{aneN}_4]^{2+} + \text{I}_2^- \quad (5)$$

$$\text{I}_2^- + [\text{Ni}^{III}rac\text{-Me}_2[14]\text{aneN}_4]^{3+} \xrightarrow{fast} [\text{Ni}^{II}rac\text{-Me}_2[14]\text{aneN}_4]^{2+} + \text{I}_2 \quad (6)$$

Further evidence of an inner-sphere mechanism is that the reaction rates appear to be too rapid to adhere to the Marcus correlation observed for other complexes with this anion. Also, the rate constants for the halides fall in the same order for all of X = Cl, Br, and I, suggesting a common initial step.

The rate of formation of the inner-sphere complex [Ni^{III}Me$_6$[14]ane-N$_4$SO$_4$]$^+$ has been investigated using pulse radiolysis techniques (*77*). Although there is evidence for hydrolysis (pK 3.7 ± 0.2) of the diaquonickel(III) complex, substitution by SO$_4^{2-}$ occurs only on the unhydrolyzed ion with $k_f = 1 \times 10^6$ M^{-1} sec^{-1} ($I = 0.03$ M). After allowance for outer-sphere ion pairing and ionic strength effects, the formation rate constant at $I = 1.0$ M is considered to be in the range 3–5 × 10^2 M^{-1} sec^{-1}, which is close to those for the halide complexes, confirming the suggestion of a dissociative interchange mechanism in these substitution processes.

In nickel(III) peptide complexes, there is a strong in-plane field provided by the deprotonated peptide linkages (*117, 118*). Two axially coordinated water molecules are present in the tetragonally distorted complexes which exchange much more rapidly than for the [14]aneN$_4$ species with a substitution rate of constant >10^6 M^{-1} sec^{-1} for the formation of the imidazole complex (*141*). However, except for the terminal peptide group, equatorial substitution is very slow. Substitution and rearrangement (*125*) reactions of these species reveal acid-induced bond breaking ($k \sim 0.1-15$ sec^{-1}) as [H$^+$] increases in the range 0.004–1.0 M. These substitutions are reversible in dilute acidic conditions. Coordination of terpyridyl (*126*) at the two axial and the free equatorial sites in a meridional manner yields complexes (log $K \sim 12.8$) which are stabilized with respect to self-redox decomposition in basic media.

2. Electron Transfer Reactions of Nickel(III)

a. Reactions of Nickel(II)/(III) Ions with Metal Complexes. Relatively few kinetic studies have been made on the oxidation of nickel(II)species. The reactions of [Ni^{II}[14]aneN$_4$]$^{2+}$ and its derivatives (*143, 144*) with aquocobalt(III) have been investigated. The mechanism may be described in terms of the following reaction scheme:

$$Co^{3+} + [Ni^{II}[14]aneN_4]^{2+} \xrightarrow{k_1} Co^{2+} + [Ni^{III}[14]aneN_4]^{3+} \quad (7)$$

$$CoOH^{2+} + [Ni^{II}[14]aneN_4]^{2+} \xrightarrow{k_2} Co(II) + [Ni^{III}[14]aneN_4]^{3+} \quad (8)$$

The overall second-order observed rate constant is of the form $k_0 = k_1 + k_2 K_h/[H^+]$. Part of the difficulty in assigning the reaction types lies in the square–planar/octahedral equilibrium of the nickel(II) ions. However, Endicott has assigned an outer-sphere mechanism for reaction Eq. (7). Confirmation of this is provided in the reactions of [Ni^{II}([9]aneN$_3$)$_2$]$^{2+}$, which retains octahedral geometry (*107*) in both the reduced and oxidized states. A rate law similar to that above is observed (*145*) in the oxidation of a nickel(II) oxime complex by cobalt(III). In this case, the nickel(III) formed is a transient and further reaction takes place to yield a nickel(IV) species. Of interest is the fact that for the CoOH^{2+} pathway with reductants (*107, 145, 146*) which are unambiguously outer sphere, the self-exchange rate constant (vide infra) for the CoOH$^{2+/+}$ couple (~ 10 M^{-1} sec^{-1}) is about 11 orders of magnitude greater than that for the aquo Co$^{3+/2+}$ system (*143*). Self-exchange data have also been derived from the reactions of nickel(II) macrocycles and oxime complexes with [Fe(phen)$_3$]$^{3+}$ complexes (*147, 148*).

b. Reactions of Nickel(III) Complexes. The stabilization of this oxidation state by macrocycles, diimines, and oximes has enabled a good deal of kinetic investigation into electron transfer processes involving these ions. Also, in many systems there is little ambiguity regarding the pseudooctahedral nature of the low-spin d^7 ions. These complexes have redox potentials (versus nhe) in the order 0.6–1.3 V (*10,*

106, 108) so that systematic studies on related species permit evaluation of self-exchange rates via a Marcus correlation. Also, as more structural information becomes available it is possible to relate these rate constants with bond extension or contraction leading to the transition state complex.

The oxidations by $[Ni^{III}([9]aneN_3)_2]^{3+}$ and $[Ni^{III}([10]aneN_3)_2]^{3+}$, of a series of metal poly(pyridine) complexes, have been investigated (107, 146, 149). A typical reaction of this type is seen in Eq. (9), where

$$[Ni^{III}([9]aneN_3)_2]^{3+} + [Co(phen)_3]^{2+} \longrightarrow [Ni^{II}([9]aneN_3)_2]^{2+} + [Co(phen)_3]^{3+} \quad (9)$$

the second-order rate constant $k_{12} = 5.6 \times 10^5 \, M^{-1} \, \text{sec}^{-1}$. There are no hydrogen ion effects on this outer-sphere reaction. The rate law and lack of $[H^+]$ dependence are a feature of these processes, which are particularly suitable for the calculation of rate constants, since no bond making or breaking occurs during electron transfer. As a result of the redox reaction, the coordination shells and immediate environments of the reactants and products will change. Using cross-reaction data of the type in Eq. (9), the rate constant for reaction, k_{12}, is related (150–152) to the individual self-exchange reaction rates k_{11}, k_{22} and to the overall equilibrium constant, K_{12}, by the expression, Eq. (10).

$$k_{12} = (k_{11}k_{22}K_{12}f_{12})^{1/2}W_{12} \quad (10)$$

where

$$\ln f_{12} = \frac{[\ln K_{12} + (W_{12} - W_{21})/RT]^2}{4\left[\ln\left(\frac{k_{11}k_{22}}{A_{11}A_{22}}\right) + \frac{W_{11} + W_{22}}{RT}\right]}$$

$$W_{12} = \exp[-(w_{12} + w_{21} - w_{11} - w_{22})/2RT] \quad (11)$$

$$w_{ij} = \frac{Z_i Z_j e^2}{D_s \sigma_{ij}(1 + \beta\sigma_{ij})} \quad (12)$$

$$A_{ii} = \left[\frac{4\pi N e^2 \nu \beta \gamma r}{1000}\right]_{ii} \quad (13)$$

where w_{ij} represents the work required to bring together ions i and j of charges $Z_i Z_j$ to the separation distance $\sigma_{ij} = r_i + r_j$ and $\beta = (8\pi N e^2/1000 D_s kT)^{1/2}$. The nuclear vibration frequency which destroys the activated complex configuration is designated $\nu\beta$ and γr is the thickness of the reaction layer. A value of $3 \times 10^{10}\ M^{-1}\ Å^{-1}\ \sec^{-1}$ has been quoted for the value of A/σ^2.

A plot of $\ln k_{12} - \ln w_{12}$ against $\ln(k_{22}K_{12}f_{12})$ should be linear with slope of 0.5 and intercept corresponding to $\frac{1}{2}\ln k_{11}$, where k_{11} is the self-exchange rate for the nickel(III)/(II) couple. Relatively few direct measurements have been made of the self-exchange in nickel(III/II) systems. The fact that the nickel(III) complexes, being low-spin d^7, are ESR active has permitted an estimate (148) of the exchange between $[\text{Ni}^{\text{III/II}}[14]\text{aneN}_4]^{3+/2+}$ complexes. Using ^{61}Ni- ($I = 3/2$) where a quartet feature observed in g_{zz} is consistent with hyperfine interaction between the unpaired electron and the ^{61}Ni nucleus, the reaction proceeds with enhancement of the quartet (Fig. 8).

$$[^{58}\text{Ni}^{\text{III}}[14]\text{aneN}_4]^{3+} + [^{61}\text{Ni}^{\text{II}}[14]\text{aneN}_4]^{2+} \rightleftharpoons$$
$$[^{58}\text{Ni}^{\text{II}}[14]\text{aneN}_4]^{2+} + [^{61}\text{Ni}^{\text{III}}[14]\text{aneN}_4]^{3+} \tag{14}$$

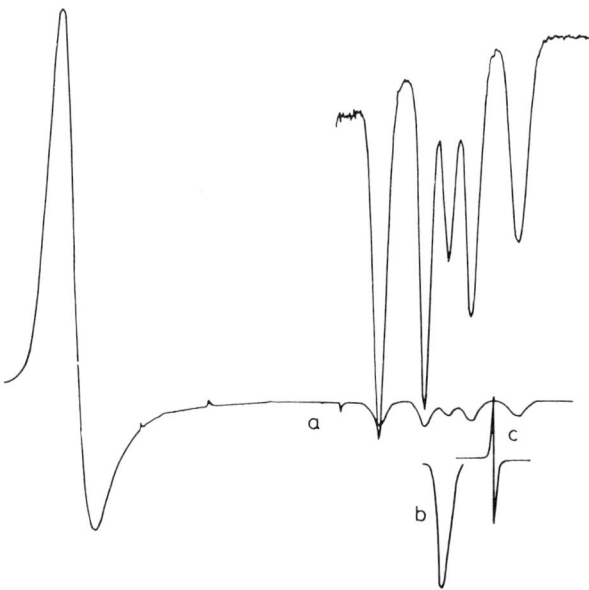

FIG. 8. (a) The EPR spectrum of $[^{61}\text{Ni}^{\text{III}}[14]\text{aneN}_4(\text{SO}_4)_2]^-$ and the corresponding spectra of (b) the ^{58}Ni complex and (c) dpph; from Ref. 148 by permission of the authors and the Royal Society of Chemistry.

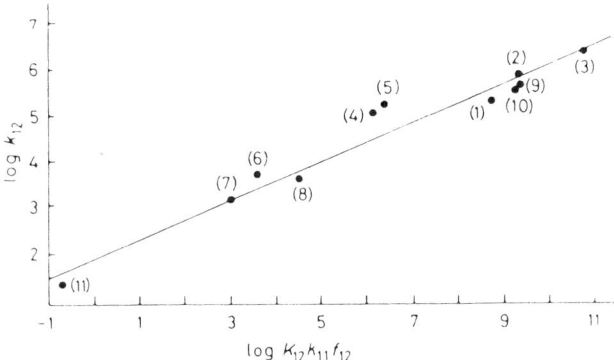

FIG. 9. Plot of $\log k_{12}$ against $\log K_{12}k_{11}f_{12}$; from Ref. 149 by permission of the authors and the Royal Society of Chemistry.

Monitoring of g_{zz} with time affords an approximate value for k_{11} ($\sim 1 \times 10^3$ M^{-1} sec^{-1}). Substantiation of this value was obtained in independent cross-reactions between [Ni^{II}[14]aneN$_4$]$^{2+}$ and [Fe(phen)$_3$]$^{3+}$. As mentioned previously, for the nickel[14]aneN$_4$ species the question of solvation in the axial positions is not completely resolved. By using [Ni$^{III/II}$([9]aneN$_3$)$_2$]$^{3+/2+}$ and the corresponding 10-membered ring complex, where octahedral geometry is retained, excellent cross-correlations have been obtained (Fig. 9), leading to calculated self-exchange rates. A feature of several of the reactions is the closely similar reduction potentials leading to an equilibrium condition,

$$[\text{Ni}^{III}([9]\text{aneN}_3)_2]^{3+} + [\text{Ni}^{II}[14]\text{aneN}_4]^{2+} \rightleftharpoons$$
$$[\text{Ni}^{II}([9]\text{aneN}_3)_2]^{2+} + [\text{Ni}^{III}[14]\text{aneN}_4]^{3+} \quad (15)$$

where $\Delta E = 0.05$ V. In such circumstances, suitable manipulation of the reagent concentrations permits both the forward and back reactions to be treated as pseudo first order. The observed rate constant k_{obs} may then be expressed in the form (149) $k_{\text{obs}} = k_f\{[\text{Ni}^{II}[14]\text{aneN}_4]^{2+}\} + k_b\{[\text{Ni}^{II}([9]\text{aneN}_3)_2]^{2+}\}$. At constant [Ni([9]aneN$_3$)$_2$]$^{2+}$ concentration, the plot of k_{obs} against [Ni^{II}[14]aneN$_4$]$^{2+}$ concentration should be linear, and the intercepts obtained under differing concentrations are a linear function of [Ni([9]aneN$_3$)$_2$]$^{2+}$. From these data, both k_f and k_b may be derived (Fig. 10).

Several isomers of [Ni^{II}Me$_2$[14]aneN$_4$]$^{2+}$, and one of [Ni^{II}Et$_2$[14]aneN$_4$]$^{2+}$, have been isolated (153). Using crystallography and ^{13}C

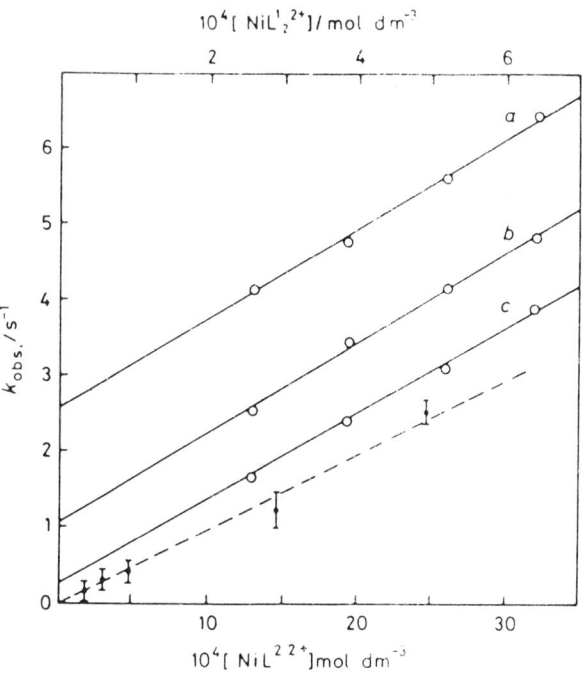

FIG. 10. Plots of k_{obs} against $[NiL']^{2+}$ at various concentrations (a–c) of $[NiL^2]^{2+}$. The dashed line represents the dependence of $[NiL^2]^{2+}$; from Ref. *149* by permission of the authors and the American Chemical Society.

and ^{1}H NMR spectroscopy, the α-$[Ni^{II}meso\text{-}Me_2[14]aneN_4]^{2+}$ and $[Ni^{II}meso\text{-}Et_2[14]aneN_4]^{2+}$ species, derived from the meso forms of the dimethyl and diethyl ligands, have equatorial distribution of the

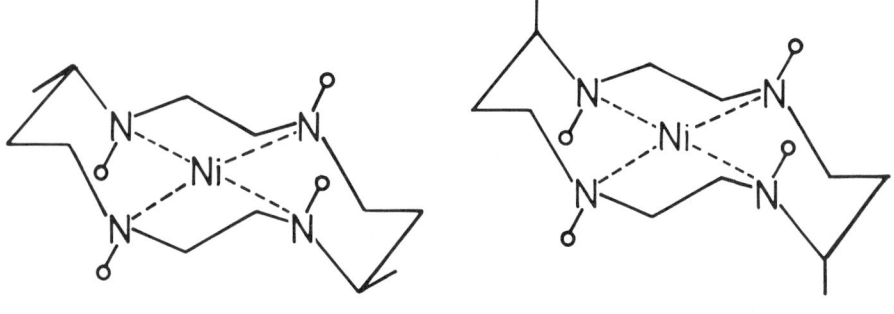

α-$[Ni^{II}meso\text{-}Me_2[14]aneN_4]^{2+}$

13

β-$[Ni^{II}meso\text{-}Me_2[14]aneN_4]^{2+}$

14

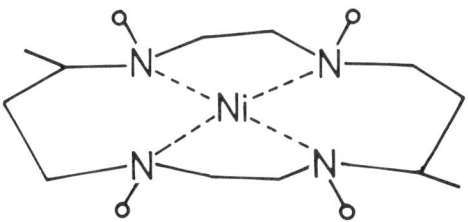

δ-[Ni^{II}*meso*-Me$_2$[14]aneN$_4$]$^{2+}$

15

substituent groups, leaving the nickel(II) center readily available for solvation. In the case of the β-[Ni^{II}*meso*-Me$_2$[14]aneN$_4$]$^{2+}$, axial configurations of the methyl groups are indicated. In δ-[Ni^{II}*meso*-Me$_2$[14]aneN$_4$]$^{2+}$, there is a modification of the 14-membered ring system to the *trans*-IV structure. Details of the stereochemistry of the ring conformation are presented in Table VI. Significant differences are observed in the solution and redox properties for the various isomers. As is known, nickel(II) macrocycles exhibit a square–planar octahedral equilibrium (K_{eq}) with water (solvent) ligating in the axial positions of the six-coordinate ion. This is accompanied by a marked decrease in the absorbance at ~450 nm. Approximate values of K_{eq} may be derived from changes in the extinction coefficient. Fabbrizzi (68) has previously obtained a value of $K_{eq} = 0.39$ for the parent [14]aneN$_4$ (Table III). The values of ~0.4 and ~0.7 for the equatorially disposed dimethyl and diethyl groups suggest the same degree of unhindered approach of the axial solvent. However, for the β isomer, significant

TABLE VI

Ring Conformation, Solution, and Redox Properties of Nickel(II) Derivatives of [14]aneN$_4$

Complex	$k_{eq}{}^a$	Substituent[b]	$E^{0\,c}$	Ring conformation	Configuration
α-*meso*-Me$_2$[14]aneN$_4$	0.4	eq, eq	0.995	trans III	S,S,R,R
meso-Et$_2$[14]aneN$_4$	0.4	eq, eq	0.990	trans III	S,S,R,R
β-*meso*-Me$_2$[14]aneN$_4$	<0.05	ax, ax	1.155	trans III	S,S,R,R
δ-*meso*-Me$_2$[14]aneN$_4$	<0.05	int, int	1.145	trans IV	S,R,R,S
rac-Me$_2$[14]aneN$_4$	0.2	int, ax	1.115	trans III	S,S,R,R

[a] Equilibrium in 0.5 HClO$_4$.
[b] eq, Equatorial; ax, axial; int, intermediate.
[c] Voltage versus a normal hydrogen electrode.

steric hindrance is anticipated, owing to the axial orientation of the ring substituents. In this instance, there is a marked decrease in the value of K_{eq}. A similar low value is found for the δ isomer, which has a trans-IV structure with groups intermediate between axial and equatorial. The ability of the ring to fold has been considered as a factor in the formation of reaction intermediates (154, 155) and this may be a feature in these systems. Of equal interest are the pronounced changes in the redox potentials of the nickel(III)/(II) couples (Table VI). Again, the E^0 values for the equatorially substituted species are similar to the $[Ni^{III/II}[14]aneN_4]^{3+/3+}$ value. However, increased strain in the system leads to more strongly oxidizing species. Using the outer-sphere reagent $[Ni^{II}([9]aneN_3)_2]^{2+}$ as reductant, a series of cross-reactions with these and other macrocycles was studied. Pronounced variations are observed (Table VII) in the rates of self-exchange for the various isomers. By examining the inner- and outer-shell reorganizational barriers, it is seen that although solvation effects must be included, inner coordination sphere reorganization is important and indeed accounts for the large variations in self-exchange rate constants. This effect is also observed in reactions of the hexamethyl derivatives where the additional methyl groups present increase the barrier to axial solvation. Again significantly lower self-exchange rates are observed. These are shown in Table VII, which lists all the self-exchange data so far derived for nickel(II)/(III) couples.

Reactions of nickel(III) complexes have also been used to examine exchange parameters for other metal ion complexes. An excellent correlation is provided in a Marcus treatment of the data for the reactions with $TiOH^{2+}$ as reductant (156). The self-exchange rate $(TiOH^{2+/+} = 9 \times 10^3 \ M^{-1} \ sec^{-1})$ is in excellent agreement with that derived by Sutin (157). A feature of this study is that the reaction rates for $[Ni^{III}[14]aneN_4(Cl)_2]^+$ and its dimethyl derivative with $TiOH^{2+}$ are $\sim 10^2$ faster than substitution at the metal center and so the reactions are postulated as outer sphere. Only in conditions of high ($>0.5 \ M$) chloride is the axial substitution complete. From the data, the rate constants for the nickel(III)/(II) self-exchanges may be evaluated as 3.4×10^4 and $9.5 \times 10^4 \ M^{-1} \ sec^{-1}$, respectively. These values are higher than for other nickel(III) species and may reflect the effects of the lower overall charges on the ions. Also, the nickel(II)/(III) exchange involves transfer of a σ^*d electron between a high-spin nickel(II) (d^8) and a low-spin (d^7) nickel(III). X-ray crystal data are available for the oxidized (81) and reduced (83) forms of the nickel complex. Whereas there is a bond length change of 0.088 Å in the Ni–N equatorial plane, the Ni–Cl differences show $\Delta d = 0.04$ Å.

TABLE VII

Self-Exchange Rate Constants for Nickel(III)/(II) Complex

Oxidant	k_{11} (M^{-1} sec^{-1})	Reference
[Ni^{III}Me$_2$[14]aneN$_4$(Cl)$_2$]$^+$	9.5×10^4	156
[Ni^{III}([10]aneN$_3$)$_2$]$^{3+}$	2.5×10^4	146
[Ni^{III}[14]aneN$_4$(Cl)$_2$]$^+$	3.4×10^4	156
[Ni^{III}Et$_2$[14]aneN$_4$(OH$_2$)$_2$]$^{3+}$	2.5×10^4	153
[Ni^{III}α-meso-Me$_2$[14]aneN$_4$(OH$_2$)$_2$]$^{3+}$	7.5×10^3	153
[Ni^{III}([9]aneN$_3$)$_2$]$^{3+}$	6×10^3	149
[Ni^{III}Me$_2$[14]dieneN$_4$]$^{3+}$	3×10^3	143
[Ni^{III}(bpy)$_3$]$^{3+}$	1.5×10^3	159[a]
[Ni^{III}[14]aneN$_4$(H$_2$O)]$^{3+}$	1×10^3	148
[Ni^{III}rac-Me$_2$[14]aneN$_4$(H$_2$O)]$^{3+}$	1.55×10^2	153
[Ni^{III}Me$_6$[14]aneN$_4$(H$_2$O)]$^{3+}$	31	—[b]
[Ni^{III}δ-meso-Me$_2$[14]aneN$_4$(H$_2$O)$_2$]$^{3+}$	26	153
[Ni^{III}β-meso-Me$_2$[14]aneN$_4$(H$_2$O)$_2$]$^{3+}$	6	153
[Ni^{III}meso-Me$_6$[14]aneN$_4$(H$_2$O)$_2$]$^{3+}$	2	143
[Ni^{III}Me$_2$L]$^+$	2×10^3	163
[Ni^{III}Me$_2$LH]$^{2+}$	4×10^2	163
[Ni^{III}Me$_4$[14]tetraeneN$_4$(H$_2$O)$_2$]$^{3+}$	0.009	—[c]
[Ni^{III}trans-Me$_6$[14]dieneN$_4$(H$_2$O)$_2$]$^{3+}$	0.009	—[c]
[Ni^{III}cis-Me$_6$[14]dieneN$_4$(H$_2$O)$_2$]$^{3+}$	0.008	—[c]
[Ni^{III}H$_{-3}$G$_3$a(H$_2$O)$_2$]	1.2×10^5	165[d]
[Ni^{III}H$_{-3}$G$_4$(H$_2$O)$_2$]$^-$	1.3×10^4	165[d]
[Ni^{III}H$_{-3}$G$_5$(H$_2$O)$_2$]$^-$	4.2×10^4	165[d]

[a] McAuley, A., and Olubuyide, O., unpublished data; reaction with [Ru(bpy)$_3$]$^{3+}$.

[b] Other tris(diimine) complexes react similarly.

[c] Fairbank, M., and McAuley, A., unpublished data; reaction with [Ni^{III}([9]aneN$_3$)$_2$]$^{3+}$.

[d] Additional data available in Refs. 164 and 165.

The relatively small bond length changes in the axial positions may account for the increased rate of electron exchange.

Applications of the Marcus theory to reactions of nickel(III) species with Fe^{2+} and VO^{2+} aquo ions (158) lead to values of 10^{-3}–10^{-2} M^{-1} sec^{-1} for Fe$^{3+/2+}$ and 10–10^3 M^{-1} sec^{-1} for VO(OH)$^{2+/+}$. These rate constants are larger than for the corresponding data derived using poly(pyridine) derivatives where there may be a contribution from the π^*–π^* interaction of the ligand orbitals.

As has been noted (29, 31), nickel(III) poly(pyridine) complexes may be prepared. Electron-exchange rates have been determined (15) from a series of cross-reactions. The rate of the electron transfer reaction has

$$[\text{Ni}^{\text{III}}(\text{bipy})_3]^{3+} + [\text{Ni}^{\text{II}}(\text{phen})_3]^{2+} \underset{k_y}{\overset{k_x}{\rightleftharpoons}} [\text{Ni}^{\text{II}}(\text{bipy})_3]^{2+} + [\text{Ni}^{\text{III}}(\text{phen})_3]^{3+} \quad (16)$$

been measured in 1.0 M H_2SO_4 in both directions ($K = 0.54$). Because of the similarities of the two species in terms of size and free energy, the values of k_x and k_y (1.1. × 10^3 and 2.0 × 10^3 M^{-1} sec^{-1}) have been used in the expression $k_{11} = (k_x k_y)^{1/2}$, yielding a self-exchange rate constant $k_{11} = 1.5 \times 10^3$ M^{-1} sec^{-1}. Other exchange rates are closely similar to this value. Although crystal-structure data are available for the $[\text{Ni}^{\text{III/II}}(\text{bipy})_3]^{3+/2+}$ couples, the evidence in terms of geometry is still the subject of discussion. Frozen solution EPR spectra (31) of the low-spin d^7 $[\text{Ni}^{\text{III}}(\text{bipy})_3]^{3+}$ are consistent with tetragonal elongation in the Jahn–Teller distorted system. However, the crystal structure (32) at 160 K shows tetragonal compression with two pairs of shorter axial bonds leading to an average difference Δd_0 in the Ni–N bond lengths of 0.12 Å. There is the possibility of a static or dynamic (33) disorder where the four longer spacings are the mean of two shorter and two longer bonds. The $\Delta d_0 \sim 0.1$ Å accounts for the relatively slow transfer ($k_{11} = 1.5 \times 10^3$ M^{-1} sec^{-1}) of a $\sigma^* d$ electron when compared to the d^8/d^7 $[\text{Co}(\text{bipy})_3]^{+/2+}$ and d^7/d^6 $[\text{Co}(\text{bipy})_3]^{2+/3+}$ exchanges. In the former $k_{11} \sim 10^8$ M^{-1} sec^{-1} there is the transfer of a πd electron between high-spin d^8 and high-spin d^7, and there is no significant change in the Co–N bond lengths. This feature and the change in spin multiplicity are responsible for the lower value of $k_{11} \sim 18$ M^{-1} sec^{-1}.

Comparison may now be made of the bond length changes in $[\text{Ni}^{\text{III/II}}([9]\text{aneN}_3)_2]^{3+/2+}$ exchange. The structure of $[\text{Ni}^{\text{II}}([9]\text{aneN}_3)_2]^{2+}$ shows an almost regular octahedron with Ni–N = 2.10 Å (109). In the corresponding nickel(III) cation, the six Ni–N bonds form two sets, two at 2.110 Å and four at 1.9711 Å, in keeping with a Jahn–Teller distorted system. Thus the two axial bonds are 0.01 Å longer than in the nickel(II) ion and the four equatorial are shorter by 0.129 Å. The short bonds are very similar to those in $[\text{Ni}^{\text{III}}(\text{bipy})_3]^{3+}$ (1.924 Å) and in $[\text{Ni}^{\text{III}}(\text{TACNTA})]$ (1.93 Å) (131). In the bis([9]aneN$_3$) system, interligand repulsions are observed in the structure and it is considered that the cavity developed by the two [9]aneN$_3$ ligands is too large for the smaller nickel(III) ion. The single-crystal EPR spectrum is consistent with the unpaired electron in the d_{z^2} orbital. Also, the self-exchange rate constant (6×10^3 M^{-1} sec^{-1}) is similar in magnitude to that for other nickel(III) species [poly(pyridyl) complexes], where bond length changes are of comparable magnitude.

Recent studies by Chakravorty have shown that ligands Me$_2$LH$_2$ containing oxime–imine and amine functions can stabilize both nickel(III) and nickel(IV) oxidation states (9, 45, 55, 160). On oxidation, one or more of the oxime protons is lost and if only one oxime group is present (160), only nickel(III) is formed. A variety of kinetic studies have been made in which the nickel(III) intermediates have been characterized (56, 161, 162). Recently (163) the kinetics of the redox equilibrium Eq. (17) have been made as a function of pH. Above pH ~5,

$$[Ni^{IV}Me_2L]^{2+} + [Ni^{II}Me_2L] \rightleftharpoons 2[Ni^{III}Me_2L]^+ \qquad (17)$$

the reductions of $[Ni^{IV}Me_2L]^{2+}$ by $[Ni^{II}Me_2L]$, $[Ni^{II}Me_2LH]^+$, and $[Ni^{II}Me_2LH_2]^{2+}$ take place with second-order rate constants of 1.24×10^6, 3.8×10^3, and 3.5×10^2 M^{-1} sec^{-1}, respectively. Below pH 5, disproportionation of $[Ni^{III}Me_2L]^+$ is dominated by the reaction with $[Ni^{III}Me_2LH]^{2+}$ ($k = 3.4 \times 10^3$ M^{-1} sec^{-1}). Protonation of nickel(III) at low pH leads to the reaction $2[Ni^{III}Me_2LH]^{2+} \rightarrow [Ni^{II}Me_2LH_2]^{2+} + [Ni^{IV}Me_2L]^{2+}$ being much slower ($k_2 \sim 4$ M^{-1} sec^{-1}), although this process is favored thermodynamically. The trend in the data may be accounted for, partly by increased charges on the species leading to lower rates. There may also be a change in the electronic structure of the doubly protonated nickel(III) species (163). Exchange data derived using a Marcus treatment are presented in Table VII.

Nickel(III) deprotonated peptide complexes are readily prepared by chemical and electrochemical oxidation of the corresponding nickel(II) complex. They are moderately stable in aqueous media (164). Electrochemical, EPR, and crystallographic studies are consistent with a low-spin d^7 nickel(III) with tetragonally distorted axial geometry (154). Stability constants for axial substitution (122) are relatively insensitive to the peptide bound, but show marked deviations (factor of 20) in the order imidazole $> NH_3 \sim N_3^- >$ py. Using the stopped-flow technique, the rate constants for 25 cross-reactions have been determined. For 16 different peptide complexes, the values of the self-exchange rates fall into different groups. For triply deprotonated peptides, $k_{11} = 1.2 \times 10^5$ M^{-1} sec^{-1}, except for the pentaglycine complex $H_{-3}G_5$ (4.2×10^4 M^{-1} sec^{-1}); for doubly deprotonated systems, $k_{11} = 1.3 \times 10^4$ M^{-1} sec^{-1}, except for α-aminoisobutyryl tripeptides (5.5×10^2 M^{-1} sec^{-1}). The reactions studied all have small driving forces ($K_{12} < 50$). Axial binding by Cl$^-$, Br$^-$, and N$_3^-$ catalyzes the exchange reactions, but pyridine acts as an inhibitor. In the latter case, there is no opportunity for functioning as a bridging ligand.

The nature of the self-exchange is still under consideration. The planned use of the copper(III)–nickel(II) and (reverse) reactions to provide information on the nickel(III)/(II) peptide exchange rates showed features such as a marked dependence on the peptide ligand; also, the self-exchange constants derived from the cross-reactions were much larger than outer-sphere nickel(III)/(II) peptide electron transfer data (166). It appears that the enhanced pathways may have some inner-sphere character. Steric hindrance of the cross-reactions are important and when bulky ligands are attached to the copper the effect is greater than when associated with the nickel complex.

3. Reactions of Nickel(IV)

The mechanistic chemistry of nickel(IV) is dominated by substitution inert complexes and outer-sphere electron transfer reactions. The diamagnetic $[Ni^{IV}(dmg)_3]^{2-}$ is relatively long lived in aqueous base but unstable below pH 8. At 35°C, the decomposition rate (37) is 2.6×10^{-7} sec^{-1}, that of its outside-protonated form $[Ni^{IV}(dmg)_3H]^-$ being 1.16×10^{-5} sec^{-1}. Acid catalysis at lower pH is proposed (37) as direct proton transfer to ligand nitrogen, but there is no evidence for general acid catalysis necessary with this mechanism. The nature of the acid-catalyzed decomposition is thought (167) to involve ligand oxidation prior to dissociation since species with partially dissociated ligands cannot be trapped with strong nucleophiles, although formation of highly reactive partially dissociated complexes cannot be ruled out. Decomposition is also catalyzed by divalent metal cations (168), particularly copper(II) (169), which forms a 1:1 adduct with $[Ni^{IV}(dmg)_3]^{2-}$.

Redox reactions involving the nickel(IV) complex are also subject to divalent metal ion catalysis (170, 171). Oxidations of the two-electron reductant ascorbate (40) and the one-electron reductant $[Fe(CN)_6]^{4-}$ (172) have been examined in some detail. Both reactions have as the rate-determining step the transfer of one electron from the reductant to nickel(IV) in an outer-sphere process to give an undetected nickel(III) transient. Spectroscopic properties of the nickel(III) species have been determined by pulse radiolysis (41).

Comparisons can be drawn between the chemistry of $[Ni^{IV}(dmg)_3]^{2-}$ and that of the sexidentate bis oxime imine complex $[Ni^{IV}Me_2L]^{2+}$, which is much better characterized from a thermodynamic point of view (45, 56). It can be optically resolved (54) and shows no indications of protonation above pH 0. The isostructural nickel(III) and nickel(II) complexes are subject to protonation and are much more labile to substitution. Protonation of the oxime–imine chromophore destabilizes the higher oxidation states.

The kinetics and mechanisms of reductions of [Ni^{IV}Me$_2$L]$^{2+}$ by both one- (*55, 56, 163, 173*) and two-electron (*162, 174, 175*) reagents have been extensively examined. Invariably biphasic behavior is observed with more rapid reduction of nickel(IV) than of the nickel(III) that is produced as a reaction intermediate. The predominant outer-sphere nature allows analysis by Marcus theory and the self-exchange rate for [Ni$^{IV/III}$Me$_2$L]$^{2+/+}$ is evaluated (*163*) as 4×10^4 M^{-1} sec^{-1} at 25°C. Attempts to explain this value in terms of Ni–N bond length changes leads to the conclusion that structural changes within the ligand may be important.

The complex [Ni^{IV}Me$_2$L]$^{2+}$ is significantly more reactive than [Ni^{IV}(dmg)$_3$]$^{2-}$. For example, in reactions with [Fe(CN)$_6$]$^{4-}$, the rates are $\geq 10^8$ M^{-1} sec^{-1} (25°C, 0.1 M ionic strength) (*41*) and 0.06 M^{-1} sec^{-1} [35°C, 0.06 M ionic strength (*172*)], respectively. The only other nickel(IV) reagent which has been studied mechanistically in any detail is the 9-molybdonickelate(IV) complex, [Ni^{IV}Mo$_9$O$_{32}$]$^{6-}$, with an intermediate rate constant (*176*) of 1.2×10^4 M^{-1} sec^{-1} (25°C, 0.10 M NaCl), although in this case the reaction is very cation dependent. This latter reagent has been used (*177*) in the oxidation of aromatic hydrocarbons and in the decarboxylation of aromatic carboxylic acids in acetic acid media.

V. Reduction of Nickel(II)

A. COORDINATION ENVIRONMENTS AND STRUCTURAL CHEMISTRY

1. Amines and Imines

There is an extensive chemistry of the nickel(I) ion generated by pulse radiolysis which is beyond the scope of this review. Complexes with saturated amines such as 1,2-diaminoethane have been studied by this method and by the γ radiolysis of aqueous glasses, but the species formed have no more than a transient existence. The imine ligands phen and bpy offer a more attractive environment for nickel(I) by allowing electron delocalization over the ligand π system (*178, 179*). A number of complexes of these ligands have been reported in γ-radiolysis studies. The EPR spectra indicate that reduction is primarily metal centered with a significant orbital contribution. Electrochemical reduction of [Ni^{II}(bpy)$_3$]$^{2+}$ in anhydrous acetonitrile results in [Ni^{I}(bpy)$_3$]$^+$, which can be detected by EPR methods. The reduction potential is reported to be -1.55 V but the complex is thermodynamically unstable with

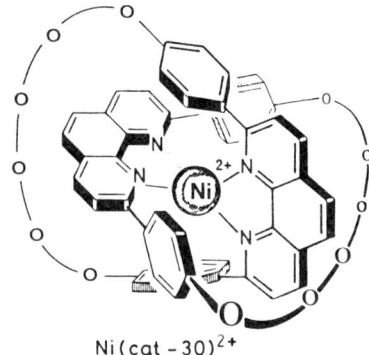

Ni(cat - 30)$^{2+}$

FIG. 11. Structure of nickel(I) catenate; from Ref. 181 by permission of the authors and the Royal Society of Chemistry.

respect to disproportionation (30, 180). It has been possible to isolate the bis complex [Ni^I(bpy)$_2$]ClO$_4$, however, indicating that nickel(I) may prefer a lower coordination number with these ligands (30). More recently (181), a four-coordinate nickel(I) catenate derived from a phenanthroline-based 30-member macrocyclic ring has been reported (Fig. 11). From consideration of the steric requirements of the ligand, a tetrahedral geometry would appear to be favored and this is consistent with the rhombic EPR. The reduced complex has an absorption maximum at 645 nm ($\varepsilon = 2400 \, M^{-1} \, cm^{-1}$) and is slow to oxidize in the presence of O$_2$.

2. Macrocycles

Reduction of a variety of square–planar nickel(II) macrocyclic complexes in acetonitrile leads to the formation of paramagnetic products (62). Saturated macrocycle complexes such as [Ni^{II}[14]aneN$_4$]$^{2+}$ are rather difficult to reduce (66, 182, 183), and the products which absorb maximally around 375 nm ($\varepsilon = 26,670 \, M^{-1} \, cm^{-1}$) are not well characterized from a structural point of view. Methyl substituents on the carbon backbone (64) and, more especially, on the ring nitrogen donor atoms (183) appear to stabilized the lower oxidation state both thermodynamically and kinetically. Substitution on the ring nitrogens retards ligand dissociation (183). Consequently, substituted products have been the subject of more extensive investigations. Reduction potential data are given in Table VIII.

Electrochemical reduction (64, 184) of [Ni^{II}Me$_6$[14]aneN$_4$]$^{2+}$ gives a product with a well-defined axial EPR signal consistent with nickel(I)

TABLE VIII

SELECTED REDUCTION POTENTIALS FOR NICKEL(II) MACROCYCLIC COMPLEXES

Complex	$E^{0\,a}$		Reference
	CH_3CN	H_2O	
$[Ni^{II}[13]aneN_4]^{2+}$	−1.12	—	64
$[Ni^{II}[14]aneN_4]^{2+}$	−1.12	−1.34	64, 183
$[Ni^{II}[15]aneN_4]^{2+}$	−0.92	—	64
$[Ni^{II}N\text{-}Me_4[14]aneN_4]^{2+}$	—	−0.91	183
$[Ni^{II}Me_6[14]aneN_4]^{2+}$	−0.99	−1.18	64, 183
$[Ni^{II}N\text{-}Me_4,Me_6[14]aneN_4]^{2+}$	—	−0.74	183

^{a} Versus a normal hydrogen electrode.

in a tetragonally elongated geometry and an absorption spectrum (183) similar to that of the unmethylated complex with a maximum at 380 nm ($\varepsilon = 26{,}320\ M^{-1}\ cm^{-1}$). As with oxidation to nickel(III), the macrocyclic hole size appears to be important in determining the reduction potential (64, 185) and as a result of the larger size of the nickel(I) ion, reduction of $[Ni^{II}[16]aneN_4]^{2+}$ is more facile than with the corresponding 15-, 14-, and 13-membered rings. Ligands such as formate ion (183) or carbon monoxide (184, 186) can bind axially to these nickel(I) species to give adducts which exhibit rhombic EPR spectra and are likely to be five-coordinate.

The nickel(II) macrocycles catalyze the decomposition of alkyl halides (187, 188) and the chemistry, which involves nickel–alkyl complexes, has been investigated recently (189) in some detail.

Incorporation of one or two imine donors into the macrocyclic framework appears to have little effect on the chemistry of the nickel(I) complexes (64, 184) unless the imine groups are in conjugation. However, reduction of $[Ni^{II}Me_4[14]1,3,8,10\text{-tetraeneN}_4]^{2+}$ results in an nickel(II)-stabilized ligand radical with $g = 2.008$ rather than a nickel(I) complex (64, 184). Addition of a second electron at lower potential is also possible and in the case the product is thought to be a nickel(I)-stabilized ligand radical (64). This abrupt change in the nature of the reduced species is not well understood, although clearly the ability of the conjugated diimine system to delocalize electron density over the ligand is an important factor. In one instance (184, 186), with a BF_2-containing ring system, both nickel(I)- and nickel(II)-stabilized radical species have been found to be in equilibrium. The proportion of the nickel(I) complex decreases with decreasing temperature.

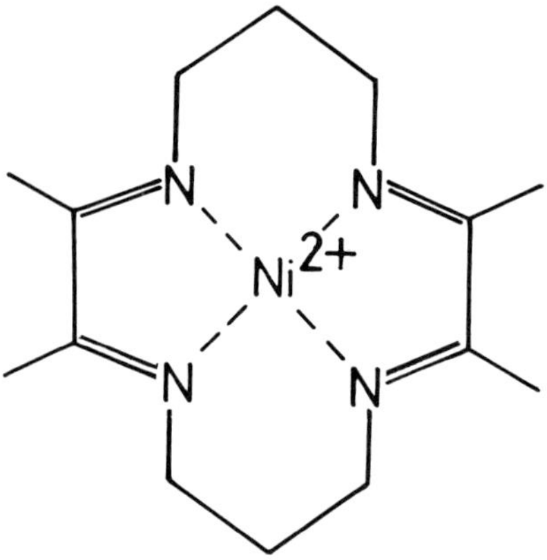

[Ni^II Me_4[14]1,3,8,10-tetraeneN_4]$^{2+}$

16

Addition of π-acceptor ligands which bind axially to the nickel(II)-stabilized radicals induces a change in the electronic configuration. The resulting species are well-defined five-coordinate nickel(I) complexes (*184, 186*), with $g_{zz} > g_{xx} = g_{yy}$ indicating a tetragonally elongated geometry. Similar results are noted for the one-electron reduced complex formed with the tetraazapyridine–diimine macrocycle (*190*), [Nicr]$^+$, which shows an EPR spectrum corresponding to a

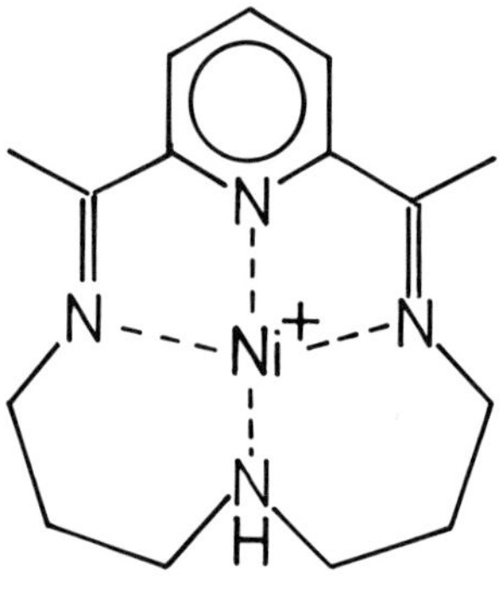

[Nicr]$^+$

17

nickel(II)-stabilized radical in the absence of π-acceptor ligands but a nickel(I) spectrum as a CO, PPh$_3$, or P(OMe)$_3$ adduct. The corresponding phenanthroline-based pentaaza macrocycle phencr forms (*191*) a pentagonal bipyramidal nickel(II) complex which is reduced to a nickel(I) species exhibiting an axial EPR spectrum with $g = 2.0575$ in acetonitrile solution. In this instance, π-acceptor ligands have a dramatic effect on the reduction potential, stabilizing nickel(I) as a six-coordinate pentagonal pyramid or seven-coordinate pentagonal bipyramid. A second pentaaza macrocycle based on terpy shows similar chemistry (*192*). In this case the nickel(I) complex can be reduced to give a nickel(I)-stabilized radical.

B. Kinetic Studies

Accessibility to the transient nickel(I) species was first gained using e_{aq}^- as reductant in pulse radiolytic experiments (*193*). More recently, Meyerstein (*185*) has shown that ring size effects can be important and that whereas the half-life of $[\text{Ni}^I[14]\text{aneN}_4]^+$ is ~2 seconds, that of the nickel(II) ion from [13]aneN$_4$ is of the order of 100 μsec. Several isomeric forms of the tetra-N-methylated (N-Me$_4$[14]aneN$_4$) nickel(II) complex ($[\text{Ni}^{II}(\text{tmc})]^{2+}$) have been prepared, and a greatly increased stability for the nickel(I) species is seen for the R,R,S,S isomer. Electrochemical reduction of $[\text{Ni}^{II}(\text{tmc})]^{2+}$ at -1.3 V (versus a standard calomel electrode), in the absence of oxygen, provides stock solutions in the millimolar concentration range which are stable for several hours at $0°$C and which may be used for kinetic studies (*189*) (vide infra).

Like other tetraaza metallo(I) complexes, the nickel(I) macrocyclic ions are powerful and labile reducing agents. A point of some interest in these systems is to design a complex couple for which the nickel(I) state is accessible at reasonable potentials. Provided the tetraaza macrocyclic ligand maintains close to planar microsymmetry, reorganizational barriers for a low-spin d^8–d^9 system might be expected to be small (*194*).

Pulse radiolysis studies of nickel(II) tetraaza macrocycles show a marked dependence on ring size. For [13]aneN$_4$, [14]aneN$_4$, and [15]aneN$_4$, the larger radius of the nickel(I) is exhibited in the redox potential and short lifetime of the 13-membered ring. Reductions of $[\text{Ni}^{II}[13]\text{aneN}_4]^{2+}$ by e_{aq}^- and CO_2^- are close to diffusion controlled in rate. The instability of the reduced ion is considered to be due to the inability of the cation to fit into the ring cavity. Of interest, however, is the reaction [Eq. (18)] in the presence of N$_2$O (*185*), where the

$$[\text{Ni}^I[13]\text{aneN}_4]^+ + \text{N}_2\text{O} + 2\text{H}_3\text{O}^+ \longrightarrow [\text{Ni}^{III}[13]\text{aneN}_4]^{3+} + \text{N}_2 + 3\text{H}_2\text{O} \quad (18)$$

product is the corresponding nickel(III) complex. Absorption maxima

($\varepsilon \sim 2\text{--}3 \times 10^3 \, M^{-1} \, cm^{-1}$) are very similar ($\lambda_{max} = 370$ nm) for all three monovalent nickel species. For the 15-membered ring system, which in the nickel(II) case is virtually exclusively octahedral (*195*), there is no reduction by CO_2^- as is observed for the other low-spin nickel(II) ions. Although thermodynamically allowed, it is concluded that reduction by CO_2^- takes place via a bridged intermediate which is not formed for the high-spin system. However, in the presence of the free radical $CH_2C(Me)_2OH$, it is postulated that a transient alkyl complex is formed which rapidly undergoes hydrolysis under the reaction conditions:

$$[Ni^I[15]aneN_4]^+ + \cdot CH_2C(Me)_2OH \longrightarrow [[15]aneN_4NiCH_2C(Me)_2OH]$$
$$\downarrow H_2O \quad (19)$$
$$[[15]aneN_4Ni^{II}(OH_2)_2]^{2+} + CH_2{=}C(Me)_2$$

Alkylnickel formation has also been extensively studied in the one-electron reductions of alkyl halides and peroxides by a nickel(I) macrocycle (*189, 196, 197*). Although two forms of the nickel(II) isomers of the tetra-N-methylated macrocycle are known, there is no ready interconversion between the $[(R,R,S,S,)Ni^{II}(tmc)]^{2+}$ and $[(R,S,S,R)]Ni^{II}(tmc)]^{2+}$. Solutions of the $[Ni^I(tmc)]^+$ complexes can be prepared in alkaline media (*183*), although within several hours isomerization of the nickel(I) forms takes place. The complexes are again strong reductants ($E^0 = -0.90$ V). Using the 1R,4R,8S,11S isomer (Fig. 12), Espenson and his co-workers have examined reactions with a variety of primary alkyl halides (*189*). For most systems studied, the overall reactions exhibit the stoichiometry according to Eq. (20):

$$2[Ni^I(tmc)]^+ + CH_3I + H_2O \longrightarrow 2[Ni^{II}(tmc)]^{2+} + I^- + CH_4 + OH^- \quad (20)$$

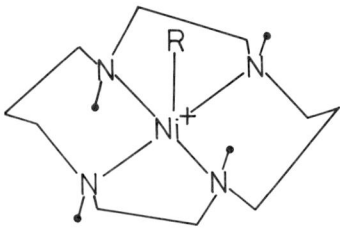

FIG. 12. [R-Ni^I(R,R,S,S-tmc)]$^+$; from Ref. *189* by permission of the authors and the American Chemical Society.

Addition of excess CH_3I to a solution of $[Ni^I(tmc)]^+$ results in the rapid loss of the absorption ($\lambda = 360$ nm, $\varepsilon = 4 \times 10^3$ M^{-1} cm^{-1}) and appearance of a less intense band at $\lambda = 346$ nm. A subsequent slower reaction gives rise to the weaker absorbance profile of $[Ni^{II}(tmc)]^{2+}$. The data are interpreted in terms of the formation of an organonickel(II) species followed by a slower hydrolysis with breaking of the Ni–C bond. Kinetic studies under conditions of excess alkyl halide show a dependence according to the equation $-d[Ni^I(tmc)^+]/dt = 2k[Ni(I)][RX]$. The data have been interpreted in terms of a rate-determining one-electron transfer from the nickel(I) species to RX, either by outer-sphere electron transfer or by halogen atom transfer, to yield the alkyl radical R. This reactive intermediate reacts rapidly with a second nickel(I) species:

$$[Ni^I(tmc)]^+ + CH_3I \xrightarrow{k} [Ni^{II}(tmc)]^{2+} + I^- + \cdot CH_3 \quad (21)$$

$$[Ni^I(tmc)]^+ + CH_3^\cdot \xrightarrow{rapid} [CH_3Ni(tmc)]^+ \quad (22)$$

The slower hydrolytic decomposition of the $[CH_3Ni(tmc)]^+$ yields CH_4 and $[Ni^{II}(tmc)]^{2+}$. Confirmation of the radical capture by nickel(I) may be illustrated using 6-bromo-1-hexene. One-electron reduction of this species yields the 1-hexanyl radical which rapidly undergoes cyclization to the cyclopentamethyl radical. Further reduction results in 1-hexane and methylcyclopentane. Reaction of $[Ni^I(tmc)]^+$ with $Br(CH_2)_4CHCH_2$ results after hydrolysis in both reduced species as forms with c-C_5H_9Me predominating. The reaction rate of the nickel(I) species with $CH_2(CH_2)_4CH=CH_2$ was evaluated as 6×10^7 M^{-1} sec^{-1}. In the case of t-butyl hydroperoxide, the presence of two, one-electron transfer steps is confirmed by the detection of $[CH_3Ni(tmc)]^+$ [from the decomposition $(CH_3)_3CO \to (CH_3)_2CO + CH_3$]. The reactivity order observed is methyl < primary < secondary.

The organonickel(II) species formed as intermediates in the reactions above react further (*196*) with alkyl halides with the formation of coupled products, alkanes and alkenes.

$$[RNi(tmc)]^+ + RX \longrightarrow [Ni^{II}(tmc)]^{2+} + (R_2 + RH + R^-H) + X^- \quad (23)$$

Kinetic dependences are first order with respect to each reactant and reactivity increases in the order shown above with Cl < Br < I. In the case of ethyl iodide, the disproportionation/combination ratio (k_d/k_c) of 0.35 ± 0.04 is identical to that observed by other methods. In the case of the acyclic valeronitrile (*198*) complex,

$$C_2H_4 + C_2H_6 \xleftarrow{k_d} 2CH_2H_5^{\cdot} \xrightarrow{k_c} C_4H_{10} \qquad (24)$$

$[NC(CH_2)_3CH_2Ni(tmc)]^+$, a facile non radical cyclization to cyclopentanone occurs as the hydrolysis reaction proceeds.

$$[NC(CH_2)_3CH_2Ni(tmc)]^+ + 2H_2O \xrightarrow{k_c}$$
$$[Ni^{II}(tmc)]^{2+} + c\text{-}(CH_2)_4C{=}O + NH_3 + HO^- \qquad (25)$$

In this system a very stable dimeric species is formed via reaction of a second nickel(I) species complexing to the cyano group. Of interest is the fact that the cyclization rate k_c is ~4 orders of magnitude slower than that for the free radical alone.

Electron transfer reactions of the organonickel species with $[Co(dmgH)_2]$ have been described (197). Reaction rates are very high ($k > 2 \times 10^6 \ M^{-1} \ sec^{-1}$) and cannot be measured by the stopped-flow technique. However, the blue organocobalt(I) products have been characterized by ^1H NMR methods. Rapid reactions have also been observed (193) in the oxidations of $[Ni^I Me_6[14]4,II\text{-dieneN}_4]^+$ and $[Ni^I Me_6[14]aneN_4]^+$ by $[M(bipy)_3]^{3+}$ (M = Co, Cr, or Fe), $[Ru(NH_3)_6]^{3+}$, and $[Co(en)_3]^{3+}$.

VI. List of Abbreviations

[9]aneN$_3$, 1,4,7-Triazacyclononane
[10]aneN$_3$, 1,4,7-Triazacyclodecane
[11]aneN$_3$, 1,4,8-Triazacycloundecane
[12]aneN$_4$, 1,4,7,10-Tetraazadodecane
[13]aneN$_4$, 1,4,7,10-Tetraazatridecane
[14]aneN$_4$, 1,4,8,11-Tetraazatetradecane
iso[14]aneN$_4$, 1,4,7,11-Tetraazatetradecane
[15]aneN$_4$, 1,4,8,12-Tetraazapentadecane
[16]aneN$_4$, 1,5,9,13-Tetraazahexadecane
[15]aneN$_5$, 1,4,7,10,13-Pentaazapentadecane
[16]aneN$_5$, 1,4,7,10,14-Pentaazahexadecane
[17]aneN$_5$, 1,4,7,11,14-Pentaazaheptadecane
[18]aneN$_6$, 1,4,7,10,13,16-Hexaazaoctadecane
Me$_2$[14]aneN$_4$, 5,12-Dimethyl-1,4,8,11-tetraazatetradecane
Et$_2$[14]aneN$_4$, 5,12-Diethyl-1,4,8,11-tetraazatetradecane
N-Me$_4$[14]aneN$_4$tmc, 1,4,8,11-Tetramethyl-1,4,8,11-tetraazacyclotetradecane
Me$_6$[14]aneN$_4$, 5,7,7,12,14,14-Hexamethyl-1,4,8,11-tetraazacyclotetradecane
N-Me$_4$Me$_6$[14]aneN$_4$, 1,4,5,7,7,8,11,12,14,14-Decamethyl-1,4,8,11-tetraazacyclotetradecane
Me$_2$[14]dieneN$_4$, 1,2-Dimethyl-1,4,8,11-tetraazacyclotetradeca-1,2-diene

$Me_4[14]1,3,8,10$-tetraeneN_4, 1,4,8,11-Tetraazacyclotetradeca-1,3,8,10-tetraene
$Me_6[14]4,11$-dieneN_4, 5,7,7,12,14,14-Hexamethyl-1,4,8,11-tetraazacyclotetradeca-4,11-diene
[14]ane$N_4CH_2CH_2$py, 6-Ethyl-2'-pyridine-1,4,8,11-tetraazacyclotetradecane
[13]ane$N_4C_6H_5$OH, 5:2'-Phenol-1,4,7,10-tetraazacyclotridecane
[14]ane$N_4C_6H_5$OH, 5:2'-Phenol-1,4,8,11-tetraazacyclotetradecane
Aib_3, Tri-α-aminoisobutyric acid

bpo, syn-2-Benzoylpyridine oxime
bpy, 2,2'-Bipyridyl
$bpyO_2$, 2,2'-Bipyridine-1,1'dioxide

$dapdH_2$, 2,6-Diacetylpyridine dioxime
$dmgH_2$, Dimethylglyoxime
(DOH)pn, 3,9-Dimethyl-4,8-diazaundeca-3,8-diene-2,10-dione dioxime
$dpgH_2$, Diphenylglyoxime

$edtaH_4$, 1,2-Diaminoethane-N,N,N',N'-tetraacetic acid
en, 1,2-Diaminoethane

G_3a, Triglycineamide
G_3, Triglycine
G_4, Tetraglycine
G_5, Pentaglycine

Me_2LH_2, 3,14-Dimethyl-4,7,10,13-tetraazahexadeca-3,13-diene-2,15-dione dioxime
$Me_2L'H$, 3-Methyl-4,7,10,13,16-pentaazahexadeca-3-ene-2-one oxime

phen, 1,10-Phenanthroline
py, Pyridine

TACNTA, 1,4,7-Triazacyclononane-N,N',N''-triacetate
terpy, 2,2':6'2''-Terpyridyl
2,2,2-tet, 1,4,7,10-Tetraazadecane
1,3,2-tet, 1,4,8,11-Tetraazaundecane
[12]dioxoane, 1,4,7,10-Tetraazadodecane-2,6-dione
[13]dioxoane, 1,4,7,10-Tetraazatridecane-11,13-dione
[14]dioxoaneN_4, 1,4,8,11-Tetraazatetradecane-5,7-dione
[15]dioxoaneN_4, 1,4,8,12-Tetraazapentadecane-9,11-dione
[16]oxoaneN_5, 1,4,7,10,13-Pentaazahexadecane-14-one
[16]dioxoaneN_5, 1,4,7,10,13-Pentaazahexadeca-14,16-dione
[13]dioxoane$N_4CH_2CH_2$py, 12-Ethyl-1'-pyridine-1,4,7,10-tetraazatridecane-11,13-dione
[13]dioxoane$N_4CH_2CH_2$pyO, 12-Ethyl-2'-pyridine-N-oxide-1,4,7,10-tetraazatridecane-11,13-dione
[14]dioxoane$N_4CH_2CH_2$py, 6-Ethyl-2'-pyridine-1,4,8,11-tetraazatetradecane-5,7-dione
[16]dioxoane$N_5CH_2XH_2$py, 15-Ethyl-2'-pyridine-1,4,7,10,13-pentaazahexadeca-4,16-dione

Acknowledgments

The authors would like to acknowledge Professor J. H. Espenson for the provision of material prior to publication. This work was supported in part by the National Science Foundation (Grant No. 84-06113), which is gratefully acknowledged.

References

1. Lancaster, J. R., *Science* **216**, 1324 (1982).
2. Albracht, S. P. J., Graf, E. C., and Thauer, R. K. *FEBS Lett* **140**, 311 (1982).
3. LeGall, J., Ljundahl, P. O., Moura, I., Peck, H. D., Xavier, A. V., Moura, J. J., Teixera, M., Huynh, B. H., and Der Vartanian, D. V., *Biochem. Biophys. Res. Commun.* **106**, 610 (1982).
4. Moura, J. J. G., Moura, I., Huynh, B. H., Krueger, H. J., Teixeira, M., DuVarney, R. C., Der Vartanian, D. V., Xavier, A. V., Peck, H. D., and Le Gall, J., *Biochem, Biophys. Res. Commun.* **108**, 1388 (1982).
5. Cammack, R., Patil, D., Aguirre, R., and Hatchikian, E. V., *FEBS Lett.* **142**, 289 (1982).
6. Albracht, S. P. J., Kalkman, M. L., and Slater, E. C., *Biochim. Biophys. Acta* **724**, 309 (1983).
7. Albracht, S. P. J., Van der Zwaan, J. W., and Fontijn, R. D., *Biochim. Biophys. Acta* **766**, 245 (1984).
8. Van der Zwaan, J. W., Albracht, S. P. J., Fontijn, R. D., and Slater, E. C., *FEBS Lett.* **179**, 271 (1985).
9. Nag, K., and Chakravorty, A., *Coord. Chem. Rev.* **33**, 87 (1980).
10. Haines, R. I., and McAuley, A., *Coord. Chem. Rev.* **39**, 77 (1981).
11. Margerum, D. W., Cayley, G. R., Weatherburn, D. C., and Pagenkopf, G. K., *Coord. Chem.* **2**, 1 (1978).
12. Wilkins, R. G., *Comments Inorg. Chem.* **2**, 187 (1983).
13. Merbach, A. E., *Pure Appl. Chem.* **54**, 1479 (1982).
14. Dwyer, F. P., and Gyarfas, E. C. J., *Proc. R. Soc. New South Wales* **83**, 232 (1949).
15. Wilkins, R. G., and Williams, M. J. G., *J. Chem. Soc.* 1763 (1967).
16. Maki, A. H., Edelstein, N., Davison, A., and Holm, R. H., *J. Am. Chem. Soc.* **86**, 4580 (1964).
17. Maki, A. H., and McGarvey, B. R., *J. Chem. Phys.* **29**, 31 (1958).
18. Drago, R. S., and Baucom, E. I., *Inorg. Chem.* **11**, 2064 (1972).
19. Peacock, R. D., and Stewart, B., *Coord. Chem. Rev.* **46**, 129 (1982).
20. Lati, J., and Meyerstein, D., *Inorg. Chem.* **11**, 2393 (1972).
21. Lati, J., and Meyerstein, D., *Inorg. Chem.* **11**, 2397 (1972).
22. Fried, I., and Meyerstein, D., *Isr. J. Chem.* **8**, 865 (1970).
23. Liu, H., Shen, W., Quayle, W. H., and Lunsford, J. H., *Inorg. Chem.* **23**, 4553 (1984).
24. Shen, W., and Lunsford, J. H., *Inorg. Chim. Acta* **102**, 199 (1985).
25. Yamashita, M., Nonaka, Y., Kida, S., Hamane, Y., and Aoki, R., *Inorg. Chim. Acta* **52**, 43 (1981).
26. Cooper, D. A., Higgins, S. J., and Levason, W., *J. Chem. Soc. Dalton Trans.* 2131 (1983).
27. Papavassiliou, G. C., and Layek, D., *Z. Naturforsch* **37B**, 1406 (1982).
28. Yamashita, M., and Murase, I., *Inorg. Chim. Acta* **97**, L43 (1985).
29. Wells, C. F., and Fox, D., *J. Chem. Soc. Dalton Trans.* 1492 (1977).

30. Prasad, R., and Scaife, D. B., *J. Electroanal. Chem.* **84**, 373 (1977).
31. Brodovitch, J. C., Haines, R. I., and McAuley, A., *Can. J. Chem.* **59**, 1610 (1981).
32. Szalada, D. J., Macartney, D. H., and Sutin, N., *Inorg. Chem.* **23**, 3473 (1984).
33. Wieghardt, K., Walz, W., Nuber, B., Weiss, J., Ozarowski, A., Stratemeier, H., and Reinen, D., *Inorg. Chem.* **25**, 1650 (1986).
34. Wada, A., Sakabe, N., and Tanaka, J., *Acta Crystallogr. Sect. B* **B32**, 1121 (1976).
35. Simek, M., *Collect. Czech. Chem. Commun.* **27**, 220 (1962).
36. Panda, R. K., Acharya, S., Neogi, G., and Ramaswamy, D., *J. Chem. Soc. Dalton Trans.* 1225 (1983).
37. Neogi, G., Acharya, S., Panda, R. K., and Ramaswamy, D., *J. Chem. Soc. Dalton Trans.* 1233 (1983).
38. Davis, D. G., and Boudreaux, E. A., *J. Electroanal. Chem.* **8**, 434 (1964).
39. Marov, I. N., Ivanova, E. K., Panfilov, A. T., and Luneva, N. P., *Z. Neorg. Khim.* **20**, 123 (1975); *J. Inorg. Chem.* **20**, 67 (1975).
40. Acharya, S., Neogi, G., Panda, R. K., and Ramaswamy, D. *J. Chem. Soc. Dalton Trans.* 1471 (1984).
41. Baral, S., and Lappin, A. G., *J. Chem. Soc. Dalton Trans.* 2213 (1985).
42. Lati, J., and Meyerstein, D., *Isr. J. Chem.* **10**, 735 (1972).
43. Baucom, E. I., and Drago, R. S., *J. Am. Chem. Soc.* **93**, 6469 (1971).
44. Sproul, G., and Stucky, G. D., *Inorg. Chem.* **12**, 2898 (1973).
45. Mohanty, J. G., Singh, R. P., and Chakravorty, A., *Inorg. Chem.* **14**, 2178 (1975).
46. Mohanty, J. G., and Chakravorty, A., *Inorg. Chem.* **15**, 2912 (1976).
47. Singh, A. N., Singh, R. P., Mohanty, J. G., and Chakravorty, A., *Inorg. Chem.* **16**, 2597 (1977).
48. Singh, A. N., *Indian J. Chem.* (1987).
49. Singh, A. N., *Synth. React. Inorg. Met. Org. Chem.* **16**, 279 (1986).
50. Singh, A. N., *Synth. React. Inorg. Met. Org. Chem.* **16**, 433 (1986).
51. Korvenranta, J., Saarinen, H., and Näsäkkälä, M., *Inorg. Chem.* **21**, 4297 (1982).
52. Saarinen, H., Korvenranta, J., and Näsäkkälä, E., *Acta Chem. Scand.* **A34**, 443 (1980).
53. Korvenranta, J., Saarinen, H., and Näsäkkälä, E., *Finn. Chem. Lett.* 81 (1979).
54. Heaney, P. J., Lappin, A. G., Peacock, R. D., and Stewart, B., *J. Chem. Soc. Chem. Commun.* 769 (1980).
55. Lappin, A. G., Laranjeira, M. C. M., and Peacock, R. D., *Inorg. Chem.* **22**, 786 (1983).
56. Lappin, A. G., and Laranjeira, M. C. M., *J. Chem. Soc. Dalton Trans.* 1861 (1982).
57. Chakravorty, A., *Isr. J. Chem.* **25**, 99 (1985).
58. Chakravorty, A., *Comments Inorg. Chem.* **4**, 1 (1985).
59. McAuley, A., and Preston, K. F., *Inorg. Chem.* **22**, 2111 (1983).
60. Singh, A. N., and Chakravorty, A., *Inorg. Chem.* **19**, 969 (1980).
61. Bentgen, J. M., Gimpert, H.-R., and Zelewsky, A., *Inorg. Chem.* **22**, 3576 (1983).
62. Olsen, D. C., and Vasilevikis, J., *Inorg. Chem.* **8**, 1611 (1969).
63. Gore, E. S., and Busch, D. H., *Inorg. Chem.* **12**, 1 (1973).
64. Lovecchio, F. V., Gore, E. S., and Busch, D. H., *J. Am. Chem. Soc.* **96**, 3109 (1974).
65. Busch, D. H., *Acc. Chem. Res.* **11**, 392 (1978).
66. Busch, D. H., Pillsbury, D. G., Lovecchio, F. V., Tait, A. M., Hung, Y., Jackels, S., Rakowski, M. C., Schammel, W. P., and Martin, L. Y., *Electrochem. Stud. Biol. Syst.* **38**, 32 (1977).
67. Hinz, F. P., and Margerum, D. W., *Inorg. Chem.* **13**, 2941 (1974).
68. Anachini, A., Fabbrizzi, L., Paoletti, P., and Clay, R. M., *Inorg. Chim. Acta* **24**, L21 (1977).
69. Fabbrizzi, L., *J. Chem. Soc. Dalton Trans.* 1857 (1979).

70. Sabatini, L., and Fabbrizzo, L., *Inorg. Chem.* **18**, 438 (1979).
71. Wilkins, R. G., Yelin, R. E., Margerum, D. W., and Weatherburn, D. C., *J. Am. Chem. Soc.* **91**, 4326 (1969).
72. Fabbrizzi, L., *J. Chem. Soc. Chem. Commun.* 1063 (1979).
73. Bencini, A., Fabbrizzi, L., and Poggi, A., *Inorg. Chem.* **20**, 2544 (1981).
74. Desideri, A., Raynor, B., and Poon, C.-K., *J. Chem. Soc. Dalton Trans.* 2051 (1977).
75. Haines, R. I., and McAuley, A., *Inorg. Chem.* **19**, 719 (1980).
76. McAuley, A., Morton, J. R., and Preston, K. F., *J. Am. Chem. Soc.* **104**, 7561 (1982).
77. Cohen, H., Kirchenbaum, L. J., Zeigerson, E., Jaacobi, M., Fuchs, E., Gingburg, G., and Meyerstein, D. *Inorg. Chem.* **18**, 2763 (1979).
78. Zeigerson, E., Ginzburg, G., Shwartz, N., Luz, Z., and Meyerstein, D., *J. Chem. Soc. Chem. Commun.* 241 (1979).
79. Zeigerson, E., Ginzburg, G., Becker, J. Y., Kirschenbaum, L. J., Cohen, H., and Meyerstein, D., *Inorg. Chem.* **20**, 3918 (1981).
80. Zeigerson, E., Bar, I., Berstein, J., Kirschenbaum, L. J., and Meyerstein, D., *Inorg. Chem.* **21**, 73 (1982).
81. Ho, T., Sugimoto, M., Toriumi, K., and Ito, H., *Chem. Lett.* 1477 (1981).
82. Yamaduta, M., Toriumi, K., and Ito, T., *Acta Crystallogr. Sect. C* **C41**, 1607 (1985).
83. Bosnich, B., Mason, R., Pauling, P. J., Robertson, G. B., and Robe, M. L., *J. Chem. Soc. Chem. Commun.* 97 (1965).
84. Barefield, E. K., and Mocella, M. T., *J. Am. Chem. Soc.* **97**, 4238 (1975).
85. Maruthamuthu, P., Patterson, L. K., and Ferraudi, G., *Inorg. Chem.* **17**, 3157 (1978).
86. Jaacobi, M., Meyerstein, D., and Lilie, J., *Inorg. Chem.* **18**, 429 (1979).
87. Morliere, P., and Patterson, L. K., *Inorg. Chem.* **20**, 1458 (1981).
88. Morliere, P., and Patterson, L. K., *Inorg. Chem.* **21**, 1837 (1982).
89. Ulman, A., Cohen, H., and Meyerstein, D., *Inorg. Chim. Acta* **64**, L127 (1982).
90. Cohen, H., Nutkovich, M., Meyerstein, D., and Shusterman, A., *Inorg. Chem.* **23**, 2361 (1984).
91. McElroy, F. C., and Dabrowiak, J. C., *J. Am. Chem. Soc.* **98**, 7112 (1976).
92. Fabbrizzi, L., and Poggi, A., *J. Chem. Soc. Chem. Commun.* 646 (1980).
93. Kodama, M., and Kimura, E., *J. Chem. Soc. Dalton Trans.* 694 (1981).
94. Hay, R. W., Bembi, R., and Sommerville, W., *Inorg. Chim. Acta* **59**, 157 (1982).
95. Fabbrizzi, L., Perotti, A., and Poggi, A., *Inorg. Chem.* **22**, 1411 (1983).
96. Kimura, E., Koike, T., Machida, R., Nagai, R., and Kodama, M., *Inorg. Chem.* **23**, 4181 (1984).
97. Fabbrizzi, L., Kaden, T. A., Perotti, A., Seghi, B., and Liselotti, S., *Inorg. Chem.* **25**, 321 (1986).
98. Fabbrizzi, L., Licchelli, M., Perotti, A., Poggi, A., and Soresi, S., *Isr. J. Chem.* **25**, 112 (1985).
99. Kimura, E., *Pure Appl. Chem.* **58**, 1461 (1986).
100. Hitaka, Y., Koike, T., and Kimura, E., *Inorg. Chem.* **25**, 402 (1986).
101. Rakowski, M. C., Rycheck, M., and Busch, D. H., *Inorg. Chem.* **14**, 1194 (1975).
102. Kimura, E., Machida, R., and Kodama, M., *J. Am. Chem. Soc.* **106**, 5497 (1984).
103. Kushi, Y., Machida, R., and Kimura, E., *J. Chem. Soc. Chem. Commun.* 216 (1985).
104. Machida, R., Kimura, E., and Kushi, Y., *Inorg. Chem.* **25**, 3461 (1986).
105. Kimura, E., and Machida, R., *J. Chem. Soc. Chem. Commun.* 499 (1984).
106. Wieghardt, K., Schmidt, W., Herrmann, W., Küppers, H. J., *Inorg. Chem.* **22**, 2953 (1983).
107. McAuley, A., Norman, P. R., and Olubuyide, O., *Inorg. Chem.* **23**, 1939 (1984).
108. Buttafava, A., Fabbrizzi, L., Perotti, A., Poggi, A., Poli, G., and Seglu, B., *Inorg. Chem.* **25**, 1456 (1986).

109. Zompa, L. J., and Margulis, T. N., *Inorg. Chim. Acta* **28**, L157 (1978).
110. Margerum, D. W., and Dukes, G. R., *In* "Metal Ions in Biological Systems" (H. Siegel, ed.), Vol. 1, p. 157. Dekker, New York, 1974.
111. Freeman, H. C., Guss, J. M., and Sinclair, R. L., *J. Chem. Soc. Chem. Commun.* 485 (1968).
112. Paniago, E. B., Weatherburn, D. C., and Margerum, D. W., *J. Chem. Soc. Chem. Commun.* 1427 (1971).
113. Bossu, F. P., and Margerum, D. W., *J. Am. Chem. Soc.* **98**, 4003 (1976).
114. Bossu, F. P., Paniago, E. B., Margerum, D. W., Kirksey, S. T., and Kurtz, J. L., *Inorg. Chem.* **17**, 1034 (1978).
115. Sakurai, T., and Nakahara, A., *Inorg. Chim. Acta* **34**, L243 (1979).
116. Bossu, F. P., and Margerum, D. W., *Inorg. Chem.* **16**, 1210 (1977).
117. Youngblood, M. P., and Margerum, D. W., *Inorg. Chem.* **19**, 3068 (1980).
118. Lappin, A. G., Murray, C. K., and Margerum, D. W., *Inorg. Chem.* **17**, 1630 (1978).
119. Sugiura, Y., and Mino, Y., *Inorg. Chem.* **18**, 1336 (1979).
120. Sakurai, T., Hongo, J.-I., Nakahara, A., and Nakao, Y., *Inorg. Chim. Acta* **46**, 205 (1980).
121. Sugiura, Y., Kuwahara, J., and Suzuki, T., *Biochem. Biophys. Res. Commun.* **115**, 878 (1983).
122. Murray, C. K., and Margerum, D. W., *Inorg. Chem.* **21**, 3501 (1982).
123. Kervan, G. E., and Margerum, D. W., *Inorg. Chem.* **24**, 3245 (1985).
124. Jacobs, S. A., and Margerum, D. W., *Inorg. Chem.* **23**, 1195 (1984).
125. Subak, E. J., Loyola, V. M., and Margerum, D. W., *Inorg. Chem.* **24**, 4350 (1985).
126. Pappenhagen, T. L., Kennedy, W. R., Bowers, C. P., and Margerum, D. W., *Inorg. Chem.* **24**, 4356 (1985).
127. Lati, J., and Meyerstein, D., *Int. J. Radiat. Phys. Chem.* **7**, 611 (1975).
128. Lati, J., and Meyerstein, D., *J. Chem. Soc. Dalton Trans.* 1105 (1978).
129. Lati, J., Koresh, J., and Meyerstein, D., *Chem. Phys. Lett.* **33**, 286 (1975).
130. Fuchs, E., Ginsburg, G., Lati, J., and Meyerstein, D., *J. Electroanal. Chem.* **73**, 83 (1976).
131. Van der Merwe, M. J., Boeyens, J. C. A., and Hancock, R. D., *Inorg. Chem.* **22**, 3489 (1983).
132. Hall, R. D., *J. Am. Chem. Soc.* **29**, 692 (1907).
133. Baker, L. C. W., and Weakley, T. J. R., *J. Inorg. Nucl. Chem.* **28**, 447 (1966).
134. Roy, A., and Chaudhury, M., *Bull. Chem. Soc. Jpn.* **56**, 2827 (1983).
135. Bhattacharya, S., Mukherjee, R., and Chakravorty, A., *Inorg. Chem.* **25**, 3448 (1986).
136. Pappenhagen, T. L., and Margerum, D. W., *J. Am. Chem. Soc.* **107**, 4576 (1985).
137. Barefield, E. K., and Busch, D. H., *J. Chem. Soc. Chem. Commun.* 523 (1970).
138. Whitburn, K. D., and Laurence, G. S., *J. Chem. Soc. Dalton Trans.* 139 (1979).
139. Fairbank, M. G., and McAuley, A., *Inorg. Chem.* **25**, 1233 (1986).
140. Fairbank, M. G., and McAuley, A., *Inorg. Chem.* **26**, 2844 (1987).
141. Margerum, D. W., *ACS Symp. Ser.* (198), 8 (1982).
142. Brodovitch, J. C., and McAuley, A., *Inorg. Chem.* **20**, 1667 (1981).
143. Endicott, J. F., Durham, B., and Kumar. K., *Inorg. Chem.* **21**, 2437 (1982).
144. Fairbank, M. G., and McAuley, A., to be published.
145. Macartney, D. H., and McAuley, A., *Can. J. Chem.* **60**, 2625 (1982).
146. Fairbank, M. G., McAuley, A., Norman, P. R., and Olubuyide, O., *Can. J., Chem.* **63**, 2983 (1985).
147. Macartney, D. H., and McAuley, A., *Can. J. Chem.* **61**, 103 (1983).
148. McAuley, A., Macartney, D. H., and Oswald, T., *J. Chem. Soc. Chem. Commun.* 274 (1982).

149. McAuley, A., Norman, P. R., and Olubuyide, O., *J. Chem. Soc. Dalton Trans.* 1501 (1984).
150. Marcus, R. A., *Annu. Rev. Phys. Chem.* **15**, 155 (1964).
151. Sutin, N., *Acc. Chem. Res.* **15**, 275 (1982).
152. Sutin, N., *Prog. Inorg. Chem.* **30**, 441 (1983).
153. Fairbank, M. G., Norman, P. R., and McAuley, A., *Inorg. Chem.* **24**, 2639 (1985).
154. Cooksey, C. J., and Tobe, M. L., *Inorg. Chem.* **17**, 1558 (1978).
155. Hay, R. W., Norman, P. R., House, D. A., and Poon, C. K., *Inorg. Chim. Acta* **48**, 81 (1981).
156. McAuley, A., Olubuyide, O., Spencer, L., and West, P. R., *Inorg. Chem.* **23**, 2594 (1984).
157. Braunschweig, B., and Sutin, N., *Inorg. Chem.* **18**, 1731 (1979).
158. Macartney, D. H., McAuley, A., and Olubuyide, O. A., *Inorg. Chem.* **24**, 307 (1985).
159. Macartney, D. H., and Sutin, N., *Inorg. Chem.* **22**, 3530 (1983).
160. Singh, A. N., Mohanty, J. G., and Chakravorty, A., *Inorg. Nucl. Chem. Lett.* **14**, 441 (1978).
161. Allan, A. E., Lappin, A. G., and Laranjeira, M. C. M., *Inorg. Chem.* **23**, 477 (1984).
162. Munn, S. F., Lannon, A. M., Laranjeira, M. C. M., and Lappin, A. G., *J. Chem. Soc. Dalton Trans.* 1371 (1984).
163. Lappin, A. G., Martone, D. P., and Osvath, P., *Inorg. Chem.* **24**, 4187 (1985).
164. Owens, G. D., Phillips, D. A., Czarnecki, J. J., Raycheba, J. M. T., and Margerum, D. W., *Inorg. Chem.* **23**, 1345 (1984).
165. Murray, C. K., and Margerum, D. W., *Inorg. Chem.* **22**, 463 (1983).
166. Margerum, D. W., *Pure Appl. Chem.* **55**, 23 (1983).
167. Acharya, S., Neogi, G., and Panda, R. K., *Inorg. Chem.* **23**, 4393 (1984).
168. Neogi, G., Acharya, S., Panda, R. K., and Ramaswamy, D., *J. Chem. Soc. Dalton Trans.* 1239 (1983).
169. Neogi, G., Acharya, S., Panda, R. K., and Ramaswamy, D., *Int. J. Chem. Kinet.* **15**, 521 (1983).
170. Acharya, S., Neogi, G., Panda, R. K., and Ramaswamy, D., *Bull. Chem. Soc. Jpn.* **56**, 2814 (1983).
171. Acharya, S., Neogi, G., Panda, R. K., and Ramaswamy, D., *Bull. Chem. Soc. Jpn.* **56**, 2921 (1983).
172. Sahu, R., Neogi, G., Acharya, S., and Panda, R. K., *Int. J. Chem. Kinet.* **15**, 823 (1983).
173. Macartney, D. H., and McAuley, A., *Inorg. Chem.* **22**, 2062 (1983).
174. Macartney, D. H., and McAuley, A., *J. Chem. Soc. Dalton Trans.* 103 (1984).
175. Lappin, A. G., Laranjeira, M. C. M., and Youde-Owei, L., *J. Chem. Soc. Dalton Trans.* 721 (1981).
176. Awanya, F. A., Thesis, University of Notre Dame, 1984.
177. Jönsson, L., *Acta Chem. Scand.* **B37**, 761 (1983).
178. Amano, C., and Fujiwara, S., *Bull. Chem. Soc. Jpn.* **46**, 1379 (1973).
179. Tanaka, N., Ogata, T., and Niizuma, S., *Inorg. Nucl. Chem. Lett.* **8**, 965 (1972).
180. Tanaka, N., and Sato, Y., *Inorg. Nucl. Chem. Lett.* **4**, 487 (1968).
181. Dietrich-Buchecker, C. D., Kern, J. M., and Sauvage, J. P., *J. Chem. Soc. Chem. Commun.* 760 (1985).
182. Jubran, N., Ginzburg, G., Cohen, H., and Meyerstein, D., *J. Chem. Soc. Chem. Commun.* 517 (1982).
183. Jubran, N., Ginzburg, G., Cohen, H., Koresch, Y., and Meyerstein, D., *Inorg. Chem.* **24**, 251 (1985).
184. Gagné, R. R., and Ingle, D. M., *Inorg. Chem.* **20**, 420 (1981).
185. Jubran, N., Cohen, H., and Meyerstein, D., *Isr. J. Chem.* **25**, 118 (1985).

186. Gagné, R. R., and Ingle, D. M., *J. Am. Chem. Soc.* **102,** 1444 (1980).
187. Becker, J. U., Kerry, J. B., Pletcher, D., and Rosas, R., *J. Electroanal. Chem. Interfacial Electrochem.* **117,** 87 (1981).
188. Gosden, C., Kerr, J. B., Pletcher, D., and Rosas, R., *J. Electroanal. Chem. Interfacial Electrochem.* **117,** 101 (1981).
189. Bakac, A., and Espenson, J. H., *J. Am. Chem. Soc.* **108,** 713 (1986).
190. Lewis, J., and Schroeder, M., *J. Chem. Soc. Dalton Trans.* 1085 (1982).
191. Ansell, C. W. G., Lewis, J., Raithby, P. R., Ramsden, J. N., and Schroeder, M. J., *Chem. Soc. Chem. Commun.* 546 (1982).
192. Constable, E. C., Lewis, J., Liptrot, M. C., Raithby, P. R., and Schroeder, M., *Polyhedron* **2,** 301 (1983).
193. Tait, A. M., Hoffman, M. Z., and Hayon, E., *Inorg. Chem.* **15,** 934 (1976).
194. Endicott, J. F., and Durham, B., *In* "Coordination Chemistry of Macrocyclic Compounds" (G. A. Melsor, ed.), p. 393. Plenum, New York, 1979.
195. Fabbrizzi, L., Micheloni, M., and Paoletti, P., *Inorg. Chem.* **19,** 535 (1980).
196. Bakac, A., and Espenson, J. H., *J. Am. Chem. Soc.* **108,** 719 (1986).
197. Ram, M. S., Bakac, A., and Espenson, J. H., *Inorg. Chem.* **25,** 3267 (1986).
198. Bakac, A., and Espenson, J. H., *J. Am. Chem. Soc.* **108,** 5353 (1986).

NICKEL IN METALLOPROTEINS

R. CAMMACK

Department of Biochemistry, King's College London (KQC), University of London, Kensington, London W8 7AH, England

I. Introduction
II. Urease
    A. Composition
    B. Spectroscopic Properties
    C. Mechanism of Catalysis
III. Hydrogenase
    A. Types of Hydrogenases
    B. EPR Spectroscopic Properties of Nickel
    C. Midpoint Redox Potentials
    D. Coordination State of Nickel
    E. Derivatives of Nickel
    F. Interactions of Nickel with Iron–Sulfur Centers
    G. Catalytic Properties
    H. Other Functions of Nickel in Hydrogenases
IV. Methyl-Coenzyme M Reductase
    A. Factor 430
    B. EPR Spectra
    C. Function
V. Carbon Monoxide Oxidoreductase
    A. Catalytic Properties
    B. Spectroscopy
VI. Concluding Remarks
VII. Abbreviations
    References

I. Introduction

Nickel has long been suspected to be an essential trace element for living organisms, but the identification of its functions in molecular terms is relatively recent. The first nickel protein to be identified was urease (urea ammonia hydrolase) ([1]). This was demonstrated 49 years after the original isolation and crystallization of the enzyme by Sumner ([2]). This enzyme is of widespread occurrence, and the specific requirement for nickel explains many of the effects of nickel deficiency in plants ([3, 4]).

Certain bacteria—in particular hydrogen bacteria, methanogens, and acetogens—were found to have a relatively high demand for nickel as a trace element. This was not recognized at first, since the requirement could be satisfied by nickel dissolved from stainless-steel culture apparatus (5). These chemolithotrophic bacteria are named from the metabolic processes by which they obtain energy for growth. All of them use hydrogen as reductant, in different ways. The hydrogen bacteria, such as *Alcaligenes* and *Nocardia,* are aerobes and catalyze the oxidation of hydrogen. Methanogens, of which the best-studied genera are *Methanobacterium* and *Methanosarcina*, are anaerobes which convert carbon dioxide to methane, using reducing agents such as hydrogen gas (6). The acetogens, such as *Clostridium thermoaceticum, Clostridium aceticum,* and *Acetobacterium woodii* (25), are also strict anaerobes and catalyze the conversion of carbon dioxide and carbon monoxide to acetic acid using hydrogen and other reductants.

The requirement for nickel was explained when it was found that the enzyme hydrogenase contains nickel. A characteristic EPR spectrum observed in membranes of methanogenic bacteria was recognized by Lancaster as being due to nickel (7, 8). This assignment was confirmed by growing the cells on the stable isotope ^{61}Ni, which has a nuclear spin $I = 3/2$, and observing the hyperfine splitting of the spectrum into four lines (9). Similar spectra have been observed in hydrogenases from sulfate-reducing bacteria (10, 11), photosynthetic bacteria (12), and hydrogen-oxidizing bacteria (13).

Hydrogenases are enzymes that catalyze the production or consumption of hydrogen gas. Not all hydrogenases contain nickel, and in some of those that contain nickel, the nickel is EPR silent (14). Properties of a few of the known Ni hydrogenases are summarized in Table I; more comprehensive listings are given in Refs. 15 and 16. The composition and biological function of hydrogenases have been reviewed recently (15–18), and this review will concentrate on the chemistry of those nickel centers.

In the methanogenic bacteria, nickel is also involved in a second process, in which a complex series of reactions leads to the release of methane gas (19–21). The enzyme involved, methyl-coenzyme M reductase, contains a cofactor F_{430} which is a nickel–porphinoid complex. The methanogens belong to an unusual group of bacteria described as Archaebacteria, which appear to be very ancient (22). Like the acetogens they are strict anaerobes. Although they are now restricted in habitat, these organisms, with their metabolism involving nickel and the related cobalamins, may have played an important part in the early phases of evolution.

TABLE I

COMPOSITION AND SPECTROSCOPIC PROPERTIES OF TYPICAL Ni HYDROGENASES

Organism	Type of hydrogenase	Center	Paramagnetic state
Methanobacterium thermoautotrophicum (methanogenic bacterium)	Soluble, deazaflavin reducing	Ni 2–3[4Fe–4S]	Oxidized, Ni(III); H_2 reduced, Ni(I)? EPR silent?
Desulfovibrio gigas (sulfate-reducing bacterium)	Soluble, periplasmic	Ni [3Fe–xS] [4Fe–4S]	Oxidized, Ni(III), Ni-A; H_2 reduced, Ni(I)?, Ni-C Oxidized Reduced
Chromatium vinosum (anaerobic photosynthetic bacterium)	Membrane bound	Ni [3Fe–xS] or [4Fe–4S]	Oxidized, Ni(III), Ni-a; H_2 reduced, Ni(I) Oxidized
Nocardia opaca (hydrogen-oxidizing bacterium)	Soluble, NAD reducing	Ni [3Fe–xS] 3[4Fe–4S] [2Fe–2S] FMN	EPR silent Oxidized Reduced Reduced Semiquinone radical

A third nickel-containing enzyme found in some strictly anaerobic bacteria is involved in the oxidation of carbon monoxide to carbon dioxide, and in the formation of acetyl-coenzyme A from carbon monoxide (23, 24). This reaction was reported to occur in sewage sludge in 1932 by Fischer (102), who was one of the cooriginators of the Fischer–Tropsch synthesis of hydrocarbons from carbon monoxide. It has been given the trivial name carbon monoxide dehydrogenase, but this seems illogical in view of the fact that carbon monoxide contains no hydrogen. In this review the alternative name carbon monoxide oxidoreductase will be used. Since the principal function of the enzyme is probably to synthesize acetyl-coenzyme A, the name "acetyl-CoA synthase" is also appropriate. It is a major enzyme of acetogenic bacteria, which have a novel pathway for fixation of CO_2 (26, 27), and it is also found in methanogens.

Both hydrogenases and carbon monoxide oxidoreductases contain iron–sulfur clusters in addition to nickel. It may be noted that in addition to the Ni hydrogenases, there is another class of Fe hydrogenases, such as those in clostridia, which contain no nickel but have a specialized type of iron–sulfur cluster (28a, 28b). Therefore, it has to be established that the nickel in Ni hydrogenases is the active site; as will be seen later, there is a considerable amount of circumstantial evidence for this.

A recent review by Hausinger (29) deals with nickel utilization by microorganisms. In addition to known effects of nickel deficiencies, nickel and its compounds are known to have toxic and carcinogenic effects. The reader is referred to recent proceedings (30a, 30b).

II. Urease

A. Composition

The first enzyme that was demonstrated to contain nickel was urease (urea amidohydrolase) from jack bean. It catalyzes the hydrolysis of urea to ammonia and carbon dioxide. The protein has a multimeric structure with a relative molecular mass of 590,000 Da. Analysis indicated 12 nickel atoms/mol. Binding studies with the inhibitors indicated an equivalent weight per active site of 105,000, corresponding to 2 nickel atoms/active site. During removal of the metal by treatment with EDTA at pH 3.7, the optical absorption and enzymatic activity correlated with nickel content. This, combined with the sensitivity of the enzyme to the chelating agents acetohydroxamic acid and phosphoramidate, indicates that nickel is essential to the activity of the enzyme (1).

B. Spectroscopic Properties

The nickel in urease is nonmagnetic and appears to be in the oxidation state Ni(II). The broad optical absorption spectrum is influenced by ligands to the metal (Fig. 1). The spectrum obtained in the presence of the competitive inhibitor mercaptoethanol, after correction for Rayleigh scattering by the protein (*31*), shows absorption peaks at 324, 380, and 420 nm, with molar absorption coefficients of 1550, 890, and 460 M^{-1} cm^{-1}, respectively. These were assigned to sulfur-to-nickel charge transfer transitions. The spectrum is changed by addition of other inhibitors, such as acetohydroxamic acid (Fig. 1B). Similar

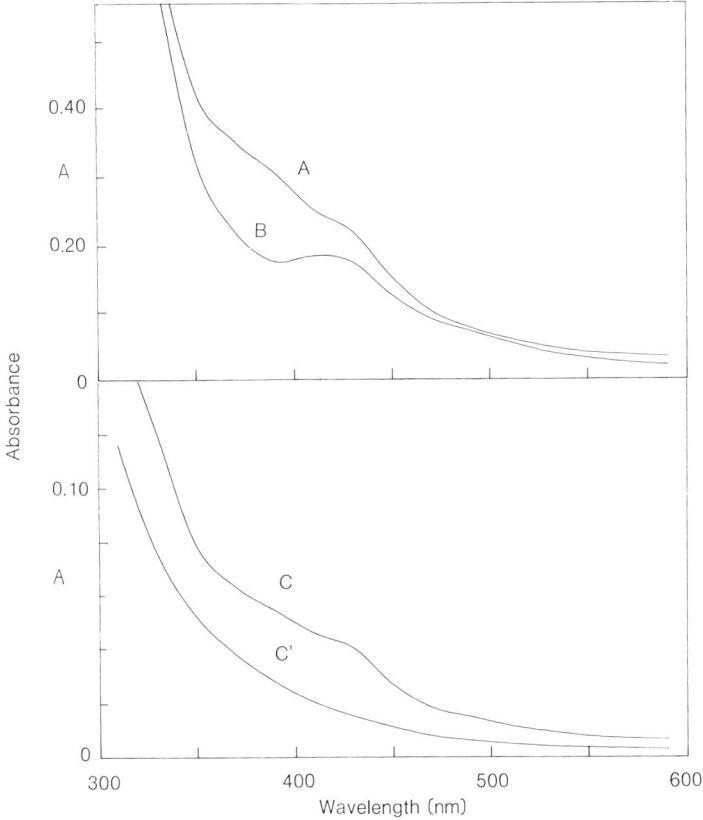

FIG. 1. Optical absorption spectra of urease from jack bean (*Canavalia ensiformis*). (A) Enzyme, 43.3 mg/ml in 1 mM β-mercaptoethanol/1 mM EDTA; (B) with 10 mM acetohydroxamic acid; (C) urease, 11.9 mg/ml after acidification, pH 3.8; (C') after 2 hours at pH 3.8; the latter enzyme retained 6.1% of its original activity. Redrawn, with permission, from Ref. *1*.

spectra have been observed in other nickel-protein complexes, including Ni(II) carboxypeptidase, which also undergoes spectra changes on addition of the inhibitor β-phenyl propionate (*32*). The spectra of urease have been interpreted in terms of an octahedral site (*31a*), although the site in carboxypeptidase, occupied by nickel, is coordinated to one cysteine sulfur, two histidine imidazoles, and the carboxylate of a glutamate residue (*32a*).

In extended X-ray absorption fine structure (EXAFS) studies of urease, Hasnain, Piggott, and co-workers (*33, 34*) demonstrated that spectra were similar to those of benzimidazole complexes, consistent

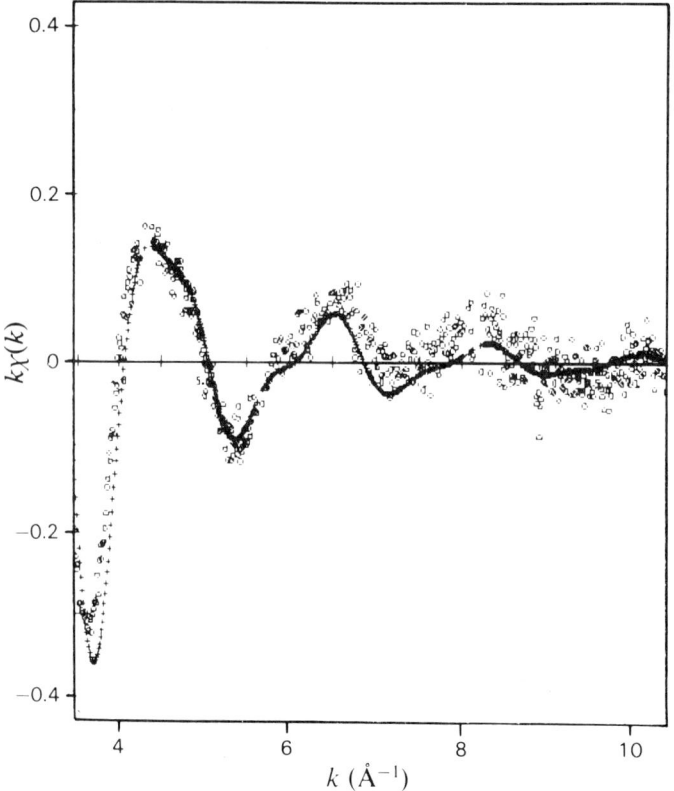

FIG. 2. Ni K-edge EXAFS spectrum of urease. The curve (+) is calculated for a single type of nickel site, and the minimization of parameters was based on those for the model complexes Ni(1-n-propyl-2-hydroxybenzylbenzimidazole)$_3$(ClO$_4$) and Ni(2-hydroxymethylbenzimidazole)$_3$Br$_2$. Atoms (with distances in nanometers given in parentheses) in the simulation were N (0.204), O (0.206), O (0.225), C (0.294), C (0.312), N (0.392), and C (0.394). Reproduced, with permission, from Ref. *34*.

with nickel coordination to histidine nitrogen and oxygen ligands. The spectra obtained for urease (Fig. 2) are presumably the average of two types of sites.

C. Mechanism of Catalysis

Dixon et al. (35) have proposed a mechanism for urease catalysis (Fig. 3) based on studies of the reactions with the poor substrates formamide, acetamide, and N-methylurea. They suggest that the two nickel ions are both in the active site, one binding urea and the other a hydroxide ion which acts as an efficient nucleophile. This implies that the nickel ions are within 0.6 nm (1 nm = 10 Å) of each other; so far it

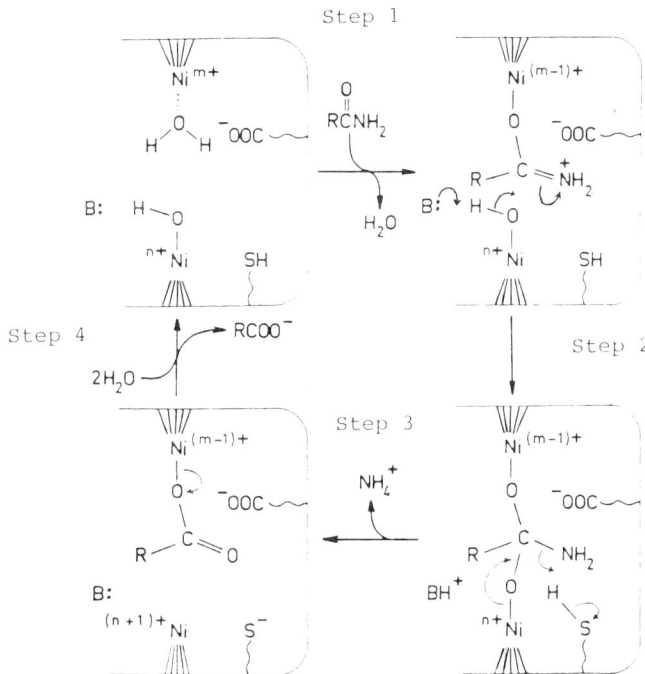

FIG. 3. Proposed reaction cycle for urease. For urea, R = —NH$_2$. Step 1: urea is activated toward nucleophilic attack by O coordination to a nickel ion; the =N$^+$H$_2$ is stabilized by interaction with a protein carboxylate. Step 2: nucleophilic attack by a hydroxide ion, coordinated to the second nickel, to form a tetrahedral intermediate. Step 3: breakdown of the tetrahedral intermediate to form a coordinated carbamate ion. Step 4: hydrolysis releases carbamate ion, the initial product of urease on urea. Reproduced, with permission, from Ref. 34.

has not been possible to confirm this spectroscopically. The reaction is analogous with peptide hydrolysis by carboxypeptidase, in which a hydroxide ion bound to a zinc ion has been implicated.

III. Hydrogenase

A. Types of Hydrogenases

In contrast to urease the nickel in other bacterial enzymes appears to have a redox function and to take up oxidation states Ni(I) and/or Ni(III). Fortunately these states have recently become better understood in inorganic systems (see the preceding review in this volume by A. G. Lappin and A. McAuley).

The Ni hydrogenases are now the most intensively studied nickel proteins. Although they have only been investigated for a few years, there is a considerable amount of evidence about the state and function of nickel in the catalytic cycle (16). In fact, "hydrogenase" is not one enzyme, but a class of enzymes. Hydrogenases may be distinguished functionally, in that they serve to produce or consume hydrogen in reactions with various physiological electron acceptors and donors. They may also be distinguished by the following molecular and catalytic properties.

1. Types of iron–sulfur clusters and other groups, such as flavin or selenium, that are present.
2. Types of EPR signals due to nickel.
3. Sensitivity to deactivation by oxygen and, in some cases, slow, reductive reactivation.
4. Different ratio of products in hydrogen isotope-exchange assays.

Among the diversity of Ni hydrogenases, there is a common pattern of protein composition, to which most conform, which consists of two protein subunits of relative molecular mass approximately 60,000 and 30,000 Da (14). There is some evidence (38) that the nickel is situated in the 60,000-Da subunit. More complex hydrogenases, such as the soluble hydrogenase of *Nocardia opaca* (Table I), contain other subunits which are concerned with the reduction of specific electron acceptors.

B. EPR Spectroscopic Properties of Nickel

The coordination and oxidation state of nickel in hydrogenase are difficult to determine, because the optical spectra are obscured by the iron–sulfur clusters and because the commonest oxidation state Ni(II) is EPR silent. However, as already noted, some hydrogenases show significant EPR signals, which are the best indicators at present of the function of nickel in these enzymes. In the following discussion, EPR spectra will be described with reference to the typical spectra from *Desulfovibrio gigas* hydrogenase, and any differences in other Ni hydrogenases will be noted.

Examples of EPR spectra of the Ni hydrogenase from *D. gigas* are shown in Fig. 4. This hydrogenase, in common with those from *Methanobacterium thermoautotrophicum* (*7, 9, 39*) and *Chromatium*

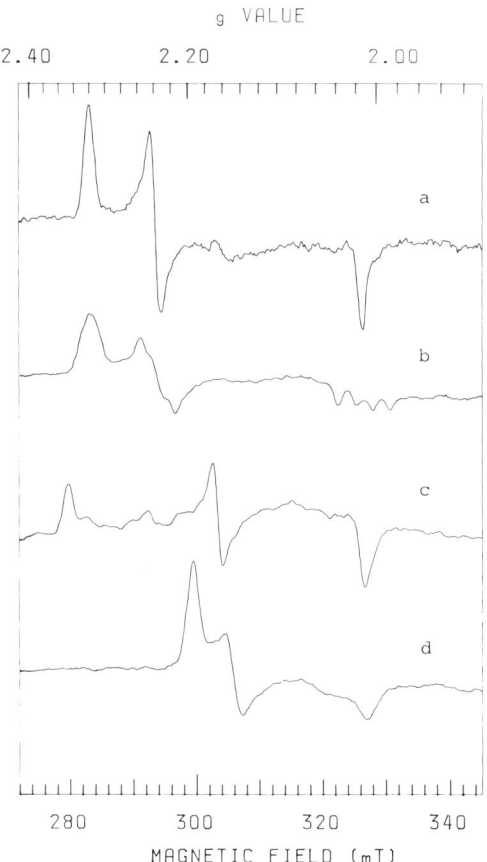

FIG. 4. EPR spectra of nickel in *D. gigas* hydrogenase: (a) Ni-A, enzyme stored under air; (b) Ni-A, enzyme enriched with ^{61}Ni; (c) Ni-B, enzyme-activated enzyme reoxidized with dichloroindophenol at pH 7.8; (d) Ni-C, enzyme activated under hydrogen. Spectra were recorded at a temperature of approximately 100 K, with microwave power 10 mW.

vinosum (*12*) shows a prominent EPR signal with g values of 2.32, 2.21, and 2.01 (Fig. 4a). In *D. gigas* this spectrum has been termed Ni-A (*40, 41*). Frequently a second minor signal is observed in oxidized hydrogenase, termed Ni-B in *D. gigas* (Fig. 4b). In *C. vinosum* hydrogenase, the corresponding signals have been termed Ni-a and Ni-b. It may be mentioned that a number of additional EPR signals have been reported, in particular hydrogenases from various bacteria, the origins of which are not clear at present, and which will not be described here.

It is generally believed that the EPR-detectable nickel in oxidized hydrogenase represents low-spin Ni(III), d^7. In support of this view, the EPR signal disappears on reduction, consistent with reduction to Ni(II). The g values of oxidized hydrogenase were interpreted by Lancaster (*7*) as due to a rhombically distorted octahedron with the unpaired electron in a $d_x{}^2$ orbital. Square-pyramidal geometry is also possible. In practice, the coordination sites in proteins are often so distorted that structural interpretations of EPR spectra based on crystal symmetry may be misleading.

When the hydrogenases of *D. gigas* (*42*), *M. thermoautotrophicum* (*39*), or *C. vinosum* (*43*) are progressively reduced by hydrogen, the EPR spectra indicate three stages: first, the Ni-A and Ni-B signals disappear; then a signal (termed Ni-C in *D. gigas*) appears (Fig. 4d); then finally this signal also disappears. The Ni-C signal represents an intermediate oxidation state since it can be restored by flushing away excess hydrogen with argon (*39, 41, 43*).

C. MIDPOINT REDOX POTENTIALS

The midpoint reduction potentials of the various EPR-detectable nickel species in hydrogenase are all less than 0 mV versus the standard hydrogen electrode (Table II). This is in contrast to synthetic inorganic complexes with amino acids (*44*), in which the oxidation of Ni(II) to Ni(III) occurs at much higher potentials (0.8–1.2 mV) and is accompanied by reorganization of the complex (*45*). This requires some explanation in view of the interpretation of the Ni-A EPR signal as Ni(III)(*7*).

There are various factors which might stabilize the Ni(III) state and thus lower the Ni(III)/Ni(II) potential, the first of these being sulfur ligation, for which there is spectroscopic evidence (Section II,D,1,2). The ligands to nickel in the above-mentioned amino acid complexes are oxygen and nitrogen. Moreover, it is well established that, for a transition metal center with a particular arrangement of ligands, a protein environment can adjust the midpoint potential over a range of

TABLE II

MIDPOINT REDOX POTENTIALS OF HYDROGENASES[a]

Hydrogenase	Redox process	E_m (mV)	E_m/pH unit (mV)
D. gigas (65)	Reduction of Ni-A	−150	−60
	Reduction of [3Fe−xS]	−35	0
	Reduction of [4Fe−4S][b]	−350	−60
	Appearance of Ni-C	−270	−120
	Disappearance of Ni-C	−390	−60
	Reductive activation (75)	−310	−60
	Oxidative deactivation (99)	−133	−60
C. vinosum	Reduction of Ni-a	−175 (43) (pH 7.3)	ND
	Reduction of [3Fe−xS][c]	−20 (100), −165 (43) (pH 7.3)	0
	Disappearance of interaction between Ni and [3Fe−xS]	180 (100), −29 (43) (pH 7.3)	−60
N. opaca (82)	Reduction of [4Fe−4S][d]	−420	ND
	Reduction of [2Fe−2S][d]	−285	ND
	Reduction of [3Fe−xS]	25	ND

[a] Potentials are values at pH 7.0, expressed relative to the standard hydrogen electrode. ND, Not determined. Numbers in parentheses are references.

[b] The redox potential of the [4Fe−4S] cluster in D. gigas hydrogenase is inferred from an extremely broad EPR signal in the reduced enzyme, which correlates with the splitting of the Ni-C signal (65).

[c] The g = 2.01 signal in C. vinosum hydrogenase is interpreted by Albracht et al. (12) as a [4Fe−4S] cluster which changes to a [3Fe−xS] cluster when interaction with the nickel center ceases.

[d] The EPR-detectable iron−sulfur clusters of N. opaca hydrogenase are associated with the NAD-reducing segment of the enzyme.

several hundred millivolts by means of electrostatic and other effects (*46*). There are other mechanisms known from chemical complexes which would also lower the potential (*47*). A confined binding site would have this effect, as has been demonstrated in Ni complexes with macrocycles (*48*). However, it may be difficult to achieve this in the more flexible environment of a protein.

The opposite problem occurs if we are to invoke Ni(I) as a participant in the reaction cycle, since the Ni(II)/Ni(I) couple is generally more negative, by about 1 V, than Ni(III)/Ni(II)(*37*). In some hydrogenases, such as in *D. gigas*, it is known that the enzyme undergoes a slow conformational change during conversion to the active state, in which the alleged Ni(I) EPR signal appears (e.g., see Ref. *49*), so the altered conformation might stabilize the Ni(I) state. The Ni(I) state is stabilized, relative to the Ni(II) state, by ligands which favor tetrahedral coordination. Dietrich-Buchecker *et al.* (*50*) have obtained stable Ni(I) complexes by using catenand ligands, containing interlocking, 30-membered, coordinating rings.

D. COORDINATION STATE OF NICKEL

In those hydrogenases in which the nickel is EPR detectable (Table I), the remarkable similarity in the lineshapes of the spectra is a strong indication that the nickel environment is highly conserved. Therefore, although there are substantial differences in the catalytic activities and specificity of hydrogenases from different organisms, it seems likely that there are, at most, only a few different types of nickel centers. It therefore seems reasonable to correlate spectroscopic information on nickel in hydrogenases from different species in order to obtain a composite picture.

It is not known at present if the nickel is coordinated directly to the protein, as in copper and iron–sulfur proteins, or to an organic cofactor, as in the molybdenum hydroxylases and hemoproteins.

1. X-Ray Absorption Spectroscopy

X-Ray absorption edge spectra and EXAFS measurements have the potential for detailed determination of the coordination geometry of the nickel sites. Nickel K-edge X-Ray absorption measurements have so far been made on two hydrogenases. At present there are few suitable model compounds containing Ni(III) or Ni(I) with sulfur ligands, and the interpretations as to coordination geometry and oxidation state should therefore be regarded as preliminary.

EXAFS studies of the F_{420}-reducing hydrogenase from *M. thermoautotrophicum* (*51*) indicated that sulfur was the principal scattering nucleus. Best fits to the data were obtained with 2.9 sulfur atoms, at a distance of approximately 0.225 nm. The spectra were refined with the aid of data from $Ni(II)(toluene-3, 4-dithiolato)_2{}^{2-}$, in which the Ni–S distance was estimated to be 0.219 nm. Lindahl *et al.* (*51*) note that because of differences between the Debye–Waller factors of the enzyme and model compound, there is uncertainty about the exact number of sulfur ligands involved. Scattering by nuclei of other ligands to nickel, such as oxygen or nitrogen, of lower atomic mass, was expected to be so weak that they could not be resolved.

In *D. gigas* hydrogenase, X-ray absorption edge measurements gave a -2-eV shift in the nickel K-edge on reduction of the sample by hydrogen (*52*). This was similar to the difference between the model compounds $[Ni(III)(maleonitriledithiolate)_2](n\text{-}Bu_4N)$ and $[Ni(II)\text{-}(maleonitriledithiolate)_2](n\text{-}Bu_4N)_2$, and was taken as evidence for reduction of Ni(III) to Ni(II) in hydrogenase. It is difficult to correlate these results with the states observed in EPR, since it was not made clear whether the hydrogen-reduced enzyme was predominantly in the EPR-silent reduced state or the EPR-detectable Ni-C state. EXAFS spectra of the nickel in oxidized *D. gigas* hydrogenase were best fitted with four or six sulfur atoms at a distance of 0.220 nm. Other low-atomic-number scatterers were not detected but the presence of a small number could not be ruled out.

2. Hyperfine Interactions

As already noted, the first application of hyperfine interactions in the EPR spectra of hydrogenase was to identify the EPR signals that are due to nickel, by using the convenient isotope ^{61}Ni. All of the signals due to nickel have been assigned in this way (Figs. 4 and 5).

The EPR spectra of oxidized hydrogenases (Fig. 4a) do not show any noticeable hyperfine splitting by ^{14}N nuclei, indicating that there is no significant delocalization onto nitrogen ligands. Neither is there any change in linewidth on exchanging H_2O by D_2O (*10, 53*).

Albracht *et al.* (*54*) investigated the hyperfine interaction of the nickel center with ^{33}S ligands in hydrogenase from *Wolinella succinogenes* grown in a medium enriched with the isotope. Hyperfine splitting was observed in the narrow-line EPR spectra of both oxidized hydrogenase, and the hydrogen-reduced form after illumination (cf. Fig. 5b). Although there was only partial (about 70%) isotopic enrichment, it was possible to estimate by spectral simulations the number of interacting

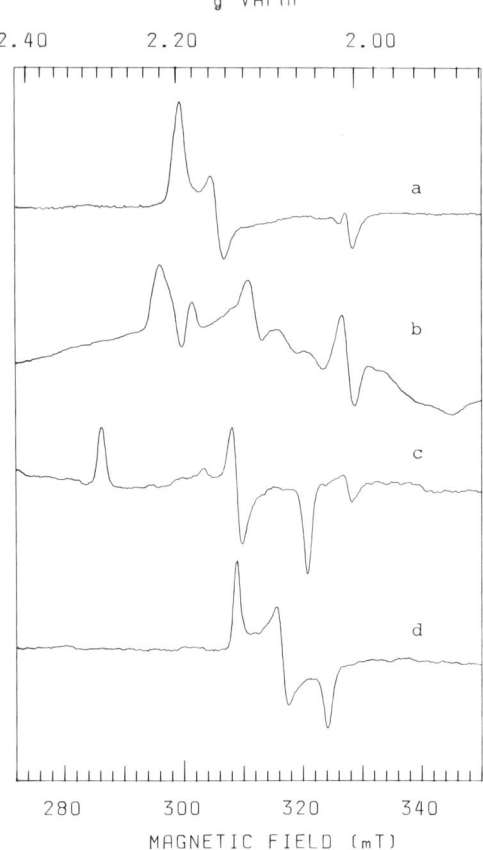

FIG. 5. EPR spectra of nickel in *D. gigas* hydrogenase: (a) Ni-C spectrum of activated hydrogenase under hydrogen, recorded at 77 K (cf. Fig. 4d); (b) same sample, recorded at 7 K, showing splitting due to interaction with reduced [4Fe–4S] cluster(s); (c) sample after illumination, recorded at 104 K; (d) hydrogen-activated hydrogenase, treated with carbon monoxide, recorded at 28 K.

sulfurs on the basis that it should be an integer value. The best fit was obtained by assuming just one interacting sulfur nucleus. The possibility cannot be discounted at present that other sulfurs are present, if the coordination geometry and electron distribution are such that the hyperfine interaction is very weak, or if the percentage enrichment of ^{33}S is overestimated. However, these results represent a significant disagreement with the estimates, from EXAFS, of three (*51*) or four (*52*) sulfurs, and further investigation is required.

The EPR spectra of Ni(III) complexes of short peptides containing thiol groups (55) were found to resemble those of hydrogenase most closely when they contained only one sulfur ligand. Sugiura et al. concluded that the most probable arrangement in hydrogenase is a single cysteine sulfur as equatorial ligand in a tetragonal geometry.

3. Electron Spin-Echo Spectroscopy

The three-pulse electron spin-echo envelope modulation (ESEEM) technique is particularly sensitive for detecting hyperfine couplings to nuclei with a weak nuclear moment, such as ^{14}N. It has been used to probe the coordination state of nickel in two hydrogenases from *M. thermoautotrophicum,* strain ΔH (56). One of these enzymes contains FAD and catalyzes the reduction of F_{420} (7,8-dimethyl-8-hydroxy-5-deazaflavin), while the other contains no FAD and has so far only been shown to reduce artificial redox agents such as methyl viologen.

The Fourier transform of the ESEEM spectra of the F_{420}-reducing hydrogenase showed a pattern of lines (Fig. 6b) which was interpreted as a hyperfine interaction of about 1.8 MHz with a ^{14}N nucleus (56). Since the spectrum was not observed in the ESEEM of the methyl viologen-reducing hydrogenase, which lacks FAD, it was suggested that the interaction might be with a flavin nitrogen. However, subsequent measurements have revealed the same interaction with nitrogen, in the hydrogenase of *Thiocapsa roseopersicina,* which contains no flavin (101). The magnitude of the hyperfine splitting suggests an indirect coordination of the nickel to nitrogen. It corresponds to a dipolar interaction over approximately 0.35 nm.

4. Magnetic Circular Dichroism

Magnetic circular dichroism (MCD) is a means of observing optical transitions due to paramagnetic species, and has been used by Johnson *et al.* (57) to observe Ni(III) against a background of iron–sulfur cluster absorption, in hydrogenases from *M. thermoautotrophicum* and *D. gigas.* Spectra of *M. thermoautotrophicum* hydrogenase were more fruitful because of the absence of paramagnetic [3Fe–xS] clusters. In the oxidized state only the nickel is paramagnetic and at low temperatures yielded MCD bands in the regions of 300–460 and 530–670 nm (Fig. 7b). These were provisionally assigned to nickel d–d transitions and sulfur-to-nickel charge transfer bands, respectively. Magnetization curves were consistent with an $S = 1/2$ ground state, as expected for low-spin Ni(III).

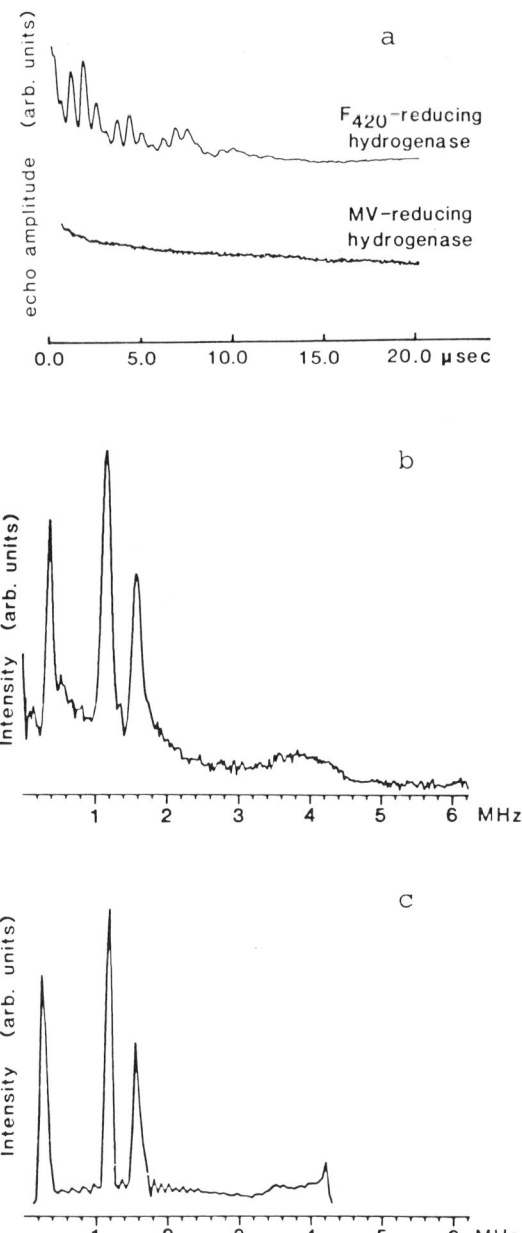

FIG. 6. ESEEM spectra of Ni in the F_{420}-reducing and methyl viologen (MV)-reducing hydrogenases from *M. thermoautotrophicum* (ΔH strain). Spectra (a) were obtained using a three-pulse stimulated-echo sequence, the time T between the second and third pulses being varied. (b) Fourier transform of FH_2ase data; (c) simulated spectra. Spectra are the average of recordings from $g = 2.0$ to 2.34. Reproduced, with permission, from Ref. 56.

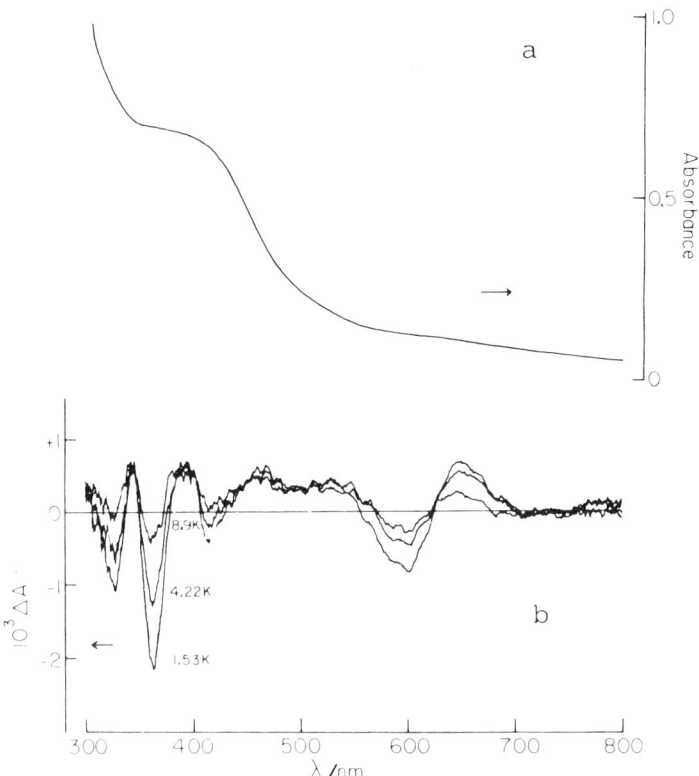

FIG. 7. Optical absorption and magnetic circular dichroism spectra of oxidized hydrogenase from *M. thermoautotrophicum* (ΔH strain), nickel concentration 120 μM. (a) Optical absorption spectrum, at room temperature; the absorption is predominantly due to iron–sulfur clusters. (b) MCD spectra recorded at 1.53, 4.22, and 8.9 K, in a magnetic field of 4.5 T; MCD is predominantly due to Ni(III), which is the only paramagnetic species in the oxidized enzyme. Reproduced, with permission, from Ref. 57.

E. Derivatives of Nickel

Altered forms of Ni hydrogenases have been observed under conditions in which substitution is expected at the hydrogen-binding site. These have been interpreted as species containing hydrogen or hydride or carbon monoxide, respectively, and are particularly relevant to the mechanism of hydrogen production and its inhibition by carbon monoxide. So far these species have only been observed by EPR. The investigation of these species and other possible diamagnetic species by other spectroscopic techniques is awaited with interest.

1. Reactions with Hydrogen

The Ni-C EPR signal is observed in some Ni hydrogenases under hydrogen. A significant property of this nickel species is that it is photosensitive; irradiation by visible or near-ultraviolet light at cryogenic temperatures produces a new type of EPR signal (Fig. 5c). The photochemical reaction is reversed at temperatures above 200 K. The rate of photolysis was shown by Van der Zwaan et al. (43) to have a kinetic isotope effect, being nearly six times slower in D_2O than in H_2O. This indicates that the photolytic process involves displacement of a hydrogen atom in the first coordination sphere, such as a nickel hydride. If that is the case, a significant hyperfine splitting would be expected. The Ni-C signal has indeed been reported (43, 53) to show broadening in ^1H_2O compared with ^2H_2O (43, 53), though this is small (0.3–0.5 mT) considering the large magnetic moment of the ^1H nucleus.

Studies of nickel protoporphyrins by resonance Raman spectroscopy show that the photochemical behavior of nickel sites may be complex and dependent on the protein environment. Nickel, in porphyrin complexes, favors a four- or six-coordinate structure (58). In nickel-substituted human hemoglobin, one type of binding site takes up four-coordinate geometry (with a γ band at 406 nm), and the other type of site, like the heme-binding site in myoglobin, is forced into five-coordination (59). After laser excitation, the four-coordinate sites undergo photoassociation with a fifth ligand, as occurs in protoporphyrin complexes. By contrast, the five-coordinate sites undergo photodissociation (60). Unlike Fe porphyrins, the recombination of the displaced ligand with the nickel site is very rapid (20 p sec).

Hydrides of Ni(I) and Ni(II) are known (37). A Ni(II) hydride appears to be an intermediate in the catalysis of olefin isomerization by phosphine complexes of nickel (61). Dilworth (62) has pointed out that stable hydride species are not obtained in model complexes with sulfur ligands. However, they may be possible within the confines of a protein chelate.

2. Reactions with Carbon Monoxide

Carbon monoxide is an inhibitor of most nickel-containing hydrogenases, an exception being the soluble hydrogenases of hydrogen bacteria (63). Kinetically, inhibition is competitive with hydrogen, which indicates that the two molecules bind to the same site. EPR spectroscopy indicates that oxidized hydrogenase, giving the Ni(III) Ni-A signal, is unreactive toward carbon monoxide. Probably it is Ni(II) or Ni(I) that binds CO. Reaction of reduced hydrogenase with CO produces both EPR-silent and EPR-detectable (Fig. 5d) states, which may represent different carbonyl species (*64, 65*).

The EPR-detectable CO derivative in *C. vinosum* hydrogenase was shown by Van der Zwaan *et al.* (*64*) to be photolysed at low temperatures, yielding a product having the same spectrum as the photolyzed hydrogen-treated enzyme (Fig. 5c). Unlike the hydrogen-reduced species however, the rate of photolysis was unaffected by deuterium isotope substitution.

3. Interactions with Oxygen

Oxygen is inhibitory toward some hydrogenases and completely destructive toward others, particularly some Fe hydrogenases (*66*). In those hydrogenases which can survive exposure to oxygen, the hydrogenase active site is unreactive toward H_2 and CO under oxidizing conditions (not necessarily the presence of oxygen). In *D. gigas* hydrogenase, DerVartanian *et al.* (*53*) have noted an increase in the spin-lattice relaxation rate of the Ni-A EPR signal under oxygen, consistent with an interaction with the triplet state of O_2. This has been interpreted (*74*) as the binding of oxygen to the nickel site, but, at present, the evidence for this is not conclusive, since the lineshape of the Ni-A signal did not change. If oxygen is associated with the Ni-A site, it must be extremely tenaciously bound, since the enzyme is unaffected by stringent removal of dissolved oxygen (*49*). It seems more likely that there is a protective mechanism for those organisms which may encounter oxygen *in vivo*, whereby the hydrogenase active site becomes inaccessible to all gases. A similar type of oxidative stabilization and reductive reactivation is observed in the Fe hydrogenase of *Desulfovibrio vulgaris* (Hildenborough strain) (*67*). It is therefore possible that the ability to form such a state depends upon some specific structure within the enzyme molecule which is distinct from the hydrogen-binding site.

F. INTERACTIONS OF NICKEL WITH IRON–SULFUR CENTERS

All the hydrogenases which have been isolated are iron–sulfur proteins. In particular, hydrogenase activity seems to be associated with at least one [4Fe–4S] cluster in addition to nickel. In some hydrogenases, including the catalytic dimer of *N. opaca* hydrogenase, both the iron–sulfur cluster and the nickel are undetectable by EPR, and it is possible that in these cases there is strong antiferromagnetic coupling. The two groups might act together to transfer two electrons to hydrogen.

Many of the Ni hydrogenases contain an iron–sulfur cluster, presumed to be of the [3Fe–xS] type, which is paramagnetic with $S = 1/2$ in the oxidized state and $S = 2$ in the reduced state (68, 69). The function of these clusters is unknown. In some hydrogenases, typified by *C. vinosum* hydrogenase and the membrane-bound hydrogenase of *Alcaligenes eutrophus* (70), the EPR spectra of the iron–sulfur clusters and the oxidized nickel center show complex lineshapes (Fig. 8). In *C. vinosum* the nickel is also EPR detectable and its spectrum also shows a

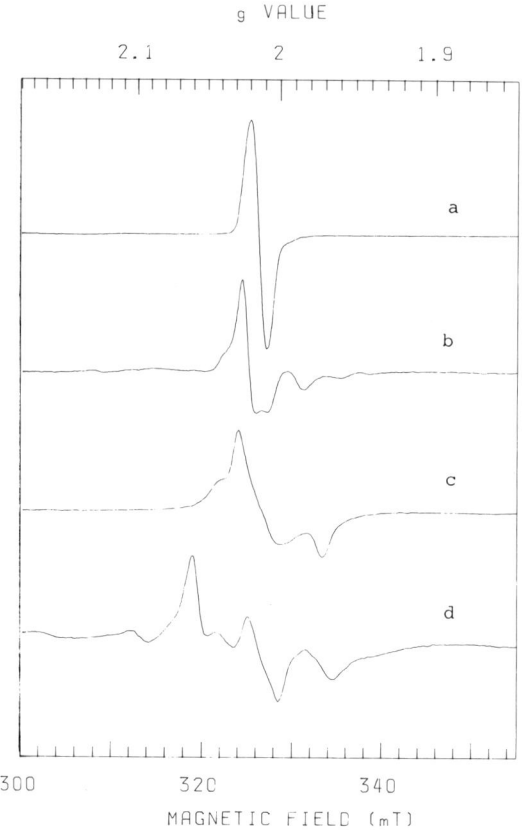

FIG. 8. EPR spectra of [3Fe–xS] clusters in oxidized hydrogenases, showing the influences of weak Ni–Fe–S electron-spin interactions. (a) *Desulfovibrio desulfuricans* (strain Norway 4) hydrogenase, showing the spectrum of an isolated [3Fe–xS] cluster; (b) *Chromatium vinosum* hydrogenase; the outer lines (Signal 2) correspond to interaction with Ni(III); (c) *Paracoccus denitrificans* hydrogenase; (d) *Alcaligenes eutrophus* membrane-bound hydrogenase. Spectra were recorded at approximately 20 K. Samples were provided by K. K. Rao, J. Serra, and K. Schneider.

splitting at temperatures where the iron–sulfur cluster is detectable (71). Spectra recorded at S-band and Q-band frequencies confirmed that these effects were due to spin–spin interaction presumably between the nickel and iron–sulfur clusters (71).

In *D. gigas* hydrogenase, splittings are not observed in the Ni-A and Ni-B signals from oxidized nickel centers (Fig. 4a), but are seen in the reduced Ni-C species at low temperatures (Fig. 5b) (41, 72). The splitting of Ni-C correlates with the reduced state of a [4Fe–4S] cluster (72). The spin–spin interactions observed in EPR are consistent with a distance between the nickel and iron-sulfur cluster of less than 1.2 nm (73).

G. Catalytic Properties

1. Activity States of Nickel-Containing Hydrogenases

The characteristic EPR signal due to Ni(III) (Fig. 4a) is observed in a number of hydrogenases, but not all (14, 15). This is consistent with the view that the signal reflects an oxidized form of the enzyme which is not directly involved in catalysis. This is supported by studies of the enzyme activity in response to oxidizing and reducing agents. A well-studied case is that of hydrogenase from *D. gigas* (41, 49). Even when extensive precautions are taken to remove oxygen, a reductant is still necessary to reduce the oxidized hydrogenase to a state which is capable of reacting with hydrogen (69).

The enzymatic activity of hydrogenases, whether measured by hydrogen production, hydrogen consumption, or hydrogen isotope exchange, is often variable, depending on the history of the sample. The hydrogenases may not manifest their full activity until they have been subjected to reactivating treatments, which vary from one hydrogenase to another. For instance, *D. gigas* hydrogenase shows a phenomenon of slow activation by hydrogen or other strong reductants (74, 75). A study of the conditions required for the slow activation of *D. gigas* hydrogenase showed that it was not induced by addition of chelating agents, or nickel ions, or reagents which in other systems have caused the conversion of [3Fe–xS] clusters to [4Fe–4S] (49). The thiol-reducing agent dithiothreitol (DTT) did not cause activation when added alone [although in certain circumstances it can have an activating effect on *C. vinosum* hydrogenase (76)]. The changes in activity of *D. gigas* hydrogenase when incubated under hydrogen are illustrated in Fig. 9a, and EPR spectra of the nickel in the enzyme are shown in Fig. 9b. The fact that the enzyme can be obtained in forms with different activity, which can all be fully reactivated, indicates that there must be more

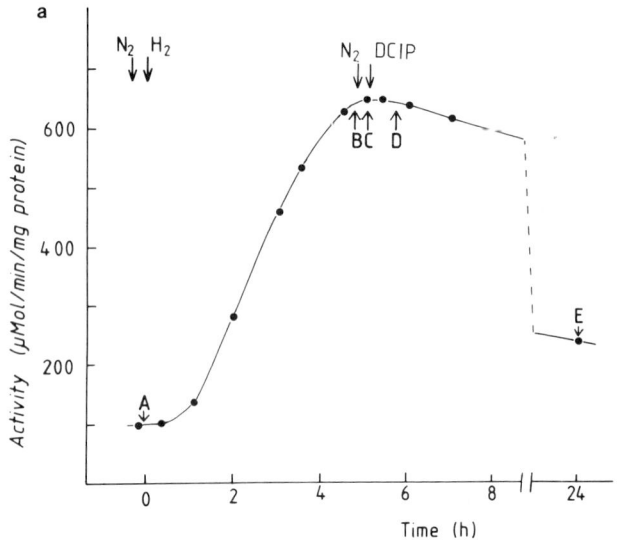

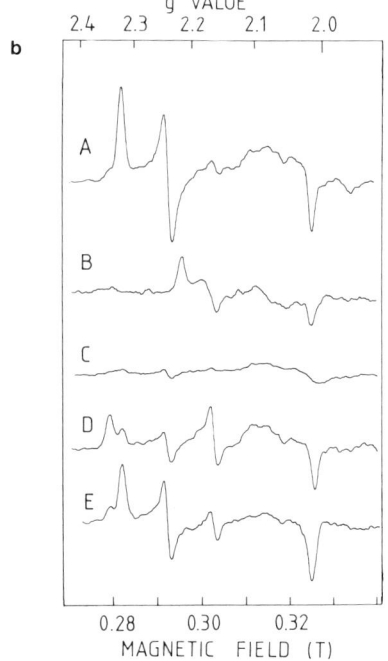

FIG. 9. (a) Activity of *D. gigas* hydrogenase (hydrogen–methyl viologen reductase) during activation under hydrogen, flushing with nitrogen, and reoxidation with dichloroindophenol (DCIP). (b) EPR spectra of samples taken at points A–E, as indicated in a, recorded at 80 K. See text for details. Adapted from Refs. *40* and *49*.

than one type of inactive oxidized enzyme. The simplest hypothesis is that there are two: one which is activated rapidly by reduction during the assay, and one which is only activated slowly by strong reductants. We have termed these forms the *ready* and *unready* states, respectively, of the enzyme. The catalytic and spectroscopic properties of the enzyme in these states are described in Table III. In the interpretation of Berlier *et al.* (74), the unready state is associated with a tightly bound oxygen molecule. Alternatively, the ready and unready states might represent different protein conformations (49).

The enzyme, which had been prepared in the presence of air (marked point A in Fig. 9a), is mostly in a state which has been described as *unready* (49). It has an activity, which corresponds to about 12% of the *active* state, in hydrogen uptake with methyl viologen, but is completely inactive in hydrogen isotope exchange (74, 77) or hydrogen uptake with a high-potential acceptor dye such as 2,6-dichloroindophenol (DCIP). We proposed that the enzyme at this stage consists of about 88% of the *unready* and 12% of the *ready* state.

After complete activation under hydrogen, a process which requires about 4 hours at 20°C (point B), the hydrogenase is fully active in all assays (49, 74). If hydrogen is then removed and the enzyme reoxidized with the dye dichloroindophenol at pH 7.8, the enzyme is principally in the *ready* state (point C), which is fully active in the hydrogen–methyl viologen reductase assay, but completely inactive in hydrogen isotope exchange (77), and is stable in air for several hours. This observation particularly indicates that the appearance or loss of activity of the enzyme is not entirely due to the removal or addition of oxygen. The

TABLE III

ACTIVITY STATES OF *D. gigas* HYDROGENASE[a]

Activity	State		
	Active	Ready	Unready
H_2–methyl viologen	Active	Active after lag	Inactive
H_2–DCIP	Active	Inactive	Inactive
H_2–isotope exchange	Active	Inactive	Inactive
Hydrogen production	Active	Active	Inactive
	EPR signals		
	Ni-C	Ni-B	Ni-A
	Broad $[4Fe–4S]_{red}$	$[3Fe–xS]_{ox}$	$[3Fe–xS]_{ox}$

[a] Conditions of formation: active state incubation under H_2, ready state anaerobic oxidation, unready state oxygen or strong reductants by DCIP, pH 8.0.

EPR signal at this stage corresponds mainly with the Ni B signal (Figs. 4d and 5a), which was therefore proposed to represent the *ready* state (*40*). After prolonged exposure to oxygen, the enzyme reverts to the *unready* state and the Ni-A signal reappears (point E).

Hydrogenases differ greatly in the rate and extent of reductive activation (*14*); *D. gigas* appears to represent a slow extreme. In other hydrogenases, such as the soluble enzymes from hydrogen bacteria, the reductive activation is always too rapid to be measurable. These hydrogenases also differ from *D. gigas* hydrogenase in that they give no Ni(III) signal in their oxidized states, and give different HD/H_2 ratios in the isotope-exchange reaction (*78*). It is not not known if these factors are correlated.

Another pattern of reductive activation has been observed in the Ni hydrogenase of *Methanobacterium formicicum* (*79*). In this enzyme, reductive activation is irreversible, requires a strong reductant, and is inhibited by the chelating agent 2,2'-bipyridyl. The conversion of a [3Fe–xS] cluster to a [4Fe–4S] has been suggested.

2. Mechanism of Catalysis

The behavior of the EPR-detectable nickel and iron–sulfur species in hydrogenase during redox-poising experiments suggests several possible reaction sequences. The Ni-C species is particularly significant since the potentials for its appearance and subsequent disappearance during reduction are comparable with the H^+/H_2 potential, at the low partial pressures of hydrogen likely to be encountered by these organisms; the potentials are also pH dependent, consistent with hydrogen production. Moreover, the low-temperature photochemical reaction of the Ni-C species shows a strong kinetic isotope effect with ^2H.

The series of steps of reduction have been variously interpreted:

$$
\begin{array}{llll}
\text{Ni(II)} & = \text{Ni(I)} & = \text{Ni(0)} & (43) \quad (1) \\
\text{EPR silent} & \text{Ni-C signal} & \text{EPR silent} & \\
\text{Ni(III)-[4Fe-4S]}^- & = \text{Ni(III)} \cdot H^- & = \text{Ni(II)} & (41) \quad (2) \\
\text{Ni(II)} & = \text{Ni(I)} & = \text{Ni(II)} \cdot H^- & (72) \quad (3) \\
\end{array}
$$

It is difficult to distinguish these possibilities on the basis of the EPR spectra since the spectra of low-spin d^7 Ni(III) and d^9 Ni(I) are similar. A disadvantage of Scheme (1) is that it requires the nickel to take up oxidation states from Ni(III) to Ni(0) over a potential range of about 240 mV, whereas in inorganic complexes they span several volts (*80*). In Scheme (2) (*41*), the nickel is coupled to an iron–sulfur cluster. On reduction of the oxidized hydrogenase, the disappearance of the Ni-A Ni(III) signal is due to reduction of the iron–sulfur cluster, which becomes paramagnetic on reduction, giving net zero spin; however, optical absorption changes on reduction of *D. gigas* hydrogenase do not support this view (*40*). Scheme (3) was proposed to explain why only a very small ^{1}H hyperfine splitting is seen in the Ni-C species, whereas a stronger coupling would be expected for a paramagnetic nickel hydride (*72*).

The hypothetical reaction mechanism shown in Fig. 10 is a variation of Scheme (3), and is consistent with the redox and pH dependence of the EPR-detectable nickel species (*65*). Hydrogen is known to undergo heterolytic cleavage (*81*); it is proposed that this is an intramolecular reaction, leading to the formation of a nickel (II) hydride and a protonated base in the enzyme (Step 1 in Fig. 10). The Ni-C species is postulated to be a protonated Ni(I) species. An alternative formulation for this state would be a dihydrogen complex, Ni(III)·H$_2$, as suggested by Crabtree (*104*). Ultimately the exact mechanism can only be determined by kinetic measurements.

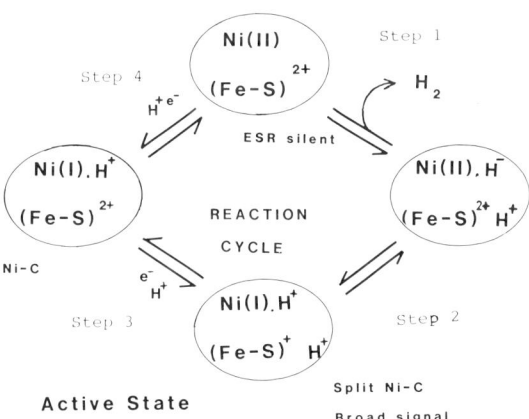

FIG. 10. Hypothetical reaction cycle for *D. gigas* hydrogenase, based on the EPR and redox properties of the nickel (Table II). Only the nickel center and one [4Fe–4S] cluster are shown. Step 1: enzyme, in the activated conformation and Ni(II) oxidation state, causes heterolytic cleavage of H$_2$ to produce a Ni(II) hydride and a proton which might be associated with a ligand to the nickel or another base in the vicinity of the metal site. Step 2: intramolecular electron transfer to the iron–sulfur cluster produces a protonated Ni(I) site (giving the Ni-C signal). An alternative formulation of this species would be Ni(III)·H$_2$. Step 3: reoxidation of the iron–sulfur cluster and release of a proton. Step 4: reoxidation of Ni and release of the other proton.

H. OTHER FUNCTIONS OF NICKEL IN HYDROGENASES

The soluble hydrogenase from the hydrogen-oxidizing bacterium *N. opaca* is one of a class of hydrogenases that contain flavin and use nicotinamide adenine dinucleotide (NAD) as electron acceptor. The protein consists of four dissimilar subunits and contains approximately four atoms of nickel, one FMN, three [Fe–4S] clusters, one [2Fe–2S] cluster, and up to one [3Fe–xS] cluster (*82*). Two of the nickel atoms were readily removed by dialysis, in contrast to the nickel in most hydrogenases. The enzyme would only catalyze electron transfer from hydrogen to NAD if cations, of which Ni^{2+} is the most effective, were added. In the absence of the cations, the enzyme could be separated as

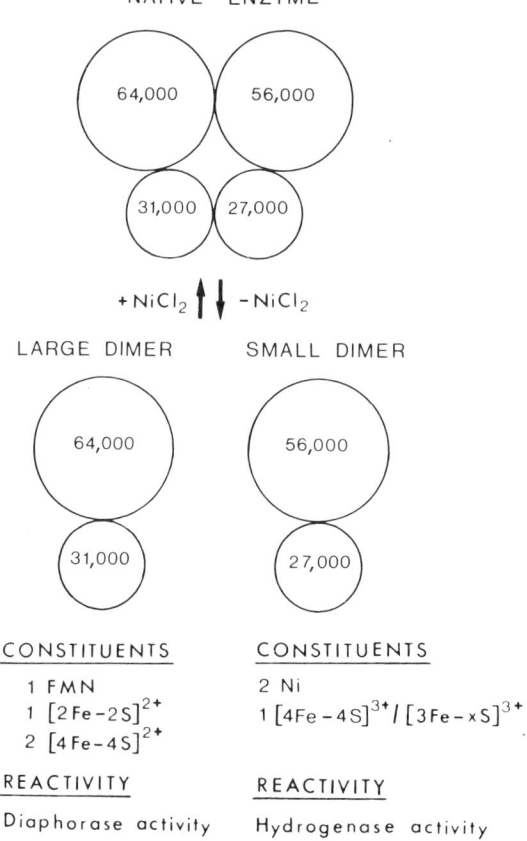

FIG. 11. Dissociation and reassociation of soluble, NAD-linked hydrogenase from *Nocardia opaca*. "Diaphorase" refers to NADH : acceptor reductase activity. Reproduced, with permission, from Ref. *82*.

two dimers (Fig. 11). One of these contained two tightly bound nickel atoms and one [4Fe–4S] cluster, and had hydrogen:acceptor reductase activity. The other contained flavin and iron–sulfur clusters and had NADH:acceptor reductase activity. Hence in *N. opaca* hydrogenase, which so far is a unique case, the nickel has two functions, in catalysis and in maintaining protein structure.

IV. Methyl-Coenzyme M Reductase

A. Factor 430

F_{430}, a yellow, water-soluble compound, was first extracted from boiled cells of methanogenic bacteria, a discovery which Wolfe (*19*) has credited to J. LeGall. Its isolation was first reported by Gunsalus and Wolfe (*83*). The cofactor has a Soret band in the visible region at 430 nm. Functionally F_{430} is a prosthetic group of the methylreductase system (*24, 84*). It is also found in the free state in cell extracts (*85*).

The presence of nickel in F_{430} was demonstrated by Diekert *et al.* (*86*) and Whitman and Wolfe (*87*). Biosynthetic incorporation of δ-aminolevulinic acid indicated that the compound has a tetrapyrrole structure. The structure of the methanolysis product of F_{430} was determined by Pfaltz *et al.* (*88*) using NMR spectroscopy. The structure of the cofactor in the cell is probably the penta acid (Fig. 12). It is a tetrahydro derivative of a porphinoid which is related to the corrins, for which Pfaltz *et al.* (*88*) have coined the term "corphin." Specificity of this macrocycle for nickel is afforded by the six-membered ring, which introduces a slight pucker into the planar structure. EXAFS measurement of the nickel environment of isolated F_{430} indicated two nitrogens at 0.191–0.192 and 0.210–0.214 nm (*89a, 89b*). This is consistent with nickel in a square–planar coordination to a puckered corphin structure. In the intact protein, protein component C of methyl-CoM reductase, the absorption-edge spectra are indicative of nickel with a coordination number greater than four, probably octahedral (*89c*).

B. EPR Spectra

Albracht *et al.* (*90*) have observed an EPR spectrum from whole cells of *M. thermoautotrophicum* (Fig. 13) which appears to arise from protein-bound F_{430}. The same type of spectrum was also observed in purified methyl-CoM reductase. The paramagnetic species on whole cells was only partially reduced by treatment with hydrogen and was unaffected by carbon monoxide. Gel filtration showed it to be part of a soluble protein. The spectrum was somewhat broadened by substitution

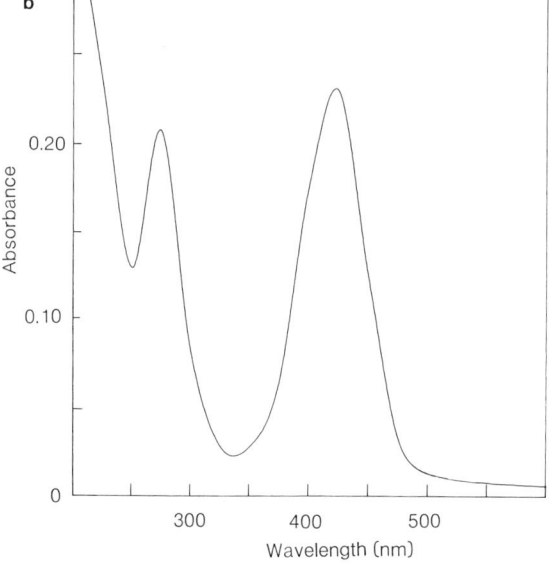

FIG. 12. (a) Structure of F_{430}, deduced from the structure of the methanolysis product (88). (b) UV/visible absorption spectrum of F_{430} in water, concentration $10^{-5} M$.

with ^{61}Ni and showed hyperfine splitting which could be simulated as interaction with four equivalent nitrogen atoms, corresponding to the four nitrogens of a tetrapyrrole. The lineshape of the spectrum is consistent with nickel in a tetragonally distorted octahedral ligand field. This confirms the assignment as a nickel species bound in a planar tetrapyrrole structure. It was not possible to determine the oxidation state of the nickel, although the observation that it was observed in isolated factor F_{430} treated with dithionite suggests that it is Ni(I).

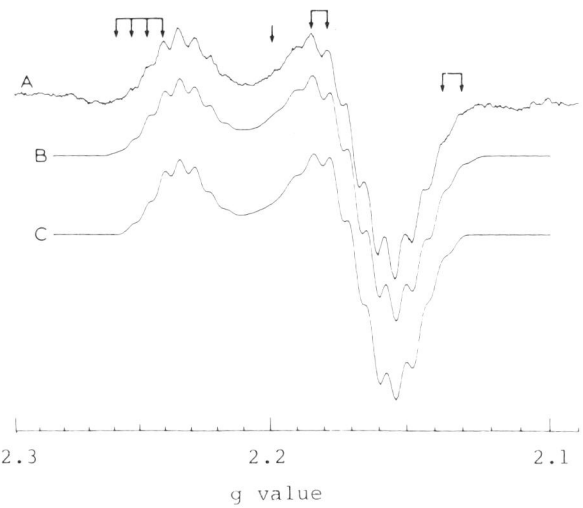

FIG. 13. (A) EPR spectrum, recorded at 70 K, of reduced F_{430}, in a hydrogen-reduced cell extract from *M. thermoautotrophicum* (Marburg strain). (B and C) Computer simulations, assuming equal interaction with four (B) or three (C) equivalent ^{14}N nuclei. The arrows indicate places where B gives a better fit. Reproduced, with permission, from Ref. 90.

C. FUNCTION

F_{430} is required for the activity of the methylreductase system, which catalyzes the reduction of the methyl group of methyl-CoM, i.e., 2-(methylthio)ethanesulfonate, to methane:

$$CH_3-S-CH_2-CH_2-SO_3^- + H_2 \longrightarrow CH_4 + HS-CH_2-CH_2-SO_3^- \quad (4)$$

This is a complex process, involving the deazaflavin-reducing hydrogenase and other proteins, provisionally named factors A-2, A-3, and C (*19*). The system requires initial activation by ATP before the reaction can proceed (*87*). Component C consists of two each of three protein subunits of 68, 45, and 38.5 kDa and contains two molecules of F_{430} per molecule, as well as stoichiometric amounts of coenzyme M (*91*). F_{430} appears to be tightly but noncovalently bound to the enzyme.

The exact role of the nickel of F_{430} in methane formation is not clear at present. Analogy with the cobalamins, and the observation of an EPR-detectable reduced state, might suggest that it is involved in either methyl group transfer, reduction, or both.

V Carbon Monoxide Oxidoreductase

A. Catalytic Properties

Carbon monoxide oxidoreductase (carbon monoxide dehydrogenase) catalyzes the interconversion of CO and CO_2 with suitable electron acceptors and donors. The reaction takes place via an enzyme-bound one-carbon intermediate:

$$CO \longrightarrow [C_1] \longrightarrow CO_2 + 2e^- + 2H^+ \qquad (5)$$

The enzyme was first isolated from *C. thermoaceticum*, *Clostridium formicoaceticum*, and *Clostridium pasteurianum* by Diekert and Thauer (92), who demonstrated that synthesis of the enzyme requires nickel (93). The enzyme is present in large amounts in acetogenic bacteria, where it is involved in an unusual pathway for fixation of CO_2 with the formation of acetate (25, 26). In acetogens this reaction is involved both in production of energy, with acetate as a waste product, and in biosynthesis of cell constituents starting from acetate. CO oxidoreductase is also present in methanogenic bacteria, where it is used in biosynthetic metabolism (25, 27).

Other reactions catalyzed by CO oxidoreductase, which may be more relevant to its physiological function, are the exchange between CO and the carbonyl group of acetyl-CoA (94)

$$CH_3-\text{*CO}-CoA + CO \longrightarrow CH_3-CO-CoA + \text{*CO} \qquad (6)$$

and the exchange between CoA and the CoA moiety of acetyl-CoA.

$$CH_3-CO-S\text{*CoA} + CoASH \longrightarrow CH_3-CO-CoA + \text{*CoASH} \qquad (7)$$

The reaction [Eq. (7)] requires a disulfide-reducing system such as dithiothreitol or disulfide reductase and a reducing agent such as NADPH or reduced ferredoxin. It is proposed [Eq. (5)] that carbon monoxide oxidoreductase binds CO as a one-carbon intermediate $[C_1]$, which can be either oxidized to CO_2 or condensed with the methyl group of a methylated corrinoid protein and CoA in the final step of acetyl-CoA synthesis.

Carbon monoxide oxidoreductase has a high molecular weight; values between 150,000 and 460,000 have been reported (25). Analysis of the enzyme from *C. thermoaceticum* indicated a composition of 2 atoms of Ni, 1 Zn, 11 Fe, and 14 inorganic sulfide per dimeric enzyme, with a relative molecular mass of approximately 150,000 Da (23).

B. Spectroscopy

Low-temperature EPR spectroscopy of carbon monoxide oxidoreductase in the reduced state gave a complex spectrum with g values 2.04, 1.94, and 1.90, and 2.01, 1.86, and 1.75, which probably represent two [4Fe–4S] clusters (95). The clusters are oxidized by CO_2 and reduced by CO.

EPR spectra of CO oxidoreductase under non reducing conditions showed a spectrum at $g = 2.21$, 2.11, and 2.02, which, by analogy with spectra observed in nickel-containing hydrogenases, was attributed to Ni(III) (96). The spectrum was of low intensity and it was not established whether it represents an active state of the enzyme.

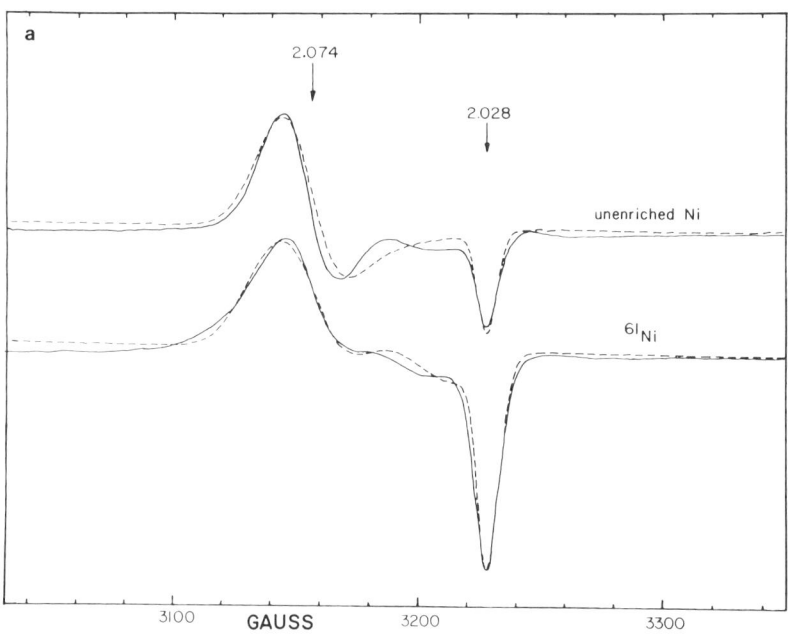

FIG. 14. EPR spectra of carbon monoxide oxidoreductase from *C. thermoaceticum*, treated with CO plus coenzyme A. Solid lines are experimental spectra, dashed lines are computer simulations, with $g_x = g_y = 2.074$, $g_z = 2.028$. Substitutions with ^{61}Ni and ^{57}Fe were made by growth of the organism on the appropriate isotopes. (a) Effect of substitution with ^{61}Ni. Simulation assumes $A_{\parallel} = 3$ MHz, $A_{\perp} = 20$ MHz. (b, p. 328) Effects of substitution with ^{57}Fe and ^{13}C. The simulation of the ^{57}Fe spectrum assumes one iron atom with $A_{\parallel} = 40$ MHz, $A_{\perp} = 60$ MHz, and two iron atoms with $A_{\parallel} = 20$ MHz, $A_{\perp} = 30$ MHz. The simulation of the ^{13}C spectrum assumes $A_{\parallel} = 26$ MHz, $A_{\perp} = 13$ MHz. Spectra provided by courtesy of Dr. S. G. Ragsdale.

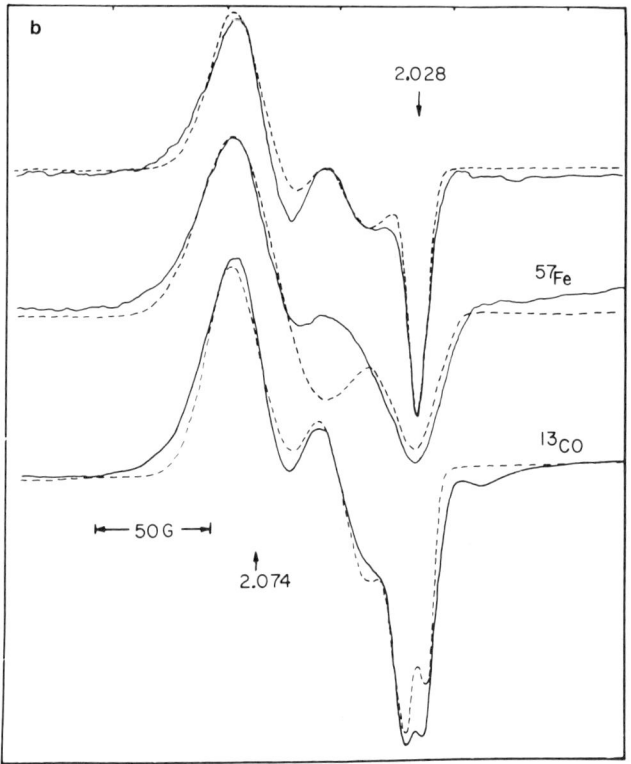

Fig. 14b. See legend on p. 327.

Two spectra which are more clearly relevant to the enzyme-catalyzed reaction were observed in the reduced enzyme after treatment with CO (Fig. 14). A rhombic spectrum, Signal 2, with g values of 2.062, 2.047, and 2.028, was observed upon reaction of carbon monoxide oxidoreductase with CO (97). Addition of coenzyme A or acetyl-CoA induced an axially symmetric spectrum, Signal 1, at $g = 2.074$ and 2.028. Spectra of the CO-treated carbon monoxide oxidoreductase were typically a mixture of the two species. It was assumed that these spectra represent different enzyme–C_1 intermediates. The spectrum showed hyperfine broadening after reaction with ^{13}CO instead of ^{12}CO and in enzyme from cells grown on ^{61}Ni (96). Subsequently it was shown (97) that the spectrum was broadened by substitution of ^{57}Fe (Fig. 14). These results demonstrated that the radical species is a complex containing nickel, carbon monoxide, and iron. Recent simulations (103) show that the spectra are consistent with a complex of one nickel and three iron atoms, analogous to a [4Fe–4S] cluster in which one iron atom is substituted by nickel.

This represents a new and fascinating type of mixed-metal cluster in biochemistry.

VI. Concluding Remarks

The four types of nickel-containing enzymes are quite distinct in the coordination sites and catalytic function of the nickel centers. In urease, the nickel appears to be bound to oxygen and nitrogen ligands and appears to remain as Ni(II), a state which favors octahedral or square–planar coordination. The function of nickel in this unique case may be analogous to that of zinc in other hydrolases such as carboxypeptidase.

In all the other nickel enzymes, there is scope for redox reactions at the nickel site, though it has not been demonstrated in all cases. The nickel in hydrogenase shows indications of at least one sulfur ligand, the others probably being oxygen; the ease with which oxidation states Ni(III) and, possibly, Ni(I) can be achieved indicates that the geometry might be distorted toward tetrahedral. In methyl-CoM reductase the nickel is held by a novel type of porphinoid macrocycle which is almost planar. In the protein, the geometry is presumably octahedral. The nickel may be capable of achieving the oxidation states Ni(II) and Ni(I). Finally, in CO oxidoreductase, hyperfine interactions suggest a mixed Ni–Fe cluster which is capable of binding a one-carbon group. The chemistry of the latter is probably the most fascinating and the least understood at present.

VII. Abbreviations

CoA, Coenzyme A
CoM, Coenzyme M

DCIP, 2,6-Dichloroindophenol

EPR, Electron paramagnetic resonance
ESEEM, Electron spin-echo envelope modulation
EXAFS, Extended X-ray absorption fine structure

FAD, Flavin adenine dinucleotide

MCD, Magnetic circular dichroism

NAD, Nicotinamide adenine dinucleotide
NADP, Nicotinamide adenine dinucleotide phosphate

References

1. Dixon, N. E., Gazzola, C., Blakeley, R. L., and Zerner, B., *J. Am. Chem. Soc.* **97,** 4131 (1975).
2. Sumner, J. B., *J. Biol. Chem.* **69,** 435 (1926).
3. Klucas, R. V., Hanus, F. J., Russell, S. A., and Evans, H. J., *Proc. Natl. Acad. Sci. U.S.A.* **86,** 2253 (1983).
4. Eskew, D. L., Welch, R. M., and Cary, E. E., *Science* **222,** 621 (1983); Eskew, D. L., Welch, R. M., and Norvell, W. A., *Plant Physiol.* **76,** 691 (1984).
5. Thauer, R. K., Diekert, G., and Schönheit, P., *Trends Biochem. Sci.* **11,** 304 (1980).
6. Balch, W. E., Fox, G. E., Magrum, L. J., Woese, C. R., and Wolfe, R. S., *Microbiol. Rev.* **43,** 260 (1979).
7. Lancaster, J. R., Jr., *FEBS Lett.* **115,** 285 (1980).
8. Lancaster, J. R., Jr., *In* "Methods in Enzymology" (S. Fleischer and L. Packer, eds.), Vol. 54, p. 412. Academic Press, New York, 1982.
9. Albracht, S. P. J., Graf. E.-G., and Thauer, R. K., *FEBS Lett.* **140,** 311 (1982).
10. Cammack, R., Patil, D., Aguirre, R., and Hatchikian, E. C., *FEBS Lett.* **142,** 289 (1982).
11. LeGall, J., Ljungdahl, P. O., Moura, I., Peck, H. D., Xavier, A., Moura, J. J. G., Teixeira, M., Huynh, B. H., and DerVartanian, D. V., *Biochem. Biophys. Res. Commun.* **106,** 610 (1982).
12. Albracht, S. P. J., Kalkman, M. L., and Slater, E. C., *Biochim. Biophys. Acta* **724,** 309 (1983).
13. Schneider, K., Patil, D. S., and Cammack, R., *Biochim. Biophys. Acta* **748,** 353 (1983).
14. Cammack, R., Fernandez, V. M., and Schneider, K., *Biochimie* **68,** 85 (1986).
15. Cammack, R., Hall, D. O., and Rao, K. K., *In* "Microbial Gas Metabolism: Mechanistic, Metabolic and Biotechnological Aspects" (R. K. Poole and C. S. Dow, eds.), p. 209, Academic Press, London, 1985.
16. Cammack, R., Fernandez, V. M., and Schneider, K., *In* "Bioinorganic Chemistry of Nickel" (J. R. Lancaster, Jr., ed.); VCH Publ., Deerfield Beach, Florida, 1988 (in press).
17. Bagyinka, C., and Kovacs, K., *Biochimie* **68,** pp. 1–221 (1986).
18. Lancaster, J. R. Jr., "Bioinorganic Chemistry of Nickel" (J. R. Lancaster, Jr., ed.). VCH Publ., Deerfield Beach, Florida, 1987 (in press).
19. Wolfe, R. S., *Trends Biochem. Sci.* **16,** 306 (1985).
20. Thauer, R. K., *Biol. Chem. Hoppe-Seyler* **366,** 103 (1985).
21. Keltjens, J. T., and Van der Drift, C., *Microbiol. Rev.* **39,** 259 (1986).
22. Woese, C. R., Magrum, L. J., and Fox, G. E., *J. Mol. Evol.* **11,** 245 (1978).
23. Ragsdale, S. G., Clark, J. E., Ljungdahl, L. G., Lundie, L. L., and Drake, H. L., *J. Biol. Chem.* **258,** 2364 (1983).
24. Diekert, G., Fuchs, G., and Thauer, R. K., *In* "Microbial Gas Metabolism: Mechanistic, Metabolic and Biotechnological Aspects" (R. K. Poole, and C. S. Dow, eds.), p. 115 Academic Press, London, 1985.
25. Zeikus, J. G., Kerby, R., and Krzycki, J. A., *Science* **227,** 1167 (1985).
26. Hemming, A., and Blotevogel, K. H., *Trends Biochem, Sci.* **16,** 198 (1985).
27. Fuchs, G., *FEMS Microbiol. Rev.* **39,** 181 (1986).
28a. Adams, M. W. W., and Mortenson, L. E., *J. Biol. Chem.* **259,** 7045 (1984).
28b. Huynh, B. H., Czechowski, M. H., Krüger, H. J., DerVartanian, D. V., Peck, H. D., and LeGall, J., *Proc. Natl. Acad. Sci. U.S.A.* **81,** 3728 (1984).
29. Hausinger, R. P., *Microbiol. Rev.* **51,** 22 (1987).
30a. Sunderman, F. W., Jr. *et al.* (eds.), "Nickel in the Human Environment." Scientific Publications, Lyon, France, 1984.

30b. Brown, S. S., and Sunderman, F. W., Jr., *Proc. Int. Congr. Nickel Metab. Toxicol. 3rd, Paris* (1985).
31. Blakeley, R. L., Dixon, N. E., and Zerner, B., *Biochim. Biophys. Acta* **744**, 219 (1983).
32. Rosenberg, R. C., Root, C. A., and Gray, H. B., *J. Am. Chem. Soc.* **97**, 21 (1975).
32a. Rees, D. C., Howard, J. B., Chakrabarti, P., Yeates, T., Hsu, B. T., Hardman, K. D., and Lipscomb, W. N., *In* "Zinc Enzymes" (I. Bertini, C. Luchinat, W. Maret, and M. Zeppezauer, eds.), p. 155. Birkhäuser Verlag, Boston, 1986.
33. Hasnain, S. S., and Piggott, B., *Biochem. Biophys. Res. Commun.* **112**, 279 (1983).
34. Alagna, L., Hasnain, S. S., Piggott, B., and Williams, D. J., *Biochem. J.* **220**, 591 (1984).
35. Dixon, N. E., Riddles, P. W., Gazzola, C., Blakeley, R. L., and Zerner, B., *Can. J. Biochem.* **58**, 1335 (1980).
36. Spratt, T. E., Sugimoto, T., and Kaiser, E. T., *J. Am. Chem. Soc.* **105**, 3679 (1983).
37. Nag, K., and Chakravorty, A., *Coord. Chem. Rev.* **33**, 87 (1980).
38. Hornhardt. S., Schneider, K., and Schlegel, H. G., *Biochimie* **68**, 15 (1986).
39. Kojima, N., Fox, J. A., Hausinger, R. P., Daniels, L., Orme-Johnson, W. H., and Walsh, C., *Proc. Natl. Acad. Sci. U.S.A.* **80**, 378 (1983).
40. Fernandez, V. M., Hatchikian, E. C., Patil, D. S., and Cammack, R., *Biochim. Biophys. Acta* **883**, 145 (1986).
41. Teixeira, M., Moura, I., Xavier, A. V., Huynh, B. H., DerVartanian, D. V., Peck, H. D., Jr., LeGall, J., and Moura, J. J. G., *J. Biol. Chem.* **260**, 8942 (1985).
42. Moura, J. J. G., Moura, I., Huynh, B. H., Krüger, H.-J., Teixeira, M., DuVarney, R. C., DerVartanian, D. V., Xavier, A. V., Peck, H. D., Jr., and LeGall, J., *Biochem. Biophys. Res. Commun.* **108**, 1388 (1982).
43. Van der Zwaan, J. W., Albracht, S. P. J., Fontijn, R. D., and Slater, E. C., *FEBS Lett.* **179**, 271 (1985).
44. Bossu, F. P., and Margerum, D. W., *Inorg. Chem.* **16**, 1210 (1977).
45. Sakurai, T., Hongo, J.-I., Nakahara, A., and Nakao, Y., *Inorg. Chim. Acta* **46**, 205 (1980).
46. Moore, G. D., Pettigrew, G. W., and Rogers, N. K., *Proc. Natl. Acad. Sci. U.S.A.* **83**, 4998 (1986).
47. Haines, R. I., and McAuley, A., *Coord. Chem. Rev.* **39**, 77 (1981).
48. Busch, D. H., *Acct. Chem. Res.* **11**, 392 (1978).
49. Fernandez, V. M., Hatchikian, E. C., and Cammack, R., *Biochim. Biophys. Acta* **832**, 69 (1985).
50. Dietrich-Buchecker, C. O., Kern, J.-M., and Sauvage, J. P., *J. Chem. Soc. Chem. Commun.* 760 (1985).
51. Lindahl, P. A., Kojima, N., Hausinger, R. P., Fox, J. A., Teo, B. K., Walsh, C. T., and Orme-Johnson, W. H., *J. Am. Chem. Soc.* **106**, 3062 (1984).
52. Scott, R. A., Wallin, S. A., Czechowski, M., DerVartanian, D. V., LeGall, J., Peck, H. D., Jr., and Moura, I., *J. Am. Chem. Soc.* **106**, 6864 (1984).
53. DerVartanian, D. V., Kruger, H. J., Peck, H. D., Jr., and LeGall, J., *Rev. Port. Quim.* **27**, 70 (1985).
54. Albracht, S. P. J., Kröger, A., Van der Zwaan, J. W., Unden, G., Böcher, R., Mell, H., and Fontijn, R. D., *Biochim. Biophys. Acta* **116** (1986).
55. Sugiura, Y., Kuwahara, J., and Suzuki, T., *Biochem. Biophys. Res. Commun.* **115**, 878 (1983).
56. Tan, S. L., Fox, J. A., Kojima, N., Walsh, C. T., and Orme-Johnson, W. H., *J. Am. Chem. Soc.* **106**, 3064 (1984).
57. Johnson, M. K., Zambrano, I. C., Czechowski, M. H., Peck, H. D. Jr., DerVartanian, D. V., and LeGall, J., *Biochem. Biophys. Res. Commun.* **128**, 220 (1985).
58. Kim, D.-H., and Holten, D., *Chem. Phys.* **75**, 305 (1983).

59. Shelnutt, J. A., Alston, K., Ho, J. Y., Yu, N. T., and Yamamoto, T., *Biochemistry* **25**, 620 (1986).
60. Findsen, E. W., Alston, K., Shelnutt, J. A., and Ondria, M. R., *J. Am. Chem. Soc.* **108**, 4009 (1986).
61. Tolman, C. A., *J. Am. Chem. Soc.* **94**, 2994 (1972).
62. Dilworth, J. R., *In* "Sulfur and its Relevance to Geo-, Bio- and Cosmosphere and Technology: Studies in Inorganic Chemistry" (A. Müller, ed.), Vol. 5. Elsevier, New York, 1984.
63. Schneider, K., Cammack, R., Schlegel, H. G., and Hall, D. O., *Biochim. Biophys. Acta* **578**, 445 (1979).
64. Van der Zwaan, J. W., Albracht, S. P. J., Fontijn, R. D., and Roelofs, Y. B. M., *Biochim. Biophys. Acta* **872**, 208 (1986).
65. Cammack, R., Patil, D. S., Hatchikian, E. C., and Fernandez, V. M., *Biochim. Biophys. Acta* **912**, 98 (1987).
66. Adams, M. W. W., Mortenson, L. E., and Chen, J.-S., *Biochim. Biophys. Acta* **594**, 105 (1981).
67. Van der Westen, H. M., Mayhew, S. G., and Veeger, C., *FEMS Microbiol. Lett.* **7**, 35 (1980).
68. Cammack, R., Lalla-Maharajh, W. V., and Schneider, K., *In* "Electron Transport and Oxygen Utilization" (C. Ho, ed.), Vol. 2. Elsevier, New York, 1982.
69. Huynh, B. H., Patil, D. S., Moura, I., Teixeira, M., Moura, J. J. G., DerVartanian, D. V., Czechowski, M. H., Prickril, B. C., Peck, H. D., Jr., and LeGall, J., *J. Biol. Chem.* **262**, 795 (1987).
70. Schneider, K., Patil, D. S., and Cammack, R., *Biochim. Biophys. Acta* **748**, 353 (1983).
71. Albracht, S. P. J., Van der Zwaan, J. W., and Fontijn, R. D., *Biochim. Biophys. Acta* **766**, 245 (1984).
72. Cammack, R., Patil, D., and Fernandez, V. M., *Biochem. Soc. Trans.* **13**, 572 (1985).
73. Coffman, R. E., and Buettner, G. R., *J. Phys. Chem.* **83**, 2392 (1979).
74. Berlier, Y. M., Fauque, G., Lespinat, P. A., and LeGall, J., *FEBS Lett.* **140**, 185 (1982).
75. Lissolo, T., Pulvin, S., and Thomas, D., *J. Biol. Chem.* **259**, 11725 (1984).
76. Albracht, S. P. J., Fontijn, R. D., and Van der Zwaan, J. W., *Biochim. Biophys. Acta* **832**, 89 (1985).
77. Hallahan, D. L., Fernandez, V. M., Hatchikian, E. C., and Cammack, R., *Biochim. Biophys. Acta* **874**, 72 (1986).
78. Lespinat, P. A., Berlier, Y., Fauque, G., Czechowski, M., Dimon, B., and LeGall, J., *Biochimie* **68**, 55 (1986).
79. Adams, M. W. W., Jin, S. L. C., Chen, J. S., and Mortenson, L. E., *Biochim. Biophys. Acta* **869**, 37 (1986).
80. Moura, J. J. G., Teixeira, M., Moura, I., and LeGall, J., *In* "Bioinorganic Chemistry of Nickel" (J. R. Lancaster, Jr., ed.). VCH Publ. Deerfield Beach, Florida, 1988 (in press).
81. Hoberman, H. D., and Rittenberg, D., *J. Biol. Chem.* **147**, 211 (1943).
82. Schneider, K., Schlegel, H. G., and Jochim, K., *Eur. J. Biochem.* **138**, 533 (1984).
83. Gunsalus, R. P., and Wolfe, R. S., *FEMS Microbiol. Lett.* **3**, 191 (1978).
84. Ellefson, W. L., Whitman, W. B., and Wolfe, R. S., *Proc. Natl. Acad. Sci. U.S.A.* **79**, 3707 (1982).
85. Hausinger, R. P., Orme-Johnson, W. H., and Walsh, K. M., *Biochemistry* **23**, 801 (1984).
86. Diekert, G., Klee, B., and Thauer, R. K., *Arch. Microbiol.* **124**, 103 (1980).
87. Whitman, W. B., and Wolfe, R. S., *Biochem. Biophys. Res. Commun.* **92**, 1196 (1980).

88. Pfalz, A., Jaun, B., Fässler, A., Eschenmoser, A., Jaenchen, R., Gilles, H. H., Diekert, G., and Thauer, R. K., *Helv. Chim. Acta* **65**, 828 (1982).
89a. Scott, R. A., Hartzell, P. L., Wolfe, R. S., LeGall, J., and Cramer, S. P., *In* "Frontiers in Bioinorganic Chemistry" (A. V. Xavier, ed.), p. 20. VCH Publ., Weinheim, FRG, 1985.
89b. Diakun, G. P., Piggott, B., Tinto, H. J., Fuchs, D. A., and Thauer, R. K., *Biochem. J.* **232**, 281 (1985).
89c. Eidsness, M. K., Sullivan, R. J., Schwartz, J. R., Hartzell, P. L., Wolfe, R. S., Flank, A. M., Cramer, S. P., and Scott, R. A., *J. Am. Chem. Soc.* **108**, 3120 (1986).
90. Albracht, S. P. J., Ankel-fuchs, D., Van der Zwaan, J. W., Fontijn, R. D., and Thauer, R. K., *Biochim. Biophys. Acta* **870**, 50 (1986).
91. Keltjens, J. T., Whitman, W. B., Caerteling, C. G., van Kooten, A. M., Wolfe, R. S., and Vogels, G. D., *Biochem. Biophys. Res. Commun.* **108**, 495 (1982).
92. Diekert, G., and Thauer, R. K., *J. Bacteriol.* **136**, 597 (1978).
93. Diekert, G., Graf, E. G., and Thauer, R. K., *Arch. Microbiol.* **122**, 117 (1979).
94. Ragsdale, S. W., and Wood, H. G., *J. Biol. Chem.* **260**, 3970 (1985).
95. Ragsdale, S. W., Ljungdahl, L. G., and DerVartanian, D. V., *Biochem. Biophys. Res. Commun.* **108**, 658 (1982).
96. Ragsdale, S. W., Ljungdahl, L. G., and DerVartanian, D. V., *Biochem. Biophys. Res. Commun.* **115**, 658 (1983).
97. Ragsdale, S. W., Wood, H. G., and Antholine, W. E., *Proc. Natl. Acad. Sci. U.S.A.* **82**, 6811 (1985).
98. Ragsdale, S. G., Wood, H. G., Morton, T. A., Ljungdahl, L. G., and DerVartanian, D. V., *In* "Bioinorganic Chemistry of Nickel" (J. R. Lancaster, Jr., ed.). VCH Publ., Deerfield Beach, Florida, 1988 (in press).
99. Mege, R.-M., and Bourdillon, C., *J. Biol. Chem.* **260**, 14701 (1985).
100. Cammack, R., Rao, K. K., Serra, J., and Llama, M. J., *Biochimie* **68**, 93 (1986).
101. Cammack, R., Kovacs, K., McCracken, J., and Peisach, J., unpublished observations.
102. Fischer, F., Lieske, R., and Winzer, K., *Biochem. Z.* **245**, 2 (1932).
103. Ragsdale, S. W., Wood, H. G., Morton, T. A., Ljungdahl, L. G., and DerVartanian, D. V., *In* "Bioinorganic Chemistry of Nickel" (J. R. Lancaster, Jr., ed.). VCH Publ., Deerfield Beach, Florida, 1988 (in press).
104. Crabtree, R. H., *Inorg. Chim. Acta* **125**, L7-L8 (1986).

NITROSYL COMPLEXES OF IRON–SULFUR CLUSTERS

ANTHONY R. BUTLER,* CHRISTOPHER GLIDEWELL,* and MIN-HSIN LI**

*Department of Chemistry, University of St. Andrews, Fife KY16 9ST, Scotland, and
**Cancer Institute, Chinese Academy of Medical Sciences, Beijing, People's Republic of China

I. Introduction
II. Synthesis
   A. $[Fe_4S_3(NO)_7]^-$ and $[Fe_2S_2(NO)_4]^{2-}$
   B. $[Fe_2(SR)_2(NO)_4]$
   C. $[Fe_4S_4(NO)_4]$ and Related Cubane-Type Clusters
   D. Other Fe–S–NO Complexes
   E. Selenium Analogs
   F. Tellurium Analogs
   G. Heterometallic Iron–Sulfur–Nitrosyl Clusters
III. Molecular Structure: X-Ray Crystallography
   A. $[Fe_4S_3(NO)_7]^-$ and $[Fe_2S_2(NO)_4]^{2-}$
   B. $[Fe_2(SR)_2(NO)_4]$
   C. $[Fe_4S_4(NO)_4]$ and Related Clusters
   D. $[Fe(NO)(S_2CNR_2)_2]$ and Related Complexes
   E. Selenium and Tellurium Analogs
   F. Heterometallic Iron–Sulfur–Nitrosyls
IV. Molecular Structure: NMR Spectroscopy
   A. ^1H NMR Spectroscopy of $[Fe_2(SR)_2(NO)_4]$
   B. ^{15}N NMR Spectroscopy of $[Fe_2(SR)_2(NO)_4]$
   C. ^{15}N NMR Spectroscopy of Tetrairon Complexes
V. Electronic Structure
   A. Oxidation States
   B. Dinuclear Complexes
   C. Tetranuclear Complexes
   D. Mononuclear Complexes
VI. Chemical Reactivity
   A. Redox Reactions
   B. Conversion of Di- and Tetrairon Complexes into Paramagnetic Monoiron Complexes
   C. Ligand Substitution Reactions
   D. Nitrosylation Reactions
   E. Analytical Applications
VII. Biological Chemistry
   A. Cancerous States and $g = 2.03$ Complexes
   B. The Antimicrobial Activity of $[Fe_4S_3(NO)_7]^-$ Salts
   C. $[Fe_2(SMe)_2(NO)_4]$ as a Natural Product
   D. $[Fe_2(SMe)_2(NO)_4]$ (Roussin Esters) and Chemical Carcinogenesis
References

I. Introduction

The first iron–sulfur–nitrosyl complexes, in salts now known to contain the anions $[Fe_2S_2(NO)_4]^{2-}$ and $[Fe_4S_3(NO)_7]^-$, were described by Roussin in 1858 (*1*). Although Roussin did not know the precise compositions of the new salts, far less their constitutions, he was convinced of their close relationship to the nitroprusside anion $[Fe(CN)_5NO]^{2-}$, which had only recently been described by Playfair (*2*). The empirical formulas of a range of salts $M_2^I[Fe_2S_2(NO)_4]$ and $M^I[Fe_4S_3(NO)_7](M^I = Na, K, Tl, or NH_4)$ were determined by Pavel in 1882 (*3*) as a result of some painstaking elemental analyses. Pavel also prepared the first organic derivative $[Fe_2(SEt)_2(NO)_4]$ by alkylation, using $K_2[Fe_2S_2(NO)_4]$, and established its empirical formula. Subsequent molecular weight determinations on a range of organic derivatives, which had been prepared by reaction of thiols and iron(II) salts with nitric oxide (*4*), showed that these compounds had dimeric molecular formulas $[Fe_2(SR)_2(NO)_4]$. Subsequent work early in the present century confirmed the dimeric nature of $[Fe_2(SR)_2(NO)_4]$ (*5*), and demonstrated that both $[Fe_2(SR)_2(NO)_4]$ and salts containing $[Fe_2S_2(NO)_4]^{2-}$ or $[Fe_4S_3(NO)_7]^-$ are diamagnetic (*6*).

Based upon the molecular formulas, but upon no other evidence, a range of structural formulas were proposed (*5, 7–9*), including structures containing *cis*-hyponitrite ligands (*5*) and bridging NO groups (*9*). Definitive structural formulas were established only by the simultaneous publication in 1958, exactly 100 years after Roussin's original description of these complexes (*1*), of X-ray structure determinations for $[Fe_2(SEt)_2(NO)_4]$ (*10*) and $Cs[Fe_4S_3(NO)_7] \cdot H_2O$ (*11*).

The recent upsurge of interest in iron–sulfur–nitrosyl complexes has been stimulated in part by the reported isolation of $[Fe_2(SMe)_2(NO)_4]$ from natural sources (*12*), by the obvious resemblances between these complexes and the naturally occurring [2Fe–2S] and [4Fe–4S] clusters of iron–sulfur proteins (*13, 14*), and by the connections between tetrairon–sulfur–nitrosyls and cubane-type clusters (*15*). Most of the work in this area has been published in the past 5 years or so, and no review has previously been made. However, a number of excellent reviews of the wider aspects of metal–nitrosyl chemistry have appeared (*16–19*).

Salts containing the anions $[Fe_2S_2(NO)_4]^{2-}$ and $[Fe_4S_3(NO)_7]^-$ are often called, from their colors, Roussin's red and black salts, respectively: similarly the organic derivatives $[Fe_2(SR)_2(NO)_4]$ are often called esters of Roussin's red salt. These names are nonsystematic and, in the case of $[Fe_2(SR)_2(NO)_4]$, scarcely accurate; nonetheless they are convenient and they carry the sanction of long usage, and will on occasion be used here.

II. Synthesis

A. $[Fe_4S_3(NO)_7]^-$ AND $[Fe_2S_2(NO)_4]^{2-}$

Salts containing these two anions, structures **1** and **2**, were obtained in impure form by Roussin in 1858 (*1*): the anion $[Fe_4S_3(NO)_7]^-$ is conveniently synthesized using a modification (*20*) of Roussin's method.

Reaction of iron(II) sulfate with a mixture of sodium nitrite and ammonium sulfide yields an intensely black solution from which $NH_4[Fe_4S_3(NO)_7]$ can be crystallized: alternatively, since salts of $[Fe_4S_3(NO)_7]^-$ are all soluble in diethyl ether, the crude reaction mixture can be extracted with ether to yield the product. Other salts can readily be prepared by metathesis.

The conversion of $[Fe(H_2O)_6]^{2+}$ into $[Fe_4S_3(NO)_7]^-$ in this reaction requires reduction of iron from iron(II) to a mean oxidation number of $-\frac{1}{2}$ (see Section V,C, for discussion of the bonding in $[Fe_4S_3(NO)_7]^-$), spontaneous self-assembly to form the tetranuclear anion, and spin coupling to yield a diamagnetic ground state. Neither the mechanism nor the stoichiometry of this reaction is clear, although Brauer (*20*) gives an equation in which 170 mol of reactants give 102 mol of products (excluding water of crystallization).

Salts of $[Fe_2S_2(NO)_4]^{2-}$ are best prepared by Pavel's modification (*3*) of Roussin's original method (*1*), which is again described by Brauer (*20*). Reaction of $NH_4[Fe_4S_3(NO)_7]$ with aqueous sodium hydroxide yields $Na_2[Fe_2S_2(NO)_4]$, from which other salts may be prepared by metathesis: $(Et_4N)_2[Fe_2S_2(NO)_4]$ is conveniently prepared from $NH_4[Fe_4S_3(NO)_7]$ by reaction with Et_4NOH (*21*). While $Na_2[Fe_2S_2(NO)_4]$ is insoluble in ether and similar solvents, unlike $M^I[Fe_4S_3(NO)_7]$, salts such as $(R_4N)_2[Fe_2S_2(NO)_4]$ and $(Ph_3PNPPh_3)_2[Fe_2S_2(NO)_4]$ are readily soluble in solvents such as ether and ketones.

The mechanism of conversion of $[Fe_4S_3(NO)_7]^-$ into $[Fe_2S_2(NO)_4]^{2-}$ almost certainly involves fragmentation to mononuclear species, followed by reassembly. At neutral pH, $[Fe_4S_3(NO)_7]^-$ gives no ESR spectrum, but upon increasing the pH to 11, an ESR spectrum appears (22), which has subsequently been identified (23) as that of the mononuclear species $[Fe(NO)_2(SH)_2]^-$. As discussed in Section VI,B, such complexes play a central role in many of the reactions of both diiron and tetrairon–sulfur–nitrosyls, particularly those involving ligand substitution and change of nuclearity.

Just as reaction of $[Fe_4S_3(NO)_7]^-$ with strong alkali gives $[Fe_2S_2(NO)_4]^{2-}$, so also reaction of $[Fe_2S_2(NO)_4]^{2-}$ with strong acids, such as CF_3COOH or HBF_4, yields (24) not only $[Fe_2(SH)_2(NO)_4]$, the "acid" corresponding to Roussin's red salt, but also $[Fe_4S_3(NO)_7]^-$. A more surprising conversion of $[Fe_2S_2(NO)_4]^{2-}$ into $[Fe_4S_3(NO)_7]^-$ is achieved (21, 25, 26) simply by dissolving a salt of $[Fe_2S_2(NO)_4]^{2-}$ in methylene chloride. Using ^{15}N-labeled material, the sole species detectable by ^{15}N NMR spectroscopy is $[Fe_4S_3(^{15}NO)_7]^-$ (21, 25), while, on a preparative scale (26), salts of $[Fe_4S_3(NO)_7]^-$ can be isolated in yields of around 60%.

Both $[Fe_4S_3(NO)_7]^-$ and $[Fe_2S_2(NO)_4]^{2-}$ can also be synthesized by carbonyl displacements from appropriate iron–carbonyl precursors. The complex $[Fe_4S_3(NO)_7]^-$ can in fact be formed from mono-, di-, or trinuclear iron–carbonylate species. Salts of $[Fe(CO)_3NO]^-$ react with elemental sulfur or with polysulfide to yield $[Fe_4S_3(NO)_7]^-$ (23, 27), while nitrosylations using sodium nitrite in aqueous ethanol of $[Fe_2S_2(CO)_6]$, $[Fe_2S_2(CO)_6]^{2-}$, and $[Fe_3S_2(CO)_9]$ all give $[Fe_4S_3(NO)_7]^-$ in yields of ~80% (27); in addition, the reaction with $[Fe_2S_2(CO)_6]^{2-}$ yields $[Fe_2S_2(NO)_4]^{2-}$ (9%).

These transformations are summarized in Scheme 1.

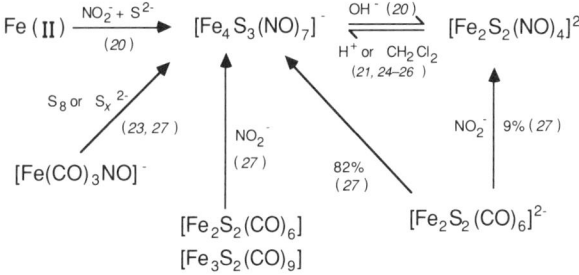

SCHEME 1. Synthetic routes to $[Fe_4S_3(NO)_7]^-$ and $[Fe_2S_2(NO)_4]^{2-}$. Numbers in parentheses are references.

B. [Fe$_2$(SR)$_2$(NO)$_4$]

Neutral [Fe$_2$(SR)$_2$(NO)$_4$] complexes, **3a** and **3b**, having a wide range of substituents R, are now known. When R is alkyl or aryl, the complexes can be synthesized from mono-, di-, or tetrairon precursors, and this forms a convenient basis for the classification of synthetic routes to [Fe$_2$(SR)$_2$(NO)$_4$].

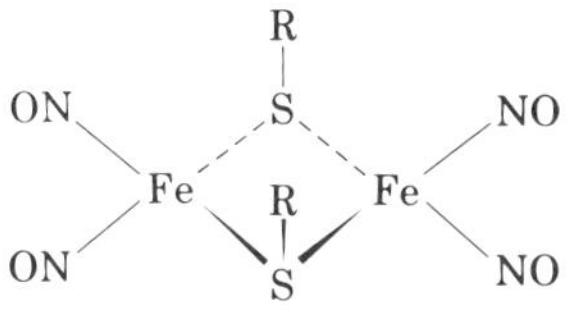

3a 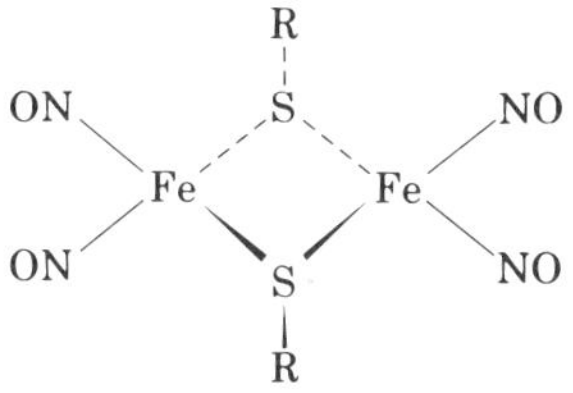 **3b**

1. Synthesis from Mononuclear Precursors

a. Iron(II) Salts. Reaction of nitric oxide with a mixture of an iron(II) salt, conveniently iron(II) sulfate, and a thiol RSH in strongly alkaline aqueous solution gives the corresponding complex [Fe$_2$(SR)$_2$(NO)$_4$] in yields around 80% (*4, 5, 12, 20*) [Eq. (1)].

$$4\text{FeSO}_4 + 8\text{KOH} + 2\text{RSH} + 4\text{NO} \longrightarrow$$
$$[\text{Fe}_2(\text{SR})_2(\text{NO})_4] + \text{Fe}_2\text{O}_3 + 4\text{K}_2\text{SO}_4 + 5\text{H}_2\text{O} \quad (1)$$

b. [Fe(CO)$_2$(NO)$_2$]. The mixed dicarbonyl–dinitrosyl reacts with PhSH to yield [Fe$_2$(SPh)$_2$(NO)$_4$] (*28*), and with Me$_2$S$_2$ to give [Fe$_2$(SMe)$_2$(NO)$_4$] [Eq. (2)] (*27*).

$$2[\text{Fe}(\text{CO})_2(\text{NO})_2] + \text{Me}_2\text{S}_2 \longrightarrow [\text{Fe}_2(\text{SMe})_2(\text{NO})_4] + 4\text{CO} \quad (2)$$

In each case, it is the carbonyl ligands which are displaced rather than the nitrosyl ligands. In a similar manner, the reaction of polysulfide (*23, 27*) with [Fe(CO)$_3$NO]$^-$ causes displacement of carbonyl, rather than nitrosyl, to yield [Fe$_4$S$_3$(NO)$_7$]$^-$ (see Section II,A).

c. [Fe(NO)(S$_2$CNMe$_2$)$_2$]. The reaction of the mononitrosyl complex with an excess of MeS$^-$ provides the dinitrosyliron product [Fe$_2$(SMe)$_2$(NO)$_4$] in 24% yield (*23*); this reaction proceeds via the mononuclear dinitrosyl [Fe(NO)$_2$(SR)$_2$]$^-$ and illustrates the ready interchange between Fe(NO)$^{2+}$ fragments and Fe(NO)$_2^+$ fragments, which will be discussed further in Section VI,B.

d. *[Fe(CO)₃NO]⁻*. In another reaction illustrating the conversion of a mononitrosyliron source into a dinitrosyliron product by the action of RS⁻, (Ph₃PNPPh₃)[Fe(CO)₃NO] gave with MeSNa a 78% isolated yield of [Fe₂(SMe)₂(NO)₄] (*23*). As with the corresponding reaction of [Fe(NO)(S₂CNMe₂)₂] (see above), the reaction proceeds via [Fe(NO)₂(SMe)₂]⁻. Similarly, reactions of [Fe(CO)₃NO]⁻ with *i*-PrS⁻ or *t*-BuS⁻ gave [Fe(NO)₂(SPr^i)₂]⁻ and [Fe(NO)₂(SBu^t)₂]⁻, respectively, as precursors (see below) of the corresponding dinuclear [Fe₂(SR)₂(NO)₄] (*23*).

e. *[Fe(NO)₂(SR)₂]⁻*. Any process which gives rise to mononuclear complexes of the type [Fe(NO)₂(SR)₂]⁻ (R ≠ H) can also yield dinuclear [Fe₂(SR)₂(NO)₄] (*23*). The mononuclear species are favored by highly polar environments such as DMF solution, but addition of toluene or a similar noncoordinating solvent of low polarity causes loss of RS⁻ and dimerization [Eq. (3)]; the reaction is reversible (*23*).

$$2[Fe(NO)_2(SR)_2]^- \underset{DMF, +2RS^-}{\overset{toluene, -2RS^-}{\rightleftharpoons}} [Fe_2(SR)_2(NO)_4] \qquad (3)$$

2. Synthesis from Dinuclear Precursors

a. *[Fe₂S₂(NO)₄]²⁻*. Probably the most versatile route to neutral [Fe₂(SR)₂(NO)₄] complexes is provided by the reaction between [Fe₂S₂(NO)₄]²⁻, acting as nucleophile, and molecular halides, RX, acting as electrophiles [Eq. (4)]. When R is an organic group this method is restricted to groups R containing a saturated α-carbon atom, but subject to this constraint has been widely employed for R = CH₃ (*24, 29*), Et (*3, 29*), C₃H₅(allyl) (*29*), PhCH₂ (*29, 30*), HC≡CCH₂, Me₃SiCH₂, and CH₃C(O)CH₂ (*29*).

$$[Fe_2S_2(NO)_4]^{2-} + 2RX \longrightarrow 2X^- + [Fe_2(SR)_2(NO)_4] \qquad (4)$$

Organometallic halides may also be employed as the electrophile in Eq. (4); thus by use of Ph₃SnCl, Me₃SnBr, Ph₃PbBr, PhHgCl (*29*), or CH₃HgCl (*31*), the corresponding [Fe₂(SMR$_x$)₂(NO)₄] complexes were obtained in yields ranging from 33% for [Fe₂(SHgPh)₂(NO)₄] to 99% for [Fe₂(SPbPh₃)₂(NO)₄] (*29*). These derivatives are all reddish crystalline solids, usually somewhat air sensitive as solids, and very air sensitive in solution.

The use of organic gem dihalides as electrophiles does not appear to have been reported, although no success was reported in attempts to react [Fe₂S₂(NO)₄]²⁻ with Me₂SnCl₂ or Ph₂PbCl₂ (*29*), presumably because the S····S bite is too large for effective chelation to tin(IV) or lead(IV).

For successful reaction of $[Fe_2S_2(NO)_4]^{2-}$ at transition metal centers a strongly electrophilic metal is required; thus while no reaction was observed (29) with $[(\eta^5\text{-}C_5H_5)Fe(CO)_2Br]$, the more electrophilic cationic complex $[(\eta^5\text{-}C_5H_5)Fe(CO)_2(THF)]BF_4$ reacted to provide a 90% yield of the air-stable $[Fe_2\{SFe(CO)_2(\eta^5\text{-}C_5H_5)\}_2(NO)_4]$.

Just as group-14 gem dihalides appear to be size limited in their reactions with $[Fe_2S_2(NO)_4]^{2-}$, so too do cis dihalide complexes of the group-10 metals. Although cis-$[(Ph_3P)_2PtCl_2]$ reacts with $[Fe_2S_2(NO)_4]^{2-}$ to provide a 95% yield of $[Fe_2\{S_2Pt(PPh_3)_2\}(NO)_4]$ (32), attempts to prepare analogous complexes containing the smaller homologs nickel and palladium failed (29). However, reaction with $[NiCl_2(Ph_2PCH_2CH_2PPh_2)]$ gave (30) $[Fe_2\{S_2Ni(Ph_2PCH_2CH_2PPh_2)\}(NO)_4]$. In contrast to the formation of this platinum complex, which proceeds simply by nucleophilic displacement of chloride by the sulfur in $[Fe_2S_2(NO)_4]^{2-}$, reaction of $[Fe_2S_2(NO)_4]^{2-}$ with the cobalt gem dihalides $[(\eta^5\text{-}C_5R_5)Co(CO)I_2]$ (R = H or Me) yielded the novel trinuclear cluster $[Fe(\mu_3\text{-}S)_2\{Co(\eta^5\text{-}C_5R_5)\}_2(NO)_2]$ (29), in which the iron is bound not only to two terminal NO ligands but to two μ_3-sulfur ligands and to two cobalt centers.

With the proton as electrophile, the primary product from $[Fe_2S_2(NO)_4]^{2-}$ is $[Fe_2(SH)_2(NO)_4]$ as noted earlier (Section II,A), but this is susceptible to decomposition, yielding $[Fe_4S_3(NO)_7]^-$ (24).

The reactions of $[Fe_2S_2(NO)_4]^{2-}$ with electrophiles are summarized in Scheme 2.

b. *[Fe_2(SR)_2(CO)_6]*. When R is an organic group, the reaction of $[Fe_2S_2(NO)_4]^{2-}$ with RX cannot always be applied; for example, if R = aryl then RX will not react. However, the carbonyl complexes

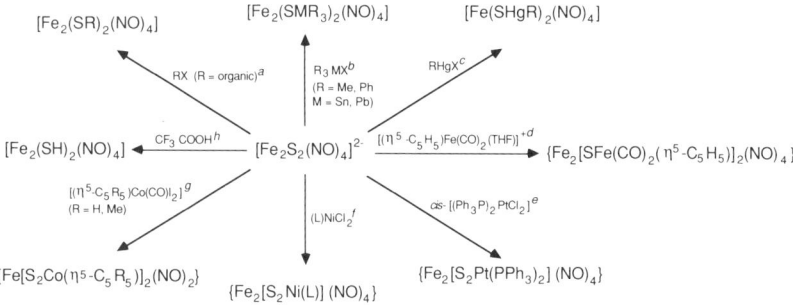

SCHEME 2. Reactions of $[Fe_2S_2(NO)_4]^{2-}$ with electrophiles. [a]Refs. 3, 24, 29, and 30. [b]Ref. 29. [c]Refs. 29 and 31. [d]Ref. 29. [e]Ref. 32. [f]Ref. 30: L = $Ph_2PCH_2CH_2PPh_2$. [g]Ref. 29. [h]Ref. 24.

$[Fe_2(SR)_2(CO)_6]$, which are readily available from the reactions of R_2S_2 with $[Fe_3(CO)_{12}]$ (33) and of RSH with $[Fe_2(CO)_9]$ (34), can be nitrosylated in reactions which replace three carbonyl ligands by two nitrosyl ligands. Nitrosylation can be effected either by use of nitric oxide gas (27, 29) or by use of sodium nitrite in aqueous ethanol (27) or, better, in dimethylformamide (DMF) (25). It is usually more convenient to employ nitric oxide if complexes of normal isotopic composition are required, but for ^{15}N labeling the use of sodium nitrite in DMF is the more convenient and economical route.

The mechanism of the reaction of $[Fe_2(SR)_2(CO)_6]$ with nitric oxide does not seem to have been investigated. The corresponding reaction with nitrite in DMF proceeds (25) via the mononuclear complex $[Fe(NO)_2(SR)_2]^-$ (see Section II,B,1), but it is not yet known how this complex is formed from $[Fe_2(SR)_2(CO)_6]$.

c. *$[Fe_2I_2(NO)_4]$*. This iodo-bridged complex reacts in tetrahydrofuran (THF) solution with thiols (RSH) in the presence of a base B (30), conveniently triethylamine, to yield the corresponding $[Fe_2(SR)_2(NO)_4]$ [Eq. (5)].

$$[Fe_2I_2(NO)_4] + 2RSH + 2B \longrightarrow [Fe_2(SR)_2(NO)_4] + 2BH^+I^- \qquad (5)$$

Study of this reaction by EPR spectroscopy (35) has shown that it proceeds via the mononuclear complexes $[Fe(NO)_2I]$ and $[Fe(NO)_2(SR)_2]^-$.

As well as introducing this fairly straightforward use of $[Fe_2I_2(NO)_4]$ and its extension, by reaction with PhELi (E = S, Se, or Te) [Eq. (6)], Rauchfuss and Weatherill have developed (30) an ingenious synthesis of $[Fe_2E_2(NO)_4]^{2-}$ [Eq. (7)].

$$[Fe_2I_2(NO)_4] + 2PhELi \longrightarrow 2LiI + [Fe_2(EPh)_2(NO)_4] \qquad (6)$$

$$[Fe_2I_2(NO)_4] + 2Li_2E \longrightarrow 2LiI + Li_2[Fe_2E_2(NO)_4] \qquad (7)$$

The products $[Fe_2E_2(NO)_4]^{2-}$ can be alkylated in the usual way (see Section II,B,2). The advantage of using $[Fe_2I_2(NO)_4]$ as starting material lies in the fact that stepwise displacement of the μ-iodo ligands is possible, by either E^{2-} or by RE^-; thus a mixed chalcogen derivative such as $[Fe_2(SCH_2Ph)(SePh)(NO)_4]$ or $[Fe_2(SCH_2Ph)(SeCH_2Ph)(NO)_4]$ can be synthesized [Eqs. (8) and (9)].

$$[Fe_2I_2(NO)_4] \xrightarrow{PhSeLi} [Fe_2I(SePh)(NO)_4] \xrightarrow{PhCH_2SH/Et_3N}$$
$$[Fe_2(SCH_2Ph)(SePh)(NO)_4] \qquad (8)$$

$$[Fe_2I_2(NO)_4] \xrightarrow{Se^{2-}} [Fe_2I(Se)(NO)_4]^- \xrightarrow{PhCH_2SH/Et_3N}$$
$$[Fe_2(SCH_2Ph)(Se)(NO)_4]^- \xrightarrow{PhCH_2Cl} [Fe_2(SCH_2Ph)(SeCH_2Ph)(NO)_4] \quad (9)$$

3. Synthesis from Tetranuclear Precursors

As noted in Section II,B,1, any process which leads to the mononuclear anionic intermediates $[Fe(NO)_2(SR)_2]^-$ can also provide, by loss of RS^-, the neutral dinuclear $Fe_2(SR)_2(NO)_4$. Both $[Fe_4S_3(NO)_7]^-$ and $[Fe_4S_4(NO)_4]$ have been shown (23) by EPR spectroscopy to react with MeS^-, yielding $[Fe(NO)_2(SMe)_2]^-$; the corresponding dinuclear $[Fe_2(SMe)_2(NO)_4]$ has been isolated (23) from these reactions in yields of 46 and 13%, respectively.

C. $[Fe_4S_4(NO)_4]$ AND RELATED CUBANE-TYPE CLUSTERS

1. $[Fe_4S_4(NO)_4]$ and $[Fe_4S_4(NO)_4]^-$

The reaction between $Hg[Fe(CO)_3NO]_2$ and elemental sulfur in refluxing toluene provides (36, 37) practical yields of the cubane-type tetranuclear $[Fe_4S_4(NO)_4]$ [Eq. (10)].

$$2Hg[Fe(CO)_3NO]_2 + \tfrac{3}{4}S_8 \longrightarrow [Fe_4S_4(NO)_4] + 2HgS + 12CO \quad (10)$$

This synthesis provides another example, comparable to that of $[Fe_4S_3(NO)_7]^-$, of the spontaneous self-assembly of a tetranuclear species 4 from a mononuclear precursor in a one-stage procedure.

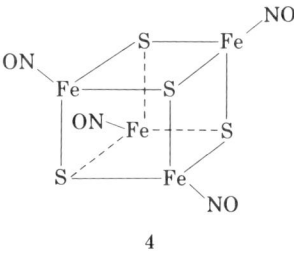

4

It is instructive, in this context, to compare the case of assembly of both nitrosylated iron–sulfur tetranuclear clusters and the related clusters with pendant thiolate groups (13), with the laborious and stepwise initial synthesis (38) of cubane C_8H_8; this comparison is, of course, simply one between thermodynamically controlled processes, in the case of the iron–sulfur systems, and kinetically controlled reactions for C_8H_8 synthesis.

The identical [Fe$_4$S$_4$(NO)$_4$] can also be synthesized more conveniently by reflux of Na[Fe$_4$S$_3$(NO)$_7$] with elemental sulfur in toluene (27); it is not yet clear whether this process involves simply the completion of the Fe$_4$S$_4$ framework by addition of one atom of sulfur to the existing Fe$_4$S$_3$ framework, or whether it proceeds by means of fragmentation, perhaps to mononuclear intermediates, followed by subsequent reformation of the iron–sulfur skeleton. Whatever the mechanism, it is clear that the formation of [Fe$_4$S$_4$(NO)$_4$] from [Fe$_4$S$_3$(NO)$_7$]$^-$ is subject to subtle factors involving both the counterion and the solubility of the products and reactants, since even on prolonged (>7 days) reflux in toluene the salt (Ph$_3$PNPPh$_3$)-[Fe$_4$S$_3$(NO)$_7$] did not react with sulfur to yield any [Fe$_4$S$_4$(NO)$_4$]. A similar contrast in fact obtains in the reactions with sulfur and polysulfide with the mercurial Hg[Fe(CO)$_3$NO]$_2$: with elemental sulfur in toluene the product is [Fe$_4$S$_4$(NO)$_4$] (36, 37), but with polysulfide in methanol [Fe$_4$S$_3$(NO)$_7$]$^-$ is formed (27).

Electrochemical reduction—or chemical reduction by [(η^5-C$_5$H$_5$)$_2$Co], sodium amalgam, or potassium benzophenone ketyl—of [Fe$_4$S$_4$(NO)$_4$] provides the monoanion [Fe$_4$S$_4$(NO)$_4$]$^-$ (37). Serious difficulty was experienced (37) in the isolation and characterization of salts of [Fe$_4$S$_4$(NO)$_4$]$^-$, and earlier attempts resulted instead in the isolation of salts of [Fe$_4$S$_3$(NO)$_7$]$^-$ (37, 39). The sequence summarized in Eq. (11) must again involve a major structural reorganization:

$$[\text{Fe}_4\text{S}_4(\text{NO})_4] \xrightarrow{e} [\text{Fe}_4\text{S}_4(\text{NO})_4]^- \longrightarrow$$
$$[\text{Fe}_4\text{S}_3(\text{NO})_7]^- + \text{unidentified by-products} \quad (11)$$

It is probable, although not proved, that the transformation of [Fe$_4$S$_4$(NO)$_4$]$^-$ to [Fe$_4$S$_3$(NO)$_7$]$^-$ proceeds by means of fragmentation to mononuclear intermediates, followed by reassembly.

The cubane-type cluster [Fe$_4$S$_4$(NO)$_4$] gives [Fe$_4$S$_3$(NO)$_7$]$^-$ not only upon reduction but also upon aerial oxidation in THF solution (37). This immediately suggests, because of the instability of [Fe$_4$S$_4$(NO)$_4$] with respect to fragmentation upon both electron addition and electron removal, that the frontier orbitals of [Fe$_4$S$_4$(NO)$_4$] are localized in an electron-precise Fe$_4$ framework; the bonding in [Fe$_4$S$_4$(NO)$_4$] and like clusters is discussed in Section V,C.

2. *[Fe$_4$S$_2$(NO)$_4$(NCMe$_3$)$_2$] and [Fe$_4$S$_2$(NO)$_4$(NCMe$_3$)$_2$]$^-$*

Formally isoelectronic with [Fe$_4$S$_4$(NO)$_4$] is the as yet unknown [Fe$_4$(NO)$_4$(NCMe$_3$)$_4$]. Attempts to prepare this latter cluster, in order to effect comparison with the known (40) stoichiometric analog

NITROSYL COMPLEXES OF IRON–SULFUR CLUSTERS 345

[Co$_4$(NO)$_4$(NCMe$_3$)$_4$], by reaction of Hg[Fe(CO)$_3$NO]$_2$ with the sulfur diimide (Me$_3$CN)$_2$S under a wide range of conditions yielded (36, 41) instead [Fe$_4$S$_2$(NO)$_4$(NCMe$_3$)$_2$], in which only two of the triply bridging sulfur ligands present in [Fe$_4$S$_4$(NO)$_4$] have formally been replaced by μ_3-NCMe ligands. For this complex also, sodium amalgam reduction yields (41) the monoanion [Fe$_4$S$_2$(NO)$_4$(NCMe$_3$)$_2$]$^-$, which appears to be less prone to rearrangement than [Fe$_4$S$_4$(NO)$_4$]$^-$.

D. Other Fe–S–NO Complexes

1. "[Fe$_3$S$_2$(NO)$_5$]$^-$"

Iron–sulfur systems containing three iron atoms are uncommon, and so the reported synthesis (42) of such a trinuclear anion [Fe$_3$S$_2$(NO)$_5$]$^-$ was of great interest, particularly as it was also reported that this anion, although an even-electron species, gave ESR spectra under a wide range of conditions.

It is therefore unfortunate that attempts to repeat the synthesis of salts containing [Fe$_3$S$_2$(NO)$_5$]$^-$ according to the published (42) procedure have all, without exception, yielded the corresponding salts of [Fe$_4$S$_3$(NO)$_7$]$^-$ (43). Neither could [Fe$_3$S$_2$(NO)$_5$]$^-$ be identified as a nitrosylation product of the carbonyl cluster [Fe$_3$S$_2$(CO)$_9$] in which the Fe$_3$S$_2$ framework is preformed (27).

Other aspects of the report (42) on [Fe$_3$S$_2$(NO)$_5$]$^-$ are surprising. Elemental analysis of the ammonium salt was reported to distinguish between iron(II) and iron(III) in [Fe$_3$S$_2$(NO)$_5$]$^-$, but to find these two types of iron present in equal numbers is most unusual for a triiron complex. Second, the molecular weight of the potassium salt was measured as 420 by mass spectrometry. This value is close to the M/Z of 421 calculated for the most abundant isotopic form of the ion-pair cation [KFe$_3$S$_2$(NO)$_5$]$^+$. Finally, the ESR spectrum reported is that of a dinitrosyliron species, which bears a remarkable resemblance to that reported (22) for a complex formed from Fe(II) and nitric oxide in aqueous alkaline solution.

Until further evidence is forthcoming it is necessary, therefore, to treat the report (42) of the preparation and properties of [Fe$_3$S$_2$(NO)$_5$]$^-$ with some caution.

2. [Fe$_2$(S$_2$O$_3$)$_2$(NO)$_4$]$^{2-}$

The synthesis of salts containing this anion was first reported in 1895 by Hofmann and Wiede (4), and subsequently confirmed by

Manchot (44) and by Brauer (45). Reaction of an iron(II) salt with potassium thiosulfate and nitric oxide yields the required anion [Eq. (12)].

$$2FeSO_4 + 4K_2S_2O_3 + 4NO \longrightarrow K_2[Fe_2(S_2O_3)_2(NO)_4] + K_2S_4O_6 + 2K_2SO_4 \quad (12)$$

This is similar to the synthesis of the neutral Roussin esters [$Fe_2(SR)_2$-$(NO)_4$] from iron(II), RSH, and nitric oxide [Eq. (1), Section II,B,1], except that in Eq. (1) the iron(II) initially present undergoes disproportionation to Fe(I), which is present in the product [$Fe_2(SR)_2(NO)_4$], and Fe(III), which is removed as oxide. In Eq. (12), by contrast, the iron(II) is reduced to Fe(I) by excess of thiosulfate, which is itself oxidized to tetrathionate. Apart from its diamagnetism (4), practically nothing is known about this anion.

3. [Fe(NO)(S₂CNR₂)₂] and Related Complexes

Reaction of an iron(II) salt, conveniently the sulfate, with sodium dialkyldithiocarbamate [R_2NCS_2]Na under nitric oxide provides excellent yields (>85%) of the iron(I) complexes [$Fe(NO)(S_2CNR_2)_2$] (structure 5) (46). Oxidation of the diethyldithiocarbamate derivative with

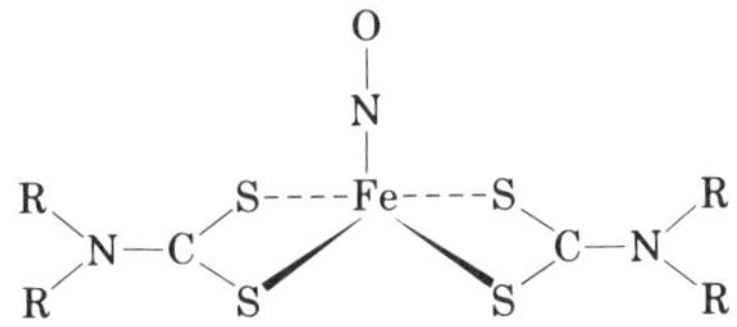

molecular bromine or iodine or with nitrogen(IV) oxide yielded cis-[$Fe(NO)X(S_2CNEt_2)_2$], for X = Br, I, or NO_2 (46), although NO did not react. Subsequent study of the oxidation of the dimethyl analog [$Fe(NO)(S_2CNMe_2)_2$] by nitrogen(IV) oxide showed (47) that at low temperature ($-60°C$, $CHCl_3$ solution) the initial product is trans-[$Fe(NO)(NO_2)(S_2CNMe_2)_2$], but that this isomer is rapidly converted to cis-[$Fe(NO)(NO_2)(S_2CNMe_2)_2$] at 5°C or above.

The iron cyclobutadiene complex [(η^4-C_4Ph_4)Fe(CO)$_2$NO]$^+$ undergoes replacement of both carbonyl ligands with $Et_2NCS_2^-$, although with retention of the nitrosyl ligand (48) to yield [(η^4-C_4Ph_4)Fe(NO)-(S_2CNEt_2)]. structure 6.

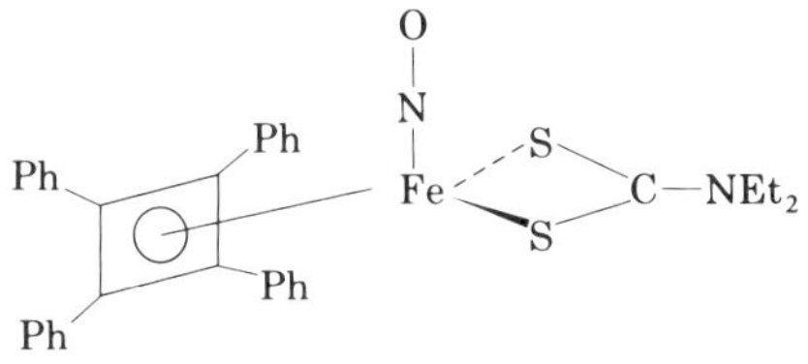

[Structure diagrams:

6: Fe complex with cyclobutadiene-like Ph₄ ring, NO, and S₂C–NEt₂ ligand

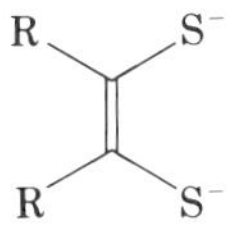

7 a R = CN
 b R = Ph
 c R = CF₃
(dithiolene with R groups on C=C, S⁻ termini)

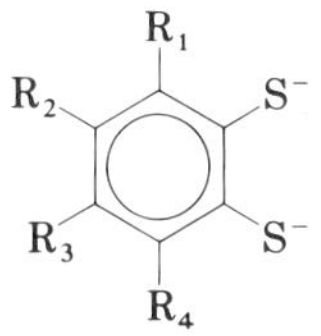

8 a R₁, R₃, R₄ = H; R₂ = CH₃
 b R₁–R₄ = Cl
(benzene-1,2-dithiolate with R₁–R₄ substituents)]

In [Fe(NO)(S$_2$CNR$_2$)$_2$] the iron is ligated by four sulfur atoms as well as by the nitrosyl ligand; similar Fe(NO)S$_4$ chromophores are found (49) in complexes of dithiolenes (**7**) and dithiols (**8**). For both dithiolene and dithiol derivatives, reaction of [FeL$_2$]$^{2-}$ (L = **7** or **8**) with NO gas yields the nitrosyl derivatives [Fe(NO)L$_2$]$^-$ (50), in which the nitrosyl ligand has effected a net oxidation. Certain of the mononegative complexes can be oxidized by iodine to neutral Fe(NO)L$_2$ or reduced with borohydride to dinegative [Fe(NO)L$_2$]$^{2-}$ (50).

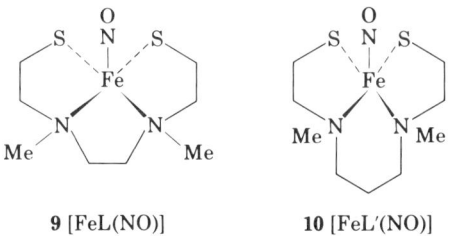

9 [FeL(NO)]   **10** [FeL'(NO)]

The Fe(NO)S$_2$N$_2$ chromophore found in complexes **9**, [FeL(NO)], and **10**, [FeL'(NO)], is almost identical to that in [Fe(NO)(S$_2$CNR$_2$)$_2$] (49). These complexes can be prepared by reaction of NO, NO$^+$, or NO$_2^+$ with the dimeric complexes [Fe$_2${(CH$_2$)$_2$(NMe CH$_2$CH$_2$S)$_2$}$_2$], i.e., (FeL)$_2$, and [Fe$_2${(CH$_2$)$_3$(NMe CH$_2$CH$_2$S)$_2$}$_2$], i.e., (FeL')$_2$, respectively. An alternative synthesis of FeL'(NO) has been described (51): reaction of

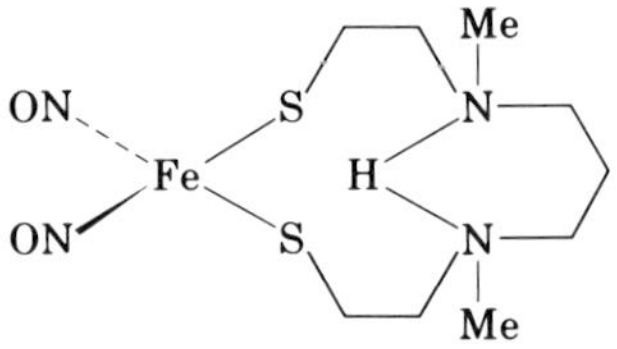

11 [Fe(NO)₂(L'H)]

[Fe(NO)$_2$Br$_2$]$^-$ with L'H$_2$ provides two products, [FeL'(NO)] and [Fe(NO)$_2$(L'H)] (**11**), which is a neutral analog of the anions [Fe(NO)$_2$(SR)$_2$]$^-$ readily available from [Fe$_2$(SR)$_2$(NO)$_4$] (*23*).

E. SELENIUM ANALOGS

The number of iron–selenium–nitrosyl complexes is substantially smaller than the iron–sulfur–nitrosyl species, as considerably less work in this area has been reported. However there are a number of differences between the sulfur systems and their analogs containing selenium or tellurium. For selenium it is convenient to divide the complexes into three classes, dependent upon the stoichiometry of the metal–chalcogen framework.

1. Fe$_2$Se$_2$ Complexes

The conversion of [Fe$_2$I$_2$(NO)$_4$] to [Fe$_2$Se$_2$(NO)$_4$]$^{2-}$ (*26, 30*) was described earlier [Eq. (7), Section II,B,2] along with the subsequent alkylation of [Fe$_2$Se$_2$(NO)$_4$]$^{2-}$; the neutral species [Fe$_2$(SeR)$_2$(NO)$_4$] can also be obtained (*26, 30*) by reaction of RSe$^-$ with [Fe$_2$I$_2$(NO)$_4$], [Eq. (6), Section II,B,2].

2. Fe$_4$Se$_4$ Complexes

Reaction of Hg[Fe(CO)$_3$NO] with elemental selenium in refluxing toluene (*52*) proceeds just like the reaction of this mercurial with elemental sulfur (*36, 37*) and gives the cubane-type tetranuclear cluster [Fe$_4$Se$_4$(NO)$_4$]. This cluster can also be readily synthesized by reaction of Na[Fe$_4$Se$_3$(NO)$_7$] with elemental selenium, also in refluxing toluene (*53*).

Reduction of [Fe$_4$Se$_4$(NO)$_4$] by a stoichiometric quantity of potassium metal in the presence of 2,2,2-cryptand yielded the salt [K(2,2,2-crypt)]$^+$[Fe$_4$Se$_4$(NO)$_4$]$^-$ (*52*), use of excess barium metal under similar conditions gave the rather unstable dinegative anion [Fe$_4$Se$_4$(NO)$_4$]$^{2-}$, which can also be made by further reduction of mononegative anion by excess potassium metal.

3. Fe_4Se_3 Complexes

Three routes have been described to the anion $[Fe_4Se_3(NO)_7]^-$, the selenium analog of the Roussin black anion. Reaction of iron(II) sulfate with sodium nitrite and sodium hydrogen selenide provides $Na[Fe_4Se_3(NO)_7]$ (27) in a reaction entirely analogous to Roussin's original synthesis of the anion $[Fe_4S_3(NO)_7]^-$ (1). Likewise, just as $[Fe_2S_2(NO)_4]^{2-}$ is converted, in CH_2Cl_2 solution, to $[Fe_4S_3(NO)_7]^-$ (Section II,A), so too the Ph_4As^+ salt of $[Fe_2Se_2(NO)_4]^{2-}$ is converted into $(Ph_4As)[Fe_4Se_3(NO)_7]$ (26) in practical yield simply by stirring it in CH_2Cl_2 for half an hour. Similarly, synthesis of $[Fe_2Se_2(NO)_4]^{2-}$ from $[Fe_2I_2(NO)_4]$ followed by acidification with glacial acetic acid, without isolation of the intermediate $Li_2[Fe_2Se_2(NO)_4]$, yields, after addition of $(Ph_3PNPPh_3)Cl$, the Fe_4Se_3 salt $(Ph_3PNPPh_3)[Fe_4Se_3(NO)_7]$ in 77% yield in a one-pot process (26).

The reaction of the selenium cubane cluster $[Fe_4Se_4(NO)_4]$ with excess of Ph_3P, which often acts as a chalcogen abstractor as well as a ligand, yields (52) a product of composition $[Fe_4Se_3(NO)_4(PPh_3)_3]$ (12), whose X-ray structure (see Section III,E) shows that it resembles $[Fe_4Se_3(NO)_7]^-$ but with the equatorial nitrosyl ligands on the basal iron atoms replaced by triphenylphosphine.

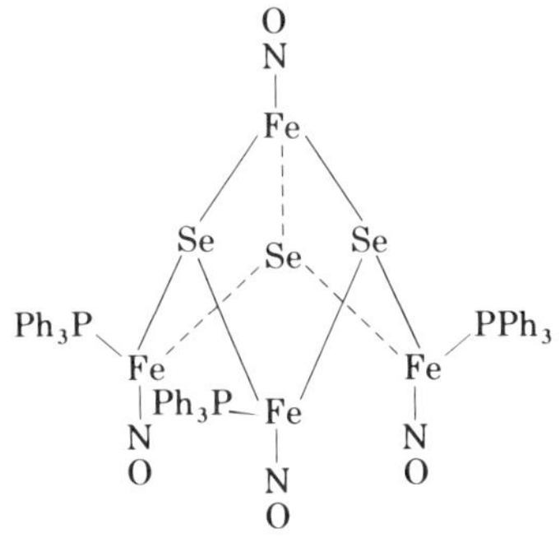

12

Reduction of $[Fe_4Se_3(NO)_4(Ph_3P)_3]$ by an excess of barium metal in THF solution in the presence of the 2,2,2-cryptand gave the $[Ba(2,2,2\text{-crypt})]^{2+}$ salt of the dinegative heptanitrosyl anion $[Fe_4Se_3(NO)_7]^{2-}$. Clearly the change from $[Fe_4Se_3(NO)_4(Ph_3P)_3]$ to $[Fe_4Se_3(NO)_7]^{2-}$

must involve major rearrangement involving not only loss of triphenylphosphine but reorganization from mononitrosyliron fragments to dinitrosyliron fragments.

F. Tellurium Analogs

In comparison with the sulfur and selenium systems, rather few iron–tellurium–nitrosyl complexes are known. One of the most productive techniques of study of this area has proved to be that employing $[Fe_2I_2(NO)_4]$ (*16, 30*). Thus reaction of PhTe$^-$ with $[Fe_2I_2(NO)_4]$ gave $[Fe_2(TePh)_2(NO)_4]$ [cf. Eq. (6), Section II,B,2], and Te^{2-} yielded $[Fe_2Te_2(NO)_4]^{2-}$ [cf. Eq. (7), Section II,B,2]. The anion $[Fe_2Te_2(NO)_4]^{2-}$ can be alkylated with electrophilic reagents such as BEt$_3$ or PhCH$_2$Cl (*26*). Just as $[Fe_2S_2(NO)_4]^{2-}$ and $[Fe_2Se_2(NO)_4]^{2-}$ are spontaneously converted in CH$_2$Cl$_2$ solution into the tetrairon anions $[Fe_4X_3(NO)_7]^-$ (X = S or Se), so also (*26*) $[Fe_2Te_2(NO)_4]^{2-}$ upon similar treatment gives $[Fe_4Te_3(NO)_7]^-$, although salts of this anion have not yet been obtained analytically pure.

In contrast to the reactions of elemental sulfur and selenium with the mercurial Hg[Fe(CO)$_3$NO]$_2$ to give cubane-type clusters $[Fe_4X_4(NO)_4]$ (X = S or Se) (*36, 37, 52*), the analogous reaction with elemental tellurium yielded (*52*) the nitrosyl-free cluster $[Fe_3Te_2(CO)_9]$. However, reaction of the mercurial either with a mixture of Ph$_2$Te$_2$ and elemental tellurium in refluxing toluene, or with PhTeBr$_3$ in cold methanol, yielded a diiron–nitrosyl $[Fe_2(TePh)_2(NO)_4]$ identical with that formed (*26, 30*) by reaction of PhTe$^-$ with $[Fe_2I_2(NO)_4]$. Reduction of $[Fe_2(TePh)_2(NO)_4]$ with an excess of barium metal yielded the corresponding dianion $[Fe_2(TePh)_2(NO)_4]^{2-}$ (*52*).

G. Heterometallic Iron–Sulfur–Nitrosyl Clusters

A variety of routes have been employed for the synthesis of heterometallic clusters containing Fe(NO)S fragments, and examples of such systems have been described in which the second metal is V, Mo, Co, Ni, and Pt; it seems probable, given the variety of potential synthetic approaches to such mixed metal systems, that many more examples await discovery.

1. Synthesis by Nitrosylation of Preformed Heterometallic Aggregates

When the heterotrimetallic anion Fe(S$_2$MoS$_2$)$_2^{3-}$ (**13**), was treated with NO gas in DMF solution, the reaction product isolated (*54*) in 50% yield was a salt of the heterometallic iron–dinitrosyl

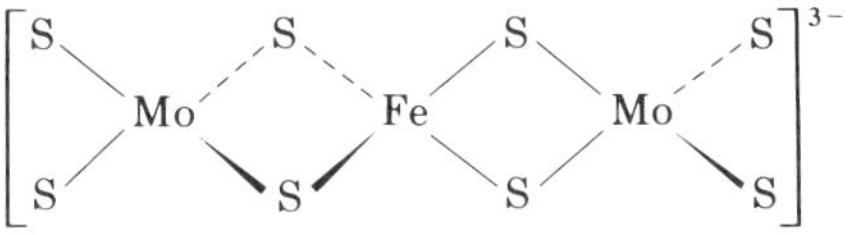

[structure 13: [S₂Mo(S₂)Fe(S₂)MoS₂]³⁻]

13

[structure 14: [S₂MoS₂Fe(NO)₂]²⁻]

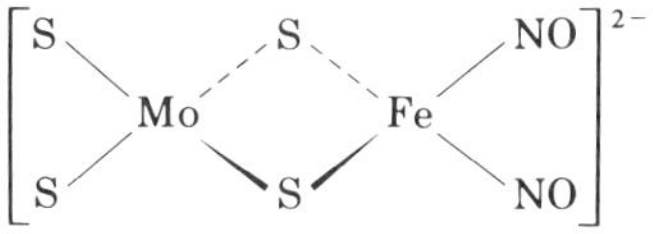

14

$[S_2MoS_2Fe(NO)_2]^{2-}$ (**14**). The tungsten analog of **14** was synthesized in a similar manner (54).

2. Synthesis by Use of Sulfur-Containing Anions as Nucleophiles

The reactions of $[Fe_2S_2(NO)_4]^{2-}$ with complexes of nickel (30) and platinum (32) to yield heterometallic Fe_2MS_2 clusters have already been described (Section II,B,2).

The tetrathiomolybdate anion $[MoS_4]^{2-}$ acts as a chelating ligand toward iron in both diiron complexes $[Fe_2(SR)_2(NO)_4]$ and tetrairon complexes $[Fe_4S_3(NO)_7]^-$ and $[Fe_4S_4(NO)_4]$ to yield two paramagnetic iron–nitrosyls which have been assigned (55) the constitutions **15** and **16**:

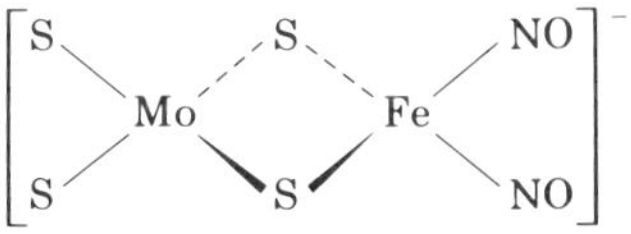

[structure 15: [S₂MoS₂Fe(NO)₂]⁻]

15

[structure 16: [S₂MoS₂Fe(NO)S₂MoS₂]²⁻ with NO on Fe]

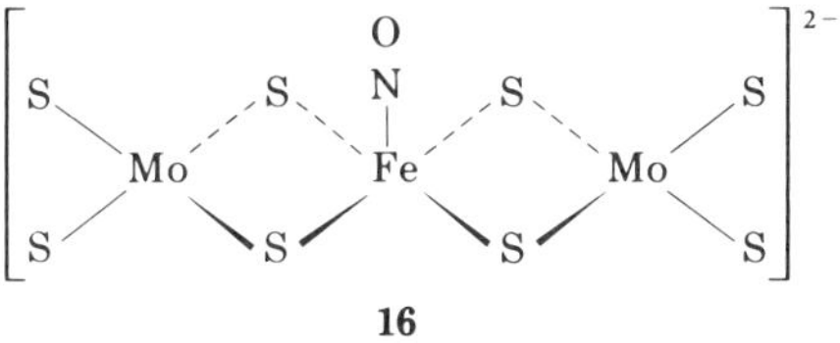

16

Complex **15** differs from **14** only in the overall oxidation level, while complex **16** is clearly related to the dithiocarbamate complex **5** (Section II,D,3).

3. Synthesis via Exchange Reactions

From the reaction of [Fe$_2$(SPh)$_2$(NO)$_4$] with the dinuclear cobalt complex [Co$_2$(SPh)$_2$(NO)$_2$(PPh$_3$)$_2$], the mixed-metal nitrosyl **17** was isolated by chromatography in 70% yield (*26*) [Eq. (13)]:

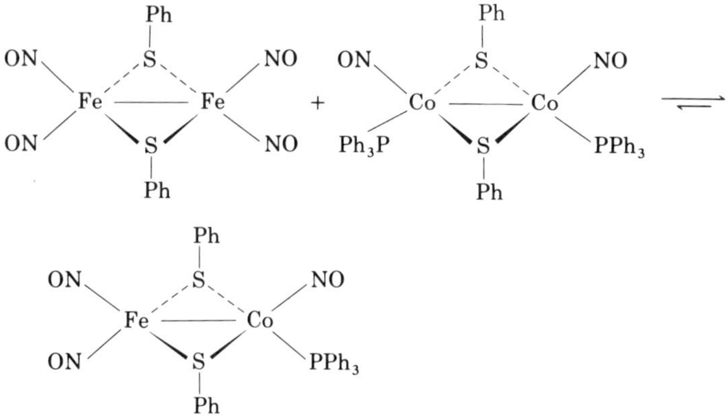

$$\tag{13}$$

Complex **17** can also be formed in low yield (*26*) along with [Fe$_2$(SPh)$_2$(NO)$_4$] by the reaction of [Co$_2$(SPh)$_2$(NO)$_2$(PPh$_3$)$_2$] with [Fe$_2$(SPh)$_2$(CO)$_6$].

4. Synthesis from Sulfur-Rich Metal Complexes

The mercurial Hg[Fe(CO)$_3$NO]$_2$ reacts with elemental sulfur S$_8$ to yield the cubane-type cluster [Fe$_4$S$_4$(NO)$_4$] (*36, 37*); it is not therefore entirely unexpected that metal complexes rich in sulfur can likewise react with this mercurial to produce novel heterometallic nitrosyls.

Thus [(η^5-C$_5$H$_4$Me)$_2$V$_2$S$_4$] reacts (*26, 56, 57*) with Hg[Fe(CO)$_3$NO]$_2$ to yield successively the complexes [(η^5-C$_5$H$_4$Me)$_2$V$_2$S$_4$Fe(NO)$_2$] (**18**)

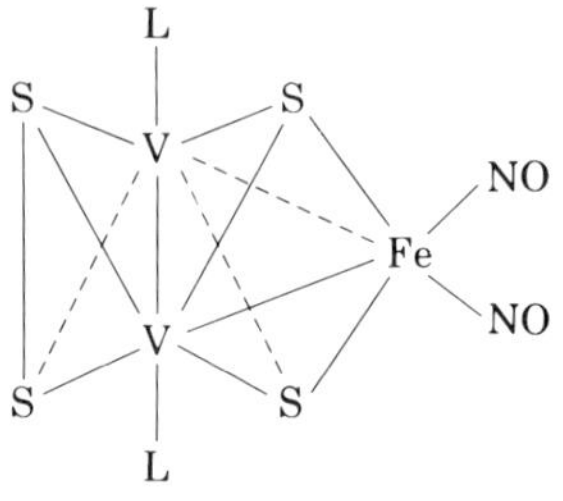

18 L = C$_5$H$_4$Me

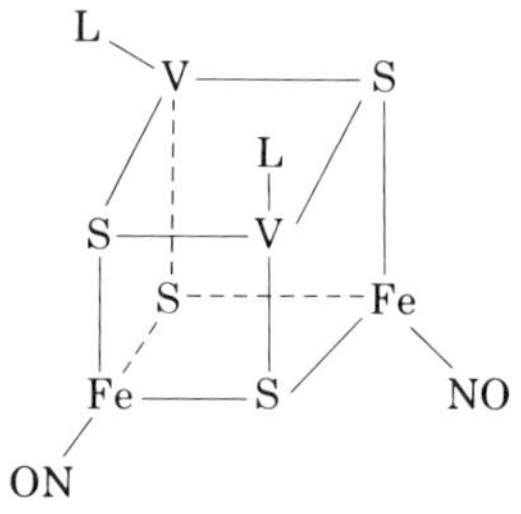

19

and the cubane-type cluster [(η^5-C$_5$H$_4$Me)$_2$V$_2$S$_4$Fe$_2$(NO)$_2$] (**19**). In like manner [(η^5-C$_5$H$_4$Me)$_2$Mo$_2$S$_4$] reacts (*58*) with Hg[Fe(CO)$_3$NO]$_2$ to give [(η^5-C$_5$H$_4$Me)$_2$Mo$_2$S$_4$Fe(NO)$_2$].

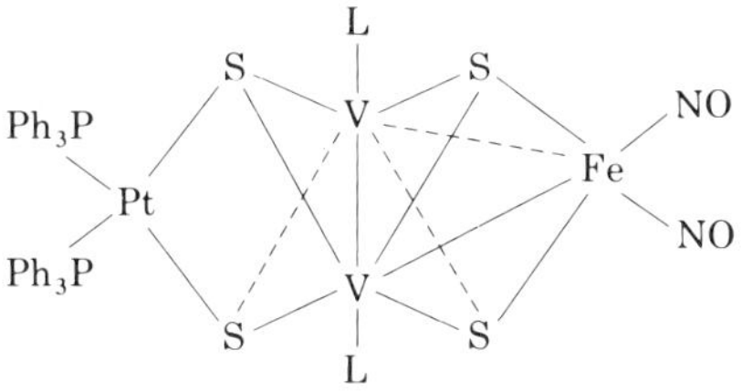

Both **18** and **19** are, furthermore, effective precursors for heterometallic nitrosyls containing three different metals. Thus (*26*) **18** reacts with [(Ph$_3$P)$_2$Pt(C$_2$H$_4$)] to give [(η^5-C$_5$H$_4$Me)$_2$V$_2$S$_4$ {Fe(NO)$_2$} {Pt(PPh$_3$)$_2$}] (**20**), while **19** reacts (*26, 56*) with [Co(NO)(CO)$_3$] in the presence of Me$_3$NO to give the cubane-type complex **21** containing a V$_2$FeCoS$_4$ core.

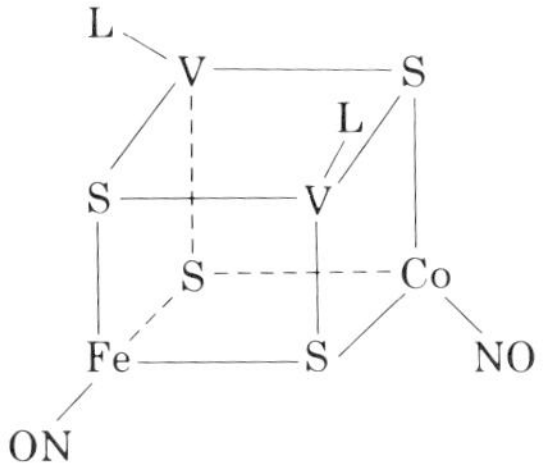

The trinuclear complex **18** reacts with [Co(NO)$_2$I]$_x$ under reducing conditions (Zn dust, MeLi, or LiBEt$_3$H) to provide a 44% yield of [(η^5-C$_5$H$_4$Me)VS$_3$FeCo$_2$(NO)$_6$] (*26*), which has been assigned the constitution **22**; this complex has the same overall electron count as [Fe$_4$S$_3$(NO)$_7$]$^-$, and it was on this basis that the structure **22**, with its strong resemblance to [Fe$_4$S$_3$(NO)$_7$]$^-$, was proposed (*26*).

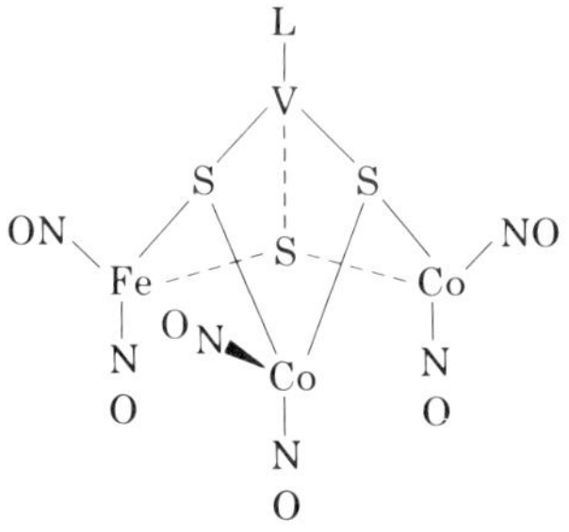

Finally, a single example of desulfurization, using a tertiary phosphine, to effect cluster transformation, has been reported (*26*) for a heterometallic nitrosyl. Reaction of complex **18** [(η^5-C$_5$H$_4$Me)$_2$V$_2$S$_4$Fe-(NO)$_2$] with tributylphosphine provided [(η^5-C$_5$H$_4$Me)$_2$V$_2$S$_3$Fe(NO)$_2$], of unknown constitution, in 80% yield.

III. Molecular Structure: X-Ray Crystallography

The principal objective of the earliest X-ray studies to be carried out on iron–sulfur–nitrosyl complexes, those on $[Fe_2(SEt)_2(NO)_4]$ (10) and $Cs[Fe_4S_3(NO)_7]$ (11), was the establishment of their gross chemical constitution. More recent X-ray studies have been concerned not only with gross structure, but additionally with detailed comparisons within series of similar species as a possible probe of electronic structure.

A. $[Fe_4S_3(NO)_7]^-$ AND $[Fe_2S_2(NO)_4]^{2-}$

The structure of $[Fe_4S_3(NO)_7]^-$ was initially determined for the cesium salt (11) and it was subsequently redetermined for the tetraphenylarsonium salt (39). In both salts the anion has the same overall structure, of essentially C_{3v} molecular symmetry, with remarkably similar dimensions. The anion contains a flattened Fe_4 tetrahedron, three faces of which are triply bridged by sulfur atoms. The unique apical iron carries a single nitrosyl ligand while the three basal irons each carry two nitrosyl ligands, which may be regarded as axial (approximately parallel to the threefold symmetry axis) and equatorial, respectively. Figure 1 shows the anion as in the Ph_4As^+ salt; in this salt the averaged Fe_a-Fe_b and Fe_b-Fe_b distances are 2.700 and 3.570 Å, respectively and the averaged Fe_a-S and Fe_b-S distances are 2.206 and 2.258 Å. All of the Fe–N–O groups fall into the category regarded as "linear," i.e., best described as based upon NO^+ rather than upon any other electronic form of the nitrosyl ligand (see Section V, for

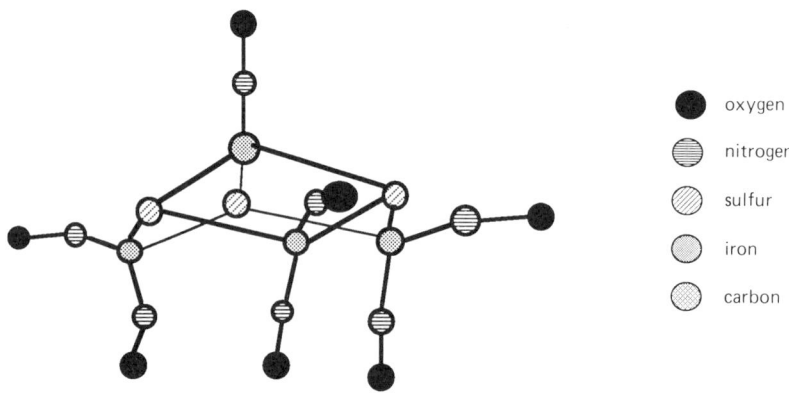

FIG. 1. The structure of the anion $[Fe_4S_3(NO)_7]^-$ as in the Ph_4As^+ salt. Redrawn from Ref. 39.

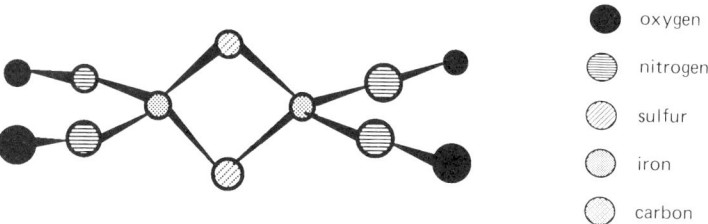

FIG. 2. The structure of the anion [Fe$_2$S$_2$(NO)$_4$]$^{2-}$ as in the Me$_4$N$^+$ salt. Redrawn from Ref. 59.

discussion of electronic structures): the Fe$_a$–N–O angle is 176.3(9)°, the equatorial and axial Fe$_b$–N–O angles are 167.5 and 166.1°, respectively.

The structure of [Fe$_2$S$_2$(NO)$_4$]$^{2-}$ has been determined only recently (59): the tetramethylammonium salt contains two crystallographically distinct anions in the asymmetric unit, each of which has approximate D_{2h} molecular symmetry, although only C_i crystallographic symmetry. Thus the Fe$_2$S$_2$ fragment is planar, with an average Fe–S distance of 2.244 Å, very similar to the S–Fe(NO)$_2$ distance, 2.258 Å in [Fe$_4$S$_3$(NO)$_7$]$^-$ (39). Likewise, the Fe–Fe distance of 2.714 Å in [Fe$_2$S$_2$(NO)$_4$]$^{2-}$ is similar to the Fe$_a$–Fe$_b$ distance, 2.700 Å, in [Fe$_4$S$_3$(NO)$_7$]$^-$. The Fe–N–O groups are again "linear" with an average angle of 165.8° (range 163.0–167.8°). Figure 2 shows one of the anions from the asymmetry unit.

Thus while none of the early formulations of the diiron structure (5, 7) proved to be correct, Seel's 1942 formulation of the structure of [Fe$_4$S$_3$(NO)$_7$]$^-$ (8), based upon the FeS structure, is remarkably close to the molecular structure found by X-ray analysis (11, 39).

B. [Fe$_2$(SR)$_2$(NO)$_4$]

When R is C$_2$H$_5$ or HgCH$_3$ the molecules are found to be centrosymmetric (10, 31), each having crystallographic C_i ($\bar{1}$) symmetry and approximate C_{2h} molecular symmetry. Thus of the two possible isomeric forms **3a** and **3b**, only the anti form **3b** is observed in the solid state for [Fe$_2$(SEt)$_2$(NO)$_4$] and [Fe$_2$(SHgCH$_3$)$_2$(NO)$_4$].

On the other hand, in [Fe$_2${S$_2$Pt(PPh$_3$)$_2$}(NO)$_4$], in which the two sulfurs of the Fe$_2$S$_2$ fragment are bonded to the same platinum, the structure necessarily has approximate C_{2v} rather than C_{2h} symmetry (32). Figures 3–5 show the molecular configuration of [Fe$_2$(SEt)$_2$(NO)$_4$], [Fe$_2$(SHgCH$_3$)$_2$(NO)$_4$], and [Fe$_2${S$_2$Pt(PPh$_3$)$_2$}(NO)$_4$], respectively.

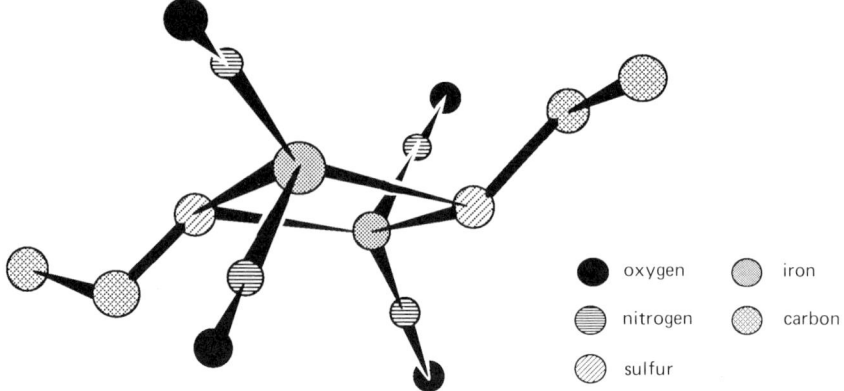

FIG. 3. The structure of [Fe$_2$(SEt)$_2$(NO)$_4$]. Redrawn from Ref. *10*.

The salient molecular dimensions for crystallographically characterized complexes [Fe$_2$(SR)$_2$(NO)$_4$] are collected in Table I, together with those for [Fe$_2$S$_2$(NO)$_4$]$^{2-}$ (*59*). The important features are (1) the smaller size of the Fe$_2$S$_2$ fragment in [Fe$_2$S$_2$(NO)$_4$]$^{2-}$; (2) the folding of the Fe$_2$S$_2$ ring about the Fe–Fe vector, in [Fe$_2$\{S$_2$Pt(PPh$_3$)$_2$\}(NO)$_4$], consequent upon the bridging by platinum of the two sulfurs; and (3) the existence in [Fe$_2$\{S$_2$Pt(PPh$_3$)$_2$\}(NO)$_4$], of two distinct configurations of the Fe–N–O fragments. The Fe–N–O groups on the same side of the molecule as the platinum atom are "linear" (FeNO angle, 172.1°), while those on the opposite side are distinctly nonlinear, with an average

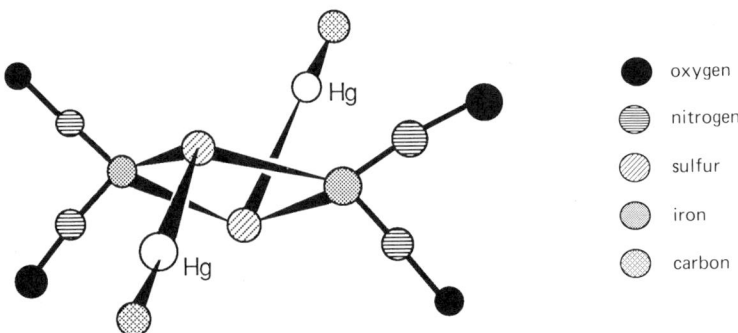

FIG. 4. The structure of [Fe$_2$(SHgCH$_3$)$_2$(NO)$_4$]. Redrawn from Ref. *31*.

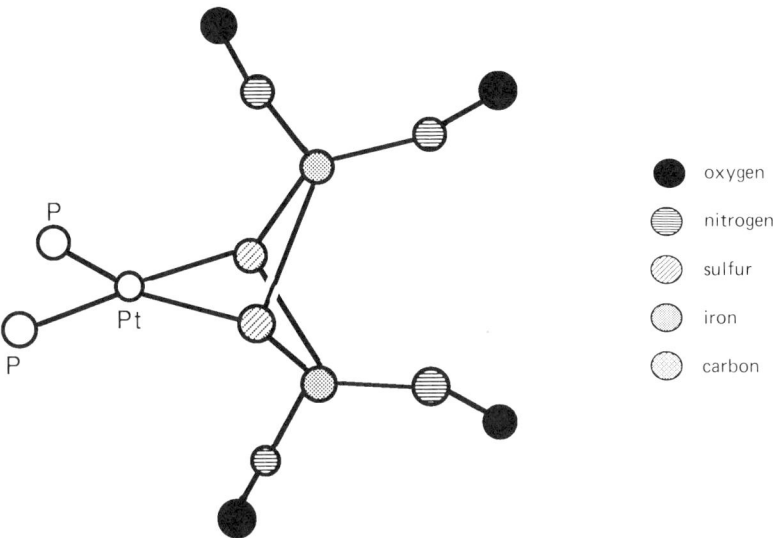

Fig. 5. The structure of $[Fe_2\{S_2Pt(PPh_3)_2\}(NO)_4]$. Redrawn from Ref. 32 (phenyl groups omitted).

FeNO angle of 150.3°. It has been suggested (32) that the bonding of the nitrosyls remote from platinum arises from the interaction of the nitrosyl π orbitals with the orbitals concentrated in the Fe–Fe bond, whose disposition in turn is determined by the puckering of the Fe_2S_2 ring; however, no theoretical support for this idea is yet available.

TABLE I

Selected Geometric Parameters for $[Fe_2S_2(NO)_4]^{2-}$ and $[Fe_2(SR)_2(NO)_4]$

Complex	Idealized molecular symmetry	d(Fe–S) (Å)[a]	d(Fe–Fe) (Å)	d(S⋯S) (Å)	(Fe–N–O) (degrees)[a]	Reference
$[Fe_2S_2(NO)_4]^{2-\,b}$	D_{2h}	2.239	2.716	3.562	165.4	59
		2.249	2.713	3.587	166.2	59
$[Fe_2(SEt)_2(NO)_4]$	C_{2h}	2.270(4)	2.720(3)	3.63	167	10
$[Fe_2(SHgCH_3)_2(NO)_4]$	C_{2h}	2.275(7)	2.771(7)	3.607	166.5	31
$[Fe_2S\{SPt(PPh_3)_2\}(NO)_4]$	C_{2v}	2.275(5)	2.802(5)	3.241(7)	150.3 (trans)	32
					172.1 (cis)	32

[a] Averaged values.
[b] Two independent molecules in an asymmetric unit.

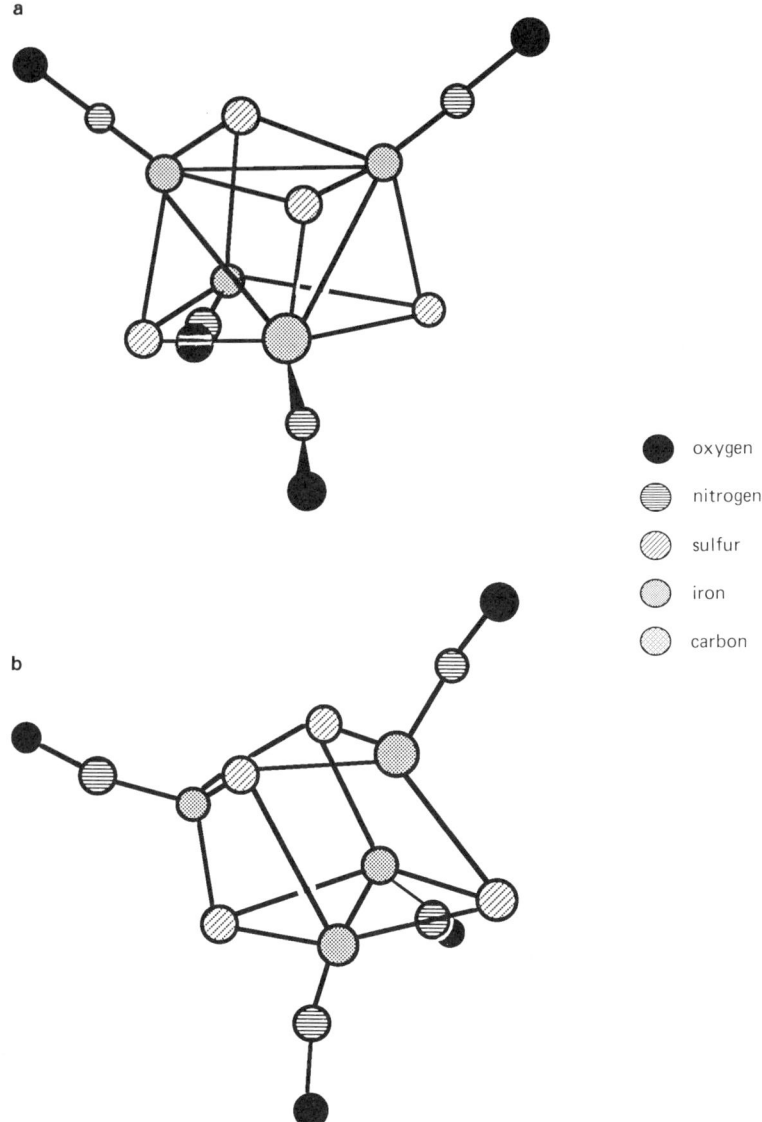

FIG. 6. The structure of (a) [Fe$_4$S$_4$(NO)$_4$] and (b) [Fe$_4$S$_4$(NO)$_4$]$^-$. Both redrawn from Ref. 37.

C. [Fe$_4$S$_4$(NO)$_4$] AND RELATED CLUSTERS

The neutral cubane-like cluster [Fe$_4$S$_4$(NO)$_4$] (4) has almost perfect tetrahedral T_d symmetry (37). The cluster has a tetrahedral Fe$_4$ core with Fe–Fe distances of 2.649 Å—slightly shorter than in [Fe$_4$S$_3$(NO)$_7$]$^-$ and [Fe$_2$(SR)$_2$(NO)$_4$]—with each face of the tetrahedron triply bridged by sulfur, such that Fe—S is 2.217 Å and S---S is 3.503 Å, again both shorter than found for [Fe$_2$(SR)$_2$(NO)$_4$]. Because of the difference between the Fe—Fe distance and the S---S distance, it is perhaps better to think of this structure as based upon two intersecting tetrahedra rather than a cube. As usual the Fe–N–O groups are linear, with an average angle of 177.6°. The structure is shown in Fig. 6a.

Addition of a single electron to [Fe$_4$S$_4$(NO)$_4$], of T_d symmetry, to provide [Fe$_4$S$_4$(NO)$_4$]$^-$, causes a reduction in symmetry to D_{2d} in the [K(2,2,2-crypt)]$^+$ salt (37). In the anion, Fig. 6b, there are two distinct types of Fe–Fe bond, of length 2.704 (twofold) and 2.688 Å (fourfold), indicative of an Fe$_4$ core which is not only reduced in symmetry, but increased in size compared with that in neutral [Fe$_4$S$_4$(NO)$_4$]. The Fe–S distances are also slightly larger in the anion, 2.231 Å, than in the neutral cluster, 2.217 Å, although the S---S distances, 3.496 (twofold) and 3.517 Å (fourfold), are very little different from the 3.503 Å in the neutral complex. From this comparison the following may be deduced (37) concerning the electronic structure: the lowest unoccupied molecular orbital (LUMO) in [Fe$_4$S$_4$(NO)$_4$] and the semioccupied molecular orbital (SOMO) in [Fe$_4$S$_4$(NO)$_4$]$^-$ (1) are localized primarily in the Fe$_4$ cage; (2) are antibonding in character; and (3) are degenerate (e or t_1 or t_2), assuming that the change of symmetry upon reduction is due to the Jahn–Teller effect.

The related complexes [Fe$_4$S$_2$(NO)$_4$(NCMe$_3$)$_2$] and [Fe$_4$S$_2$(NO)$_4$(NCMe$_3$)$_2$]$^-$, Fig. 7, each have C_{2v} symmetry in the solid state (41); a detailed comparison of the Fe$_4$S$_2$N$_2$ core geometry in neutral and anionic forms shows (41) that the LUMO in [Fe$_4$S$_2$(NO)$_4$(NCMe$_3$)$_2$] and the SOMO in [Fe$_4$S$_2$(NO)$_4$](NCMe$_3$)$_2$]$^-$ are localized in the Fe$_4$ cage, as for [Fe$_4$S$_4$(NO)$_4$] and its anion (37) with Fe–Fe antibonding character localized primarily in the unique Fe$_2$S$_2$ and Fe$_2$N$_2$ faces, rather than in the fourfold Fe$_2$SN faces.

D. [Fe(NO)(S$_2$CNR$_2$)$_2$] AND RELATED COMPLEXES

The complexes [Fe(NO)(S$_2$CNMe$_2$)$_2$] (60), Fig. 8, and [Fe(NO)(S$_2$CNEt$_2$)$_2$] (61) both have square-pyramidal geometry with a linear, apical, Fe–N–O group. The nitrogen(IV) oxide oxidation product

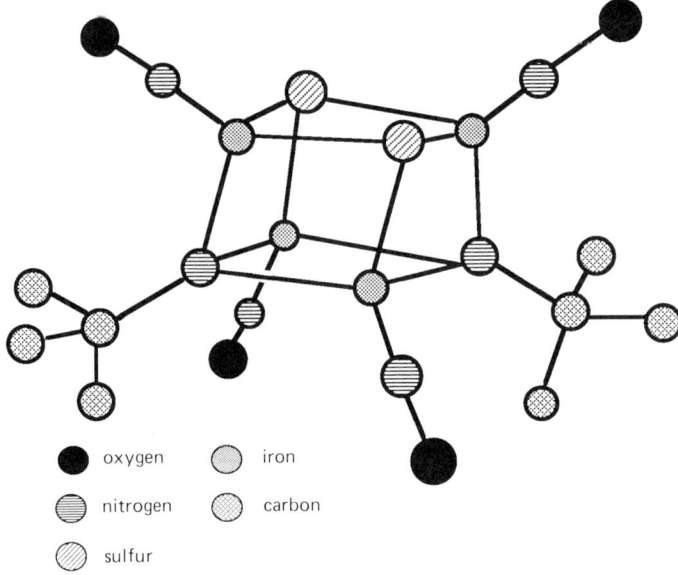

FIG. 7. The structure of [Fe$_4$S$_2$(NO)$_4$(NCMe$_3$)]. Redrawn from Ref. *41*.

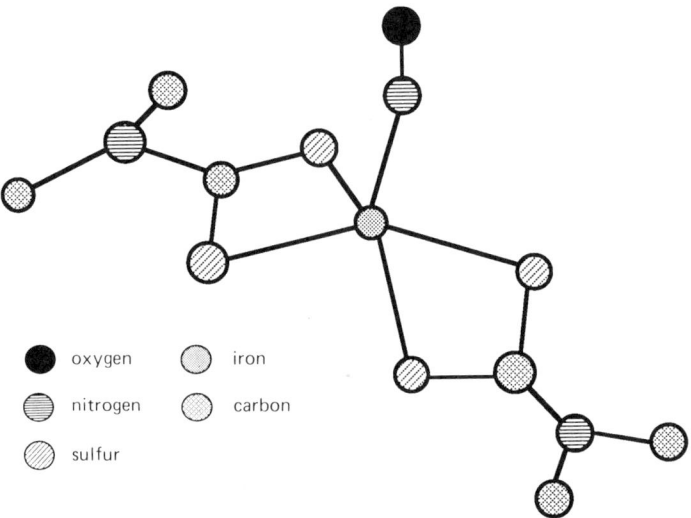

FIG. 8. The structure of [Fe(NO)(S$_2$CNMe$_2$)$_2$]. Redrawn from Ref. *60*.

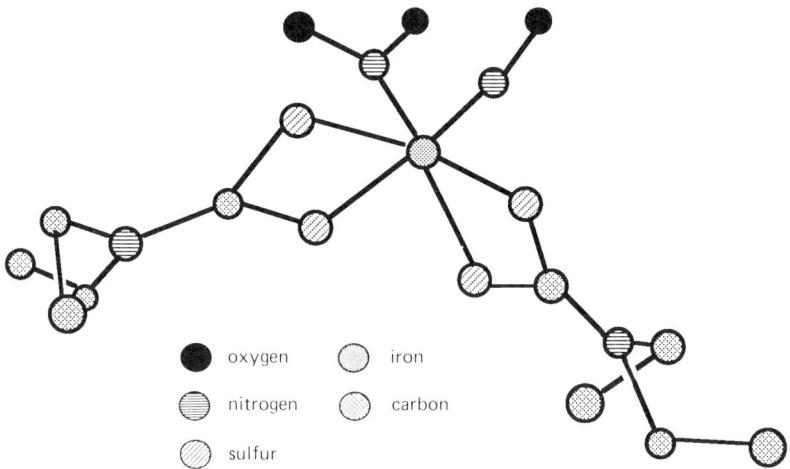

FIG. 9. The structure of cis-[Fe(NO)(NO$_2$)(S$_2$CNEt$_2$)$_2$]. Redrawn from Ref. 62.

from [Fe(NO)(S$_2$CNEt$_2$)$_2$] (46, 47) has been shown (62) to be cis-[Fe(NO)(NO$_2$)(S$_2$CNEt$_2$)$_2$], Fig. 9, again with a linear iron–nitrosyl fragment. The dithiocarbamate complex **6** also has a "linear" Fe–N–O fragment (48).

In complexes **9** and **10**, the Fe–N–O groups are distinctly bent (49, 51) although the nitrosyl ligand still occupies the apical site in an approximately square–pyramidal complex. Complex **9** crystallizes with two independent molecules in the asymmetric unit for which the bond angles Fe–N–O are 155.2(5) and 158.4(5)°, respectively (**49**), Fig. 10a, while in complex **10** the angle is (51) 155.2(9)°, Fig. 10b. The values of the Fe–N–O angle in complexes **9** and **10** are thus very similar to those of the trans nitrosyl ligands in [Fe$_2${S$_2$Pt(PPh$_3$)$_2$}(NO)$_4$] (32).

In the closely related complex **11**, the four-coordinate iron adopts nearly regular tetrahedral geometry with "linear" iron–nitrosyl groups (51), Fig. 11.

E. SELENIUM AND TELLURIUM ANALOGS

The iron–selenium cubane-type cluster [Fe$_4$Se$_4$(NO)$_4$] has been shown (52) to adopt a molecular structure of almost exact T_d symmetry, whereas in the corresponding anion [Fe$_4$Se$_4$(NO)$_4$]$^-$ the symmetry is lowered to D_{2d}, just as for the sulfur analogs (37). Upon one-electron reduction the size of the Fe$_4$ cage increases: in the neutral molecule the Fe–Fe and Fe–Se distances are 2.703 and 2.342 Å, but in the anion the

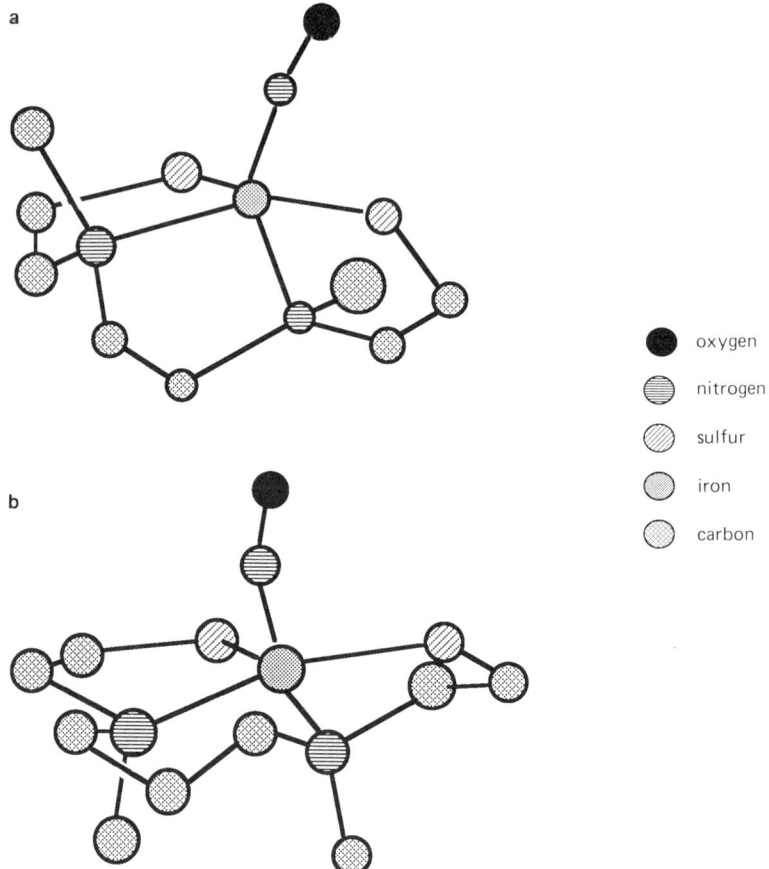

FIG. 10. The structure of (a) complex 9 and (b) complex 10. Redrawn from Refs. *49* and *51*, respectively.

Fe–Fe distances are increased to 2.770 (twofold) and 2.75 Å (fourfold), while the Fe–Se distances are all 2.354 Å: the Se ···· Se distances are scarcely affected, so that, as before, the LUMO in $[Fe_4Se_4(NO)_4]$ and the SOMO in $[Fe_4Se_4(NO)_4]^-$ are antibonding and localized primarily in the Fe_4 cage.

The triphenylphosphine complex $[Fe_4Se_3(NO)_4(PPh_3)_3]$ (**12**) has (*52*) a structure which resembles that of $[Fe_4S_3(NO)_7]^-$ (*11, 39*) in which the equatorial nitrosyl ligands on the basal iron are replaced by triphenylphosphine ligands. However, the molecular symmetry is reduced from C_{3v} to C_s, and the basal Fe_3 triangle contains one Fe–Fe distance of 2.695 and two of 2.845 Å. The structure of $[Fe_4Se_3(NO)_7]^-$

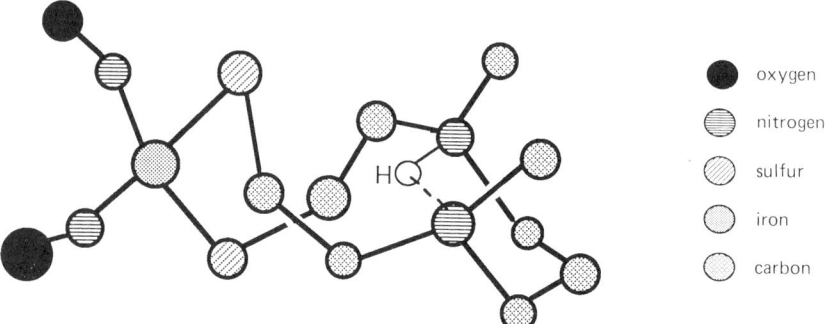

FIG. 11. The structure of complex 11. Redrawn from Ref. *51*.

appears not to have been determined, but by analogy with $[Fe_4S_3(NO)_7]^-$, C_{3v} symmetry is to be expected. In the dianion $[Fe_4Se_3(NO)_7]^{2-}$, however, the symmetry is not C_{3v} but C_s (*52*), suggesting that the LUMO in $[Fe_4Se_3(NO)_7]^-$ is of a type giving a Jahn–Teller distortion upon reduction. In the basal Fe_3 triangle of $[Fe_4Se_3(NO)_7]^{2-}$ the unique Fe–Fe distance is 3.637 Å, the others 3.750 Å, a difference of magnitude comparable to that in $[Fe_4Se_3(NO)_4(PPh_3)_3]$, although the distances themselves are very much greater. In all of these selenium complexes, the Fe–N–O fragments are linear. The only iron–tellurium–nitrosyl complex which has so far been characterized by X-ray crystallography is $[Fe_2(TePh)_2(NO)_4]$ (*52*): this complex is centrosymmetric, i.e., analogous to structure **3b** with "linear" Fe–N–O groups, and an Fe–Fe distance of 2.801 Å, very little longer than in $[Fe_2(SR)_2(NO)_4]$ complexes (Table I).

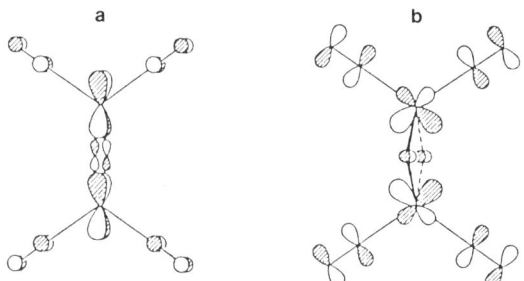

FIG. 12. Frontier orbitals in $[Fe_2(SH)_2(NO)_4]$ (a) LUMO (B_g) and (b) HOMO (B_u). Reproduced with permission from *Inorg. Chem.* **24**, 3858 (1985).

F. HETEROMETALLIC IRON–SULFUR–NITROSYLS

The structure of the heterometallic complexes **14, 17, 18,** and **19** have all been established by X-ray crystallography, (*26, 54, 56*). The principal points of interest arising from these studies, aside from the definitive establishment of the constitutions of these complexes, are as follows: (1) all the Fe–N–O groups are linear and (2) complex **17** adopts (*26*) the syn configuration analogous to **3a**, with both phenyl groups on the side of the FeS_2Co ring remote from the triphenylphosphine, and it is presumably the repulsive steric interaction between the several phenyl groups which distinguish the configuration of structure **17** from those of $[Fe_2(SR)_2(NO)_4]$ complexes (Section II,G).

IV. Molecular Structure: NMR Spectroscopy

A. ^{1}H NMR SPECTROSCOPY OF $[Fe_2(SR)_2(NO)_4]$

The ^{1}H NMR spectra of solutions of $[Fe_2(SR)_2(NO)_4]$ is noncoordinating solvents for a wide range of organic substituents R [R = Me, Et, *n*-Pr, *i*-Pr, *n*-Bu, *i*-Bu, *t*-Bu, *n*-C_5H_{11}, CH_2=$CHCH_2$, HC≡CCH_2, $CH_3C(O)CH_2$, Me_3SiCH_2, Ph, and $PhCH_2$] show (*27, 30, 63, 64*) the presence of two isomeric forms: these spectra have been interpreted in terms of the forms **3a** and **3b**, although the ^{1}H spectra do not establish these forms. For all substituents except R = *t*-Bu, the two isomers are present in virtually equal abundance, implying that in solution the ΔG_f^0 values are esentially identical. When R = *t*-Bu (*64*), the isomer ratio is 1:3 in favor of the syn isomer **3a** (rigorously established by use of ^{15}N NMR; see below). The existence of two forms has been confirmed by ^{13}C NMR in a number of cases (*27, 63*).

B. ^{15}N NMR SPECTROSCOPY OF $[Fe_2(SR)_2(NO)_4]$

The ^{15}N spectra of a number of complexes $[Fe_2(SR)_2(^{15}NO)_4]$ enriched to 99% in ^{15}N, have been studied in noncoordinating solvents (*25, 63, 64*). In every case the spectrum comprises a singlet and a pair of doublets: the singlet is assigned to the anti isomer **3b** and the two doublets, which comprise an AX system, to the syn isomer **3a**. In **3b** all the nitrosyl ligands are equivalent in C_{2h} symmetry, while in **3a**, although the two irons are equivalent and the two substituents R are also equivalent, the two nitrosyl ligands in each $Fe(^{15}NO)_2$ group are nonequivalent and the coupling $^2J(^{15}NFe^{15}N)$ between the ^{15}N nuclei yields the observed AX spectra. No molecular symmetries for **3a** and **3b** other than C_{2v} and C_{2h}, respectively, can yield the observed ^{15}N spectra. Except when R = t-Bu, the spectra show the two isomers to be of equal abundance: if R = t-Bu, the C_{2v} isomer dominates and the isomer ratio is constant at 2.85 throughout the temperature range 220–298 K: this implies that for the syn $\rightleftharpoons$ anti equilibrium, ΔH^0 is zero and ΔS^0 is 8.7 J K^{-1} mol^{-1}. One possible explanation (*64*) for the behavior of [Fe$_2$(SBu-t)$_2$(NO)$_4$] lies in differential specific solvation of the two forms: specific solvation was invoked (*65*) to rationalize some of the solvent shift properties of the ^{1}H spectrum of [Fe$_2$(SMe)$_2$(NO)$_4$], where the overall behavior was satisfactorily correlated with Taft's solvatochromic parameter, π^* (*66, 67*). The existence in solution of two forms corresponding to structures **3a** and **3b** for such a wide range of substituents is remarkable in view of the fact that only isomer **3b** has been observed in the solid state (see Section III,B).

As expected, the ^{15}N spectrum of [Fe$_2$S$_2$(^{15}NO)$_4$]$^{2-}$ consists of a singlet, as all the nitrosyl ligands are equivalent (*63*). However, in methylene chloride solution, [Fe$_2$S$_2$(NO)$_4$]$^{2-}$ is rapidly and cleanly converted to [Fe$_4$S$_3$(NO)$_7$]$^-$, which is then the sole species detectable in solution using ^{15}N NMR (*25, 63*); and by-products remain uncharacterized.

For each complex [Fe$_2$(SR)$_2$(NO)$_4$] examined, the ^{15}N chemical shift is in the range characteristic (*68–70*) of linear M–N–O groups.

C. ^{15}N NMR Spectroscopy of Tetrairon Complexes

The ^{15}N NMR of [Fe$_4$S$_3$(^{15}NO)$_7$]$^-$ (99% ^{15}N) consists of a singlet and a pair of doublets (*25*); the relative intensities allow the assignment of the singlet to the apical nitrosyl group and the two doublets, which comprise an AX system, to the axial and equatorial nitrosyl ligands on the basal irons. Since the two ligands in each basal Fe(NO)$_2$ fragment are distinct, the coupling $^2J(^{15}$NFe^{15}N) is observed. The spectrum shows that [Fe$_4$S$_3$(NO)$_7$]$^-$ retains its structure in solution. The spectrum of [Fe$_4$Se$_3$(^{15}NO)$_7$]$^-$, which has not yet been characterized by X-ray crystallography, is very similar to that of [Fe$_4$S$_3$(^{15}NO)$_7$]$^-$; this and the observation of only a single ^{77}Se resonance (*25*) show that [Fe$_4$Se$_3$(NO)$_7$]$^-$ also has C_{3v} symmetry in solution.

Each of [Fe$_4$X$_4$(^{15}NO)$_4$] (99% ^{15}N; X = S or Se) exhibits a singlet only in its ^{15}N NMR spectrum in CD$_2$Cl$_2$ solution, demonstrating structural similarity and structural integrity in solution in this solvent.

All of the tetrairon complexes have ^{15}N chemical shifts characteristic of linear Fe–N–O groups.

V. Electronic Structure

A. OXIDATION STATES

Bulk susceptibility measurements (6) and the observation of sharp, unshifted resonances in the NMR spectra of both diiron and tetrairon complexes, and of cosolutes (37), show that $[Fe_2S_2(NO)_4]^{2-}$ and all $[Fe_2(SR)_2(NO)_4]$, $[Fe_4X_3(NO)_7]^-$, and $[Fe_4X_4(NO)_4]$ (X = S or Se) species are diamagnetic. On the other hand, the mononuclear complexes 5 (18, 23), 9 and 10 (49), and 11 (51) are all paramagnetic, each containing a single unpaired electron per molecular unit. Likewise, $[Fe_4S_4(NO)_4]^-$ (37) and $[Fe_4Se_3(NO)_4(PPh_3)_3]$ (52) are paramagnetic, containing, respectively, one and probably two unpaired electrons per molecular unit.

Furthermore, Mössbauer spectroscopy has shown that the two iron atoms in $[Fe_2S_2(NO)_4]^{2-}$ are equivalent (71), and that the four sites in each of $[Fe_4S_4(NO)_4]$ (72) and $[Fe_4S_4(NO)_4]^-$ (37) are equivalent, as expected from the observed (37) T_d and D_{2d} molecular symmetries. However, Mössbauer studies of isomorphous alkali metal or ammonium $[Fe_4S_3(NO)_7]^-$ salts have provided contradictory conclusions; two studies (72, 73) concluded that the spectrum contained only a single quadrupole doublet, despite the clear crystallographic demonstration (11, 39) of two quite different iron environments (cf. Fig. 1), while in two further studies (71, 74) the spectra were interpreted in terms of two iron sites whose relative populations were refined as 2.85:1 (71) and 3.03:1 (74). It is probably best to conclude that the two geometrically different iron types in $[Fe_4S_3(NO)_7]^-$ are in similar electronic environments. Likewise, the two geometrically distinct iron types in $[Fe_4S_2(NO)_4(NCMe_3)_2]$ (41) are in very similar electronic environments, as this compound also gives a Mössbauer spectrum containing only a single quadrupole doublet (72). For each of $NH_4[Fe_4S_3(NO)_7]$, $[Fe_4S_4(NO)_4]$, and $[Fe_4S_2(NO)_4(NCMe_3)_2]$, the Mössbauer spectra are indicative of diamagnetic ground states (72), consistent with bulk susceptibility and NMR data.

Since in all iron–sulfur–nitrosyl complexes hitherto characterized by either X-ray crystallography or ^{15}N NMR spectroscopy, the Fe–N–O fragments are approximately linear, the nitrosyl ligands are considered to be bound formally as NO^+. This then implies that the formal oxidation state of the iron atoms in $[Fe_2S_2(NO)_4]^{2-}$ and in $[Fe_2(SR)_2(NO)_4]$ is Fe(−I), d^9; in $[Fe_4S_4(NO)_4]$ and $[Fe_4S_2(NO)_4(NCMe_3)_2]$ it is Fe(I), d^7, while in $[Fe_4S_3(NO)_7]^-$ the apical iron is formally Fe(I) and the three basal irons are formally Fe(−I). In the paramagnetic mononuclear complexes 5, 9, 10, and

16, the $\{Fe(NO)\}^7$ (*18*) fragments contain formally Fe(I), while in complexes **11** and **15**, the $\{Fe(NO)_2\}^9$ fragments contain formally Fe(−I).

Adoption of the oxidation state Fe(−I) for $Fe(NO)_2$ fragments leads to the assignment of Co(O) in the binuclear heterometallic complex **17**, $[(ON)_2Fe(SPh)_2Co(NO)(PPh_3)]$, while adoption of Fe(I) for the Fe(NO) fragments in the heterometallic cubane **19** leads to a formal oxidation state V(III) for vanadium.

The diamagnetic behavior of the diiron and tetrairon complexes, despite the presence of formally d^7 and/or d^9 iron centers, indicates very strong coupling between the individual paramagnetic centers: all theoretical treatments of polynuclear iron–sulfur–nitrosyl complexes to date have been based on the assumption of diamagnetism in even-electron species and have employed molecular orbital methods at various levels of approximation.

B. DINUCLEAR COMPLEXES

Extended Hückel molecular orbital (EHMO) calculations on $[Fe_2S_2(NO)_4]^{2-}$ and $[Fe_2(SH)_2(NO)_4]$, using geometries derived from that found for $[Fe_2(SEt)_2(NO)_4]$ by X-ray analysis, showed (*75*) that the frontier orbitals are concentrated in the Fe–Fe bond in each case, with the HOMO bonding and the LUMO antibonding in the Fe–Fe bond. The frontier orbitals in *anti*-$[Fe_2(SH)_2(NO)_4]$ are shown in Fig. 12. The electron-precise nature of the Fe–Fe interaction suggests that both electron addition to and electron removal from $[Fe_2(SR)_2(NO)_4]$ will weaken the Fe–Fe bond, possibly leading to cleavage into mononuclear fragments. It is noteworthy in this connection that the stoichiometrically analogous cobalt complexes $[Co_2(SR)_2(NO)_4]$, in which there is an extra pair of electrons compared with $[Fe_2(SR)_2(NO)_4]$ and which lack a metal–metal bond, are much more reactive than the iron complexes and very readily form mononuclear complexes (*26*). Thus $[Co_2(SPh)_2(NO)_4]$ reacted with triphenylphosphine to give an 88% yield of mononuclear $[Co(NO)_2(SPh)(PPh_3)]$, in 15 minutes at room temperature, while $[Fe_2(SPh)_2(NO)_4]$ gave a 50% yield of $[Fe(NO)_2(PPh_3)_2]$ only after reaction for 3 days (*26*).

The complexes $[Fe_2(SR)_2(NO)_4]$ contain a total of 34 valence electrons and are isoelectronic with $[Fe_2(SR)_2(CO)_6]$ and with $[Fe_2(CO)_9]$; with a single Fe–Fe bond, each obeys the 18-electron rule. The cobalt complexes $[Co_2(SR)_2(NO)_4]$, on the other hand, contain 36 valence electrons and thus obey the 18-electron rule in the absence of any metal–metal bond, while the heterometallic complex **17** is another 34-electron species containing a Co–Fe bond.

C. Tetranuclear Complexes

In both [Fe$_4$X$_4$(NO)$_4$](X = S or Se) and [Fe$_4$S$_2$(NO)$_4$(NCMe$_3$)$_2$] the total valence electron count is 60. This is the number characteristic of tetrahedral tetranuclear metal clusters, such as [Ir$_4$(CO)$_{12}$], in the Wade and Mingos skeletal-electron counting schemes (76, 77) and, furthermore, each iron atom in these clusters obeys the 18-electron rule, provided that it forms single Fe–Fe bonds to each of the other iron atoms in the tetrahedron.

However, the counting schemes which have proved to be so valuable in the rationalization of stoichiometry and structure in both metal–carbonyl clusters and posttransition metal clusters cannot play more than a marginal role for metal–sulfur–nitrosyl systems, since the total valence electron counts in currently known Fe$_4$ and related systems range between 54 and 72; thus, for example, [Fe$_4$S$_4$(SR)$_4$]$^{2-}$ (78, 79) has a total valence electron count of 54, [Fe$_4$S$_4$(NO)$_4$] and [Fe$_4$S$_2$(NO)$_4$(NCMe$_3$)$_2$] have 60, [Fe$_4$Se$_3$(NO)$_4$(PPh$_3$)$_3$] (52) has 62, [Fe$_4$S$_3$(NO)$_7$]$^-$ has 66, [Fe$_4$S$_4$(η^5-C$_5$H$_5$)$_4$] (80) has 68, and [Fe$_4$S$_4$(CO)$_{12}$] (81) has 72. Among heterometallic M$_4$ systems, complexes **19** and **21** (26, 56) have 58 and 59 valence electrons, respectively. In addition, many stoichiometries occur in a range of oxidation states, for example, [Fe$_4$S$_4$(SR)$_4$]$^{2-/3-}$ (78, 79, 82, 83), [Fe$_4$S$_4$(NO)$_4$]$^{0/1-/2-}$ (37), [Fe$_4$S$_2$(NO)$_4$(NCMe$_3$)$_2$]$^{0/1-}$ (41), [Fe$_4$S$_4$(η^5-C$_5$H$_5$)$_4$]$^{1-/0/1+/2+/3+}$ (80, 84, 85), [(η^5-C$_5$H$_4$CH$_3$)$_2$V$_2$Fe$_2$S$_4$(NO)$_2$]$^{0/1-/2-}$ (26), and [(η^5-C$_5$H$_4$CH$_3$)$_2$V$_2$FeCoS$_4$(NO)$_2$]$^{0/1-/2-}$ (26). In virtually all of the complexes of type L$_4$M$_4$X$_4$, the M$_4$X$_4$ core has either T_d or D_{2d} symmetry, regardless of the valence electron count, and hence these systems must be regarded as outside the scope of Wade's rules.

In tackling problems of molecular and electronic structure such as those posed by these sulfur-bridged tetrametallic clusters, two approaches are possible. One is the construction, often heavily reliant upon symmetry arguments, of a general, essentially qualitative model intended to provide a broad description of a series of molecules; the other is a detailed quantum mechanical study, at an appropriate level of theory, of individual molecules, followed by a search for generalization or patterns. A combination of both approaches is usually the most valuable for chemical understanding, and much effort along these lines has been expanded on cubane-type clusters and their derivatives.

Dahl has employed (37, 85) a qualitative molecular orbital model for [Fe$_4$S$_4$L$_4$] complexes based on T_d symmetry, the essential features of which are summarized in Fig. 13. The 20 iron d orbitals form molecular orbitals which can be divided into three major groups. The first group,

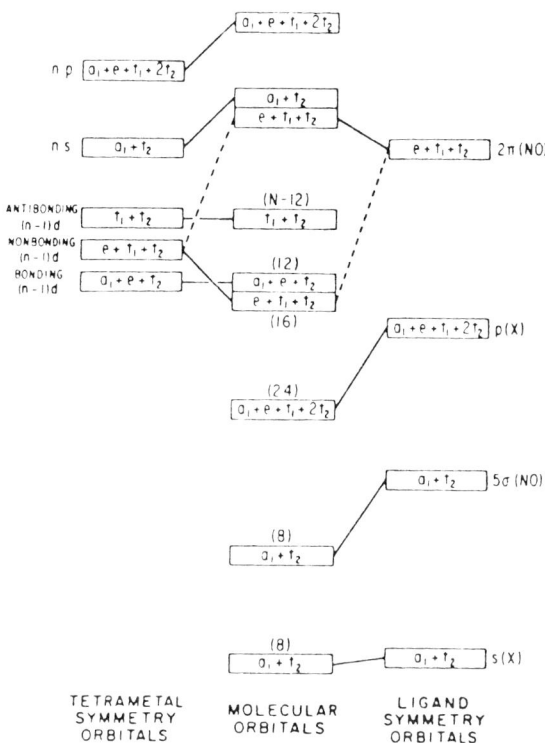

FIG. 13. Qualitative molecular orbital scheme for [Fe$_4$S$_4$L$_4$]. Complexes of T_d symmetry. Reproduced with permission from *J. Am. Chem. Soc.* **104**, 3409 (1982).

of 8 orbitals of symmetry types ($e + t_1 + t_2$), is primarily involved in π bonding with ligands L, and is essentially nonbonding in the Fe$_4$ cage; the second group, 6 orbitals of types ($a_1 + e + t_2$), is bonding in the Fe$_4$ cage; and the third group, also of 6 orbitals, of symmetry types ($t_1 + t_2$), is antibonding in the Fe$_4$ cage. While the Fe$_4$ antibonding orbitals are the highest in energy, the model retains some flexibility in the ordering of the bonding and nonbonding sets, dependent upon the pendant ligands L.

Table II shows the distribution of electrons among the molecular orbitals for a range of representative Fe$_4$S$_4$ complexes. According to this scheme, both [Fe$_4$S$_4$(NO)$_4$] and [Fe$_4$S$_4$(CO)$_{12}$] have closed-shell configurations, and hence should show exact T_d symmetry, as observed (*37, 81*); on the other hand, in T_d symmetry, both [Fe$_4$S$_4$(SR)$_4$]$^{2-}$ and [Fe$_4$S$_4$(η^5-C$_5$H$_5$)$_4$] have open-shell configuration, and hence are expected, via the Jahn–Teller effect, to undergo a lowering of

TABLE II

ELECTRON DISTRIBUTIONS IN REPRESENTATIVE Fe_4S_4 COMPLEXES UNDER ASSUMED T_d SYMMETRY

Orbitals	$[Fe_4S_4(SR)_4]^{2-}$	$[Fe_4S_4(NO)_4]$	$[Fe_4S_4(\eta^5\text{-}C_5H_5)_4]$	$[Fe_4S_4(CO)_{12}]$
Fe_4 antibonding $(t_1 + t_2)$	0	0	8	12
Fe_4 nonbonding $(e + t_1 + t_2)$	10	16	16	16
Fe_4 bonding $(a_1 + e + t_2)$	12	12	12	12
Fe–X $(a_1 + e + t_1 + 2t_2)$	24	24	24	24
Fe–L $(a_1 + t_2)$	8	8	8	8
Total valence electrons	54	60	68	72

their symmetry, again exactly as observed (78–80). Reduction of $[Fe_4S_4(NO)_4]$ to $[Fe_4S_4(NO)_4]^-$ yields an open-shell species, which according to this scheme is Jahn–Teller sensitive; the observed (37) structure for $[Fe_4S_4(NO)_4]^-$ has D_{2d} symmetry, as do $[Fe_4S_4(SR)_4]^{2-}$ (78, 79) and $[Fe_4S_4(\eta^5\text{-}C_5H_5)_4]$ (80). Furthermore, this model neatly rationalizes the steady increase in the Fe–Fe distances as the number of Fe_4 antibonding electrons increases from zero in $[Fe_4S_4(NO)_4]$ (37) to six in $[Fe_4S_4(\eta^5\text{-}C_5H_5)_4]^{2+}$ (85), seven in $[Fe_4S_4(\eta^5\text{-}C_5H_5)_4]^+$ (84), and eight in neutral $[Fe_4S_4(\eta^5\text{-}C_5H_5)_4]$ (80).

According to Dahl's model (37), $[Fe_4S_4(NO)_4]$ with 60 valence electrons has 12 electrons which are bonding in the Fe_4 cage, equivalent to six single Fe–Fe bonds, one along each edge of the cage. In the 66-electron anion $[Fe_4S_3(NO)_7]^-$ (39), application of the same basic model with the overall symmetry lowered from T_d to C_{3v} provides a Fe_4 cage having 12 bonding and 6 antibonding electrons, equivalent to a total Fe_4 bond order of 3, or a single Fe–Fe bond along only three edges of the Fe_4 cage; experimentally (39), there are three Fe–Fe distances of 2.70 Å (bonds) between apical and basal irons and three of 3.57 Å (nonbonds) between pairs of basal irons.

Quantitative calculations (75) on $[Fe_4S_4(NO)_4]$ using the extended Hückel molecular orbital (EHMO) method have provided results in broad agreement with the Dahl scheme (37) (cf. Fig. 13 and Table II): the 4 molecular orbitals arising from the NO orbitals lie at -15.61 (a_1) and -15.54 (t_2) eV; the 12 Fe–S orbitals are grouped between -15.30 and -14.31 eV; and the 14 occupied Fe–Fe orbitals are grouped between -13.28 and -11.41 eV. However, the orbitals which are, respectively, bonding or nonbonding for the Fe_4 cage are not found in separate energy regions, but are interleaved. The highest occupied molecular orbital (HOMO) of t_2 type is nonbonding in Fe_4, while the

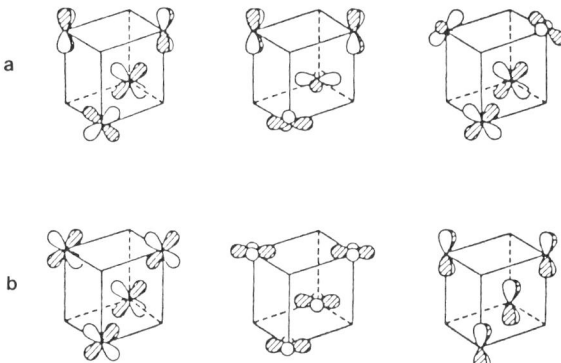

FIG. 14. Principal contributing orbitals to the (a) LUMO (T_1) and (b) HOMO (T_2) of [Fe$_4$S$_4$(NO)$_4$]. Reproduced with permission from *Inorg. Chem.* 24, 3858 (1985).

LUMO of t_1 type is antibonding in Fe$_4$; the principal components of these orbitals are shown in Fig. 14. Thus addition of electrons to [Fe$_4$S$_4$(NO)$_4$] should weaken the cluster bonding, as observed (37), but removal of one or two electrons should not seriously influence the cluster bonding. Similar calculations on [Fe$_4$S$_3$(NO)$_7$]$^-$ show a broadly similar order of the molecular orbitals, with the uppermost 17 occupied orbitals concentrated in the Fe$_4$ cage. Here bonding, nonbonding, and antibonding orbitals are all interleaved: the HOMO is a bonding Fe$_4$ orbital of a_1 type and the LUMO an antibonding Fe$_4$ orbital e type; the main components of these orbitals are shown in Fig. 15. Of the 17 occupied Fe$_4$ orbitals, 7 are bonding, 6 nonbonding, and 4 antibonding, giving a total net Fe$_4$ bond order of 3, essentially as required by the Dahl model (37). Since the HOMO and LUMO in [Fe$_4$S$_3$(NO)$_7$]$^-$ are, respectively, bonding and antibonding in Fe$_4$, either addition or subtraction of an electron is expected to weaken the cage bonding.

Both [Fe$_4$S$_4$(NO)$_4$] and [Fe$_4$S$_3$(NO)$_7$]$^-$ were calculated (75) to have degenerate LUMOs of t_1 and e symmetry, respectively, so that Jahn–Teller ions are expected to result, in each case, from addition of an electron. In [Fe$_4$S$_3$(NO)$_7$]$^{2-}$, the system must distort along an e vibrational mode to give a product of C_s symmetry [cf. [Fe$_4$Se$_3$(NO)$_7$]$^{2-}$ (52)], while in [Fe$_4$S$_4$(NO)$_4$]$^-$ the system distorts either along an e mode to give a product of D_{2d} symmetry [as observed (37)], or along a t_2 mode to give a product of C_{3v} symmetry.

The electronic structures of the related complexes [(η^5-C$_5$H$_5$)$_4$M$_4$X$_4$] have been discussed by Bottomley (86).

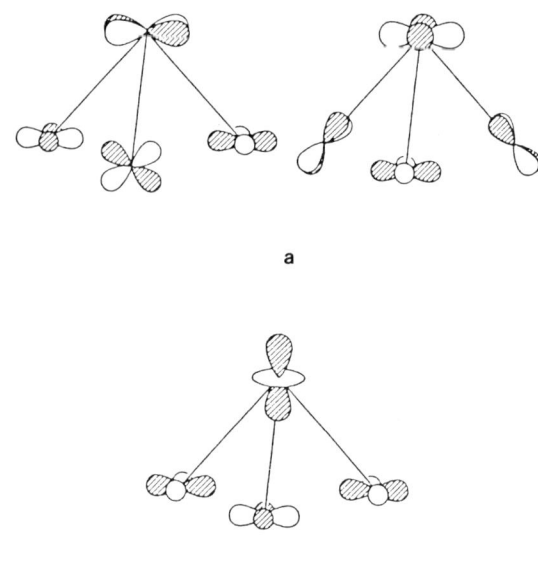

FIG. 15. Principal contributing orbitals to the (a) LUMO (E) and (b) HOMO (A_1) of $[Fe_4S_3(NO)_7]^-$. Reproduced with permission from *Inorg. Chem.* **24**, 3858 (1985).

D. MONONUCLEAR COMPLEXES

For the paramagnetic monoiron complexes **5** (R = Me) and **16**, having approximately square-pyramidal coordination of iron, EHMO calculations have shown (55) that the SOMO is of σ symmetry, directed along the Fe–N–O vector (Fig. 16a): this appears to be generally so for square–pyramidal iron- and nitrosyl-containing $\{Fe(NO)\}^7$ fragments, including also $[Fe(NO)(SH)_4]^{2-}$ (55) and $[Fe(CN)_4NO]^{2-}$ (87). On the other hand, in paramagnetic dinitrosyliron complexes of the general type $[Fe(NO)_2X_2]$, containing $\{Fe(NO)_2\}^9$ fragments, the SOMO is (55) of π symmetry with respect to the iron–nitrosyl directions (Fig. 16b). For linear Fe–N–O fragments with the nitrosyl ligand bound as NO^+, the tetrahedral $\{Fe(NO)_2\}^9$ complexes contain d^9 Fe(−I) and therefore have $S = \frac{1}{2}$ ground states. The square–pyramidal $\{Fe(NO)\}^7$ species contain d^7 Fe(I), but the d electron configuration $(e)^4(b_2)^2(a_1)^1(b_1)^0$ in approximate C_{4v} symmetry ensures spin pairing and hence an $S = \frac{1}{2}$ ground state.

The importance of the different nodal properties of the two types of SOMO shown in Fig. 16 lies in their influence upon the nitrogen hyperfine coupling $A(^{14}N)$ in the electron spin resonance spectra in

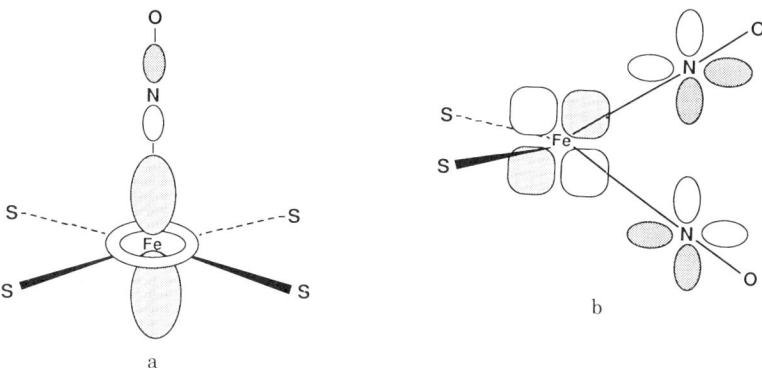

FIG. 16. Principal contributing orbitals to the SOMO in (a) square–pyramidal {Fe(NO)}^{7} complexes and (b) tetrahedral {Fe(NO)$_2$}^{9} complexes.

paramagnetic monoiron complexes. Those complexes having a σ SOMO of the type shown in Fig. 16a have $A(^{14}N)$ values in the range 12–15 G, while those having a π SOMO as shown in Fig. 16b have $A(^{14}N)$ values in the range 2–6 G (23, 55, 87). Hence the ESR spectra of paramagnetic monoiron–nitrosyl complexes give immediate information on the electronic structure, provided only that $A(^{14}N)$ can be resolved. Thus, for example, the complexes **9** and **10** each have an ESR spectrum with $A(^{14}N)$ of 12 G (49, 51), indicative of a σ SOMO in each case, while the series of complexes [Fe(NO)$_2$(SR)$_2$]$^-$ and [Fe(NO)(SR)$_3$]$^-$ (R = alkyl; see Section VI,B) have (23, 55) $A(^{14}N)$ values clearly indicative of a π SOMO in every case.

VI. Chemical Reactivity

The reactions of the anion [Fe$_2$S$_2$(NO)$_4$]$^{2-}$ with electrophiles have already been described (Section II,B,2). Apart from these reactions, most of the chemistry so far reported for iron–sulfur–nitrosyl systems involves the dinuclear complexes [Fe$_2$(SR)$_2$(NO)$_4$], the tetranuclear [Fe$_4$S$_3$(NO)$_7$]$^-$ and [Fe$_4$S$_4$(NO)$_4$], and the paramagnetic mononuclear species [Fe(NO)$_2$(SR)$_2$]$^-$ and [Fe(NO)(SR)$_3$]$^-$, which prove to be important reactive intermediates in a wide range of reactions.

A. Redox Reactions

Most of the reactions of the [Fe$_4$S$_3$(NO)$_7$]$^-$ salts described by Roussin in his original report (1) on [Fe$_4$S$_3$(NO)$_7$]$^-$ and [Fe$_2$S$_2$(NO)$_4$]$^{2-}$ were redox reactions: with a number of simple metal salts such as AgNO$_3$ or

Hg(NO$_3$)$_2$, oxidation of [Fe$_4$S$_3$(NO)$_7$]$^-$ occurred with liberation of nitrogen oxides and presumably complete breakup of the cluster; oxidants such as KMnO$_4$, PbO, HgO, and elemental halogens also liberated nitrogen oxide from [Fe$_4$S$_3$(NO)$_7$]$^-$. None of these investigations was other than qualitative, since Roussin had no adequate knowledge of the composition of his salt. More recently (88) it was reported that the reaction between [Fe$_4$S$_3$(NO)$_7$]$^-$ and [Fe(CN)$_6$]$^{3-}$ involved the transfer of three electrons per cluster, but no reaction products were characterized.

Cyclic voltammetry on [Fe$_4$S$_4$(NO)$_4$] showed (37) two reversible one-electron reduction waves corresponding to reduction of the neutral complex to [Fe$_4$S$_4$(NO)$_4$]$^-$ and thence to [Fe$_4$S$_4$(NO)$_4$]$^{2-}$, rather than to oxidation to [Fe$_4$S$_4$(NO)$_4$]$^+$ and reduction to [Fe$_4$S$_4$(NO)$_4$]$^-$ as reported in a preliminary communication (89); chemical reduction with, for example, [(η^5-C$_5$H$_5$)$_2$Co], sodium amalgam, or potassium metal in the presence of 2,2,2-crypt yields salts of [Fe$_4$S$_4$(NO)$_4$]$^-$ (37), whose structural characterization was described in Section III,C. Isolation and characterization of the dianion [Fe$_4$S$_4$(NO)$_4$]$^{2-}$ have not yet been reported. However, in the analogous selenium system, where cyclic voltammetry again showed (52) two reversible one-electron reduction waves corresponding to reduction of [Fe$_4$Se$_4$(NO)$_4$] to [Fe$_4$Se$_4$(NO)$_4$]$^-$ and thence to [Fe$_4$Se$_4$(NO)$_4$]$^{2-}$, salts of both anions were isolated; the dianion appears to be rather labile, and only the monoanion has been characterized (52).

In like manner, cyclic voltammetry indicates a one-electron reduction of [Fe$_4$S$_2$(NO)$_4$(NCMe$_3$)$_2$] (41), and by use of sodium amalgam as reductant the monoanion [Fe$_4$S$_2$(NO)$_4$(NCMe$_3$)$_2$]$^-$ was isolated as its (Ph$_3$PNPPh$_3$)$^+$ salt. The heterometallic cubanes **19** and **21** each show two reversible one-electron reduction waves in cyclic voltammetry (26), indicative of the formation of [(LV)$_2$Fe$_2$(NO)$_2$S$_4$]$^-$ and [(LV)$_2$Fe$_2$(NO)$_2$S$_4$]$^{2-}$ and of [(LV)$_2$FeCo(NO)$_2$S$_4$]$^-$ and [(LV)$_2$FeCo(NO)$_2$S$_4$]$^{2-}$, respectively (L = η^5-MeC$_5$H$_4$), although the isolation of these anionic products has not yet been reported.

In contrast to the stepwise reduction of [Fe$_4$S$_4$(NO)$_4$], [Fe$_4$Se$_4$(NO)$_4$], and the heterometallic cubanes **19** and **21**, cyclic voltammetry of the diiron complex [Fe$_2$(TePh)$_2$(NO)$_4$] showed (52) a single reversible two-electron wave corresponding to reduction to [Fe$_2$(TePh)$_2$(NO)$_4$]$^{2-}$, salts of which were subsequently isolated by chemical reduction. This dianion is a 36-electron species, and hence is isoelectronic with the neutral cobalt complexes [Co$_2$(SR)$_2$(NO)$_4$] (26); it thus lacks an Fe–Fe bond.

A further synthetically useful chemical reduction has been found in the reactions of $[Fe_4X_4(NO)_4]$ (X = S or Se) with carbon monoxide under pressure (81): the 60-electron clusters $[Fe_4X_4(NO)_4]$ were thereby converted to the 72-electron clusters $[Fe_4X_4(CO)_{12}]$.

In contrast, synthetically useful oxidations are so far few in number, but reaction of the 66-electron species $[Fe_4X_3(NO)_7]^-$ with elemental X (X = S or Se) provides (27, 53) good yields of the 60-electron cubanes $[Fe_4X_4(NO)_4]$; although excellent routes exist to $[Fe_4X_4(NO)_4]$ from $Hg[Fe(CO)_3NO]_2$ (37, 52), synthesis from $[Fe_4X_3(NO)_7]^-$ is probably the most convenient method when isotopically enriched cubanes are required.

B. Conversion of Di- and Tetrairon Complexes into Paramagnetic Monoiron Complexes

1. Diiron Complexes

Although the neutral diiron species $[Fe_2(SR)_2(NO)_4]$ give sharp, unshifted ^1H NMR spectra in noncoordinating or weakly coordinating solvents, in solvents which are good coordinating agents such as dimethylformamide (DMF) or dimethyl sulfoxide (DMSO), only very broad, unresolved absorptions were observed in the ^1H NMR spectra (65) of $[Fe_2(SMe)_2(NO)_4]$. Subsequent examination (23) by ESR spectroscopy, of solutions of $[Fe_2(SMe)_2(NO)_4]$ in a range of coordinating solvents, gave clear evidence of the formation of paramagnetic mononuclear dinitrosyliron complexes. Two types of solvocomplex were identified: $[Fe(NO)_2(L)_2]^+$ (L = solvent), in which both RS^- groups have been displaced, and $[Fe(NO)_2(SMe)L]$, in which the RS groups remain bound to iron. The principal ESR parameters for a range of solvocomplexes formed from $[Fe_2(SMe)_2(NO)_4]$ are summarized in Table III; a similar range of solvocomplexes from $[Fe_2(SPr^i)_2(NO)_4]$ was also characterized. The case of formation of solvocomplexes was further demonstrated by the formation (23) of such a complex—having $g = 2.030$, but having no resolvable hyperfine coupling—simply by dissolving $[Fe_2(SCH_2COOH)_2(NO)_4]$ in water. The solvocomplexes all have g values close to 2.03, and, in general, π-acceptor solvents form complexes of the type $[Fe(NO)_2(L)_2]^+$, while solvents with little or no π-bonding capacity as ligands form neutral solvocomplexes of the type $[Fe(NO)_2(SR)(L)]$. With DMF or dimethylacetamide as solvent, both possible solvocomplexes are observed.

By use of a more strongly nucleophilic reagent RS^-, again in DMF as solvent, very ready conversion of $[Fe_2(SR)_2(NO)_4]$ to $[Fe(NO)_2(SR)_2]^-$ occurs (23) [Eq. (14)].

TABLE III

ESR Characteristics of Solvocomplexes Formed from $[Fe_2(SMe)_2(NO)_4]$

Solvent	Complex	g	$A(^{14}N)^a$	$A(^1H)(SR)^a$	$A(X)^a$
DMF	$[Fe(NO)_2(DMF)_2]^+$	2.033	2.4(2N)	—	4.0(2H)
	$[Fe(NO)_2(SMe)(DMF)]$	2.027	2.4(2N)	4.5(3H)	4.6(1H)
DMSO	$[Fe(NO)_2(SMe)(DMSO)]$	2.032	6.0(2N)	3.2(3H)	—[b]
Pyridine (py)	$[Fe(NO)_2(py)_2]^+$	2.031	2.2(2N)	—	4.5(2N)
2,6-Dimethylpyridine	$[Fe(NO)_2(S)_2]^{+\,c}$	2.031	2.3(2N)	—	4.6(2N)
Quinoline	$[Fe(NO)_2(S)_2]^{+\,c}$	2.032	2.2(2N)	—	4.4(2N)
Et_2NH	$[Fe(NO)_2(SMe)(Et_2NH)]$	2.030	4.0(2N)	2.0(3H)	—[b]
Pyrrolidine (pyr)	$[Fe(NO)_2(SMe)(pyr)]$	2.029	3.9(2N)	2.0(3H)	—[b]
Piperidine (pip)	$[Fe(NO)_2(SMe)(pip)]$	2.029	4.2(2N)	2.1(3H)	—[b]

[a] A values in gauss; X represents ^1H or ^{14}N in coordinated solvent molecules.
[b] Hyperfine coupling not resolved.
[c] S represents 2,6-dimethylpyridine or quinoline.

$$[Fe_2(SR)_2(NO)_4] + 2RS^- \longrightarrow 2[Fe(NO)_2(SR)_2]^- \qquad (14)$$

Alternatively, base can be added to a solution of $Fe_2(SR)_2(NO)_4$ in RSH [Eq. (15)].

$$[Fe_2(SR)_2(NO)_4] + 2RSH \xrightarrow{\text{base}} 2[Fe(NO)_2(SR)_2]^- \qquad (15)$$

These mononuclear species give ESR spectra characterized in every case by hyperfine coupling to two ^{14}N nuclei and to the α hydrogens, when present, of the SR groups (23, 55) (Table IV). If base is added

TABLE IV

ESR Characteristics of $[Fe(NO)_2(SR)_2]^-$ and $[Fe(NO)(SR)_3]^-$

	$[Fe(NO)_2(SR)_2]^-$			$[Fe(NO)(SR)_3]^-$	
R	g	$A(^{14}N)$ (G)	$A(^1H)$ (G)	g	$A(^{14}N)$ (G)
H	2.028	2.7	0.5(2H)	2.020	5.0
Me	2.028	2.1	2.1(6H)	2.021	4.5
Et	2.027	2.0	2.6(4H)	2.021	4.6
i-Pr	2.027	2.5	1.3(2H)	2.021	5.0
t-Bu	2.027	2.7	—(0H)	2.021	5.0
Ph	2.027	2.5	—(0H)	—	—
$PhCH_2$	2.028	2.4	1.4(4H)	2.021	4.8

to a dilute solution of [Fe$_2$(SR)$_2$(NO)$_4$] in a different thiol R'SH, or if excess of R'S$^-$ is added to a DMF solution of [Fe$_2$(SR)$_2$(NO)$_4$], the product detected by ESR is always the ligand-exchanged complex [Fe(NO)$_2$(SR')$_2$]$^-$, rather than [Fe(NO)$_2$(SR)$_2$]$^-$ [Eq. (16)].

$$[Fe_2(SR)_2(NO)_4] + 2R'SH \xrightarrow{base} [Fe(NO)_2(SR')_2]^- \qquad (16)$$

The reaction of [Fe$_2$(SR)$_2$(NO)$_4$] with RS$^-$ proves to be the prototype of a wide range of similar reactions of [Fe$_2$(SR)$_2$(NO)$_4$] with other nucleophiles Y$^-$ to yield [Fe(NO)$_2$(Y)$_2$]$^-$. Examples of Y$^-$ so far found to yield [Fe(NO)$_2$(Y)$_2$]$^-$ include Y$^-$ = Br$^-$ or I$^-$ (*35*), NCO$^-$ or NCS$^-$ (*23*), NO$_2^-$ (*21*), and [MoS$_4$]$^{2-}$, which yields the dimetallic complex **15**, [Fe(NO)$_2$(S$_2$MoS$_2$)]$^-$ (*55*).

In the formation of each of these complexes, the incoming nucleophile becomes bound to iron. This is in contrast to the reactions of the only other iron–nitrosyl complex whose reactivity toward nucleophiles has been extensively investigated (*19, 90*), the nitroprusside ion, [Fe(CN)$_5$NO]$^{2-}$. In this anion, the Fe–N–O fragment is linear, both in the solid state (*91*) and in solution (*25*), just as for [Fe$_2$(SR)$_2$(NO)$_4$], but in all its reactions with nucleophiles X$^-$ (X$^-$ = OH$^-$ (*92*), SH$^-$ (*93*), RS$^-$ (*94*), amines (*95, 96*), NO$_2^-$ (*25*), N$_3^-$ and NH$_2$OH (*97*), and carbanions (*98, 99*), the incoming nucleophile is bonded to the nitrogen atom of the nitrosyl ligand; [SO$_3^{2-}$ is probably bonded to the oxygen of the nitrosyl ligand (*100–102*)]. Empirical correlations have been described (*19, 103, 104*) which indicate that metal–nitrosyl complexes having v(NO) > 1880 cm^{-1} undergo addition of nucleophiles to the nitrogen of the nitrosyl ligand; that complexes having 1880 cm^{-1} > v(NO) > 1800 cm^{-1} undergo no reactions with nucleophiles; and that complexes having 1800 cm^{-1} > v(NO) undergo addition of nucleophiles to the metal. In [Fe(CN)$_5$NO]$^{2-}$, v(NO) is 1938 cm^{-1} (*19*), whereas in [Fe$_2$(SR)$_2$(NO)$_4$], v(NO) always falls below 1800 cm^{-1} (*63*); hence the difference between Fe$_2$(SR)$_2$(NO)$_4$ and [Fe(CN)$_5$NO]$^{2-}$ in their reactions with nucleophiles falls neatly into the correlation between reactivity and v(NO).

2. Tetrairon Complexes

It was first observed (*22*) some years ago that although aqueous solutions of [Fe$_4$S$_3$(NO)$_7$]$^-$ gave no ESR spectra at neutral pH, raising the pH gave the spectrum of a mononitrosyl species characterized by $g = 2.027$ and $A(^{14}N) = 4.7$ G. On the other hand, in DMF solution

[Fe$_4$S$_3$(NO)$_7$]$^-$ gives (23) a complex ESR spectrum indicative of the presence of at least three paramagnetic species. These presumably include solvocomplexes.

When [Fe$_4$S$_3$(NO)$_7$]$^-$ reacts with RS$^-$ in DMF solution, then, for a range of substituents R, the products are (23) [Fe(NO)$_2$(SR)$_2$]$^-$, as formed from [Fe$_2$(SR)$_2$(NO)$_4$], together with [Fe(NO)(SR)$_3$]$^-$; in this latter series (Table IV), although the $A(^{14}N)$ value depends upon R, no hyperfine coupling to the α-hydrogen atoms in R was resolved, so that all the spectra of [Fe(NO)(SR)$_3$]$^-$ comprise three lines only. The mononitrosyl complexes are fairly short lived, and within 2 days only the dinitrosyl complexes were detectable. Since the formal oxidation states of iron in [Fe(NO)$_2$(SR)$_2$]$^-$ and [Fe(NO)(SR)$_3$]$^-$ are Fe(−I) and Fe(I), respectively, it is likely that [Fe(NO)$_2$(SR)$_2$]$^-$ arises from the basal iron atoms in [Fe$_4$S$_3$(NO)$_7$]$^-$, and that [Fe(NO)(SR)$_3$]$^-$ arises from the apical iron [Eq. (17)].

$$[Fe_4S_3(NO)_7]^- + 9RS^- \longrightarrow 3[Fe(NO)_2(SR)_2]^- + [Fe(NO)(SR)_3]^- + 3S^{2-} \quad (17)$$
$$\text{3Fe(−I) + Fe(I)} \qquad \text{3Fe(−I)} \qquad \text{Fe(I)}$$

The conversion of [Fe(NO)(SR)$_3$]$^-$ into [Fe(NO)$_2$(SR)$_2$]$^-$ has been ascribed (23) to a disproportionation [Eq. (18)].

$$2[Fe^I(SR)_3NO]^- \longrightarrow [Fe^{-I}(NO)_2(SR)_2]^- + [Fe^{III}(SR)_4]^- \quad (18)$$

Reaction of K[Fe$_4$S$_3$(NO)$_7$] with KOH in molten Ph$_2$S$_2$ as solvent leads to the isolation in modest yield of [Fe(NO)$_2$(SPh)$_2$]$^-$; X-ray analysis of the (Et$_4$N)$^+$ salt confirmed the presence of the monomeric anion with linear Fe−N−O groups and an approximately tetrahedral iron center (104a).

Mononuclear complexes are formed (23) from [Fe$_4$S$_3$(NO)$_7$]$^-$ also by reaction with Me$_2$NCS$_2^-$, when [Fe(NO)(S$_2$CNMe$_2$)$_2$] can be isolated in 88% yield based upon total iron, so that in addition to the apical Fe(NO) group, also the basal Fe(NO)$_2$ groups are incorporated into the product. In contrast, complete and rapid conversion of all the iron−nitrosyl fragments in [Fe$_4$S$_3$(NO)$_7$]$^-$ into Fe(NO)$_2$ groups is effected (23) by reaction of Na[Fe$_4$S$_3$(NO)$_7$]·H$_2$O with NaNO$_2$ in DMF, yielding [Fe(NO)$_2$(SH)$_2$]$^-$, but no mononitrosyl complexes. With [MoS$_4$]$^-$, [Fe$_4$S$_3$(NO)$_7$]$^-$ forms not only [Fe(NO)$_2$(SH)$_2$]$^-$ and [Fe(NO)(SH)$_3$]$^-$ but also [Fe(NO)(S$_2$MoS$_2$)$_2$]$^{2-}$ (55).

It appears that mononitrosyliron fragments {Fe(NO)}^{7} are favored in the presence of chelating ligands such as [Me$_2$NCS$_2$]$^-$ or [MoS$_4$]$^{2-}$, but that dinitrosyliron fragments $\{Fe(NO)_2\}^9$ are favored by the nonchelating ligands RS^-.

The formation of mononuclear complexes from $[Fe_4S_4(NO)_4]$ is broadly similar to formation from $[Fe_4S_3(NO)_7]^-$; thus (23) the anions RS^- yield the mononitrosyls $[Fe(NO)(SR)_3]^-$ [Eq. (19)], which are converted into the corresponding dinitrosyls, while $Me_2NCS_2^-$ yields $[Fe(NO)(S_2CNMe_2)_2]$.

$$[Fe_4S_4(NO)_4] + 12RS^- \longrightarrow 4[Fe(NO)(SR)_3]^- + 4S^{2-} \quad (19)$$

The mononuclear dinitrosyl complexes $[Fe(NO)_2(SR)_2]^-$ are readily formed in DMF solution from both diiron and tetrairon precursors; when a DMF solution of such a complex is made less polar by addition of a large volume of benzene or toluene, the green $[Fe(NO)_2(SR)_2]^-$ is rapidly and smoothly converted to red $[Fe_2(SR)_2(NO)_4]$ [Eq. (20)].

$$2[Fe(NO)_2(SR)_2]^- \xrightarrow[\text{solvent}]{\text{nonpolar}} 2RS^- + [Fe_2(SR)_2(NO)_4] \quad (20)$$

The process described in Eq. (20) is effectively the reverse of that described by Eq. (15).

C. Ligand Substitution Reactions

Mononuclear solvocomplexes of types $[Fe(NO)_2(L)_2]^+$ and $[Fe(NO)_2(SR)(L)]$ (L = solvent) are intermediates (23) in the thiol exchange reaction [Eq. (21)], and the proposed exchange mechanism is shown in Scheme 3. With the exception of $[Fe(NO)_2(SR)(SR')]^-$, all of the complexes in Scheme 3, and all of the individual equilibria, were observed (23) for at least one combination of R,R' and solvent L. In aromatic solvents such as pyridine the exchange proceeds (23) solely via $[Fe(NO)_2(L)_2]^+$, in solvents such as Et_3N solely via $[Fe(NO)_2(SR)(L)]$, and in solvents such as DMF, the exchange proceeds via both routes.

$$[Fe_2(SR)_2(NO)_4] + 2R'SH \longrightarrow 2RSH + [Fe_2(SR')_2(NO)_4] \quad (21)$$

In addition to rapid exchange of thiolate ligands with R'SH and for $R'S^-$, the mononuclear complexes $[Fe(NO)_2(SR)_2]^-$ also undergo (23) rapid exchange of the nitrosyl ligands in the presence of isotopically labeled nitrite; a W-shaped N_2O_3 intermediate was proposed (23) [Eq. (22)].

$$[L_xFeNO] \xrightarrow{*NO_2^-} \left[L_xFe-N\begin{matrix}O\\ \diagdown\\ O\\ *N\\ |\\ O\end{matrix}\right]^- \longrightarrow \left[L_xFe\begin{matrix}O\\ \diagdown N\diagup\\ \diagup \quad \diagdown O\\ *N\\ |\\ O\end{matrix}\right]^- \xrightarrow{-NO_2^-} [L_xFe*NO] \quad (22)$$

Similar fast exchange occurs with [Fe(NO)(S$_2$CNMe$_2$)$_2$] (23).

The very ready interconversion of Fe(NO)$^{2+}$ and Fe(NO)$_2^+$ fragments has already been described (Section VI,B).

One of the most striking and surprising of the substitution reactions of [Fe$_4$S$_3$(NO)$_7$]$^-$ is its conversion into nitroprusside [Fe(CN)$_5$NO]$^{2-}$. In his original paper (1) describing the synthesis of [Fe$_4$S$_3$(NO)$_7$]$^-$ and [Fe$_2$S$_2$(NO)$_4$]$^{2-}$, Roussin suggested a close connection between these anions and the nitroprusside ion, even though he did not know the constitutions, or even the accurate compositions, of any of them; he claimed also to have interconverted [Fe$_4$S$_3$(NO)$_7$]$^-$ and

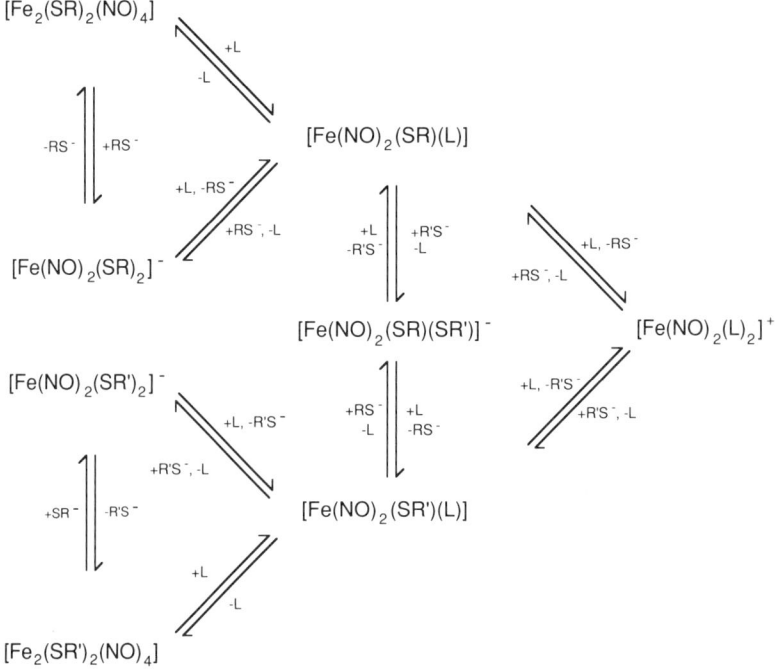

SCHEME 3. Proposed mechanisms of thiolate exchange in [Fe$_2$(SR)$_2$(NO)$_4$]. Ref. 23.

$[Fe(CN)_5NO]^{2-}$. This surprising claim has recently been substantiated (105). Reaction of $[Fe_4S_3(NO)_7]^-$ with excess cyanide in aqueous solution gives $[Fe(CN)_5NO]^{2-}$, in quantitative yield based upon iron, while reaction of excess H_2S with a hot aqueous solution of $[Fe(CN)_5NO]^{2-}$ provides, after extraction, $[Fe_4S_3(NO)_7]^-$ in quantitative yield, based upon nitrosyl ligands as the limiting factor. The formation of $[Fe(CN)_5NO]^{2-}$ must necessarily involve the fragmentation of the Fe_4S_3 core of $[Fe_4S_3(NO)_7]^-$, whose formation from $[Fe(CN)_5NO]^{2-}$ must in turn involve the spontaneous self-assembly of the cluster; otherwise, no mechanistic detail is known about these remarkable transformations. The formation of $[Fe_4S_3(NO)_7]^-$ in the reaction between $[Fe(CN)_5NO]^{2-}$ and dithioacetoin $CH_3C(S)CH(SH)CH_3$ (106) is presumably dependent upon the formation of H_2S by hydrolysis of the dithioacetoin.

The important reactions giving rise to mononuclear intermediates and some of the important ligand substitution reactions are summarized in Scheme 4. The interconversions summarized in Schemes 3 and 4 point to the central role played by the paramagnetic mononuclear complexes $[Fe(NO)_2(SR)_2]^-$ is much of the chemistry of iron–sulfur–nitrosyl systems. Regardless of whether the initial reacting complex is mononuclear, e.g., $[Fe(CO)_3NO]^-$ or $[Fe(NO)(S_2CNMe_2)_2]$, dinuclear, e.g., $[Fe_2(SR)_2(NO)_4]$, or tetranuclear, e.g., $[Fe_4S_4(NO)_4]$ or $[Fe_4S_3(NO)_7]^-$, many reactions proceed via common intermediates. All of the substitution reactions of $[Fe_2(SR)_2(NO)_4]$, $[Fe_4S_3(NO)_7]^-$, and $[Fe_4S_4(NO)_4]$ which have so far been studied in any detail appear, with the exception of the reactions of $[Fe_2S_2(NO)_4]^{2-}$ with electrophiles, to proceed via paramagnetic mononuclear intermediates. So also do a number of synthetic routes to the

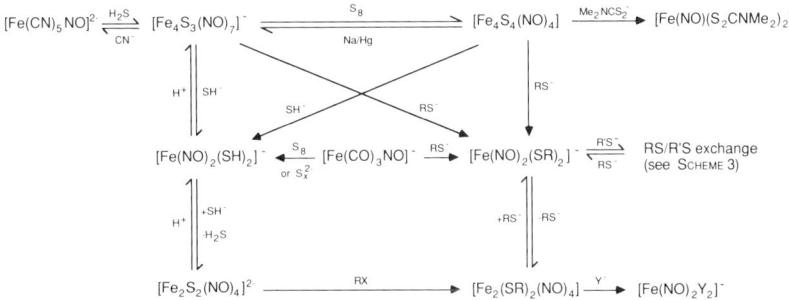

SCHEME 4. Important substitution reactions of iron–sulfur–nitrosyl systems.

diiron and tetrairon complexes, especially those from [Fe(CO)$_3$NO]$^-$ and [Fe$_2$I$_2$(NO)$_4$] (*23, 35*).

D. Nitrosylation Reactions

As well as being readily transferred from one iron center to another, in the interconversion of Fe(NO)$^{2+}$ and Fe(NO)$_2{}^+$ fragments, nitrosyl groups can also be transferred fairly readily to iron from external sources.

The preformed synthetic models (*13*) **23** and **24** for the [2Fe–2S] and [4Fe–4S] clusters, respectively, of natural redox proteins can be nitrosylated (*107*) under mild conditions with either nitric oxide or nitrite.

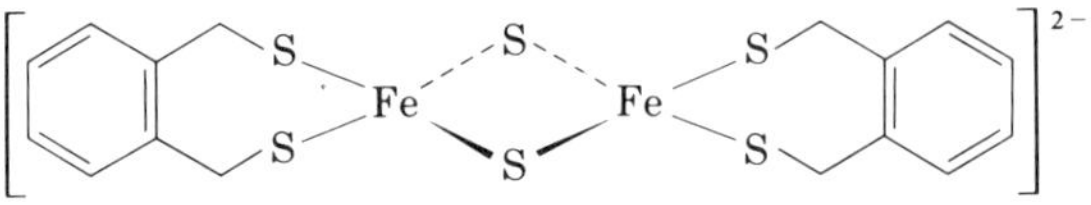

23

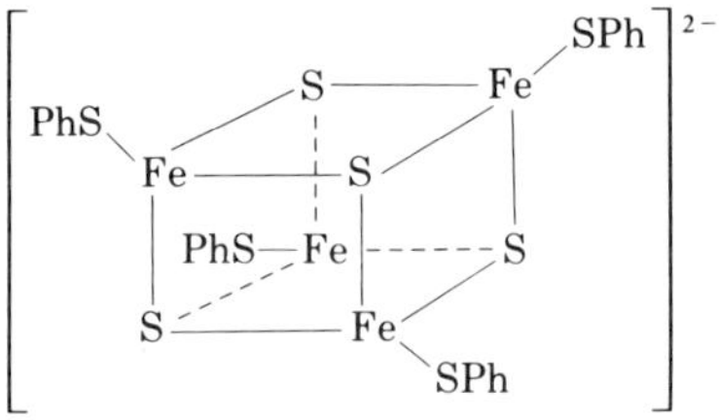

24

While nitric oxide in DMF solution yields mononuclear solvo-complexes of types [Fe(NO)$_2$(SR)(L)] and [Fe(NO)$_2$(L)$_2$]$^+$, nitrite in aqueous solution leads via [Fe(NO)$_2$(SH)$_2$]$^-$, detected by ESR spectroscopy, to isolable [Fe$_4$S$_3$(NO)$_7$]$^-$ in yields of 44 and 38% from compounds **23** and **24**, respectively.

More important than the demonstration (*107*) of nitrosylation of synthetic model clusters is the demonstration (*108*) of nitrosylation by nitrite of the [4Fe–4S] center in vegetative cells of *Clostridium botulinum*. Treatment of a reduced culture of *C. botulinum* with nitrite in the presence of ascorbate resulted in loss of the ESR signal at $g = 1.94$, characteristic of the reduced iron–sulfur protein, and its replacement by a signal at $g = 2.035$, characteristic of [Fe(NO)$_2$X$_2$] complexes, most likely of the [Fe(NO)$_2$(SR)$_2$]$^-$ type.

Although many of the details remain to be clarified, transfer of nitrosyl groups onto preformed iron–sulfur frameworks of the natural type is well established (*107, 108*). The transfer of nitrosyl groups from iron–sulfur–nitrosyl complexes to atoms other than iron is more problematical, as reported work on the ability of these complexes to nitrosate secondary amines RR'NH, with formation of secondary nitrosamines RR'NNO, has provided conflicting conclusions.

Initial reports (*12, 109*) indicated the ready nitrosation of a range of secondary amines by [Fe$_2$(SMe)$_2$(NO)$_4$]; however, in a more recent, thorough study on morpholine and pyrrolidine it was found (*110*) that although these underwent slow nitrosylation by Fe$_2$(SMe)$_2$(NO)$_4$ in air, under anaerobic conditions no nitrosylation occurred. However, these reactions were carried out in methylene chloride solution, and it is possible that in a coordinating solvent where solvocomplexes [Fe(NO)$_2$(SMe)(L)] and [Fe(NO)$_2$(L)$_2$]$^+$ are present, the rate of nitrosyl transfer may be higher: this possibility awaits investigation. In buffered aqueous acetone, when solvocomplexes are probably only present in very low concentration, the pH profiles reported (*110*) for the nitrosation of morpholine and pyrrolidine by [Fe$_2$(SMe)$_2$(NO)$_4$] show, for morpholine, an essentially constant rate over the pH range 4–11, with faster rates at extremes of pH; for pyrrolidine the rate increases steadily over the pH range 1–13. A similar *in vivo* study in rate on the nitrosation of proline by [Fe$_2$(SMe)$_2$(NO)$_4$] showed (*110*) slow formation of *N*-nitrosoproline, at a rate much faster than the background nitrosation found in the absence of [Fe$_2$(SMe)$_2$(NO)$_4$].

For both the *in vitro* studies with morpholine and pyrrolidine and the *in vivo* study with proline, it was observed that "partially decomposed" [Fe$_2$(SMe)$_2$(NO)$_4$] was a much more potent nitrosation agent than the pure complex (*110*). The nature of the decomposition was not reported, and it is not clear what the significance of this report may be, although nitrite was probably present in the decomposed material.

In view of these disparate findings (*12, 109, 110*) there is a need for a systematic product and rate study of the nitrosating ability of [Fe$_2$(SMe)$_2$(NO)$_4$] toward amines. It is likely that this will be substantially less than that of [Fe(CN)$_5$NO]$^{2-}$ (*95*), judged by the respective NO stretching frequencies (*19, 103, 104*).

E. Analytical Applications

The formation by [Fe$_4$S$_3$(NO)$_7$]$^-$ of water-insoluble salts with protonated amines has led (*111*) to its use as a gravimetric reagent for the microdetermination of naturally occurring nitrogenous bases such as nicotinamide. The $[Fe_4S_3(NO)_7]^-$ ion has also been extensively employed as a structural probe of the active sites of certain enzymes, especially the zinc enzyme carbonic anhydrase (112–117).

VII. Biological Chemistry

A. Cancerous States and $g = 2.03$ Complexes

Iron-centered paramagnetic complexes formed by reactions between iron salts and nitric oxide in the presence of anionic ligands, and characterized by $g = 2.03$, were first reported over 20 years ago (22); similar complexes, of the general type $[Fe(NO)_2X_2]^{x+}$, have subsequently been produced by reactions of iron salts and nitric oxide in the presence of halides and pseudohalides (118), alcohols and alkoxides (119), mercaptides (120, 121), and mercaptopurines and mercaptopyrimidines (122).

Similar paramagnetic iron–nitrosyl complexes, characterized by usually anisotropic g values around 2.03, have been observed (123–125) in extracts of rat liver following the administration of specific chemical carcinogens, such as aminoacetylfluorene (126) and N-methyl-N'-nitro-N-nitrosoguanidine (127), and also in extracts from many of the organs of several species of experimental animals maintained on a normal diet supplemented only by sodium nitrite and iron(II) sulfate (128, 129). In each of these biological studies (123–125, 128, 129), the identification of the $g = 2.03$ species as being of the general type $[Fe(NO)_2X_2]^{x+}$ was far from definitive, in that in general the substituent X was not positively identified; identifications were based primarily upon comparisons with the complexes originally observed (22) by Phillips and his colleagues. In every case the complexes were assigned constitutions of the type $[Fe(NO)_2(SR)_2]^-$, where R was unidentified but where SR may be taken either to be cysteine or to be a peptide or protein containing cysteine at the iron-binding site.

These assignments have been placed upon a much firmer basis with the definitive characterization (23) of the complexes $[Fe(NO)_2(SR)_2]^-$ synthesized from $[Fe_2(SR)_2(NO)_4]$ (Section VI,B,2), and the demonstration of the formation of such $g = 2.03$ complexes from synthetic and natural $[Fe_2S_2(SR)_4]^{2-}$ and $[Fe_4S_4(SR)_4]^{2-}$ complexes under mild conditions (107, 108).

Thus paramagnetic monoiron complexes $[Fe(NO)_2(SR)_2]^-$ are both readily formed *in vivo* (108, 123–125, 128, 129) and are associated with cancerous states in experimental animals (123–125), although this association does not of itself imply any causal relationships.

B. The Antimicrobial Activity of $[Fe_4S_3(NO)_7]^-$ Salts

The potassium salt $K[Fe_4S_3(NO)_7]$ was many years ago shown to exhibit bacteriostatic action against a range of microorganisms, including both aerobic and anaerobic types (130–132); its antiseptic action was shown (130) to provide a good disinfectant of contaminated drinking water.

When nitrite, a well-known and widely used food preservative, is heated either with meat products or with appropriate culture media, an inhibitor is formed active against several *Clostridium* and *Salmonella* species (133–137); several types of iron–sulfur–nitrosyl complexes have been investigated as the possible active agents (138–141). The complex $[Fe_4S_3(NO)_7]^-$ was shown (138–140) to inhibit *Clostridium* species and to be formed (140) when nitrite was autoclaved with a test medium containing iron(II) sulfate and cysteine: at the same time it was shown (140) that neither S-nitrosocysteine nor $[Fe_2S_2(NO)_4]^{2-}$ could be produced under conditions which produced $[Fe_4S_3(NO)_7]^-$. The complex isolated from a culture medium of acid-hydrolyzed casein, ascorbate, and nitrite was shown (140) chemically and spectroscopically to have a very close resemblance to an alkali metal salt of $[Fe_4S_3(NO)_7]^-$, although identity was not proved.

While these observations indicate that $[Fe_4S_3(NO)_7]^-$ both acts as a clostridial inhibitor and can be isolated from nitrite-treated proteins, it does not follow that the intact anion is the active species. As discussed in Section VI,D, the action of nitrite upon preformed iron–sulfur clusters leads to the formation of paramagnetic mononuclear dinitrosyliron complexes (107, 108), while (Section VI,B) similar complexes are also readily formed (23) from intact $[Fe_4S_3(NO)_7]^-$, and hence complexes of the type $[Fe(NO)_2(SR)_2]^-$ are plausibly the active species in the antimicrobial action of both nitrite and $[Fe_4S_3(NO)_7]^-$. Consistent with this view, it has been found (142) that both $[Fe_4S_3(NO)_7]^-$ and the mononuclear complex $[Fe(NO)_2(SCH_2CHNH_2COOH)_2]^-$, derived from cysteine, were inhibitors of *Clostridium sporogenes* in culture medium; although the required dosage of the mononuclear complexes (180 μmol dm^{-1}) was higher than that of $[Fe_4S_3(NO)_7]^-$ (42 μmol dm^{-3}), these figures reflect not only the potency of the complexes but also their lability during administration.

The central role of the complexes $[Fe(NO)_2(SR)_2]^-$ in the reaction chemistry of iron–sulfur–nitrosyl complexes and their very ready formation both in *in vitro* and *in vivo* (108, 123–125, 128, 129) suggest that the antimicrobial activity of nitrite depends not only upon the disruption of respiration [by destruction of the natural iron–sulfur clusters of redox proteins (108)] but also specifically upon the formation of $[Fe(NO)_2(SR)_2]^-$ species, whose exact role is, however, as yet undefined. This would provide a ready interpretation of the antimicrobial action of $[Fe_4S_3(NO)_7]^-$.

C. $[Fe_2(SMe)_2(NO)_4]$ AS A NATURAL PRODUCT

During a search for abnormal components in the diet of peasants living in Linxian, Henan Province, China (cf. Section VII,D), $[Fe_2(SMe)_2(NO)_4]$ was detected by gas chromatography/mass spectrometry (GC/MS) in vegetables which had been preserved by storage in water (12, 143, 144). Although $[Fe_2(SMe)_2(NO)_4]$ cannot be detected in the freshly harvested plants, during the preservation period its concentration builds up to levels in the range 0.1–0.45 mg/kg of plant matter (145). It is not found in similarly preserved vegetables from localities when nitrite levels in the local water supply are significantly lower than those in Linxian, nor where preservation in brine is practiced.

On the assumption that the Fe_2S_2 framework of naturally occurring $[Fe_2(SMe)_2(NO)_4]$ originated from either a [2Fe–2S] or a [4Fe–4S] cluster in a redox protein, a plausible biosynthetic route for the formation of $[Fe_2(SMe)_2(NO)_4]$ based upon known chemistry is that shown in Scheme 5.

Step (i), the formation of $[Fe(NO)_2(SH)_2]^-$ by reaction of nitrite (e.g., from groundwater or by reduction of nitrate) with preformed iron–sulfur clusters, is known to proceed readily (107, 108), and step (ii), the conversion of $[Fe(NO)_2(SH)_2]^-$ to $[Fe_4S_3(NO)_7]^-$ under appropriate conditions of pH, has also been demonstrated (23) (cf. also Scheme 4).

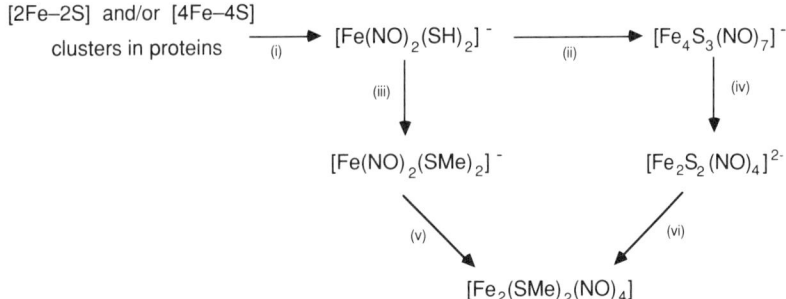

SCHEME 5. Possible biosynthetic pathways to $[Fe_2(SMe)_2(NO)_4]$: individual steps (i)–(vi) are discussed in the text.

Step (iv), the conversion of $[Fe_4S_3(NO)_7]^-$ to $[Fe_2S_2(NO)_4]^{2-}$, normally requires high pH if no external source of sulfur is provided (1, 3, 20, 21); an attractive alternative for near-neutral pH conditions is enzyme-mediated sulfur transfer, but this has not yet been investigated. Likewise, methylation of $[Fe(NO)_2(SH)_2]^-$ to provide $[Fe(NO)_2(SMe)_2]^-$, step (iii), has not been investigated yet.

Step (v), conversion of $[Fe(NO)_2(SMe)_2]^-$ to the product $[Fe_2(SMe)_2(NO)_4]$, occurs spontaneously under the appropriate conditions of solvent polarity/polarizability (23) [cf. Eq. (20) and Schemes 3 and 4]. Step (vi) requires the methylation of $[Fe_2S_2(NO)_4]^{2-}$; while this has been demonstrated for methylation by methyl halides (3, 24, 29), methylation by biological methyl transfer, e.g., from methionine or S-adenosylmethionine, has yet to be investigated. Thus, although neither of the biosynthetic routes suggested in Scheme 5 has yet been fully investigated, several of the steps [(i), (ii), and (vi)], are established, while others [(iv) and (vi)] are already known to occur under nonbiological conditions.

D. $[Fe_2(SMe)_2(NO)_4]$ (ROUSSIN ESTERS) AND CHEMICAL CARCINOGENESIS

The results of a major epidemiological study of the geographical variations in the occurrence throughout China of a wide variety of different types of cancer have recently been published in the form of an atlas (146). One result of this survey was the revelation of a very high, but very localized, incidence of esophageal cancer in the Linxian valley of Henan province in northern China (146, 147). Esophageal cancer also exhibits very high incidence in a number of other localities in both Asia and Africa (148), but the Chinese occurrence has so far attracted the most international attention.

In Linxian, where the adult incidence of esophageal cancer can be as high as one in four, the condition develops through mild and severe stages of epithelial hyperplasia, although reversible changes or recurring processes are frequently encountered (149).

The probable role of local foodstuffs in the development of esophageal cancer in Linxian was deduced from a study of domestic poultry. Poultry reared in Linxian commonly suffer from cancer of the gullet, analogous to human esophageal cancer. When part of the Linxian human population was rehoused some distance away and was provided with poultry free of gullet cancer, feeding of these birds with food scraps prepared by methods normal in Linxian caused rapid development of gullet cancers.

Among the dietary staples in the rural communities of the Linxian valley are millet, cornbread, and the pickled vegetables previously mentioned (Section VII,C), prepared by storing fresh green leaves in water over several weeks. The water supply is rather high in both nitrate and nitrite.

When consumed, the cornbread is often contaminated with the mold *Fusarium moniliforme*, and it has been shown (*150, 151*) that this mold, in the presence of nitrite, will elaborate a range of *N*-nitroso compounds: Me_2NNO, Et_2NNO, $Me(PhCH_2)NNO$, and $(Me_2CHCH_2CH_2)$-$(CH_3COCHMe)NNO$. Cornbread contaminated in this way induces tumors in rats (*150*).

Just as the cornbread is often contaminated with *F. moniliforme*, so too the preserved vegetables are generally contaminated with *Geotrichum candidum*. The formation of these molds is not necessarily indicative of poor domestic hygiene, rather they are regarded by some as an enhancement (cf. certain Western European cheeses). Extracts of these preserved vegetables were found (*152*) to induce stomach cancers in mice and rats, and $[Fe_2(SMe)_2(NO)_4]$ could be isolated from vegetables preserved in this way (*12, 143–145, 152*).

The complex $[Fe_2(SMe)_2(NO)_4]$ has been found to have only weak mutagenic properties when employed alone (*149, 152–154*), although it potentiated the action of 3-methylcolanthrene (*154*). A recent study of the effects of $[Fe_2(SMe)_2(NO)_4]$ and $Me(PhCH_2)NNO$ upon mice found (*155*) that while $[Fe_2(SMe)_2(NO)_4]$ alone caused no lesions in the esophagus or the forestomach, it markedly enhanced the lesion count induced by $Me(PhCH_2)NNO$; thus it promotes the tumorigenic properties of $Me(PhCH_2)NNO$.

Hence although $[Fe_2(SMe)_2(NO)_4]$ is formed in the preservation of vegetables, presumably under the action of *G. candidum*, only when these are consumed along with moldy cornbread (or some other source of nitrosamines) is the complex likely to prove harmful. The action of the mold in the formation of $[Fe_2(SMe)_2(NO)_4]$ has not been fully established. It is perhaps significant that in urban Beijing, vegetables preserved in the same way contained no more that 5×10^{-3} mg kg^{-1} of $[Fe_2(SMe)_2(NO)_4]$ (*144*), and the incidence of esophageal cancer is low.

A further dietary component which may potentiate the carcinogenic action of *N*-nitrosamines is millet, which contains silica fibers (*156*), known to promote the carcinogenic action of 11-methyl-15-cyclopenta[*a*]phenanthrene (*157*), and presumably of certain other carcinogens also.

Thus, although the weak mutagenicity of $[Fe_2(SMe)_2(NO)_4]$ is hardly consistent with its being the sole cause of esophageal cancer in Linxian, its presence in food there (and, as far as is known, nowhere else), and the very high incidence of just one kind of cancer in Linxian, strongly suggest some connection between the two. Possibly the lipid solubility of $[Fe_2(SMe)_2(NO)_4]$ and its nitrosating activity give it a unique *in vivo* activity. Compounds of this type are a challenge not only to our understanding of iron–sulfur cluster chemistry, but also to a full appreciation of chemical carcinogenesis.

REFERENCES

1. Roussin, F. Z., *Ann. Chim. Phys.* **52**, 285 (1858).
2. Playfair, L., *Ann. Chem. Pharm.* **74**, 340 (1850).
3. Pavel, O., *Ber. Dtsch. Chem. Ges.* **15**, 2600 (1882).
4. Hofmann, K. A., and Wiede, O. F., *Z. Anorg. Allgem. Chem.* **9**, 295 (1895).
5. Reihlen H., and von Friedolsheim, A., *Liebigs Ann. Chem.* **457**, 71 (1927).
6. Cambi, L., and Szegö, L., *Atti Accad. Lincei* **13**, 168 (1931).
7. Manchot, W., and Linckh, E., *Ber. Dtsch. Chem. Ges.* **59**, 412 (1926).
8. Seel, F., *Z. Anorg. Allgem. Chem.* **249**, 308 (1942).
9. Addison, C. C., and Lewis, J., *Q. Rev. Chem. Soc.* **9**, 115 (1955).
10. Thomas, J. T., Robertson, J. H., and Cox, E. G., *Acta Crystallogr.* **11**, 599 (1958).
11. Johansson, G., and Lipscomb, W. N., *Acta Crystallogr.* **11**, 594 (1958).
12. Wang, G. H., Zhang, W. X., and Chai, W. G., *Acta Chim. Sin.* **38**, 95 (1980).
13. Holm, R. H., *Chem. Soc. Rev.* **10**, 455 (1981).
14. Spiro, T. G., ed., "Iron-Sulphur Proteins." Wiley (Interscience), New York, 1982.
15. Garner, C. D., *In* "Transition Metal Clusters" (B. F. G. Johnson, ed.), p. 265. Wiley (Interscience), New York, 1980.
16. Johnson, B. F. G., and McCleverty, J. A., *Prog. Inorg. Chem.* **7**, 277 (1966).
17. Connelly, N. G., *Inorg. Chim. Acta Rev.* **6**, 48 (1972).
18. Enemark, J. H., and Feltham, R. D., *Coord. Chem. Rev.* **13**, 339 (1974).
19. McCleverty, J. A., *Chem. Rev.* **79**, 53 (1979).
20. Brauer, G. (ed.), "Handbuch der Präparativen Anorganischen Chemie," 2nd Ed., Vol. 2, p. 1526. Enke, Stuttgart, 1960.
21. Butler, A. R., Glidewell, C., and Johnson, I. L., unpublished work.
22. McDonald, C. C., Phillips, W. D., and Mower, H. F., *J. Am. Chem. Soc.* **87**, 3319 (1965).
23. Butler, A. R., Glidewell, C., Hyde, A. R., and Walton, J. C., *Polyhedron* **4**, 797 (1985).
24. Beck, W., Grenz, R., Götzfried, F., and Vilsmaier, E., *Chem. Ber.* **114**, 3184 (1981).
25. Butler, A. R., Glidewell, C., Hyde, A. R., and McGinnis, J., *Inorg. Chem.* **24**, 2931 (1985).
26. Weatherill, T. D., Ph.D. thesis, University of Illinois at Urbana-Champaign (1985).
27. Butler, A. R., Glidewell, C., Hyde, A. R., McGinnis, J., and Seymour, J. E., *Polyhedron* **2**, 1045 (1983).
28. Hieber, W., and Beck, W., *Z. Anorg. Allg. Chem.* **305**, 274 (1960).
29. Seyferth, D., Gallagher, M. K., and Cowie, M., *Organometallics* **5**, 539 (1986).
30. Rauchfuss, T. B., and Weatherill, T. D., *Inorg. Chem.* **21**, 827 (1982).
31. Mak, T. C. W., Book, L., Chung Chieh, Gallagher, M. K., Li-cheng Song, and Seyferth, D., *Inorg. Chim. Acta* **73**, 159 (1983).

32. Mazany, A. M., Fackler, J. P., Jr., Gallagher, M. K., and Seyferth, D., *Inorg. Chem.* **22**, 2593 (1983).
33. King, R. B., *J. Am. Chem. Soc.* **84**, 2460 (1962).
34. De Beer, J. A., and Haines, R. J., *J. Organomet. Chem.* **24**, 757 (1970).
35. Butler, A. R., Glidewell, C., Hyde, A. R., and Walton, J. C., *Polyhedron* **4**, 303 (1985).
36. Gall, R. S., Chu, C. T.-W., and Dahl, L. F., *J. Am. Chem. Soc.* **96**, 4019 (1974).
37. Chu, C. T.-W., Lo, F. Y.-K., and Dahl, L. F., *J. Am. Chem. Soc.* **104**, 3409 (1982).
38. Eaton, P. E., and Cole, T. W., *J. Am. Chem. Soc.* **86**, 3157 (1964).
39. Chu, C. T.-W., and Dahl, L. F., *Inorg. Chem.* **16**, 3245 (1977).
40. Gall, R. S., Connelly, N. G., and Dahl, L. F., *J. Am. Chem. Soc.* **96**, 4017 (1974).
41. Chu, C. T.-W., Gall, R. S., and Dahl, L. F., *J. Am. Chem. Soc.* **104**, 737 (1982).
42. Dymicky, M., *Proc. Schevchenko Sci. Soc. Sect. Chem. Biol. Med.* **8**, 70 (1980).
43. Butler, A. R., Glidewell, C., Hyde, A. R., and McGinnis, J., *Polyhedron* **2**, 1399 (1983).
44. Manchot, W., *Ber. Dtsch. Chem. Ges.* **59**, 2445 (1926).
45. Brauer, G., Ref. 20, p. 1528.
46. Büttner, H., and Feltham, R. D., *Inorg. Chem.* **11**, 971 (1972).
47. Ileperuma, O. A., and Feltham, R. D., *J. Am. Chem. Soc.* **98**, 6039 (1976).
48. Lalor, F. J., Brookes, L. H., Ferguson, G., and Parvez, M., *J. Chem. Soc. Dalton Trans.* 245 (1984).
49. Karlin, K. D., Rabinowitz, H. N., Lewis, D. L., and Lippard, S. J., *Inorg. Chem.* **16**, 3262 (1977).
50. McCleverty, J. A., Atherton, N. M., Locke, J., Wharton, E. J., and Winscom, C. J., *J. Am. Chem. Soc.* **89**, 6082 (1967).
51. Baltusis, L. M., Karlin, K. D., Rabinowitz, H. N., Dewan, J. C., and Lippard, S. J., *Inorg. Chem.* **19**, 2627 (1980).
52. Nelson, L. L., Ph.D. thesis, University of Wisconsin, Madison, (1981).
53. Butler, A. R., Glidewell, C., Hyde, A. R., and McGinnis, J., *Polyhedron* **3**, 1165 (1984).
54. Coucouvanis, D., Simhon, E. D., Stremple, P., and Baenziger, N. C., *Inorg. Chem. Acta* **53**, L135 (1981).
55. Butler, A. R., Glidewell, C., Johnson, I. L., and Walton, J. C., unpublished work.
56. Rauchfuss, T. B., Weatherill, T. D., Wilson, S. R., and Zebrowski, J. P., *J. Am. Chem. Soc.* **105**, 6508 (1983).
57. Bolinger, C. M., Weatherill, T. D., Rauchfuss, T. B., Rheingold, A. L., Day, C. S., and Wilson, S. R., *Inorg. Chem.* **25**, 634 (1956).
58. Brunner, H., Kauermann, H., and Wachter, J., *Angew, Chem. Int. Ed. Engl.* **22**, 549 (1983).
59. Lin Xianti., Zheng An, Lin Shanhao, Huang Jinling, and Lu Jiaxi, *J. Struct. Chem. (Wuhan)* **1**, 79 (1982).
60. Davies G. R., Jarvis, J. A. J., Kilbourn, B. K., Mais, R. H. B., and Owston, P. G., *J. Chem. Soc. (A)* 1275 (1970).
61. Colapietro, M., Domenicano, A., Scaramuzza, L., Vaciago, A., and Zambonelli, L., *J. Chem. Soc. Chem. Commun.* 583 (1967).
62. Ileperuma, O. A., and Feltham, R. D., *Inorg. Chem.* **16**, 1876 (1977).
63. Glidewell, C, and Johnson, I. L., unpublished work.
64. Butler, A. R., Glidewell, C., and Johnson, I. L., *Polyhedron* **6**, 1147 (1987).
65. Glidewell, C., and Hyde, A. R., *Polyhedron* **4**, 1155 (1985).
66. Abboud, J.-L. M., Taft, R. W., and Kamlet, M. J., *J. Chem. Res(S)* 98 (1984).
67. Kamlet, M. J., and Taft, R. W., *Acta. Chem. Scand.* **B39**, 611 (1985).
68. Bell, L. K., Mingos, D. M. P., Tew D. G., Larkworthy, L. F., Sandell, B., Povey, D. C., and Mason, J., *J. Chem. Soc. Chem. Commun.* 125 (1983).

69. Bell, L. K., Mason, J., Mingos, D. M. P., and Tew, D. G., *Inorg. Chem.* **22,** 3497 (1983).
70. Evans, D. H., Mingos, D. M. P., Mason, J., and Richards, A., *J. Organomet. Chem.* **249,** 293 (1983).
71. Wang, Y., *Huaxue Tongbao* **6,** 18 (1983); *Chem. Abstr.* **99,** 203021 (1983).
72. Sedney, D., and Reiff, W. M., *Inorg. Chim. Acta* **34,** 231 (1979).
73. Kerler, W., Nenwirth, W., Fluck, E. Kuhn, P., and Zimmerman, B., *Z. Phys.* **173,** 321 (1963).
74. Kostiner, E., Steger, J., and Rea, J. R., *Inorg. Chem.* **9,** 1939 (1970).
75. Sung, S.-S., Glidewell, C., Butler, A. R., and Hoffmann, R., *Inorg. Chem.* **24,** 3856 (1985).
76. Wade, K., *Adv. Inorg. Chem. Radiochem.* **18,** 1 (1976).
77. Mingos, D. M. P., *J. Chem. Soc. Chem. Commun.* 706 (1983).
78. Averill, B. A., Herskovitz, T., Holm, R. H., and Ibers, J. A., *J. Am. Chem. Soc.* **95,** 3523 (1973).
79. Que, L., Bobrik, M. A., Ibers, J. A., and Holm, R. H., *J. Am. Chem. Soc.* **96,** 4168 (1974).
80. Wei, C. H., Wilkes, G. R., Treichel, P. M., and Dahl, L. F., *Inorg. Chem.* **5,** 900 (1966).
81. Nelson, L. L., Lo, F. Y.-K., Rae, A. D., and Dahl, L. F., *J. Organomet. Chem.* **225,** 309 (1982).
82. Berg, J. M., Hodgson, K. O., and Holm, R. H., *J. Am. Chem. Soc.* **101,** 4586 (1979).
83. Laskowski, E. J., Reynolds, J. G., Frankel, R. B., Foner, S., Papaefthymiou, G. C., and Holm, R. H., *J. Am. Chem. Soc.* **101,** 6562 (1979).
84. Trinh-Toan, Fehlhammer, W. P., and Dahl, L. F., *J. Am. Chem. Soc.* **99,** 402 (1977).
85. Trinh-Toan, Teo, B. K., Ferguson, J. A., Meyer, T. J., and Dahl, L. F., *J. Am. Chem. Soc.* **99,** 408 (1977).
86. Bottomley, F., and Grein, F., *Inorg. Chem.* **21,** 4170 (1982).
87. Glidewell, C., and Johnson, I. L., *Inorg. Chim. Acta* **132,** 145 (1987).
88. Treadwell, W. D., and Huber, D., *Helv. Chim. Acta* **26,** 18 (1943).
89. Gall, R. S., Chu, C. T.-W., and Dahl, L. F., *J. Am. Chem. Soc.* **96,** 4019 (1974).
90. Swinehart, J. H., *Coord. Chem. Rev.* **2,** 385 (1967).
91. Manoharan, P. T., and Hamilton, W. C., *Inorg. Chem.* **2,** 1043 (1963).
92. Swinehart, J. H., and Rock, P. A., *Inorg. Chem.* **5,** 573 (1966).
93. Rock, P. A., and Swinehart, J. H., *Inorg. Chem.* **5,** 1078 (1966).
94. Morando, P. J., Borghi, E. B., de Schteingart, L. M., and Blesa, M. A., *J. Chem. Soc. Dalton Trans.* 435 (1981).
95. Maltz, H., Grant, M. A., and Navaroli, M. C., *J. Org. Chem.* **36,** 363 (1971).
96. Butler, A. R., Glidewell, C., Reglinski, J., and Waddon, A., *J. Chem. Res.* (*S*) 279, (*M*) 2768 (1984).
97. Wolfe, S. K., Andrade, C., and Swinehart, J. H., *Inorg. Chem.* **13,** 2567 (1974).
98. Wolfe, S. K., and Swinehart, J. H., *Inorg. Chem.* **7,** 1855 (1968).
99. Butler, A. R., Glidewell, C., Chaipanich, V., and McGinnis, J., *J. Chem. Soc. Perkin Trans.* **2,** 7 (1986).
100. Moser, W., Chalmers, R. A., and Fogg, A. G., *J. Inorg. Nucl. Chem.* **27,** 831 (1965).
101. Fogg, A., Norbury, A. H., and Moser, W., *J. Inorg. Nucl. Chem.* **28,** 2753 (1966).
102. Andrade, C., and Swinehart, J. H., *Inorg. Chem.* **11,** 648 (1972).
103. Bottomley, F., Brooks, W. V. F., Clarkson, S. G., and Tong, S. B., *J. Chem. Soc. Chem. Commun.* 919 (1973).
104. Bottomley, F., *Acc. Chem. Res.* **11,** 158 (1978).
104a. Strasdeit, H., Krebs, B., and Henkel, G., *Z. Naturforsch.* **41b,** 1357 (1986).
105. Glidewell, C., and McGinnis, J., *Inorg. Chim. Acta* **64,** L171 (1982).
106. Wharton, E. J., and McCleverty, J. A., *Inorg. Nucl. Chem. Lett.* **6,** 549 (1970).

107. Butler, A. R., Glidewell, C., Hyde, A. R., and Walton, J. C., *Inorg. Chim. Acta* **106**, L7 (1985).
108. Reddy, D., Lancaster, J. R., Jr., and Cornforth, D. P., *Science* **221**, 769 (1983).
109. Wang, M. Y., Li, M. H., Jiang, Y. Z., Sun, Y. H., Li, G. Y., Zhang, W. X., Chai, W. G., and Wang, G. H., *Cancer Res. Prevent. Treat.* **10**, 145 (1983).
110. Croisy, A., Ohshima, H., and Bartsch, H., *I.A.R.C. Sci. Publ.* **57**, 327 (1984).
111. Dobry, A., *Trav. Membres Soc. Chim. Biol.* **23**, 1438 (1941).
112. Dobry-Duclaux, A., *Biochim. Biophys. Acta* **39**, 33 (1960).
113. Dobry-Duclaux, A., *Makromol. Chem.* **44**, 155 (1961).
114. Dobry-Duclaux, A., *Biochim. Biophys. Acta* **54**, 76 (1961).
115. Dobry-Duclaux, A., *Biochim. Biophys. Acta* **77**, 1 (1963).
116. Dobry-Duclaux, A., *Biochim. Biophys. Acta* **89**, 1 (1964).
117. Dobry-Duclaux, A., *Bull. Soc. Chim. Biol.* **48**, 887 (1966).
118. Burlamacchi, L., Martini, G., and Tiezzi, E., *Inorg. Chem.* **8**, 2021 (1969).
119. Martini, G., and Tiezzi, E., *Trans. Faraday Soc.* **67**, 2538 (1971).
120. Jezowska-Trzebiatowska, B., and Jezierski, A., *J. Mol. Struct.* **19**, 635 (1973).
121. Boyer, M. P., Morton, J. R., and Preston, K. F., *J. Phys. Chem.* **84**, 2989 (1980).
122. Basoni, R., Gaggelli, E., Tiezzi, E., and Valensin, G., *J. Chem. Soc. Perkin Trans.* **2**, 423 (1975).
123. Woolum, J. C., and Commoner, B., *Biochim. Biophys. Acta* **201**, 131 (1970).
124. Chiang, R. W., Woolum, J. C., and Commoner, B., *Biochim. Biophys. Acta* **257**, 452 (1972).
125. Nagata, C., Ioki, Y., Kodama, M., Tagashira, Y., and Nakadate, M., *Ann. N.Y. Acad. Sci.* **222**, 1031 (1973).
126. Commoner, B., Woolum, J. C., Senturia, L. H., and Ternberg, J. L., *Cancer Res.* **30**, 2091 (1970).
127. Lawley, P. D., *Nature (London)* **218**, 580 (1968).
128. Vanin, A. F., and Varich, V. Ya., *Stud. Biophys.* **86**, 177 (1981).
129. Varich, V. Ya., and Vanin, A. F., *Biophysics* **28**, 1125 (1983).
130. Dobry, A., and Boyer, F., *Ann. Inst. Pasteur* **71**, 455 (1945).
131. Candeli, A., and Mancini, M., *Boll. Soc. Med. Chir. Modena* **48**, 3 (1948).
132. Candeli, A., *Boll. Soc. Ital. Biol. Sper.* **25**, 495 (1949).
133. Perigo, J. A., Whiting, E., and Bushford, T. E., *J. Food Technol.* **2**, 377 (1967).
134. Perigo, J. A., and Roberts, T. A., *J. Food Technol.* **3**, 91 (1968).
135. Johnston, M. A., Pivnich, H., and Samson, J. M., *Can. Inst. Food Sci. Technol. J.* **2**, 52 (1969).
136. Roberts, T. A., and Garcia, C. E., *J. Food Technol.* **8**, 463 (1973).
137. Pivnich, H., and Chang, P. C., *Proc. Int. Symp. Nitrite Meat Prod.* 111 (1973).
138. Van Roon, P. S., *Proc. Int. Symp. Nitrite Meat Prod.* 117 (1973).
139. Mirna, A., and Coretti, K., *Fleischwirtschaft* **54**, 507 (1974).
140. Ashworth, J., Didcock, A., Hargreaves, L. L., Jarvis, B., Walters, C. L., and Larkworthy, L. F., *J. Gen. Microbiol.* **84**, 403 (1974).
141. Moran, D. M., Tannenbaum, S. R., and Archer, M. C., *Appl. Microbiol.* **30**, 838 (1975).
142. Van Roon, P. S., *Proc. Eur. Meet. Meat Res. Work., 26th* **2**, 227 (1980).
143. Wang, G. H., Zhang, W. X., and Chai, W. G., *Adv. Mass Spectrom.* **8B**, 1369 (1980).
144. Zhang, W. X., Xu, M. S., Wang, G. H., and Wang, M. Y., *Cancer Res.* **43**, 339 (1983).
145. Wang, M. Y., Lu, S. H., Ji, C., Wang, Y. L., and Li, M. H., *Cancer Res. Prevent. Treat.* **10**, 12 (1983).
146. "Atlas of Cancer Maps of the People's Republic of China." China Map Press, Beijing, 1980.

147. Yang, C. S., *Cancer Res.* **40,** 2633 (1980).
148. Cook-Mozaffari, P. J., *Nutr. Cancer* **1,** 5 (1979).
149. Li, M. H., and Cheng, S. J., *Chin. Med. J.* **97,** 311 (1984).
150. Li, M. H., *Proc. Conf. Cancer Res. People's Rep. China, 1st* 131 (1980); *Chem. Abstr.* **94,** 186612v (1981).
151. Li, M. H., Lu, S. H., Ji, C., Wang, M., Wang, Y., Cheng, S., and Tiam, G., *In* "Genetic and Environmental Factors in Experimental and Human Cancer", (H. V. Galboln, ed.). Japan Science Society Press, Tokyo, 1980.
152. Li, M. H., Lu, S. H., Ji, C., Wang, Y., Wang, M., Cheng, S., and Tiam, G., *Proc. Int. Symp. Princess Takamatsu Cancer Res. Fund* 139 (1979); *Chem. Abstr.* **97,** 180397 (1982).
153. Lu, S. H., Camus, A. M., Tomatis, L., and Bartsch, H., *J. Natl. Cancer Inst.* **66,** 33 (1981).
154. Cheng, S. J., Sala, M., Li, M. H., Courtois, I., and Chouroulinkov, I., *Carcinogenesis* **2,** 313 (1981).
155. Lu, S., Lin, P., Lu, F., Wang, Y., and Wang, M., *Zhonghua Zhongliu Zazhi* **7,** 241 (1985); *Chem. Abstr.* **104,** 63958e (1986).
156. O'Neill, C., Jordon, P., Bhatt, T., and Newman, R., *Ciba Found. Symp.* **121,** 214 (1986).
157. Bhatt, T., Coombs, M., and O'Neill, C., *Int. J. Cancer* **34,** 519 (1984).

INDEX

A

Abbreviations
  for ligands and solvents, 158–160
  for metalloproteins, 329
  for nickel redox chemistry, 288–290
Acetogens, 298
Acetyl-CoA synthase, 300
Acid-base equilibria, of hydroxo-bridged complexes, 106–118
Acid-base titration, 58–59
Acid cleavage, of polynuclear hydroxo-bridged complexes, 121–122
Acid hydrolysis kinetics, 133, 135
  of cobalt(III) dihydroxo-bridged complexes, 141–144
  of dinuclear monohydroxo-bridged complex cleavages, 123–130
Adiabaticity, porphyrins and, 48
μ-Amido-di-μ-hydroxo complex, acid cleavage, 155
μ-Amido-μ-hydroxo complex, hydroxo bridge cleavage, 149–155
Amines
  hydroxo-bridged complexes, absorption spectra, 71
  nickel(II) oxidation, 245–254
Aminotroponeimines, 29
Ammine hydroxo-bridged complexes, absorption spectra of, 71
Ammonia ligands, 131
Angular Overlap Model, 72–75
Aniono erythro ions, acid-base equilibria of, 107
Antiferromagnetic coupling, in dinuclear system, 73
Aqua-bridged hydroxo complexes
  acid-base equilibria, 110
  of dihydroxo-bridged species, 114–115
  intermolecular hydrogen-bonded pairs, 87
  terminally coordinated, 110–118
Cis Aquaerythro ion cleavage, 124
8-Azaadenine, 205
8-Azapurines, 173
2-Azidopyridine, 206

B

Base hydrolysis kinetics
  of cobalt(III) dihydroxo-bridged complexes, 143–144
  of hydroxo-bridged complexes, 130–131
Benzotriazole
  copper(II) halide adducts, 194–195
  derivatives, spectroscopic spectra, 185–186
  formulation of, 173
  iron adducts, 182
  nickel adducts, 189, 191
  palladium(II) salt adducts, 192–193
Bimolecular reactions, spin state changes and, 46
Bis(bidentate) chelate complex, 32
Bis(chelate) complexes, 29
Bis(methylcyclopentadienyl) manganese(II), 44
Cis-Bis(phenanthroline)iron(II) complex, 45
Blue species
  acid-base monohydroxo-bridged equilibria, 107
  dinuclear, 58
  chromium(III) complexes, formation by oxidation, 89–90
Boltzmann equilibrium, 48
Bridging ligands, 56

C

Cadmium tetrazole and tetrazole complexes, 229–230
Carbon monoxide dehydrogenase, 300
Carbon monoxide oxidoreductase
  catalytic properties, 325–326
  spectroscopy of, 326–328
Catenated nitrogen ligands, triazole and triazolate complexes. *See* specific complexes
Chiral ligands, 67
Chromium
  tetrazole and tetrazole complexes, 214–215
  triazole and triazolate complexes, 180
Chromium(II)-catalyzed hydrolysis, 131

Chromium(II)-charcoal catalysis, 82
Chromium(III) hydroxo-bridged complexes, 56–57
  acid-base equilibria
    bridging hydroxide, 106–107
    dihydroxo-bridged species, 108–110
  ammine and amine oligomers, 81–83
  aqua oligomers, reactions of, 95–96
  aqua system
    kinetics, 133
    monohydroxo-bridged, acid-base equilibria of, 113
    stability constant, 105–106
  conclusions, 156–158
  crystallographic data, 59–69
  dihydroxo-bridged, 58-59
  dinuclear
    ammine and amine, 58
    formation by hydrolysis, 76–81
    monohydroxo-bridged cleavage, acid hydrolysis kinetics, 123–126
  electronic spectra, 12
  formation
    by hydrolysis, 76–81
    by hydroxo bridge cleavage or formation, 93–97
    by isomerization, 97–98
    by oxidation, 89–90
    solid-state reactions, 86–88
  hexaaquachromium(III) ion oligomers, 83–85
  mixed bridge systems, 64, 65
  mono- and dihydroxo-bridged dinuclear equilibria, 131–137
    acid-catalyzed cleavage, 138
    base-catalyzed bridge cleavage, 140
    bridge formation, 138–140
  monohydroxo-bridged, stability constants, 100–103
  oligomers with nuclearity higher than three, 81–86
  paramagnetic, magnetic measurements of, 59
  polynuclear, cleavage by strong acids, 121–122
  polynuclear ethylenediamine, 82
  stability complexes, trihydroxo-bridged species, 104–105
  stability constant determination, 99–100
  structure determination, 57–59
  tri- and dihydroxo-bridged equilibria, 148–149
  tri- and tetranuclear aqua species, acid-base equilibria, 118
Cleavage reactions
  of amido-bridged dicobalt(III) complexes, 149-155
  bridge-cleavage, 58
  formation kinetic parameters, 135–137
  of polynuclear hydroxo-bridged complexes, 121–122
Clostridium thermoaceticum, 326
Cobalt
  tetrazole and tetrazole complexes, 217–220
  triazole and triazolate complexes, 187–189
Cobalt(II) complexes
  bond length changes in spin state transition, 8–9
  crystallographic evidence, 42
  discrete spin state isomers, 28
  octahedral, spin equilibria in solution, 39
  oxidation of, 91
  Raman laser temperature-jump experiment, 18
  spin equilibria
    electron transfer and, 45–46
    octahedral, 39
    relaxation times compared to iron(III), 42
    in solution, 27–28
  volumes of activation, 11
Cobalt(III) hydroxo-bridged complexes, 57
  bridge cleavage of mixed bridge complexes
    amido–bridged dicobalt(III) complexes, 149-155
  conclusions, 156–158
  crystallographic data, 59–69
  dihydroxo-bridged, kinetics of hydrolysis, 141–144
  dinuclear
    formation by hydrolysis, 80–81
    monohydroxo-bridged cleavage, acid hydrolysis kinetics, 123, 126–127
  formation
    by hydroxo bridge cleavage or formation, 93–97
    by hydroxysis, 80–81
    by isomerization, 97–98
    by oxidation, 90–91
    by redox reactions, 98
    solid-state reactions, 86–88

mixed bridge systems, 64, 65
oligomers with nuclearity higher than
    three, formation by hydrolysis, 85–86
paramagnetic planar-diamagnetic
    octahedral equilibria, 44
polynuclear, cleavage by strong acids, 122
spin equilibria, electron transfer and, 45–46
stability complexes, trihydroxo-bridged
    species, 104–105
stability constant determination, 99–100
structure determination, 57–59
tri- and hydroxo-bridged equilibria,
    145–148
Copper
  tetrazole and tetrazole complexes,
    226–229
  triazole and triazolate complexes,
    194–200
Curie temperature dependence, of
    paramagnetic species in NMR, 15

D

Desulfovibrio gigas hydrogenase
  catalytic properties
    activity state, 317–320
    mechanism of catalysis, 320–321
  EPR spectra, 305–306
  midpoint redox potential, 306–308
  X-ray absorption, 309
Diamagnetic states, 16
Dibromobis(phosphine)nickel(II) complex,
    bond length changes in spin state
    transition, 6
Dichromium(III) complex, bridged, 57
Dihalobis(phosphine)nickel(II) complexes,
    3
Dihydroxo-bridged complexes
  acid-base equilibria, 108–110
    quantitative considerations, 115–118
  binuclear, 66–67
  crystallographic data, 61, 63
  stability complexes, 103–104
Diiron nonacarbonyl, 184
Dimerization of hydroxo-bridged complexes,
    kinetics of, 119–121
Dimethylmanganocene, 44
Dinuclear hydroxo-bridged complexes
  formation by hydrolysis, 76–81
  monohydroxo-bridged, stability constants,
    101

Diphosphinedihalonickel(II) complexes
  planar-tetrahedral spin equilibrium,
    29–30
  reaction coordinate profile, 31

E

Electronic spectra, of spin equilibrium
    complexes, 12–13
Electron paramagnetic resonance (EPR)
  carbon monoxide oxidoreductase,
    326–328
  discrete spin state isomers and, 28
  lifetime limits of spin state
    interconversions, 38
  of methanogenic bacteria membranes,
    298
  of methyl-coenzyme M reductase, 323,
    325
  of nickel, 244–245
  of nickel(III), 317
  of nickel in D. gigas hydrogenase, 310
  of organometallic complexes, 44
  vs. NMR, 16
Electron spin-echo spectroscopy, of
    hydrogenase, 311, 312
Electron transfer, spin equilibria and, 45–46
Enzymes, nickel-containing, 329
EPR. *See* Electron paramagnetic resonance
Ethylenediamine complex, 67
Ethylenediamine ligands, 131
Ethylene glycol, 4
Excited states, spin equilibria implications,
    47–48

F

$Factor_{430}$, 323, 324
Ferric myoglobin, 48–49
Ferricytochrome *c*, 49
Franck-Condon factors, 47

G

Geometric structure, of spin-equilibrium
    complexes, 6–11
Gold
  tetrazole and tetrazole complexes,
    226–229
  triazole and triazolate complexes, 194–200
Gouy method, 4, 10

H

Hafnium, triazole and triazolate complexes, 180
Heisenberg Hamiltonian, 72
Heme proteins, spin equilibria implications, 48–49
Hexaaquachromium(III) ion oligomers, formation by hydrolysis, 83–85
Hexazine complexes, 230–232
Hexazines, 172, 230–232
Hexols, 86
Hydrogenases, 298, 299
  catalytic properties
    activity states, 317-320
    mechanism of catalysis, 320-321
  coordination state of nickel
    hyperfine interactions, 309–311
    x-ray absorption spectroscopy, 308–309
  derivatives of nickel, 313
    carbonyls, 314-315
    hydrides, 314
    interactions with oxygen, 315
    other functions of, 321–322
  EPR spectroscopic properties of nickel, 304–306
  interactions with iron-sulfur centers, 315-317
  magnetic circular dichroism, 311, 313
  midpoint redox potentials, 306–308
  types, 304
Hydrogen bacteria, 298. *See also* specific bacteria
Hydrogen bonds
  intramolecular, 67–68
  in monohydroxo-bridged species, 101–103
  in terminally coordinated water, monohydroxo-bridged species, 112–114
Hydrolysis
  chromium(II)-catalyzed, 131
  of cis nitro erythro cation, 92
  of hydroxo-bridged polynuclear complexes, 59, 75–86
  photochemical, 131
Hydroxide, bridging of, 106–107
Hydroxo-bridged complexes
  acid-base equilibria, 106–118
    bridging hydroxide, 106–107
      dihydroxo-bridged species, 108–110
      monohydroxo-bridged species, 107–108
      trihydroxo-bridged species, 110
    bridging water, 110
    higher polynuclear species, 118
    terminally coordinated water, 110–118
      quantitative considerations, 115–118
      vs. mononuclear species, 111–112
    aqua hydroxo species, acid-base equilibria, dihydroxo-bridged species, 114–115
  bridge cleavage of mixed bridge complexes
    hydroxo-bridge cleavage of amido-bridged dicobalt(III) complexes, 149–155
    other, 156
  chromium(III). *See* Chromium(III) hydroxo-bridged complexes
  cleavage
    of polynuclear complexes in strong acids, 121–122
  cleavage reactions, base hydrolysis kinetics, 130–131
  conclusions, 156–158
  crystal structures, 87, 88
  formation
    by hydroxo bridge cleavage or formation, 93–97
    by isomerization, 97–98
    from other polynuclear species, 91–98
    by oxidation, 88–91
    by photochemical reactions, 98
    of polynuclear complexes by hydrolysis reactions, 75–86
    by redox reactions, 98
    solid-state reactions, 86–88
  α interactions, 68–69
  β interactions, 69
  kinetics
    of cleavage of dinuclear monohydroxo-bridged complexes, acid hydrolysis, 123–130
    for condensation reaction, 119–121
  magnetic properties, 70–75
  mono- and dihydroxo-bridged dinuclear equilibria, 131–137
    acid-catalyzed cleavage, 138
    base-catalyzed bridge cleavage, 140
    bridge formation, 138–140
    uncatalyzed cleavage, 137–138
  oligomers of nuclearity higher than three, 81–86

polynuclear
  cleavage in strong acids, 121–122
  formation by hydrolysis reactions, 75–86
spectroscopic properties, 70–75
stability complexes
  dihydroxo-bridged species, 103–104
  trihydroxo-bridged species, 104–105
stability constants
  determination of, 98–100
  for monohydroxo-bridged complexes, 100–103
structure
  crystallographic data, 59–69
  determination of, 57–59
tri- and dihydroxo-bridged equilibria chromium(III), 148–149
tri- and hydroxo-bridged equilibria cobalt(III), 145–148
Di-μ-Hydroxo dichromium(III) complexes, 73–74
Cis Hydroxo erythro ion, 91–92

I

Imines, nickel(II) oxidation, 245–254
Inelastic neutron scattering experiments, 74–75
Ion-exchange chromatography, 99
Iridium
  tetrazole and tetrazole complexes, 217–220
  triazole and triazolate complexes, 187–189
Iridium(III) hydroxo-bridged complexes
  acid-base equilibria, dihydroxo-bridged species, 108–110
  aqua hydroxo species
    acid-base equilibria, monohydroxo-bridged species, 113
  conclusions, 156–158
  crystallographic data, 59–69
  dinuclear, formation by hydrolysis, 80–81
  formation
    by hydroxo bridge cleavage or formation, 93–97
    by oxidation, 91
    solid-state reactions, 86–88
  mono- and dihydroxo-bridged dinuclear equilibria, 131–137
    acid-catalyzed cleavage, 138
    base-catalyzed bridge cleavage, 140
    bridge formation, 138–140
  polynuclear, cleavage by strong acids, 122
  spectroscopic and magnetic properties, 70–75
Iron
  $[Fe(CO)_2(NO)_2]$, $[Fe_2(SR)_2(NO)_4]$ synthesis from, 339
  $[Fe(CO)_3NO]^-$, $[Fe_2(SR)_2(NO)_4]$ synthesis from, 340
  $[Fe_2I_2(NO)_4]$, $[Fe_2(SR)_2(NO)_4]$ synthesis from, 342–343
  $[Fe(NO)(S_2CNMe_2)_2]$, $[Fe_2(SR)_2(NO)_4]$ synthesis from, 339
  $[Fe(NO)_2(SR)_2]^-$, $[Fe_2(SR)_2(NO)_4]$ synthesis from, 340
  $[Fe_2S_2(NO)_4]^{2-}$, $[Fe_2(SR)_2(NO)_4]$ synthesis from, 340–341
  $[Fe_4S_4(NO)_4]$, synthesis, 343
  $[Fe_4S_4(NO)_4]^-$, synthesis, 343–343
  $[Fe_4S_4(NO)_4]$ and related cubane-type clusters, 343–345
  $[Fe_4S_4(NO)_4(NCMe_3)_2]$ and $[Fe_4S_2(NO)_4(NCMe_3)_2]^-$, synthesis, 344–345
  $[Fe_2(SR)_2(CO)_6]$, $[Fe_2(SR)_2(NO)_4]$ synthesis from, 341–342
  tetrazole and tetrazole complexes, 215–217
Iron(II) complexes
  bond length changes in spin state transition, 6, 7, 9
  electronic spectra, 12
  octahedral spin equilibrium, reaction coordinate profile, 24–25
  Raman laser temperature-jump experiment, 18
  spin equilibria
    octahedral, 39
    porphyrins and, 48
    in solution, 22–26
  spin state interconversions, 27
  vibrational spectra, 13
  volumes of activation, 11, 45
Iron(III) complexes
  bond length changes in spin state transition, 7–9
  dithiocarbamate, lifetime limits on solid-state spin state interconversions, 37–38
  electronic spectra, 12
  Mössbauer spectroscopy, 22
  spin equilibria

octahedral, 39, 41
porphyrins and, 48
relaxation times compared to cobalt(II), 42
in solid state, lifetime limits, 37-38
in solution, 26-27
volumes of activation, 11
Iron(II) salts, [Fe$_2$(SR)$_2$(NO)$_4$] synthesis from, 339
Iron(II) Schiff base complex, 13
Iron(II) spin-equilibrium complex, 16
Iron-selenium complexes
  Fe$_2$Se$_2$ complexes, 348
  Fe$_4$Se$_3$ complexes, 349-350
  Fe$_4$Se$_4$ complexes, 348-349
Iron-selenium-nitrosyl complexes, 348-350
Iron-sulfur centers
  interactions with Ni hydrogenases, 315-317
  nitrosyl complexes of, introduction, 336
Iron-sulfur-nitrosyl complexes, 336
  biological chemistry
    antimicrobial activity of [Fe$_4$S$_3$(NO)$_7$]$^-$ salts, 385-386
    cancerous states and $g = 2.03$ complexes, 384
    [Fe$_2$(SMe)$_2$(NO)$_4$] and chemical carcinogenesis, 387-389
    [Fe$_2$(SMe)$_2$(NO)$_4$] as natural product, 386-387
  chemical reactivity, 373-384
    analytical applications, 383-384
    conversion of di- and tetrairon complexes into paramagnetic monoirons
    diiron complexes, 375-377
    tetrairon complexes, 377-379
    ligand substitution, 379-382
    nitrosylation, 382-383
    redox reactions, 373-375
  electronic structure
    dinuclear complexes, 367
    mononuclear complexes, 372-373
    oxidation states, 366-367
    tetranuclear complexes, 368-371
  [Fe(NO)(S$_2$CNR$_2$)$_2$] and related complexes, 346-348
  [Fe$_3$S$_2$(NO)$_5$]$^-$, 345
  [Fe$_4$S$_4$(NO)$_4$] and related cubane-type clusters, 343-345
  [Fe$_2$(S$_2$O$_3$)$_2$(NO)$_4$]$^{2-}$, 345-346
  [Fe$_2$(SR)$_2$(NO)$_4$]
    ^{1}H NMR spectroscopy, 364
    ^{15}N NMR spectroscopy, 364-365
  heterometallic cluster synthesis
    by nitrosylation of performed heterometallic aggregates, 350-351
    from sulfur-rich metal complexes, 352-353
    by use of sulfur-containing anions as nucleophiles, 351
    via exchange reactions, 352
  molecular structure
    by NMR, 364-365
    by X-ray crystallography, 354-365
  selenium analogs, 348-350
  synthesis
    [Fe$_2$S$_2$(NO)$_4$]$^-$, 337-338
    [Fe$_2$S$_2$(NO)$_7$]$^-$, 337-338
    [Fe$_4$S$_3$(NO)$_7$]$^-$, 337-338
    [Fe$_2$(SR)$_2$(NO)$_4$], 339-343
      from dinuclear precursors, 340-343
      from mononuclear precursors, 339-340
      from tetranuclear precursors, 343
  tellurium analogs, 350
  tetrairon complexes, ^{15}N NMR spectroscopy, 365
  x-ray crystallography
    [Fe(NO)(S$_2$CNR$_2$)$_2$] and related clusters, 359-361
    [Fe$_4$S$_3$(NO)$_7$]$^-$ and [Fe$_2$S$_2$(NO)$_4$]$^{2-}$, 354-355
    [Fe$_4$S$_4$(NO)$_4$] and related clusters, 359
    [Fe$_2$SR$_2$(NO)$_4$], 355-358
    heterometallics, 364
    selenium analogs, 361-363
    tellurium analogs, 361-363
Iron-tellurium-nitrosyl complexes, 350
Isomerization
  formation of hydroxo-bridged complexes, 97-98
  spin state equilibria and, 44-45

J

Jahn-Teller distortion, 27, 42, 46

K

β-Ketoimines, 29

L

Landé interval rule, 73
Laser photoperturbation. *See* Photoperturbation
Ligands
  abbreviations for, 158–160
  bridging hydroxo, 106–107
  ethylenediamine, 131
  macrocytic nickel(II), 254–261
  nonbridging, substitution of, in formation of hydroxo-bridged complexes, 91–92
  terminally coordinated, acid-base equilibria of, 110–118
  with tetrazole ring system. *See* Tetrazole and tetrazole complexes
Light-induced excited spin state trapping, 21, 38–39

M

Magnetic measurements, 58
Magnetic resonance, spin equilibria dynamics investigations, 14–16
Magnetic susceptibility, of spin-equilibrium complexes, 4–6
Manganese
  tetrazole and tetrazole complexes, 215
  triazole and triazolate complexes, 180–181
Manganese(II) complexes, 16
Manganese(III) complexes, 44
Mercury, tetrazole and tetrazole complexes, 229–230
Metal-ligand bonds, changes from spin state transition, 6–11
Metalloproteins, nickel containing, carbon monoxide oxidoreductase, 325–328
Metal spin-equilibria complexes, 2–3. *See also* specific complexes
  four-coordinated d type, 2
  implications, 43–44
    excited states, 47–48
    porphyrins and heme proteins, 48–49
    reaction mechanisms
      electron transfer, 45–46
      racemization and isomerization, 44–45
      substitution, 46
  in solid state, 36–39
    lifetime limits, 37–38
    measured rates, 38–39
  in solution, 22–36
  static properties
    electronic spectra, 12–13
    geometric structure, 6–11
    magnetic susceptibility, 4–6
    vibrational spectra, 13
  summary and interpretation
    of octahedral $\Delta S = 1$ equilibria, 41–42
    octahedral $\Delta S = 2$ equilibria, 39–41
    planar-octahedral $\Delta S = 1$ equilibria, 43
    planar-tetrahedral $\Delta S = 1$ equilibria, 42–43
  techniques
    magnetic resonance, 14–16
    Mössbauer spectroscopy, 21–22
    photoperturbation, 20–21
    temperature-jump relaxation, 16–18
Metal spin-equilibrium complexes
  techniques
    ultrasonic relaxation, 18–20
Meta-vanadate, 83
Methanobacterium formicicum, 320
Methanobacterium thermoautotrophicum
  hydrogenase
    electron spin-echo spectroscopy, 311, 312
    EPR spectrum, 323, 325
    EXAFS studies, 309
Methanogens, 298
Methanol, 4
Methyl-coenzyme M reductase, 323–325
  EPR spectra, 323, 325
  F_{430} and, 323–324
  function, 324–325
Methyl-CoM reductase, 329
Methylmercury(II)hydroxide, 205
Molybdenum
  tetrazole and tetrazole complexes, 214–215
  triazole and triazolate complexes, 180
Molybdenum(II) complex, singlet-triplet equilibrium, 44
Monohydroxo-bridged complexes
  acid-base equilibria, 107–108
    in terminally coordinated water, hydrogen bond interactions, 112–114
  acid–base equilibria, quantitative considerations, 115–118

crystallographic data, 61, 62, 66
stability constants, 100–103
Mössbauer spectroscopy
  of iron complexes, 3
  lifetime limits of spin state
    interconversion and, 37–38
  of spin-equilibrium complexes, 21–22

N

NAD-linked hydrogenase, 321–322
Naphthotriazole, 173
Nickel, 241–242
  bacterial demand for, 298
  in metalloproteins, 297–300, 329
    abbreviations, 329
    carbon monoxide oxidoreductase, 325–328
    hydrogenase, 304–322
    methyl-coenzyme M reductase, 323–325
    urease
      composition, 300
      mechanism of catalysis, 303–304
      spectroscopic properties, 301–303
  redox chemistry
    probes of structure, 243–245
    steric and electronic requirements, 242–243
    tetrazole and tetrazole complexes, 220–226
    triazole and triazolate complexes, 189–193
Nickel carbonyl hydrogenases, 314–315
Nickel hydride hydrogenases, 314
Nickel hydrogenases. *See* Hydrogenases
Nickel(II) complexes
  bond length changes in spin state transition, 8–10
  oxidation coordination environments and structural chemistry
    amines, imines, and oximes, 245–254
    amino acids and peptides, 261–265
    macrocycles, 254–261
    other ligands, 265–266
    remarks on structural probes of higher oxidation-state nickel, 266
  photoperturbation, 20–21
  planar-tetrahedral equilibria, 2
  Raman laser temperature-jump experiment, 18
  reduction
    coordination environments and structural chemistry
      amines and imines, 281–282
      macrocycles, 282–285
    kinetic studies, 285–288
  spin equilibrium
    of planar-five-coordinate equilibria, 34–36
    in solution, 28–36
    planar-octahedral equilibria, 32–36
    planar-tetrahedral equilibria, 29–32
    tetrahedral, NMR studies, 14
  volumes of activation, 11
Nickel(III) complexes
  kinetic studies
    complex formation reactions, 266–270
    electron transfer reactions, 270–280
    nickel(II)/(III) ions with metal complexes, 270
Nickel(II) porphyrin complexes, 13
Nickel(IV), kinetic studies of, 280–281
Niobium, triazole and triazolate complexes, 180
Nitrogen ligands, catenated
  hexazine complexes, 230–232
  pentazolate complexes, 230–232
Nitrosyl complexes of iron-sulfur clusters, introduction, 336
NMR. *See* Nuclear magnetic resonance
Nocardia opaca hydrogenase, 315, 321–322
Nuclearity, 58
Nuclear magnetic resonance (NMR)
  Evans method, 4–5, 10
  of hydroxo-bridged complexes, 59
  interconversion between planar and tetrahedral isomers, 15
  of iron-sulfur-nitrosyl complexes, 364–365
  low-temperature spectra, 29
  spin equilibrium, 3, 14–16

O

Octahedral transition metal complexes, spin equilibrium, 2
Oligomerization, 99
Oligomers, condensation, in formation of hydroxo-bridged complexes, 91
Organometallic complexes, spin equilibria, 44
Osmium, tetrazole and tetrazole complexes, 215–217
Oxidation reactions, in formation of hydroxo-bridged complexes, 88–91

Oximes, nickel(II) oxidation, 245–254
Oxo bridge complexes, 64, 65
Oxygen, interactions with nickel hydrogenases, 315

P

Palladium
  tetrazole and tetrazole complexes, 220–226
  triazole and triazolate complexes, 189–193
Paramagnetic states, 16
Pentaazadienes, structure, 171–172
1,5-Pentamethylenetetrazole, 206–207
Pentazine, 172
Pentazolate complexes, 230–232
Pentazole, 172
pH, potentiometric measurement of, 58, 99
Photochemical reactions
  formation of hydroxo-bridged complexes, 98
  hydrolysis, 131
Photoperturbation
  of complexes with larger driving force, 24
  of iron(III) complexes in solution, 26–27
  measurement of interconversion rates of metal complexes in solid state, 38–39
  for planar-tetrahedral equilibria of nickel(II) complexes, 29
  of spin-equilibria complexes, 20–21
Platinum
  tetrazole and tetrazole complexes, 220–226
  triazole and triazolate complexes, 189–193
Polydentate ligands, 56
Polymerization, in condensation of monohydroxo-bridged species, 119
Polynuclear complexes, 56
Porphyrins, spin equilibria implications, 48–49
Pulse ultrasonic relaxation method, 18
Pyridine, addition, dissociation and substitution to planar nickel(II) complex, 34

R

Racemization, spin state equilibria and, 44–45
Radiationless deactivation processes, of excited state to ground state, 47
Raman laser temperature-jump technique, 17–18
  for cobalt(II) complexes in solution, 27–28
  of iron(III) complexes in solution, 26
  limitations, 24
  spectra of spin-equilibrium complexes, 13, 16–18
Reaction coordinate profiles
  planar-octahedral equilibria of nickel(II), 33
  planar-tetrahedral spin equilibrium of diphosphinedihalonickel(II) complex, 31
Relaxation times, 40
  for planar-tetrahedral equilibria in nickel(II) complexes in solution, 31
Resonator ultrasonic relaxation method, 18
Rhenium
  tetrazole and tetrazole complexes, 215
  triazole and triazolate complexes, 180–181
Rhodium
  azides, 190
  tetrazole and tetrazole complexes, 217–220
  triazole and triazolate complexes, 187–189
Rhodium(III) hydroxo-bridged complexes
  acid-base equilibria, dihydroxo-bridged species, 108–110
  acid hydrolysis, 156
  aqua hydroxo species, acid-base equilibria, monohydroxo-bridged species, 113
  conclusions, 156–158
  crystallographic data, 59–69
  dinuclear
    formation by hydrolysis, 80–81
    monohydroxo-bridged cleavage, acid hydrolysis kinetics, 123, 127–130
  formation
    by hydroxo bridge cleavage, 93–97
    by isomerization, 97–98
    by redox reactions, 98
    solid-state reactions, 86–88
  mixed bridge systems, 64, 65
  mono- and dihydroxo-bridged dinuclear equilibria, 131–137
    acid-catalyzed cleavage, 138
    base-catalyzed bridge cleavage, 140
    bridge formation, 138–140
  monohydroxo-bridged, stability constants, 100–103

polynuclear, cleavage by strong acids, 122
spectroscopic and magnetic properties, 70–75
stability complexes, trihydroxo-bridged species, 104–105
tri- and dihydroxo-bridged equilibria, 149
Rhodo complexes
  dichromium(III) bridged complex, 57
  oxo-bridge of, 64, 65
  spectroscopic and magnetic properties, 72–73
Rhodo ion, 91
  acid-base equilibria, 107–108
  cleavage, 124
Rhodoso complex, 57, 82
Roussin esters, chemical carcinogenesis and, 387–389
Ruthenium, tetrazole and tetrazole complexes, 215–217

S

Salicylaldimines, 29
Sexadentate iron(II) complex, racemization, 45
Silver
  tetrazole and tetrazole complexes, 226–229
  triazole and triazolate complexes, 194–200
Solvents
  abbreviations for, 158–160
  NMR temperature calibrant, 4–5
  relaxation time and, 24
Spectroscopy
  of benzotriazole derivatives, 185–186
  EPR. *See* Electron paramagnetic resonance
  NMR. *See* Nuclear magnetic resonance
  visible/ultraviolet, 58
Spin-crossover, definition, 2
Spin equilibria dynamics in metal complexes. *See* Metal complexes, spin equilibria
Spin-forbidden nonadiabaticity, 25
Spin-forbidden transitions, 25–26
Spin-orbit coupling, for planar-tetrahedral isomerization of nickel(II) complexes, 32
Spin restrictions, electron transfer and, 46

Stability complexes
  dihydroxo-bridged species, 103–104
  trihydroxo-bridged species, 104–105
Substitution, spin equilibria implications, 46
Superconducting quantum interference detector (SQUID), 4

T

Tanning process, hydroxo-bridged aqua chromium(III) oligomers and, 95–96
Tantalum, triazole and triazolate complexes, 180
Technetium
  tetrazole and tetrazole complexes, 215
  triazole and triazolate complexes, 180–181
Tellurium analogs, of iron-sulfur-nitrosyl complexes, 350
Temperature dependence
  of nickel(II) complex interconversion rates, 29
  of paramagnetic species in NMR, 15
  of spin state populations in solids, 36–37
Temperature-jump relaxation. *See* Raman laser temperature-jump technique
Tetraaminenickel(II) complexes, 34–36
Tetraazadienes, structure, 171–172
Tetranuclear species, bridge cleavage in acidic solutions, 84
Tetrazenes structure, 171–172
Tetrazine structure, 172
Tetrazole and tetrazolate complexes, 205–207
  group survey, 213
  cadmium, 229–230
  chromium, 214–215
  cobalt, 217–220
  copper, 226–229
  gold, 226–229
  iridium, 217–220
  iron, 215–217
  manganese, 215
  mercury, 229–230
  molybdenum, 214–215
  nickel, 220–226
  osmium, 215–217
  palladium, 220–226
  platinum, 220–226
  rhenium, 215
  rhodium, 217–220
  ruthenium, 215–217

silver, 226–229
technetium, 215
tungsten, 214–215
zinc, 229–230
spectroscopic studies
  electronic spectra, 211–212
  electron paramagnetic resonance
    spectra, 213
  Mössbauer spectra, 213
  nuclear magnetic resonance spectra,
    212–213
  vibrational spectra, 210–211
structural properties, 208–210
structure, 172
synthesis, 207–208
1,2,3,4-Tetrazoline-5-thione, structure, 172
Thermal equilibrium, between different spin
  states. *See* Spin-crossover; Spin
  equilibrium
Thermodynamics, of equilibria between
  mono- and dihydroxo-bridged
  complexes, 132
1,2,3,4-Thiatriazoline-5-thione, structure,
  172
1,2,3,4-Thiatriazoline-5-thionate anion, 173
1,2,3,4-Thiatriazoline-5-thione, derivatives,
  175
Thiocapsa roseopersicina hydrogenase, 311
Titanium, triazole and triazolate complexes,
  180
Transmission coefficient, 25, 45
Triazenes, structure, 171–172
Triazolate complexes, 173–174
  bonding modes, 176–177
  group survey
    cadmium, 200–204
    chromium, 180
    cobalt, 187–189
    copper, 194–200
    gold, 194–200
    hafnium, 180
    iridium, 187–189
    iron, 181–186
    manganese, 180–181
    mercury, 200–204
    molybdenum, 180
    nickel, 189–193
    niobium, 180
    osmium, 181–186
    palladium, 189–193
    platinum, 189–193
    rhenium, 180–181
    rhodium, 187–189
    ruthenium, 181–186
    silver, 194–200
    tantalum, 180
    technetium, 180–181
    titanium, 180
    tungsten, 180
    vanadium, 180
    zinc, 200–204
    zirconium, 180
  spectroscopic studies
    electronic spectra, 178
    electron paramagnetic resonance
      spectra, 179
    nuclear magnetic resonance spectra,
      178
    vibrational spectra, 178
  structural properties, 175–178
  synthesis, 174–175
  x-ray diffraction studies, 177
Triazole complexes, 173–174
  group survey
    cadmium, 200–204
    chromium, 180
    cobalt, 187–189
    copper, 194–200
    gold, 194–200
    hafnium, 180
    iridium, 187–189
    iron, 181–186
    manganese, 180–181
    mercury, 200–204
    molybdenum, 180
    nickel, 189–193
    niobium, 180
    osmium, 181–186
    palladium, 189–193
    platinum, 189–193
    rhenium, 180–181
    rhodium, 187–189
    ruthenium, 181–186
    silver, 194–200
    tantalum, 180
    technetium, 180–181
    titanium, 180
    tungsten, 180
    vanadium, 180
    zinc, 200–204
    zirconium, 180
  numbering system, 174

spectroscopic studies
  electronic spectra, 178
  electron paramagnetic resonance spectra, 179
  nuclear magnetic resonance spectra, 178
  vibrational spectra, 178
 structural properties, 175–178
 structure, 172
 synthesis, 174–175
 x-ray diffraction studies, 177
1-N-(1,2,3-Triazole)cyclopentadienyl)dicarbonyliron bisuate, 182–183
Trihydroxo-bridged complexes
 crystallographic data, 61, 64
 stability complexes, 104–105
Tris(dithiocarbamate)cobalt(III) complex, racemization, 45
Tris(N,N-dialkyldithiocarbamato)iron(III) complexes, 2
Tungsten
 tetrazole and tetrazole complexes, 214–215
 triazole and triazolate complexes, 180

U

Ultrasonic relaxation
 of iron(III) complexes in solution, 26–27
 limitations of method, 24
 of spin-equilibrium complexes, 18–20
Urease, 297
 composition, 300
 mechanism of catalysis, 303–304
 spectroscopic properties, 301–303

V

Vanadium, triazole and triazolate complexes, 180
Van Vleck Hamiltonian function, 74
Vibrational (infrared and Raman) spectroscopy, of hydroxo-bridged complexes, 59
Vibrational spectra, spin equilibrium complexes, 13
Volumes of activation
 for iron(II) complex, 45
 between spin states, 10–11, 40

W

Werner's brown salt, 67, 82, 91

X

X-ray crystallography
 changes in metal-ligand bond lengths from spin state transition, 6
 cobalt(III) complexes, 90–91
 diffraction studies, triazolate, 177
 of hydroxo-bridged complexes, 59–69
 hydroxo-bridged complexes, α-type interactions in solid state, 68
 iron-sulfur-nitrosyl complexes, 354–365
  [Fe(NO)(S$_2$CNR$_2$)]$_2$ and related clusters, 359–361
  [Fe$_4$S$_3$(NO)$_7$]$^-$ and [Fe$_2$S$_2$(NO)$_4$]$^{2-}$, 354–355
  [Fe$_4$S$_4$(NO)$_4$] and related clusters, 359
  [Fe$_2$SR$_2$(NO)$_4$], 355–358
  heterometallics, 364
  selenium analogs, 361–363
  tellurium analogs, 361–363
 structure determinations, 56
 for hydroxo-bridged polynuclear complexes, 57–58
 tetrazole and tetrazole complexes, 210
 triazole moieties, 176–177

Z

Zinc, tetrazole and tetrazole complexes, 229–230
Zirconium, triazole and triazolate complexes, 180

CONTENTS OF RECENT VOLUMES

VOLUME 22

Lattice Energies and Thermochemistry of Hexahalometallate(IV) Complexes, A_2MX_6, Which Possess the Antifluorite Structure
H. Donald B. Jenkins and Kenneth F. Pratt

Reaction Mechanisms of Inorganic Nitrogen Compounds
G. Stedman

Thio-, Seleno-, and Tellurohalides of the Transition Metals
M. J. Atherton and J. H. Holloway

Correlations in Nuclear Magnetic Shielding, Part II
Joan Mason

Cyclic Sulfur-Nitrogen Compounds
H. W. Roesky

1,2-Dithiolene Complexes of Transition Metals
R. P. Burns and C. A. McAuliffe

Some Aspects of the Bioinorganic Chemistry of Zinc
Reg H. Prince

SUBJECT INDEX

VOLUME 23

Recent Advances in Organotin Chemistry
Alwyn G. Davies and Peter J. Smith

Transition Metal Vapor Cryochemistry
William J. Power and Geoffrey A. Ozin

New Methods for the Synthesis of Trifluoromethyl Organometallic Compounds
Richard J. Lagow and John A. Morrison

1,1-Dithiolato Complexes of the Transition Elements
R. P. Burns, F. P. McCullough, and C. A. McAuliffe

Graphite Intercalation Compounds
Henry Selig and Lawrence B. Ebert

Solid-State Chemistry of Thio-, Seleno-, and Tellurohalides of Representative and Transition Elements
J. Fenner, A. Rabenau, and G. Trageser

SUBJECT INDEX

VOLUME 24

Thermochemistry of Inorganic Fluorine Compounds
A. A. Woolf

Lanthanide, Yttrium, and Scandium Trihalides: Preparation of Anhydrous Materials and Solution Thermochemistry
J. Burgess and J. Kijowski

The Coordination Chemistry of Sulfoxides with Transition Metals
J. A. Davies

Selenium and Tellurium Fluorides
A. Engelbrecht and F. Sladky

Transition-Metal Molecular Clusters
B. F. G. Johnson and J. Lewis

INDEX

VOLUME 25

Some Aspects of Silicon-Transition-Metal Chemistry
B. J. Aylett

The Electronic Properties of Metal Solutions in Liquid Ammonia and Related Solvents
Peter P. Edwards

Metal Borates
J. B. Farmer

Compounds of Gold in Unusual Oxidation States
Hubert Schmidbaur and Kailash C. Dash

Hydride Compounds of the Titanium and Vanadium Group Elements
G. E. Toogood and M. G. H. Wallbridge

INDEX

VOLUME 26

The Subhalides of Boron
A. G. Massey

Carbon-Rich Carboranes and Their Metal Derivatives
Russell N. Grimes

Fluorinated Hypofluorites and Hypochlorites
Jean'ne M. Shreeve

The Chemistry of the Halogen Azides
K. Dehnicke

Gaseous Chloride Complexes Containing Halogen Bridges
Harald Schäfer

One-Dimensional Inorganic Platinum-Chain Electrical Conductors
Jack M. Williams

Transition-Metal Alkoxides
R. C. Mehrotra

Transition-Metal Thionitrosyl and Related Complexes
H. W. Roesky and K. K. Pandey

INDEX

VOLUME 27

Alkali and Alkaline Earth Metal Cryptates
David Parker

Electron-Density Distributions in Inorganic Compounds
Koshiro Toriumi and Yoshihiko Saito

Solid-State Structures of the Binary Fluorides of the Transition Metals
A. J. Edwards

Structural Organogermanium Chemistry
K. C. Molloy and J. J. Zuckerman

Preparations and Reactions of Inorganic Main-Group Oxide Fluorides
John H. Holloway and David Laycock

The Chemistry of Nitrogen Fixation and Models for the Reactions of Nitrogenase
Richard A. Henderson, G. Jeffery Leigh, and Christopher J. Pickett

Trifluoromethyl Derivatives of the Transition Metal Elements
John A. Morrison

INDEX

VOLUME 28

Fast-Atom Bombardment Mass Spectrometry and Related Techniques
Jack Martin Miller

The Chemistry of Berkelium
J. R. Peterson and D. E. Hobart

Preparations and Reactions of Oxide Fluorides of the Transition Metals, the Lanthanides, and the Actinides
John H. Holloway and David Laycock

Chemical Effects of Nuclear Transformations
G. A. Brinkman

Homocyclic Selenium Molecules and Related Cations
Ralf Steudel and Eva-Maria Strauss

The Element Displacement Principle: A New Guide in p-Block Element Chemistry
A. Haas

Compounds of Pentacoordinated Arsenic(V)
R. Bohra and H. W. Roesky

Perchlorate Ion Complexes
N. M. N. Gowda, S. B. Naikar, and G. K. N. Reddy

INDEX

VOLUME 29

Inorganic Silylenes. Chemistry of Silylene, Dichlorosilylene, and Difluorosilylene
Chao-Shiuan Liu and Tsai-Lih Hwang

Trifluorophosphine Complexes of Transition Metals
John F. Nixon

Solvent Extraction of Metal Carboxylates
Hiromichi Yamada and Motoharu Tanaka

Alkyne-Substituted Transition Metal Clusters
Paul R. Raithby and Maria J. Rosales

Organic Superconductors: Synthesis, Structure, Conductivity, and Magnetic Properties
Jack M. Williams and Kim Carneiro

Where Are the Lone-Pair Electrons in Subvalent Fourth-Group Compounds?
S.-W. Ng and J. J. Zuckerman

INDEX

VOLUME 30

Catenated Nitrogen Ligands Part I. Transition Metal Derivatives of Triazenes, Tetrazenes, Tetrazadienes, and Pentazadienes
David S. Moore and Stephen D. Robinson

The Coordination Chemistry of 2,2':6',2"-Terpyridine and Higher Oligopyridines
E. C. Constable

High-Nuclearity Carbonyl Clusters: Their Synthesis and Reactivity
Maria D. Vargas and J. Nicola Nicholls

Inorganic Chemistry of Hexafluoroacetone
M. Witt, K. S. Dhathathreyan, and H. W. Roesky

INDEX

VOLUME 31

Preparation and Purification of Actinide Metals
J. C. Spirlet, J. R. Peterson, and L. B. Asprey

Astatine: Its Organonuclear Chemistry and Biomedical Applications
I. Brown

Polysulfide Complexes of Metals
A. Müller and E. Diemann

Iminoboranes
Peter Paetzold

Synthesis and Reactions of Phosphorus-Rich Silylphosphanes
G. Fritz

INDEX